MATHEMATICA®
BEYOND MATHEMATICS

THE WOLFRAM LANGUAGE
IN THE REAL WORLD

MATHEMATICA®
BEYOND MATHEMATICS
THE WOLFRAM LANGUAGE IN THE REAL WORLD

José Guillermo Sánchez León
Universidad de Salamanca, Spain

CRC Press
Taylor & Francis Group
Boca Raton London New York

CRC Press is an imprint of the
Taylor & Francis Group, an **informa** business
A CHAPMAN & HALL BOOK

CRC Press
Taylor & Francis Group
6000 Broken Sound Parkway NW, Suite 300
Boca Raton, FL 33487-2742

Printed and bound in India by Replika Press Pvt. Ltd.

Printed on acid-free paper
Version Date: 20170202

International Standard Book Number-13: 978-1-4987-9629-3 (Hardback)

Visit the Taylor & Francis Web site at
http://www.taylorandfrancis.com

and the CRC Press Web site at
http://www.crcpress.com

Contents

Preface

About *Mathematica* and the Wolfram Language

If you have used *Mathematica* occasionally or heard of it, you may have the false impression that it is a program for performing complicated calculations, usually for academic purposes. However, this idea is far from the truth. Actually, *Mathematica* is much more than that. By putting together the computational power and ease of use of the Wolfram Language, *Mathematica*'s high-level general-purpose programming language, the program can be used in any scientific or technical field: Aerospace engineering, environmental sciences, financial risk management, medical imaging and many others. (You can get an idea by visiting: http://www.wolfram.com/solutions).

Mathematica can be considered a tool that empowers non-professional programmers to develop applications although if you are a professional programmer, you will see that the software provides a development environment similar to the ones available for C++ or FORTRAN. You can even use the program to control external devices.

About the Book

Although many books have been written about *Mathematica,* very few of them cover the new functionality added to the most recent versions of the program. This text introduces the new features using real-world examples, based on the experience of the author as a consultant. In the process, you will also learn more about the Wolfram Language and how you can use it to solve a wide variety of problems. Both are the most important objectives of the book. To accomplish that, the author raises questions from a wide range of topics and answers them by taking full advantage of *Mathematica's* latest features. For example: What is the hidden image in "The Ambassadors" painting by Holbein? What sources of energy does the world really use? How can we calculate tolerance limits in manufacturing processes? Are our cities getting warmer? Is the novel "El Quijote" written in Pi? How do we know how old our planet is? How can we find planets outside our solar system? How big is our galaxy? And the universe? How do we know it? How can we model the distribution of radioactive isotopes in the human body? And a tsunami? What are and how can we create Mandelbrot fractals? How can we measure the genetic distance between species? How can we perform financial calculations in real time? How do we value financial derivatives? How can we make entertaining simulations for teaching mathematics, physics, statistics, ... ? Why are there no free quarks?

The answers to the previous questions will not only help you master *Mathematica*, but also to become more familiar with the corresponding topics themselves.

About the Book Objectives

This book will not only be useful to newcomers but also to those familiar with the program and interested in learning more about the new functionality included in the latest versions.

Those readers with minimal or no knowledge of *Mathematica* are strongly advised to read chapter 1 along with Stephen Wolfram's book *An Elementary Introduction to the Wolfram Language* available from within the program documentation: **Help ► Wolfram Documentation ► Introductory Book ≫**

This text will also make it easy to start programming using the Wolfram Language and to learn how to take full advantage of its capabilities.

The final objective of *Mathematica Beyond Mathematics* is to help you avoid feeling overwhelmed by the software's vast capabilities. The author has explored a significant part of them choosing the most relevant parts and illustrating them with examples from many different sources including the program documentation. Links to additional resources will also be provided. The main aim of all this is to reduce significantly the amount of time required to master the tool. The commands used will be explained using short sentences and simple examples.

About the Use of the Book

Although this book is not a manual, inexperienced readers should at least read the first four chapters in consecutive order to gain a solid understanding of how *Mathematica* works. Regarding the rest of the chapters, you should keep in mind the following: a) Chapter 5 covers the most innovative features of the program such as how to access data from many different fields and how to use natural language to interact with the software; b) Chapter 6 deals with relatively advanced probability and statistics topics; c) Chapters 7 to 11 explore topics related to a single area of knowledge (mathematics, astronomy, nuclear physics, biokinetic modeling and economics and finance). You can read them according to your preferences; and d) Chapter 12 is for those readers facing problem requiring big computational resources (parallel computing, grid-enabled calculations, etc.) and/or the ones interested in other programs related to *Mathematica* such as web*Mathematica* for adding computational capabilities to websites or Workbench, a complete software development environment using the Wolfram Language.

The principal theme of each chapter is used as the motivation to illustrate certain features of *Mathematica* that you may find useful for solving a great variety of problems. For example: Chapter 6 teaches you how to build a *Mathematica* package; in Chapter 8, related to astronomy, you learn how to create dynamic images; Chapter 10, covering the modeling of biological systems, discusses the resolution of differential equations in *Mathematica* using concrete examples.

About the First English Edition and the Supplementary Materials

Everything shown in the text has been done using *Mathematica*, including the access to information sources.

The book has been written using *Mathematica* 10 and 11 and edited using *Mathematica* 11, although it should still be useful with future versions of the program as well.

Mathematica Beyond Mathematics is based on *Mathematica más allá de las matemáticas, 2ª Edición* but its contents have been improved and updated. In this endeavor, Ruben Garcia Berasategui, a lecturer in Jakarta International College, has played a fundamental role, not only by translating the text but also by making insightful comments and suggestions.

The text doesn't display "In[*n*]:=" and "Out[*n*]:=", the default symbols that *Mathematica* adds to Input (where the commands are entered) and Output (where the results are shown) cells.

In a few cases, the book uses specific files available for downloading from the author's website: http://diarium.usal.es/guillermo.

At the end of each chapter there is a section containing additional resources carefully selected that can be accessed most of the time from either *Mathematica* or through the Internet.

The author would appreciate any comments or suggestions. Please send them to: guillermo2046@gmail.com, with the subject: '*Mathematica* Beyond Mathematics'.

Salamanca (Spain), January 2017

About the Author

J. Guillermo Sánchez León (http://diarium.usal.es/guillermo). Engineer, physicist and mathematics PhD holder, is an associate professor in the University of Salamanca and works in the energy industry. He has conducted research in a variety of fields: Modeling, optimization, medical physics, astronomy, finance and others. In 1999 he was awarded a research grant at Wolfram Research Inc. headquarters in Champaign (Illinois, USA) after his statistical applications with *Mathematica* project won a competition sponsored by the company. Since then he has been an active *Mathematica* and web*Mathematica* alpha and beta tester. He's also a Wolfram certified instructor and has extensive experience in teaching and developing programs with *Mathematica* and web*Mathematica*. Some of them are cited as references in *Mathematica*'s official web page and are available online in both English and Spanish. Among his more than 100 articles, there are several where *Mathematica* and web*Mathematica* have been used extensively.

About the Translator

Rubén García Berasategui is a math, finance and statistics lecturer in Jakarta International College. He holds a bachelor's degree in business administration and an MBA. He's been using *Mathematica* since 1997 and in 2012 became the first Wolfram certified instructor in Southeast Asia. He has been training hundreds of newcomers to *Mathematica* and sharing his passion for the program with them ever since.

Acknowledgments

The author would like to express his gratitude to Sarfraz Khan, Editor of CRC Press (Taylor & Francis Group) for his vision in realizing the need in the market for a book about *Mathematica* like this one, and for his encouragement and support during the production process.

The author would also like to say thank you to Addlink Software Científico (http://www.addlink.es), sponsor of the first Spanish edition; Antonio Molina and Juan

Antonio Rubio for their editorial guidance; and J. M. Cascón (USAL), J. López-Fidalgo (UCLM), S. Miranda (Solventis), R. Pappalardo (US), J. M. Rodríguez (USAL) and C. Tejero (USAL) for their comments and corrections.

1

Getting Started

In this chapter we will take a brief tour of Mathematica, a program for modern technical computing based on the Wolfram Language. It is important to read this chapter carefully since its contents are essential to be able to follow the rest of the book. Even if you are already a Mathematica user, you may find the following pages useful to get familiar with the new functionality included in the most recent program releases.

1.1 *Mathematica,* an Integrated Technical Computing System

Mathematica is a software application that goes beyond numeric and symbolic calculations. It is a modern technical computing system based on a high-level general-purpose programming language, the *Wolfram Language*. You will be able to start using it after a few minutes but, if you want to take full advantage of its capabilities, you will have to spend time to master its language although probably less than with other programming languages.

Mathematica took a revolutionary step forward in Version 8 with the introduction of the **free-form linguistic input**, that consists of writing in plain English a request that the program will try to answer. Using this format ensures that newcomers can get started very quickly. For experienced users, it means help in finding the most appropriate functions to perform a desired calculation. To get an idea of *Mathematica*'s capabilities before beginning to use the program, let's take a look at the following examples (for now, you just need to read what follows, later we will show you how to write inputs and run commands):

- To solve an equation such as $3x^2 + 2x - 4 = 0$, you can literally type what you want *Mathematica* to do. The output shows the correct *Mathematica* input syntax and the answer, which in this case includes both, the exact solution and the decimal approximation.

```
Solve 3 x^2 + 2 x - 4 = 0
↳ Results (1 of 2)
{Reduce[-4 + 2*x + 3*x^2 == 0, x],
 N[Reduce[-4 + 2*x + 3*x^2 == 0, x]]}
```

$$\left\{x = \frac{1}{3}\left(-1 - \sqrt{13}\right) \vee x = \frac{1}{3}\left(\sqrt{13} - 1\right),\ x = -1.53518 \vee x = 0.868517\right\}$$

- To add quantities in different currencies, just enter them and the result is shown in the currency that was used first; if you execute the command, the result will most likely be different due to changes in the exchange rates from the time the input was first executed.

```
$500 + £307 + 127 € + 1520 ¥ »
$500. + £307. + €127. + ¥1520
```

$1 061.78

- We can even ask very specific questions: To learn more about the temperature in a location, just write down the name of the place followed by “temperature”. The most important advantage is that the information from the answer can be used inside the *Mathematica* environment.

```
Salamanca temperature »
AirTemperatureData[ Salamanca (city) ]
```

17.3 °C

The *Mathematica* free-form input is integrated with Wolfram|Alpha, http://www.wolframalpha.com, a type of search engine that defines itself as a computational knowledge engine and that instead of returning links like Google or Bing, provides an answer to a given question in English. If you want, you can access Wolfram|Alpha directly from within *Mathematica,* and we will show you how to do that later on.

- To represent the Do major music scale, you just need to write:

```
Do major
```

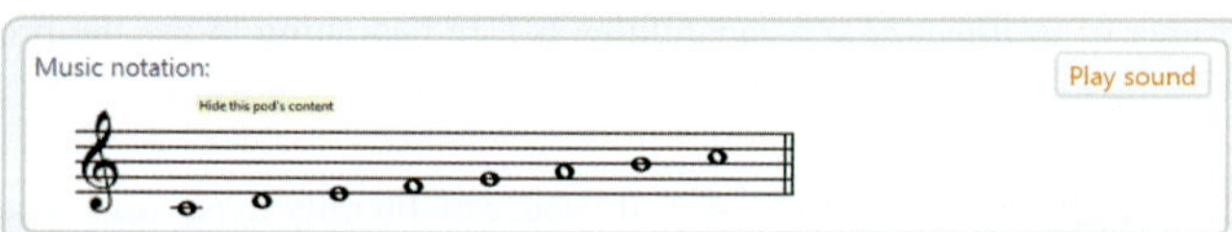

Another example of the new functionality recently incorporated in *Mathematica* is the possibility of accessing scientific and technical databases through the Internet from different fields already formatted by Wolfram Research for their direct manipulation inside the application.

- In this example, we obtain the value of the Euro in US dollars after executing the command.

```
FinancialData["EUR/USD"]
```

1.0661

- We can choose a chemical compound and get many of its properties. In this case, we get the plot of the caffeine molecule.

```
ChemicalData["Caffeine", "MoleculePlot"]
```

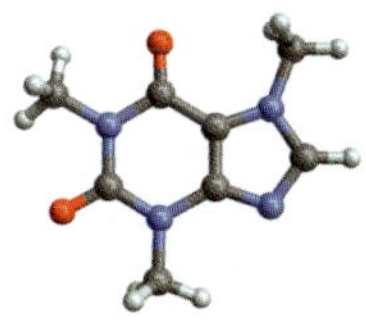

With very brief syntax we can build functions that in other programming languages would require several lines of code.

- With just two lines of commands we can generate a dynamic clock that is synchronized with the time in our computer and in a different time zone.

```
Dynamic[Refresh[Row[
   {ClockGauge[AbsoluteTime[], PlotLabel -> Style["Local", Large, Bold]],
    ClockGauge[AbsoluteTime[TimeZone → +9],
     PlotLabel -> Style["Tokyo", Large, Bold]]}], (UpdateInterval -> 1)]]
```

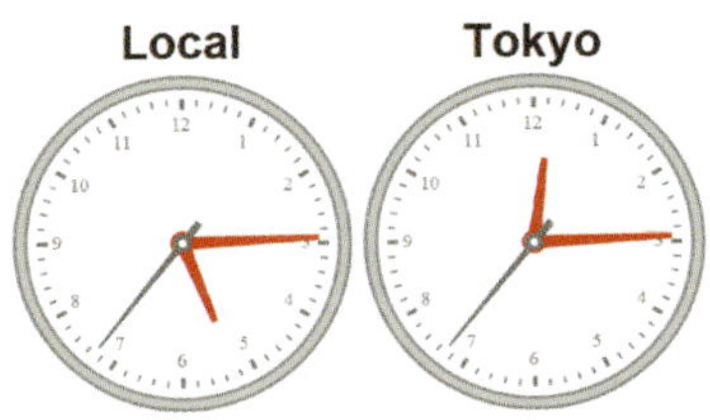

- In less than half a line we can write the necessary instructions to make an interactive model of the Julia fractal set.

```
Manipulate[JuliaSetPlot[0.365 - k ⅈ,
  PlotLegends → Automatic, ImageSize → Small], {k, 0.4, 0.5}]
```

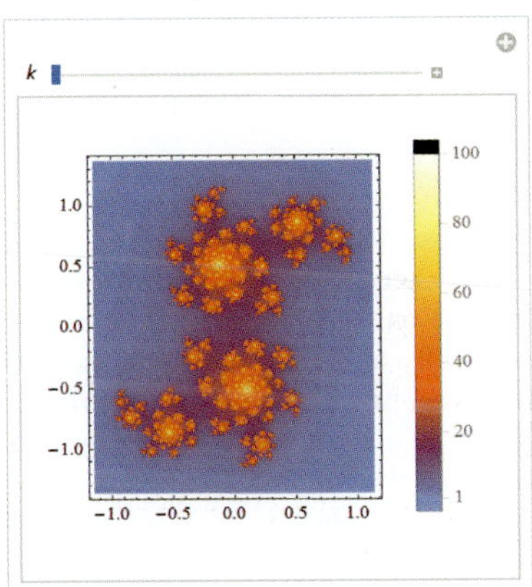

- We find the intersection points of a circumference and a parabola and graphically represent them. To write special symbols such as "∧" , the best thing to do is to use Palettes, which we will discuss later on.

```
pts = Solve[x^2 + y^2 = 1 ∧ y - 2 x^2 + 3/2 = 0, {x, y}]
```

$$\left\{\left\{x \to -\frac{1}{2}\sqrt{\frac{1}{2}(5-\sqrt{5})},\ y \to \frac{1}{4}(-1-\sqrt{5})\right\}, \left\{x \to \frac{1}{2}\sqrt{\frac{1}{2}(5-\sqrt{5})},\ y \to \frac{1}{4}(-1-\sqrt{5})\right\}, \left\{x \to -\frac{1}{2}\sqrt{\frac{1}{2}(5+\sqrt{5})},\ y \to \frac{1}{4}(\sqrt{5}-1)\right\}, \left\{x \to \frac{1}{2}\sqrt{\frac{1}{2}(5+\sqrt{5})},\ y \to \frac{1}{4}(\sqrt{5}-1)\right\}\right\}$$

```
Show[
 {ContourPlot[{x^2 + y^2 == 1, y - 2 x^2 + 3/2 == 0}, {x, -1.5, 1.5}, {y, -1.5, 1.5}],
  Graphics[{Red, PointSize[Medium], Point[{x, y} /. pts]}]}]
```

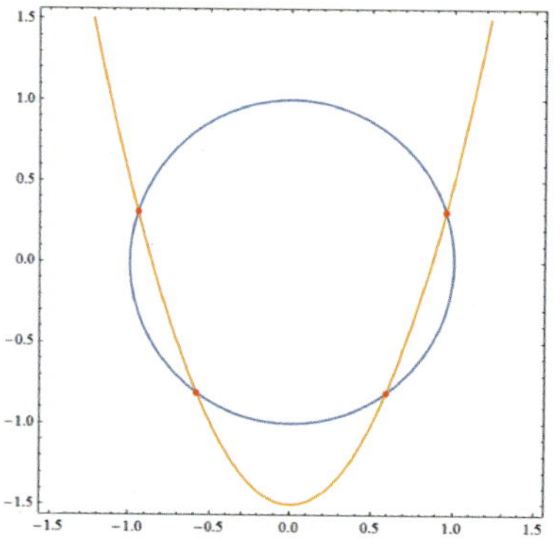

Apart from the examples given above showcasing some of *Mathematica*'s new functionality, it is also worth mentioning: (i) The *Wolfram Predictive Interface*, that makes suggestions about how to proceed after executing a command, (ii) The context-sensitive input assistant to help us choose the correct function, (iii) The possibility of running the program from the browser (WolframCloud.com), including specific cloud-computing related commands, (iv) The machine learning capabilities that, for example, enable us to analyze a handwritten text and keep on improving its translation through repetition, (v) Very powerful functions for graphics and image processing. We will refer to these and other new capabilities later on.

1.2 First Steps

The program can be run locally or in the cloud accessing the **Wolfram Cloud** (http://www.wolframcloud.com/) with a browser. From here on, except mentioned otherwise, it will be assumed that you have installed *Mathematica* locally and activated its license. When installing it, an assistant will guide you through the activation process. If this is the first time, it will take you to the user portal (https://user.wolfram.com), where you will have to sign up. The website will ask you to create a Wolfram ID, use your email address, and your own password. It will also automatically generate an activation key. Remember your Wolfram ID and password because they will be useful later on if, for example, you would like to install the program in a different computer or have an expired license. In both cases, you will need a new activation key. The Wolfram ID and password will also be necessary to access the Wolfram

Cloud, and we will give a specific example at the end of the chapter.

To start the program in Windows, under the programs menu in your system, click on the *Mathematica* icon. Normally there are two icons; choose *Mathematica* and not *Mathematica* Kernel.

Under OS X, the program is located in the applications folder and you can launch it either locating it with Finder, using the Dock or from the launchpad (OS X Lion or more recent versions).

By default, a welcome screen (Figure 1.1) similar to this one will appear:

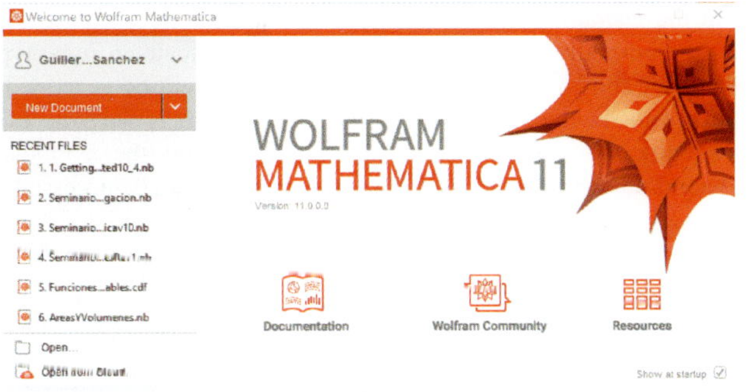

Figure 1.1 The Welcome Screen in *Mathematica* 11.

This screen contains several external links (hyperlinks). For example, by clicking on **Resources**, you will be able to access a broad collection of reference materials about how to use *Mathematica,* including videos: http://www.wolfram.com/broadcast/. You can always go back to this screen from the menu bar: **Help ▶ Welcome Screen...** . In the upper left-hand corner (in *Mathematica* 10/11) you will see this icon: 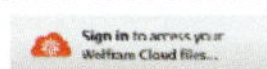 . If you are a registered user, chances are you already have a Wolfram ID. In that case, click on the icon and sign in. Once the process has been completed, the icon will be replaced with the name that you used when you signed up, as shown in the previous image. The advantage of this is to gain access from within *Mathematica* to your files in the cloud (WolframCloud) but for now, you can skip this step if desired. If you click on `NewDocument`, below the icon shown at the beginning of the paragraph, a new blank *Notebook* will be created. Notebooks are the fundamental way to interact with *Mathematica.* You can also create them by clicking on **File ▶ New ▶** Notebook.nb.

A *notebook* is initially a blank page. We can write on it the text and instructions that we want. It is always a good idea to save the file before we start working on it and, as in most programs, this can be done in the menu bar: **File ▶ Save As ▶** "Name of the File" (*Mathematica* by default will assign it the extension .nb).

It is always convenient to have access to the toolbar from the beginning of the session; to do that, select in the menu bar **Window** and check **Toolbar ▶ Formatting** (or **Show Toolbar** in Version 9 and earlier). Please note the drop-down box to the left of the toolbar (Figure 1.2), that we'll refer to as: *style box.* It will be useful later on.

Figure 1.2 *Mathematica*'s Formatting Toolbar.

Once you begin to type, a new cell will be created. It will be identified by a blue right-bracket square (]) ("cell marker") that appears on the right side of the *notebook.* Each cell constitutes a unit with its own properties (format, evaluable cell, etc.). To see the cell type, place the cursor inside or on the cell marker. The type will appear in the *style box* (Figure, Input, Text, Title, Program...). We can also change the format directly inside the box. Type in a text cell and try to change its style.

```
This is a cell formatted using the Program style.
```

When a blank *notebook* is created, it has a style (cell types, fonts, etc.) that assigns to each type of cell certain properties (evaluable, editable, etc). The first time a *notebook* is created, *Mathematica* assigns a style named *Default* (by default). Later on we will see how to choose different styles. In the *Default* style, when a new cell is created, it will be of the *Input* type, which is the one used normally for calculations. When we evaluate an *Input* type cell, "In[*n*]:= *our calculation*" will be shown and a new cell of type *Output* will be generated where you will see the calculation result in the form of "Out[*n*]:= *result*" (*n* showing the evaluation order). However, in this book we sometimes use an option to omit the symbols "In[*n*]:=" and "Out[*n*]:=". These and other options to customize the program can be accessed through **Edit ▸ Preferences ...**.

> To execute an *Input* cell select the cell (placing the cursor on the cell marker) and press SHIFT + ENTER or ENTER on the numeric keypad. You can also access the contextual menu by right-clicking with the mouse and selecting "Evaluate Cell".

```
2 + 2
```

4

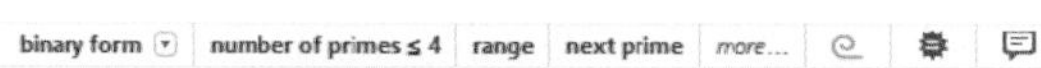

Figure 1.3 The Suggestions Bar.

Since *Mathematica* 9, when a cell is executed, a toolbar appears below its output as shown above (Figure 1.3). This bar, named the **Suggestions Bar**, provides immediate access to possible next steps optimized for your results. For example, If you click on **range** you will see that a new input is being generated and a new suggestions bar will appear right below the result:

```
Range[4]
```

{1, 2, 3, 4}

The **Suggestions Bar** is part of the predictive interface. It tries to guide you with suggestions to simplify entries and ideas for further calculations. It is one more step to reduce the time it takes to learn how to use the program.

All written instructions inside a cell will be executed sequentially. If at the end of an instruction we type semicolon ";", the instruction will be executed but its output will not be shown. This is useful to hide intermediate results. In an evaluation cell (*Input*) we can include one or more comments, writing them in the following format: (* comment *). Nevertheless, it is recommended to include the comments in separate cells in text format. Remember that you can do that by choosing *Text* in the *style box*.

To create a new cell we just need to place the cursor below the cell where we are and then start writing. We can also create a new cell where the cursor is located by pressing ALT + ENTER.

To facilitate writing, *Mathematica* provides **Palettes**. These can be loaded by clicking on **Palettes** in the menu bar. There are several palettes to make it easier to type mathematical symbols such as the **Basic Math Assistant** palette or **Other ▶ Basic Math Input**. It would be useful from now on to keep one of them open.

Alternatively, you can use keyboard shortcuts to write special symbols, indexes, subindexes, etc. The most useful ones are: subindex CTRL + _ , superindex CTRL + ^ or CTRL + 6, fraction CTRL + / and the square root symbol CTRL + 2. Depending on your keyboard configuration, to use some of the previous symbols you may have to press first + SHIFT (key CAPS) for example: superindex CTRL + SHIFT + ^.

- Using a palette or the keyboard shortcuts write and execute the following expression. Please remember: an empty space is the equivalent of a multiplication symbol: that is, instead of 4×5 you can write 4 5.

$$\frac{4\times 5}{5} + \sqrt{4} - 3^2$$

−3

The previous *Input* cell was written in the standard format (*StandardForm*). It is the one used by default in *Input* cells. To check it, position the cursor on the cell marker and in the menu bar choose: **Cell ▶ Convert To ▶**. *StandardForm* will appear with a check mark in front of it.

Expressions can also be written without the need for special symbols by using the form: *InputForm*. It is very similar to the one used by other programming languages such as C, FORTRAN or BASIC. For instance: multiplication = " * ", division = " / ", addition = " + ", subtraction " = " , exponentiation =" ^ ", and square root of a = "`Sqrt[a]`". We can replace the multiplication symbol by an empty space. It is very important to keep that in mind since *Mathematica* interprets differently "**a2**" and "**a 2**": "**a2**" is a unique symbol and "**a 2**" is **a*2**.

- In *InputForm* the previous expression can be written as:

```
(4*5)/5 + Sqrt[4] - 3^2
```

−3

Initially, it was its symbolic calculation capabilities that made *Mathematica* popular in the academic world.

- As an example, write and execute the following in the standard form in an input cell (use a palette to help you enter the content).

$$\partial_{x,y}\text{Sin}\left[x\sqrt{y}\right]$$

$$\frac{\cos(x\sqrt{y})}{2\sqrt{y}} - \frac{1}{2}x\sin(x\sqrt{y})$$

A customized style, to which we will refer later, has been defined in this notebook so that the outputs are shown in a format that corresponds to traditional mathematical notation. In *Mathematica* it is called *TraditionalForm*. We could have defined a style so that the inputs also use the traditional notation as well, but in practice it is not recommended. However, you can convert a cell to the *TraditionalForm* style anytime you want.

- Copy the *input* cell from above to a new cell and in the menu bar select **Cell ▶ Convert To ▶ TraditionalForm**. Notice that the cell bracket appears with a small vertical dashed line. You will be able to use the same menu to convert to other formats.

$$\frac{\partial^2 \sin\left(x\sqrt{y}\right)}{\partial x\,\partial y}$$

$$\frac{\cos\left(x\sqrt{y}\right)}{2\sqrt{y}} - \frac{1}{2}x\sin\left(x\sqrt{y}\right)$$

For Windows, since Windows 7, you can use the **Math Handwriting Input**, available under accessories, to write the entries manually using a tablet or an electronic board.

The **linguistic** or **free-form** format (available since *Mathematica* 8) allows us to write in plain English the operation we want to execute or the problem we want to solve. To start using it press = at the beginning of the cell (input type) and you will notice that will appear. After that write your desired operation. Your instruction will automatically be converted into standard *Mathematica* notation, and after that the answer will be displayed. Let's take a look at some examples.

- To add from 1 to 10 you can write "sum 1 to 10" or other similar English expressions such as "sum from 1 to 10".

```
sum 1 to 10
↳ Sum
  Sum[i, {i, 1, 10}]
```

55

If you follow the previous steps, the entry proposed may be different from the one shown here. The reason behind this is that when using the free-form format, the program connects to Wolfram|Alpha, Wolfram's computation knowledge engine, that is constantly evolving.

Move the cursor over `Sum[i,{i, 1, 10}]` and the message "Replace cell with this input" will appear. If you click on the symbol + located in the upper right-hand corner, additional information about the executed operation will be shown.

If you have received any message related to connection problems, it is because the free-form format and the use of data collections that we will refer to later on, requires that your computer is connected to the Internet. If it is and you still have connectivity problems it may be related to the *firewall* configuration in your network. In that case you may have to configure the *proxy* (see http://support.wolfram.com/kb/12420) and ask for assistance from your network administrator.

> When you don't know the correct function syntax, you can use the free-form format to write what you want to do. *Mathematica* probably will show you the correct syntax immediately, click on it and only the input in the correct notation will remain with the free-form format entry disappearing.

The free-form input format generally works well when it's dealing with brief instructions in English. In practice it is very useful when one doesn't know the specific syntax to perform certain operations in *Mathematica*.

Many more examples can be found in:

http://www.wolfram.com/mathematica/new-in-8/free-form-linguistic-input/basic-examples.html.

It's also very useful to use **Inline Free-form Input**.

- In this example, we assume we don't know the syntax to define an integral. In an input cell we click on **Insert ▸ Inline Free-form Input** and ⊟ will appear. Inside this box we type **integrate x^2**.

```
⊟ Integrate x^2
```

$$\frac{x^3}{3}$$

- The entry is automatically converted to the standard syntax in *Mathematica*.

```
⊟ Integrate[x^2, x] ✓
```

$$\frac{x^3}{3}$$

- By clicking on the above *input* over the ⊟ symbol, the cell will go back to its free-form format ⊟ .

```
⊟ Integrate x^2
```

$$\frac{x^3}{3}$$

- In the following example we have used **Inline Free-form Input** writing **integrate cos x from 0 to x**.

```
⊟ Integrate[Cos[x], {x, 0, x}] ✓
```

$\sin(x)$

- By clicking in ✓ the input is automatically converted to the standard syntax in *Mathematica.* It can be used as a template.

```
Integrate[Cos[x], {x, 0, x}]
```

$\sin(x)$

- The command below returns the average acceleration of the gravity *g* on earth at sea level.

```
⊟ earth gravity
```

$9.80\ \text{m/s}^2$

By default the units used are the ones from the international system (SI):

- Here we combine a standard function inside the **Inline Free-form Input** with `QuantityMagnitude` that shows the numeric value, without including the units:

```
QuantityMagnitude[⊟ earth gravity ]
```

9.80

As we have indicated in the introduction, you can directly access Wolfram|Alpha from *Mathematica*. To do that just press the equal sign on your keyboard twice ("==") at the beginning of an input cell. The symbol will appear, then type what you want.

Pi with 200 decimals

Input interpretation:

π	200 decimal places

Result:

3.141592653589793238462643383279502884197169399375105820974944592307816406·.
286208998628034825342117067982148086513282306647093844609550582231725359·.
4081284811174502841027019385211055596446229489549303819 6

WolframAlpha

- Let's go back to one of our introductory examples:

  Do major

  From the output, not shown, we selected the part that we were interested in ("Music notation") by clicking on the symbol and choosing the desired option ("Formatted pod"). A new entry was generated showing only the musical scale.

```
WolframAlpha["Do major",
 IncludePods → "MusicNotation", AppearanceElements → {"Pods"},
 TimeConstraint → {30, Automatic, Automatic, Automatic}]
```

Music notation:

Play sound

If you click on **Play sound** you will hear the sound.

> The most important difference between using the free-form input notation and the Wolfram|Alpha one, is that in the first case the entry that we write is sent by the program to a Wolfram server that will return the syntax in *Mathematica* and it will be executed locally, while by using the Wolfram|Alpha notation, all the processing is done in the server and our computer only displays the output. If we have *Mathematica* it is better to use the free-form notation to get the *Mathematica* syntax and then we will be able to use the output easily in subsequent calculations.

Wolfram|Alpha is very useful from a browser since we don't need to have the program installed: There are applications *-Apps-* for smart phones and tablets that facilitate its use and *Widgets* developed with it.

Besides all the previously described ways to define the cell type, there is another one that consists of placing the cursor immediately next to the cell to which you want to add another one. You will see "+" below the cell to its left. Click on it and choose the cell type that you want to define: *Input*, , , Text, etc.

1.3 The Help System

If you have doubts about what function to employ, the best approach in many cases is to use the free-form input format and see the proposed syntax. However, if you would like to deepen your knowledge about the program you will have to use its help system.

The help system consists of different modules or guides (**Help ▸Wolfram Documentation**) from where you can access an specific topic with its corresponding instructions (**guide/...**). Additionally, you will find tutorials and external links. Clicking on any topic will show its contents.

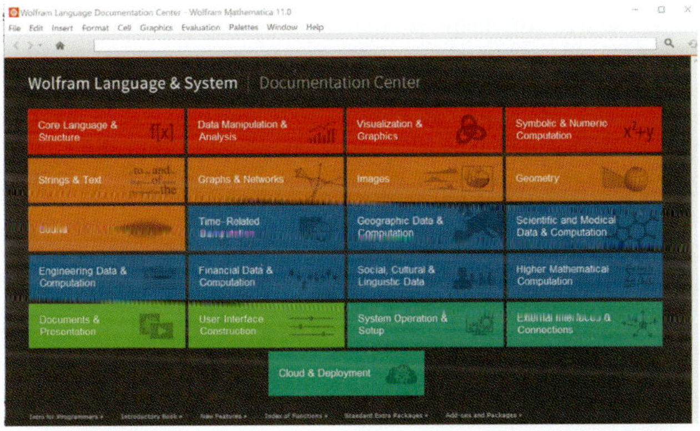

Figure 1.4 The Wolfram Language Documentation Center.

In the screen-shot shown above (Figure 1.4), after clicking on **Data Manipulation & Analysis ▸ Importing & Exporting** *Mathematica* displays the window below (Figure 1.5). Notice that besides giving us the functions related to the chosen topic, we can also access tutorials and even external resources such as video presentations.

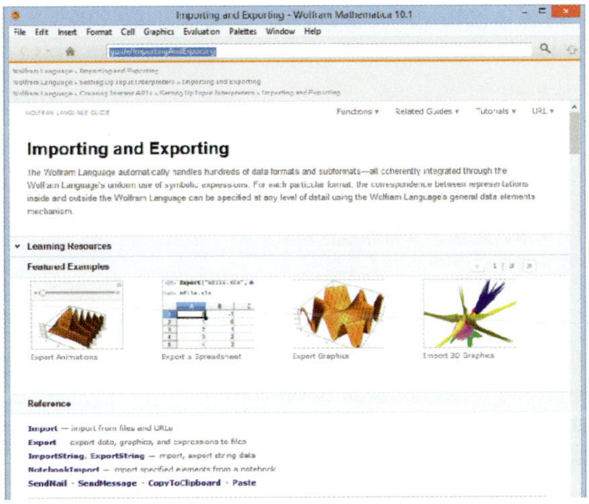

Figure 1.5 How to import and export using *Mathematica*.

If we don't have a clear idea of the function we want to use or we remember it vaguely, we can use the search toolbar (Figure 1.6) with two or three words related to what we are looking for.

Figure 1.6 The Search Toolbar.

Another type of help is the one given when a mistake is made. In this case we should pay attention to the text or sound messages that we may receive. For example, if we try to evaluate the following text cell, we will receive a beep.

"This is a text entry"

With **Help ▸ Why the Beep?...** we will be able to know the reason behind the beep.

Another additional source of help is the Input Assistant with its context-sensitive autocompletion and function templates features: when you start writing a function, *Mathematica* helps you automatically complete the code (similar to the system used by some mobile phones for writing messages). For example: If you start typing **Pl** *Mathematica* will display all the functions that start with those two letters right below them (Figure 1.7). Once you find an appropriate entry, click on it and a template will be added.

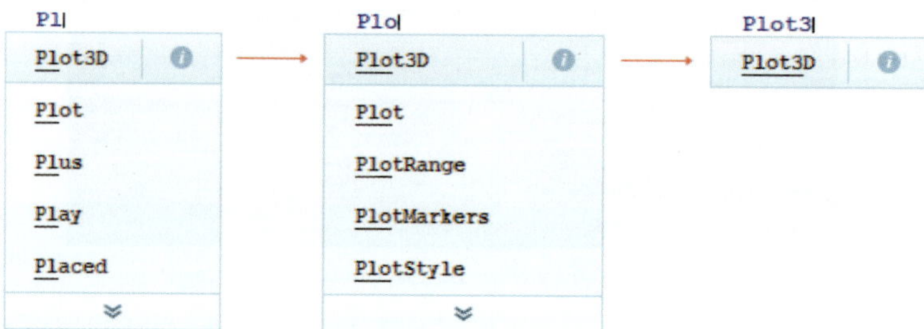

Figure 1.7 The Input Assistant in action.

Sometimes a down arrow will be shown at the end of the function offering us several templates (Figure 1.8). Choose the desired one.

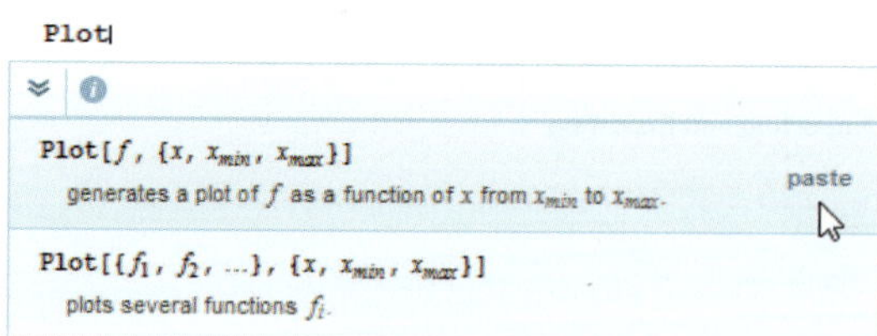

Figure 1.8 Function templates for the **Plot** function.

Additionally, a color code enables syntax error detection. If you have written a function and it is shown in blue, probably it has not been typed correctly. All *Mathematica* and user-defined functions will be shown in black. For example: Write in an input cell **Nmaximize**, it will be shown in blue, write **NMaximize** and you will see that it is shown in black and a template is displayed.

1.4 Basic Ideas

Users with prior knowledge of other programming languages such as C or FORTRAN, very often try to apply the same style (usually called procedural) in *Mathematica.* The program itself makes it easy to do so since it has many similar commands such as: For, Do, or Print. Although this programming style works, we should try to avoid it if we want to take full advantage of *Mathematica*'s capabilities. For that purpose we should use a *functional* style, that basically consists of building operations in the same way as classical mathematics.

This section is for both the uninitiated into *Mathematica* and those *Mathematica* users that still use procedural language routines. Because of that, if you have ever programmed in other languages please try not to use the same programming method.

In this brief summary we will refer to some of the most frequently used commands and ideas. We will usually not explain their syntax or we'll do it in a very concise way. For additional details, just type the command, select it with the cursor and press <F1>. If you don't understand everything that you read, don't worry, you will be given more details in later chapters.

> When you don't know the syntax of a function, type it, select it with the cursor and press the <F1> key.

1.4.1 Some Initial Concepts

- Let's begin by writing the expression below.

```
expr1 = 5 ! 6^2/∛2 π Cos[3 π] ⅇ^-7 Sin[1]
```

$$-\frac{2160 \times 2^{2/3}\,\pi \sin(1)}{e^7}$$

We have assigned it the name **expr1**, to be able to use the result of the calculations later on. To write the transcendental numbers π and e we have used special symbols (e is not the same as e, e is a letter without any other meaning) located in one of the palettes. If we prefer to use the keyboard we can write Pi for π and E^ or Exp[] for e. Notice that we use a single equal sign "=" to indicate sameness or equivalence in equations. For comparisons, we will use the double equal sign "==".

- The same operations can be written as in the cell below, using the typical *InputForm* notation.

```
(5 !  6^2/2^(1/3) Pi Cos[3 Pi] Sin[1]) / E^7 (*This is a commment*)
```

$$-\frac{2160 \times 2^{2/3}\,\pi \sin(1)}{e^7}$$

We have included a comment inside the actual cell "(* comment *)". This comment will be neither considered in the computation nor shown in the output.

Naturally, in both cases we obtain the same result. There is something that should draw our attention in this output. Apparently, there are terms that have not been evaluated and that appear again as such: π, Sin[1], y e^7. This is because *Mathematica* does not simplify or make approximations if that means losing precision. It literally works with infinite precision. Nevertheless, we can force it to compute the decimal approximation, the usual approach in other programs. To do that we will use the function **N** as shown in the next function. A similar result can be obtained by including decimals in some of the numbers.

- In this example we use the % symbol that recalls the last output. **//N** or **N**[*expr*] will use in the calculations machine precision even though the output may show fewer decimals.

```
% // N
```

−8.26548

- Another option is to substitute the number with a decimal approximation. You can try in the previous example using **Sin[1.0]** instead of **Sin[1]**.
- By default 5 decimals are shown, but with **N[*expr*, n]** we can use a precision **n** as large as desired.

```
N[expr1, 30]
```

−8.26547962656679448059436738241

Mathematica contains thousands of prebuilt commands or functions (Functions) that are always capitalized. The program differentiates between lower and upper cases. For example: `NMinimize` is not the same as Nminimize. Arguments are written inside square brackets [arg]. However, users can define a function using lower case.

Lists are a fundamental *Mathematica* concept, frequently used in many contexts. In a list elements are inside curly brackets {}. Later on we will refer to them.

> Remember: All functions in *Mathematica* are capitalized: **Log**; arguments are written inside square brackets: **Sin[2 x + 1]**; list elements are inside curly brackets: $\{a_1, ..., a_n\}$; ranges are written as lists: **Plot[x, {x, 0, 2 Pi}]** ; parentheses are used to group operations: **(Log[a x] + Sin[b x])/(a - b)**.

Use the help to investigate the various functions and their arguments. Remember that you can select a function with the cursor and press <F1> to get information about it. Besides its syntax, examples, tutorials and other related functions are displayed.

- Write **Pl** in an *input* cell. The context-sensitive input assistant will show you all the functions that start with Pl. Select Plot. Next to Plot a new down arrow will appear. Click on it and a template like the one shown below will be created.

```
Plot[f, {x, xmin, xmax}]
```

Figure 1.9 Template for the **Plot** function.

The cell above is an *input* type cell. If you'd like to avoid its evaluation: **Cell ▶ Cell Properties** and uncheck Evaluable. That is what we have done in this case in the original document since the idea is to see the cell content but not to execute it.

- In the previous template, all the basic function arguments are displayed. Now you can fill them in using the options as the next example shows. Execute the command.

```
Plot[Cos[x], {x, 0, 30}]
```

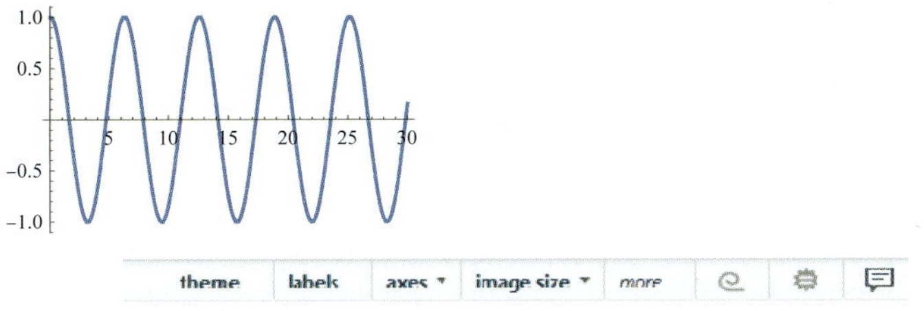

Figure 1.10 The Suggestions Bar for plots.

- The resulting output is a plot with the suggestion bar (Figure 1.10) appearing right below it. Since Version 9, thanks to the WolframPredictiveInterface, the Image Assistant and Drawing Tools provide point-and-click image processing and graphics editing. Click on **theme...** and a menu with several choices to customize the plot will unfold (Figure 1.11).

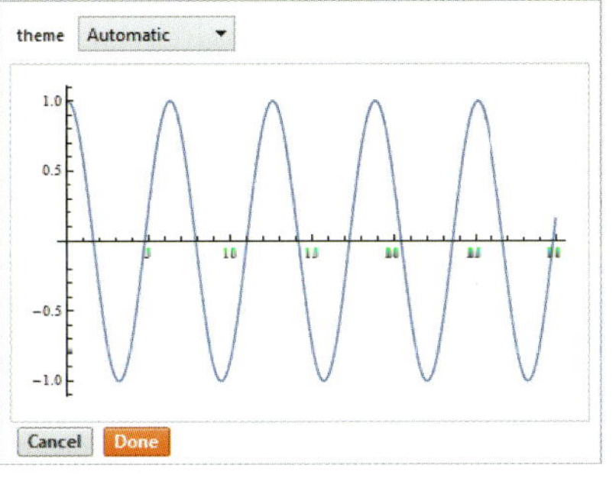

Figure 1.11 Customizing a plot using the "**theme...**" option in the Suggestions Bar.

- Alternatively, or if you are using a version prior to 9, you can customize the plot using the Plot options directly (type Plot, select it with the mouse and press <F1> to see them).

Besides arguments, commands also have options. One way of seeing those options directly is `Options`[*func*]:

- `Solve` options:

```
Options[Solve]
```

{Cubics → True, GeneratedParameters → *C*, InverseFunctions → Automatic, MaxExtraConditions → 0, Method → Automatic, Modulus → 0, Quartics → True, VerifySolutions → Automatic, WorkingPrecision → ∞}

Next, we'll show some examples. Replicate them using the **Basic Math Assistant** that includes templates for hundreds of commands.

- Definite integral example:

$$\int \mathtt{Sin[x]}^4\, \mathtt{Cos[x]}^3\, dx$$

$$\frac{3\sin(x)}{64} - \frac{1}{64}\sin(3x) - \frac{1}{320}\sin(5x) + \frac{1}{448}\sin(7x)$$

- Use `Simplify` anytime you need to simplify a complex expression.

```
Simplify[x^8 - 4 x^6 y^2 + 6 x^4 y^4 - 4 x^2 y^6 + y^8]
```

$(x^2 - y^2)^4$

There are also other commands to manipulate expressions. You can find them in the palette **Other ▶ Algebraic Manipulation**. Note that you can manipulate an entire expression or just part of it.

- Try to use it with this example simplifying the Sin and Cos arguments separately.

```
Sin[x^8 - 4 x^6 y^2 + 6 x^4 y^4 - 4 x^2 y^6 + y^8] / Cos[x^4 + 2 x^2 y^2 + y^4]
```

- To get:

```
Sin[((x^2 - y^2)^4)] / Cos[(x^2 + y^2)^2]
```

- In this example a random distribution of spheres in a three-dimensional space is shown.

```
Graphics3D[
 {Yellow, Sphere[RandomInteger[{-5, 5}, {10, 3}]], ImageSize → 250}]
```

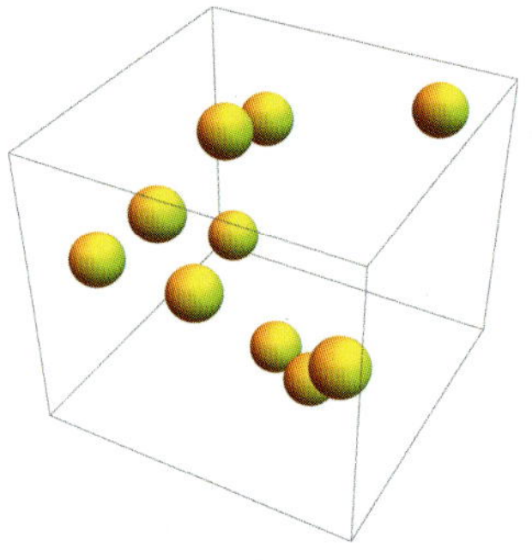

You can rotate the image to see it from different angles by clicking with the mouse cursor inside.

If you haven't used *Mathematica* previously, the last command may seem strange. Let's analyze it step by step:

- Highlight `RandomInteger` and press <F1>. The help page for the command will be displayed. In this case with **`RandomInteger[{-5, 5}, {10, 3}]`** we are generating 10 sublists with 3 elements, each element being a random number between -5 and 5. We are going to use them to simulate the coordinates $\{x, y, z\}$.
- The previous command is inside `Sphere` (do the same, highlight `Sphere` and press <F1>) a command to define a sphere or a collection of spheres with the coordinates $[\{\{x_1, y_1, z_1\}, \{x_2, y_2, z_2\}, \ldots\}, r]$, where r is the radii of the spheres. If omitted, as in this example, it's assumed that $r = 1$.

```
Sphere[RandomInteger[{-5, 5}, {10, 3}]]
```

$$\text{Sphere}\left[\begin{pmatrix} 2 & 3 & 0 \\ 4 & 1 & -1 \\ 4 & -4 & -3 \\ -5 & 2 & 4 \\ -2 & 3 & -2 \\ -4 & -5 & 2 \\ -1 & -3 & 3 \\ 1 & 5 & -5 \\ -4 & -2 & -1 \\ 5 & -4 & -1 \end{pmatrix}\right]$$

Note: In this particular notebook, list outputs are sometimes shown as matrices.

We use the command `Graphics3D` to display the graphs. We have added `Yellow` to indicate that the graphs, the spheres in this case, should be yellow and the option `ImageSize` to fix the graph size and make its proportions more adequate for this example. Finally we arrive at the command: **`Graphics3D[{Yellow, Sphere[RandomInteger[{-5, 5}, {10, 3}]], ImageSize → 250}]`**.

> When you don't understand an instruction consisting of several different commands, decompose it into its individual instructions and analyze each one separately using the help system.

- In the next cell we use `Rotate` to show a rotated output, in this case by 1 radian, compared to the usual display.

```
Rotate[1/Sqrt[1 + x], 1]
```

$$\frac{1}{\sqrt{1+x}}$$

Type inside the output and you'll see that it retains its properties.

- Practically anything can be used as a symbol. Generate a small red sphere.

```
Graphics3D[{Red, Sphere[]}, ImageSize → 30]
```

- Substitute the symbol thus obtained, 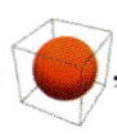, pasting it below replacing b and evaluating the cell.

```
Expand[(1 + b)^3]
```

$b^3 + 3\,b^2 + 3\,b + 1$

Expand[(1 + )^3]

^{3} + 3 ^{2} + 3  + 1

1.4.2 Replacements

A replacement consists of a rule to substitute one or several symbols with other symbols or values. To do that you apply the following syntax: "*exp* /. *rule*" that will replace *exp* with the contents of *rule*.

- In the expression $a\,x + b\,y$, a is replaced with 3 and b with 5, using a replacement rule.

```
exp1 = a x + b y /. {a → 3, b → 5};
```

- From now on anytime we call **exp1** we will see that the replacement has taken place:

```
exp1
```

$3\,x + 5\,y$

If you apply several replacement rules to the same expression in succession, the replacement takes places consecutively:

- We first apply the rule (**/.a→b**) to the expression $a\,x + b\,y$ and then the second rule (**/.b→c**) is applied to the previous result.

```
a x + b y /. a → b /. b → c
```

$c\,x + c\,y$

- In the expression below we make an assignment that consists of assigning the symbol **sol** to the solution of $a\,x^2 + b\,x + c == 0$.

```
sol = Solve[a x^2 + b x + c == 0, x];
```

- The output is a list.

```
sol
```

$$\left\{\left\{x \to \frac{-\sqrt{b^2 - 4\,a\,c} - b}{2\,a}\right\}, \left\{x \to \frac{\sqrt{b^2 - 4\,a\,c} - b}{2\,a}\right\}\right\}$$

> Remember that for assignments a single equal sign "=" is used, while to indicate an equality in an equation or a comparison, we will use a double equal sign "==" that when typed will be automatically converted to "==".

- To verify that the previous result represents the solutions to the equation $a\,x^2 + b\,x + c == 0$ we use the replacement "*exp* /. *rule*" that will replace in *exp* whatever is specified in *rule*.

```
a x^2 + b x + c == 0 /. sol
```

$$\left\{\frac{\left(-\sqrt{b^2 - 4\,a\,c} - b\right)^2}{4\,a} + \frac{1}{2\,a} b \left(-\sqrt{b^2 - 4\,a\,c} - b\right) + c = 0, \frac{\left(\sqrt{b^2 - 4\,a\,c} - b\right)^2}{4\,a} + \frac{1}{2\,a} b \left(\sqrt{b^2 - 4\,a\,c} - b\right) + c = 0\right\}$$

We can see that the replacement has taken place, but it's convenient to simplify it. For that we'll do as follows: we use `Simplify [%]` where % enables us to call the result (*Out*) of the last entry and evaluate the cell.

- We will check that effectively the equality holds and therefore the solution is correct. In this case it was evident, but we can apply the same method to more complex situations.

```
Simplify[%]
```

{True, True}

When an assignment is not going to be used later on, it may be convenient to delete it using the command `Clear`.

- The following command removes the assignment previously associated to **sol**.

```
Clear[sol]
```

- Now, when we type **sol** we see that it has no assignment.

```
sol
```

sol

1.4.3 Functions

> *Mathematica* makes it easy to define functions in a way very similar to regular mathematics. The usual syntax in the case of a function of one variable is: **f[x_]:= expr[x]**. The *blank* ("_") next to the variable is used to specify the independent function variable.

- Let's see it with an example.

```
f[x_] := 0.2 Cos[0.3 x^2]
```

- Now we can assign a value to the independent variable and we'll get the function value.

```
f[3]
```

−0.180814

- We could have also typed the previous two operations in the same cell (don't forget in this case to enter ";" at the end of each function). However, it's better to use one cell for each function until you have enough practice. It will help you find mistakes.

```
f[x_] := 0.2 Cos[0.3 x^2];
f[3]
```

−0.180814

- Now we are ready to visualize the Derivative (') of f[x] in a specific interval.

```
Plot[f'[x], {x, 0, 10}]
```

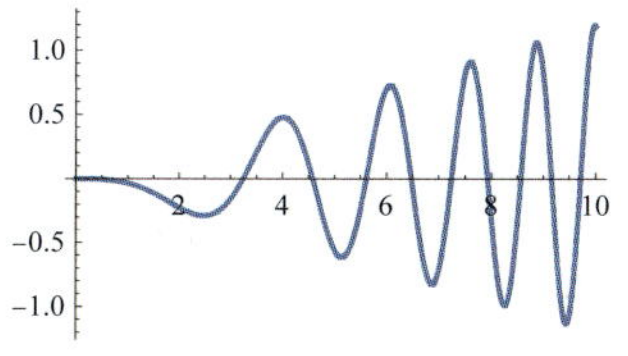

- The same approach can be extended to several variables. In this example we use the previously defined $f(x)$ function.

```
g[x_, y_] := f[x] 2.3 Exp[-0.3 y^2]
```

- Now we can present the result in a 3D graph using Plot3D. We use the option **PlotRange->All** to display all the points in the range for which the function is calculated, remove the option to see what happens.

```
Plot3D[g[x, y], {x, -2 π, 2 π}, {y, -2 π, 2 π}, PlotRange -> All]
```

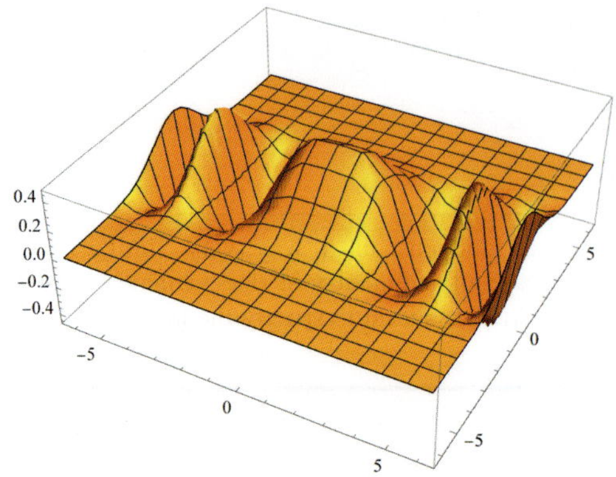

1.4.4 Dynamic Assignment of Variables

We can make dynamic assignments in which the symbol (make sure that in the menu bar **Evaluation ▶ Dynamic Updating Enable** is checked) returns an entry that changes dynamically as we make new assignments to it.

```
Clear[a]
```

- Let's create a variable.

```
Dynamic[a]
```

a

- If we now assign to **a** different values or expressions, you'll see that the output above keeps on changing.

```
a = √25
```

5

```
a = Integrate[x^2, x]
```

$\frac{x^3}{3}$

- We can also create a box with `InputField` and write any *input* in it.

```
InputField[Dynamic[a]]
```

a

- Try to use any f[x] in the box above, for example **Sin[x]**, and you will see how the next graph is updated. The process will repeat itself anytime you write a new function in the box.

```
Dynamic[Plot[a, {x, 0, 5}, Filling→ Axis]]
```

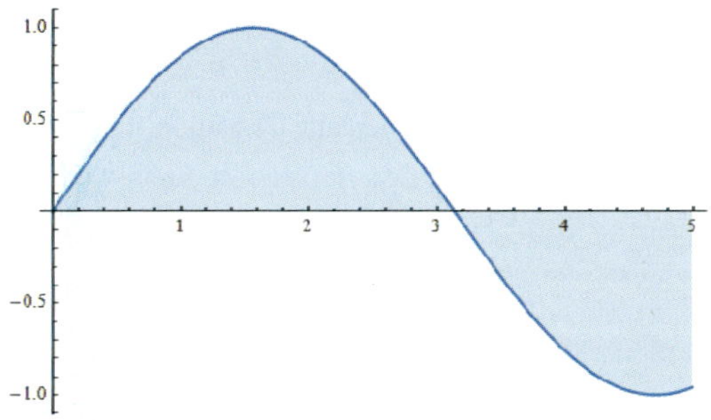


```
Clear[a]
```

1.4.5 List and Matrices

> Lists (List) are a fundamental concept in *Mathematica.* Lists or domains are written inside curly brackets with the elements separated by commas: {a, b, c}. For a description of list operations see the documentation: guide/ListManipulation. We'll describe them in more detail in later chapters.

- In the example below we generate a list with `Table`.

```
list1 = Table[{Sin[n], Cos[n]}, {n, 20}];
```

- We see that these are pairs of numbers corresponding to an ellipse that we can use to represent it graphically.

```
ListPlot[list1]
```

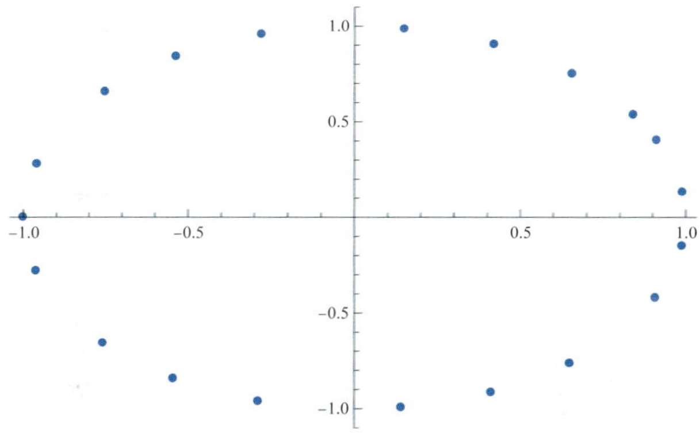

Matrices have list format and consist of lists of sublists.

- Let's create a matrix of random numbers.

```
mat = RandomInteger[10, {5, 5}];
```

- We add **`InputForm`** so you can see the output displayed in the internal format used by *Mathematica.*

```
mat // InputForm
```

```
{{4, 6, 3, 9, 8}, {3, 7, 0, 8, 6}, {5,
6, 10, 8, 10}, {3, 1, 10, 3, 8}, {0, 5,
0, 4, 5}}
```

- If desired, matrices can be presented in standard mathematical notation.

```
MatrixForm[mat]
```

$$\begin{pmatrix} 4 & 6 & 3 & 9 & 8 \\ 3 & 7 & 0 & 8 & 6 \\ 5 & 6 & 10 & 8 & 10 \\ 3 & 1 & 10 & 3 & 8 \\ 0 & 5 & 0 & 4 & 5 \end{pmatrix}$$

- Since they are lists, matrices can be handled as such. For example, here we extract the second element of the first sublist:

```
mat[[1, 2]]
```

6

For further details about matrix operations you can consult the documentation: guide/MatrixOperations.

1.4.6 Graphics

One of the most remarkable aspects of *Mathematica* is its graphical capabilities. The program's tool for 2D graphics can be accessed by pressing CTRL+D or by selecting in the menu bar **Graphics ▶ Drawing Tool**.

The most commonly used functions to represent graphics are `Plot`, `Plot3D` and `ListPlot`, of which we have already seen some examples. However, there are many more.

- Since *Mathematica* 9, when typing **`Plot`** the context-sensitive input assistant will show you all the words that include the word Plot, browse them. Alternatively type the following command and execute it (its output is omitted):

```
?*Plot*
```

The names that are shown can give you a hint about the type of plot they generate. Click on the chosen name to get more detailed information. In some cases you will be referred to the plot options.

- The graphical functions don't end there. The ones that include the word **`Chart`** are usually related to statistical graphs. Similarly to the previous case you can use the command:

```
?*Chart*
```

- Graphics-related functions include numerous options, as you can see in the case of **`Plot`**. Use the following command (its output is omitted):

```
Options[Plot3D]
```

- The next cell displays the surface $\cos(x)\sin(y)$ bounded by the ring $3 \leq x^2 + y^2 \leq 30$. Note that to set the region where the function exists we use the `RegionFunction` option.

```
Plot3D[Cos[x] Sin[y], {x, -2 π, 2 π},
  {y, -2 π, 2 π}, RegionFunction → (3 ≤ #1^2 + #2^2 ≤ 30 &),
  BoxRatios → Automatic, PlotStyle → Blue, ImageSize → Small]
```

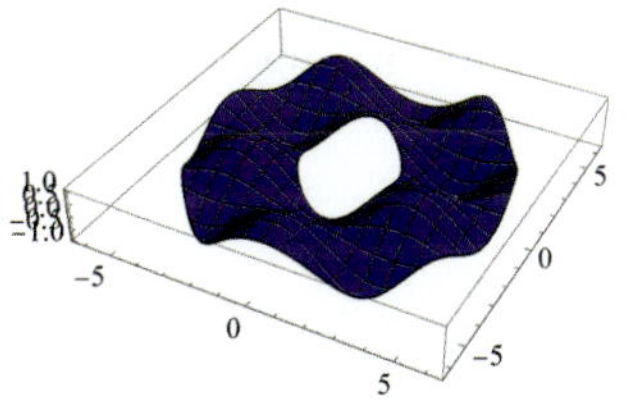

You can interact with 3D graphics: Click on the graph and you'll see that you can rotate it, with CTRL + click (o ⌘-click) you can zoom, and with SHIFT+click you can move it around.

- The code below generates the famous Lissajous curves. Using `Grid` we can show a two-dimensional mesh of objects, in this case graphs. We also use the `Tooltip` function to display labels related to objects (not only graphics) that will be shown when the mouse pointer is in the area of such objects. Here, when the mouse pointer hovers over the chosen curve, you'll see the function that generated it.

```
Grid[Table[Tooltip[ParametricPlot[{Sin[n t], Sin[m t]}, {t, 0, 2 Pi},
    ImageSize → 70, Frame → True, FrameTicks → None, Axes → False],
   {Sin[n t], Sin[m t]}], {m, 3}, {n, 3}]]
```

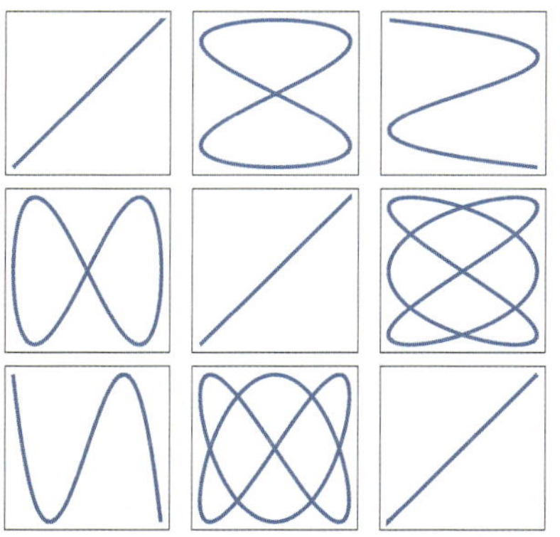

Other functions that do not include **`Plot`** or **`Chart`** but that are also useful for creating graphs are `Graphics`, `Graphics3D` and `ColorData`, as the following example shows.

- Representation of the water molecule.

```
Graphics3D[{Specularity[White, 50], ColorData["Atoms", "H"],
  Sphere[{0, 0, 0}, .7], Sphere[{1.4, 0, 0}, .7], ColorData["Atoms", "O"],
  Sphere[{.7, 0, .7}]}, Lighting → "Neutral", ImageSize → Small]
```

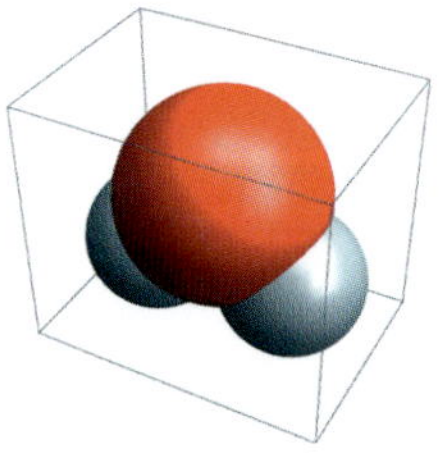

If you want to learn more about graphics, you can select the text with the cursor and press <F1>. For example: Select the text shown below and press <F1>:

tutorial/GraphicsAndSoundOverview

Graphics can be completed with legends using `PlotLegends` (if you are using a version prior to *Mathematica* 9 you will need to first execute **`Needs["PlotLegends`"]`**)

- In the following example we use **`AxesLabel`** to add labels to the axes and **`PlotLegends → "Expressions"`** to show the represented functions.

```
Plot[{Sin[x], Cos[x], Tan[x]}, {x, 0, 2 Pi},
 AxesLabel → {"x", "f(x)"}, PlotLegends → "Expressions"]
```

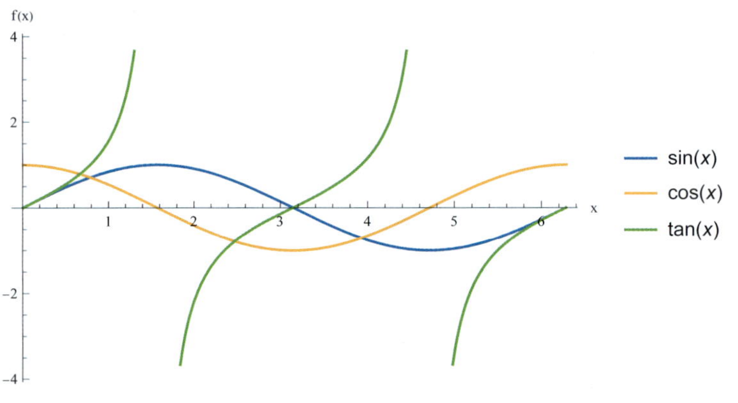

1.4.7 Images

Image processing has also experienced a significant improvement in the most recent versions of *Mathematica*. Starting with *Mathematica* 9, when clicking on an image you will see a menu with numerous options (Figure 1.12).

- The following command loads an image so we can play with it.

```
ExampleData[{"TestImage", "Lena"}]
```

Figure 1.12 The Image Assistant.

- Since version 9, a row of icons, the Image Assistant, appears below the image. Click on any of them. For example, after clicking on **more...** the following menu (Figure 1.13) will be deployed:

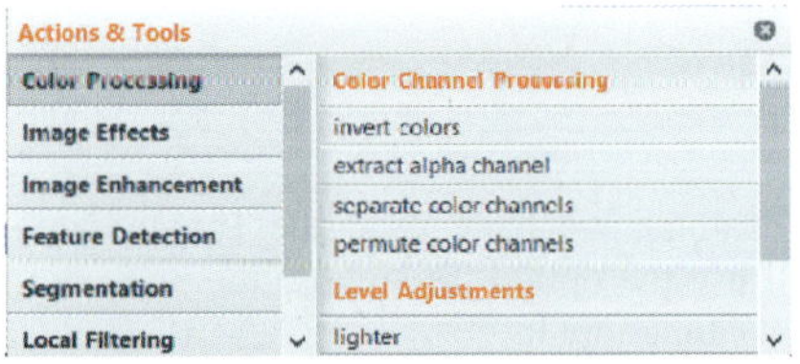

Figure 1.13 Additional capabilities available from the "**more... option**" in the Image Assistant.

- Play with the controls. When you click on one of them, a miniature version of the image will be shown where you will be able to check the effect on the original image. For example: When clicking on **invert colors** you'll see Figure 1.14.

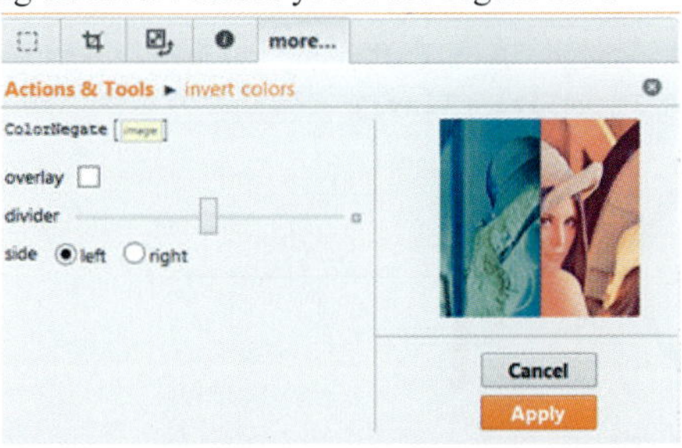

Figure 1.14 Inverting the colors using the Image Assistant.

After getting the desired effect press **Apply** and the change will be applied to the image. An alternative way is to copy the command that appears in the menu: **`ColorNegate`**[*image*] to obtain the same result. Copy it to an *input* cell and execute it. This last procedure is the appropriate one if you want to create a program or explain how the image has been modified. We will be using this method when we refer to image processing in later chapters.

1.4.8 Manipulate

Among the improvements in the most recent versions of *Mathematica* is the addition of instructions to enable dynamic and interactive operations. One of the most significant examples of this is the `Manipulate` command, as can be seen in the following examples.

- Type the command shown below. Note how sliders are created for each of the parameters being defined. Click on the + symbol included in the output to show the buttons created by `Manipulate`.

```
Manipulate[Plot[Sin[frequency x + alpha], {x, -2 π, 2 π},
  ImageSize -> Small], {frequency, 1, 5}, {alpha, 0, Pi/2}]
```

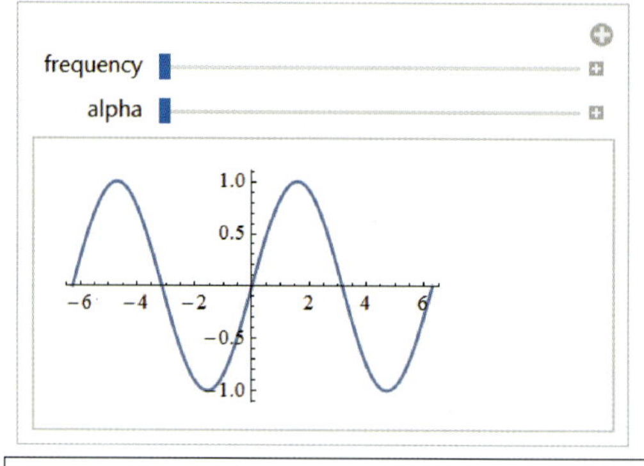

In the previous example we have used the **ImageSize → Small** option for printing purposes. We will use the option throughout the book but if you repeat the example in your computer you can ignore it.

1.4.9 Gauges

Since *Mathematica* 9 different types of gauges have been included that can behave dynamically and be customized:

```
Through[{AngularGauge, VerticalGauge, ThermometerGauge, HorizontalGauge}[
  42, {0, 100}, ImageSize→Tiny]]
```

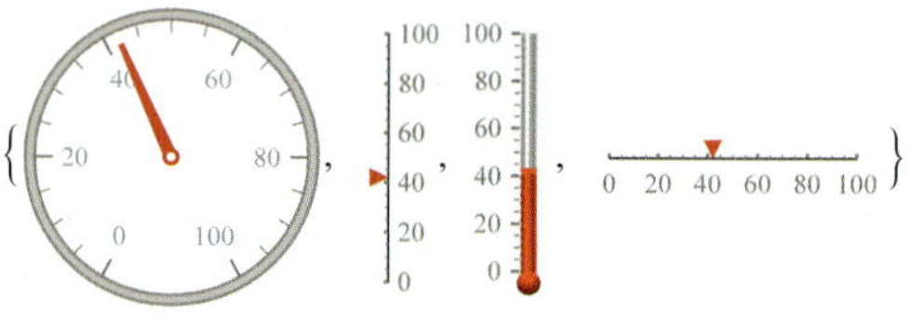

1.4.10 Handwritten Text Recognition

Among the latest new features is the inclusion of machine learning functions. Let's see an example.

- Handwrite different numbers on a white piece of paper, scan them or type them in a touch screen (in MS Windows you can use the **Crop** application) and establish their equivalence as shown in the next function.

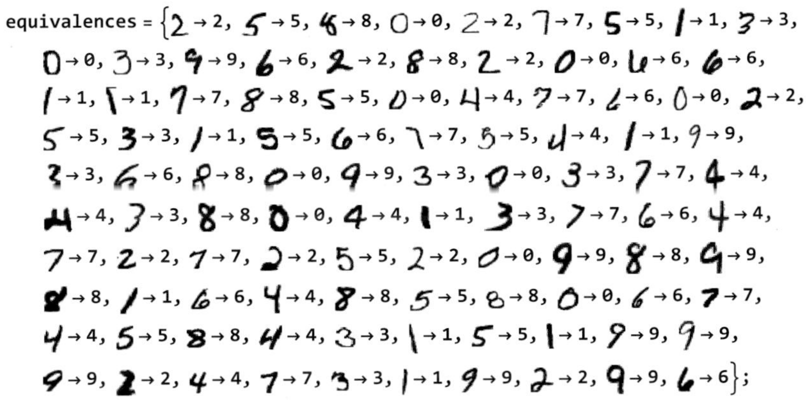

- Use the command `Classify` to establish equivalences based on the examples. Note that for a given number, its handwritten equivalent may not be the same. This function uses statistical criteria to assign a weight to each established equivalence.

```
digits = Classify[equivalences]
```

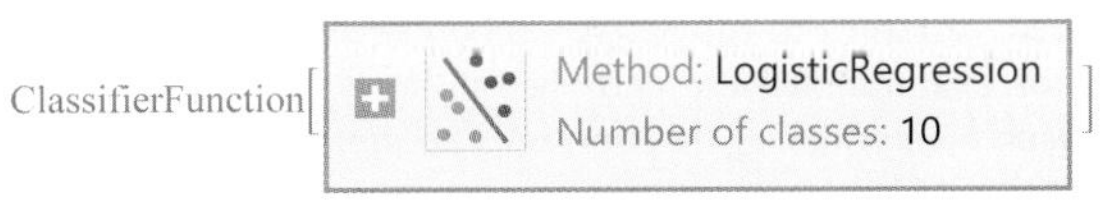

- Copy from above the numbers 0 to 10 in a list like the following:

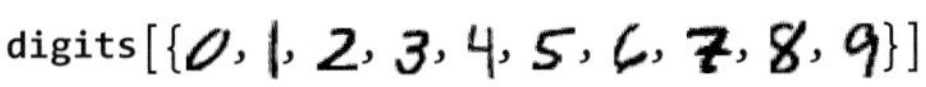

{0, 1, 2, 3, 4, 5, 4, 7, 8, 9}

- Note that the identification has been done correctly except in the case of 6 that has been mistaken with 4. The program assigns probabilities based on the stored data. You can check that the probability assigned to the symbol 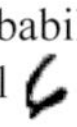 is higher for 4 than for 6. Even a person may not be sure whether the handwriting represents a 6 or a 4.

digits[, "TopProbabilities"]

{4 → 0.375915, 6 → 0.354657, 0 → 0.235305}

- If 6 is written in a less equivocal way the probability of a correct identification improves substantially.

digits[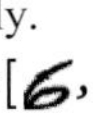, "TopProbabilities"]

{6 → 0.992954}

1.4.11 Things to Consider

Don't forget that for everything to work correctly, it's necessary that all the commands in the same session are executed sequentially. Cells are executed based on the order of evaluation and not on the order they are shown on the screen. This means that if a function calls previous functions, those functions should have been executed in advance. It's not enough to see them already typed. If you modify an assignment that is used in a later function, you will have to execute again all the related cells.

- Execute sequentially the following three cells:

```
a = 3 (*first*)
```

3

```
b = x^a (*second*)
```

x^3

```
a b(*third*)
```

$3x^3$

Now type **a = 2** in the first cell. Execute it and then the third one (**a b**) What has happened?

> ***Mathematica* retains information about all the variables defined in a session, even after starting a new *notebook*.** Because of that, when starting a new section that will not be using the variables previously defined it would be a good idea to clear them first.

- One way to remove all the values and definitions is as follows:

```
Clear["Global`*"]
```

Another way to achieve the previous result is to limit the application of the variables to a certain context through the menu bar: **Evaluation ▶ Notebook Default Context**. This offers the possibility to limit the defined variables to the active *notebook* or to certain cells grouped following a specific criterion, for example: all the cells belonging to the same section. As a matter of fact this chapter has been created using this option. The *input* counter will reset itself when a new group of cells is created (here it cannot be seen because we have suppressed the symbol In[*n*]=). Nevertheless, this method is less straightforward than **Clear["Global`*"]**. We should evaluate the most appropriate way based on the situation.

If you really want to remove all the information and not only the variables, the best way is to quit the session. This requires that you exit *Mathematica* (actually it requires that you exit the program *Kernel*). You can do that in several ways: With **File ▶ Exit** or if you don't want to exit the notebook use either **Evaluation ▶ Quit Kernel** or in an input cell type **Quit[]**. Once you have exited, the program removes from memory all the assignments and definitions. When you execute the next command the kernel will be loaded again.

Although *Mathematica* tries to maintain compatibility with *notebooks* created in previous versions, the compatibility is not always complete so you may have to make some modifications. When you open a notebook created in a previous *Mathematica* version for the first time, the program will offer you the possibility of checking the file automatically; Accept it and read the comments that you may receive carefully. In many cases *Mathematica* will modify expressions directly and let you know about the changes made giving you the option to accept or reject them. In other cases it will offer suggestions.

> Remember: If you don't understand a function use the Documentation Center.

1.5 Computational Capabilities

Early versions of *Mathematica* were mainly oriented toward symbolic computations. However, the latest versions of the program, apart from having significantly increased those capabilities, have complemented them with powerful functions for numeric calculations. An example of this combination is the large collection of probability related functions that provide seamlessly either symbolic or numeric answers depending on the inputs.

1.5.1 Equation Solving

We are going to show several available functions for solving different types of equations. In the examples we will use only one variable but the same functions can be used with a higher number of variables. Don't forget to use "==" to indicate equality in an equation.

- It is interesting to compare `Solve` and `Reduce`.

```
Solve[{x + a y + 3 z == 2, x + y - z == 1,
2 x + 3 y + a z == 3}, {x, y, z}]
```

$$\left\{\left\{x \to 1,\ y \to -\frac{1}{-a-3},\ z \to -\frac{1}{-a-3}\right\}\right\}$$

- `Reduce` generates all the solutions depending on the value of the parameter a and uses a more formal notation in its output.

```
Reduce[{x + a y + 3 z == 2, x + y - z == 1,
2 x + 3 y + a z == 3}, {x, y, z}]
```

$$\left(a = 2 \wedge y = \frac{1}{5}(5 - 4x) \wedge z = \frac{x}{5}\right) \vee \left(a - 2 \neq 0 \wedge x = 1 \wedge a + 3 \neq 0 \wedge y = \frac{1}{a+3} \wedge z = y\right)$$

- `Reduce` and `Solve` accept constraints regarding the variables domain.

```
Reduce[ x√(27^(2 x-1)) == √(9^(2 x-1)) , x , Reals]
```

$$x = \frac{1}{2} \vee x = 3$$

- `Solve` can solve geometrically formulated problems. In this case the solution to the intersection of a line passing through the points {0,0},{2,1}, with a circumference of radius $r = 1$ centered at {0, 0}(*Mathematica* uses `Circle` to refer to the circumference and `Disk` to refer to the circle, by default the commands assume a circle or circumference of $r = 1$, and centered at {0, 0}):

```
Solve[{x, y} ∈ InfiniteLine[{{0, 0}, {2, 1}}] && {x, y} ∈ Circle[], {x, y}]
```

$$\left\{\left\{x \to -\frac{2}{\sqrt{5}},\ y \to -\frac{1}{\sqrt{5}}\right\},\ \left\{x \to \frac{2}{\sqrt{5}},\ y \to \frac{1}{\sqrt{5}}\right\}\right\}$$

```
Graphics[{{Blue, InfiniteLine[{{0, 0}, {2, 1}}], Circle[]},
  {Red, Point[{x, y}] /. %}}]
```

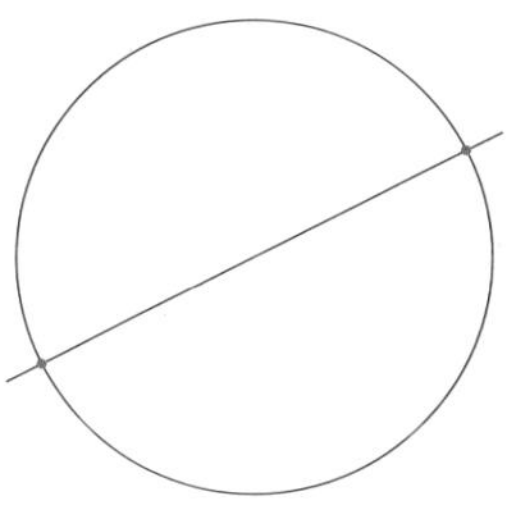

An analytical solution may not exist in many cases or we may be interested in a numeric result. In those situations we can use `FindRoot` or `NSolve`.

- With `FindRoot` you must define, for a given equation, the variable whose value you want to find and an initial starting point at which the search for the solution starts.

```
FindRoot[Cos[x] == x + Log[x], {x, 1}]
```

$\{x \to 0.840619\}$

```
NSolve[x^5 - 6 x^3 + 8 x + 1 == 0, x]
```

$\{\{x \to -2.05411\}, \{x \to -1.2915\}, \{x \to -0.126515\}, \{x \to 1.55053\}, \{x \to 1.9216\}\}$

1.5.2 Integration

- The symbolic integration capabilities of *Mathematica* are very powerful as shown in the example below.

```
Integrate[Exp[1 - x^2], x]
```

$\frac{1}{2} e \sqrt{\pi} \operatorname{erf}(x)$

- However, there are times when symbolic integration is not possible and is necessary to integrate using numerical methods (`Log` refers to the base e logarithm, to indicate the base 10 logarithm type Log[10, *expr*]).

```
NIntegrate[Log[x + Sin[x]], {x, 0, 2}]
```

0.555889

- It is also possible to limit the integration to a certain region by using the `Boole` function. In this example this region is a cone with radius 1.

```
Integrate[ (x² + y²) Boole[0 ≤ z ≤ 1 && x² + y² ≤ z²],
 {x, -1, 1}, {y, -1, 1}, {z, 0, 1} ]
```

$\frac{\pi}{10}$

- With `RegionPlot` we can represent the integration region, a feature that from an educational point of view may be of interest.

```
RegionPlot3D[0 ≤ z ≤ 1 && x^2 + y^2 ≤ z^2,
 {x, -1, 1}, {y, -1, 1}, {z, 0, 1}, ImageSize → Small]
```

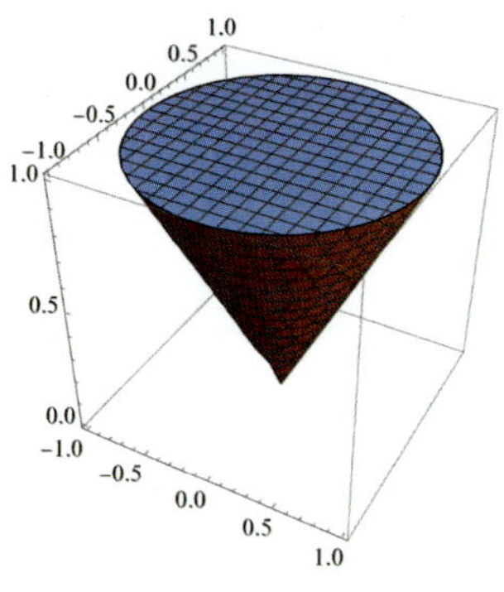

We can even specify assumptions, with `Assumptions`, about the parameters of the function that we wish to integrate.

- In this example we specify that when integrating x^n, n is greater than 1.

```
Integrate[x^n, {x, 0, 1}, Assumptions -> n > 1]
```

$$\frac{1}{n+1}$$

- We can also integrate over a region. In this example, over a sphere centered at {0,0,0} and with radius r>0. Obviously, we are calculating the volume of the sphere.

```
Integrate[1, {x, y, z} ∈ Ball[{0, 0, 0}, r], Assumptions -> r > 0]
```

$$\frac{4 \pi r^3}{3}$$

1.5.3 Sums, Logical Operations, Simplifications and Differential Equations

- We can evaluate sums and products with finite and infinite terms.

$$\sum_{k=1}^{\infty} \frac{1}{k^6}$$

$$\frac{\pi^6}{945}$$

- We can also perform logical operations.

```
Log[2] < 1.4 < √2
```

True

- To check whether two expressions are equivalent "===" can be used. Note that 3 y 3.0 in *Mathematica* are different numbers.

```
1 === 3/3
```

True

```
1 === 3.0/3
```

False

- Next, we solve a differential equation for a pendulum with damping constant k. Using `Manipulate` we can make the value of k vary from 0 to 3 and plot the solution. By using `Evaluate` we are forcing *Mathematica* to solve the differential equation first, and then substitute x and k with the given values. Note that the use of the pattern "**y[x]/.** *something*". " /." means "replace with" (in the example: y[x] is replaced by the equation solution).

```
Manipulate[Plot[Evaluate[y[x] /.
    DSolve[{y''[x] + k y'[x] + 40 y[x] == 0, y[0] == 1, y'[0] == 1/3}, y[x], x]],
  {x, 0, 4}, ImageSize → Small], {k, 0, 3}]
```

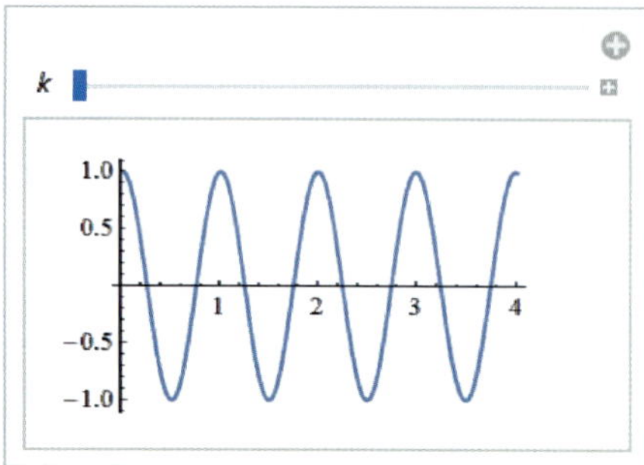

1.5.4 Application: The Lorenz System

There are many functions available for numerical methods, such as `NDSolve`, used when solving differential equations numerically.

- An example of the use of `NDSolve` is its application to the well-known Lorenz system, a simplified model to study atmospheric convection movements that exhibit chaotic behavior.

```
eqs = {x'[t] == -3 (x[t] - y[t]),
   y'[t] == -x[t] z[t] + 26.5 x[t] - y[t], z'[t] == x[t] y[t] - z[t]};
ics = {x[0] == z[0] == 0, y[0] == 1};
sol = NDSolve[{eqs, ics}, {x, y, z}, {t, 0, 200}, MaxSteps → ∞]
```

{{x → InterpolatingFunction[Domain: {{0., 200.}} Output: scalar],

y → InterpolatingFunction[Domain: {{0., 200.}} Output: scalar],

z → InterpolatingFunction[Domain: {{0., 200.}} Output: scalar]}}

- The solution to the previous equation in the phase space is displayed using `ParametricPlot3D`.

```
ParametricPlot3D[{x[t], y[t], z[t]} /. sol[[1]], {t, 0, 200},
 PlotPoints → 1000, ColorFunction → (Hue[#4] &), ImageSize → Small]
```

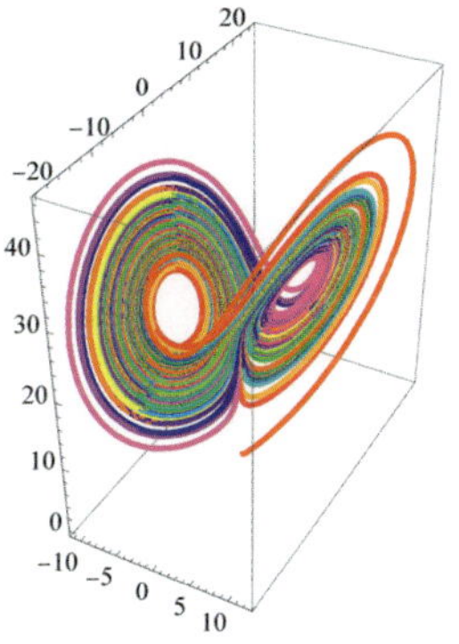

1.5.5 Statistical Calculations

Mathematica probably includes as many functions, if not more, than other specialized statistical software programs. The downside is that users must have a deep understanding of what they are trying to accomplish.

- This example shows the probability density function (pdf) and the cumulative density function (cdf) of a Poisson distribution with mean $\mu = 10$. An option in `PlotLegends` is included to place the legend in the desired position.

```
pdf = PDF[PoissonDistribution[μ], x]
```

$$\begin{cases} \frac{e^{-\mu}\mu^x}{x!} & x \geq 0 \\ 0 & \text{True} \end{cases}$$

```
cdf = CDF[PoissonDistribution[μ], x]
```

$$\begin{cases} Q(\lfloor x \rfloor + 1, \mu) & x \geq 0 \\ 0 & \text{True} \end{cases}$$

```
DiscretePlot[Evaluate[{pdf, cdf} /. μ → 10], {x, 0, 15}, PlotRange → All,
 PlotLegends → Placed[{"Probability", "Cumulative Probability"}, Below]]
```

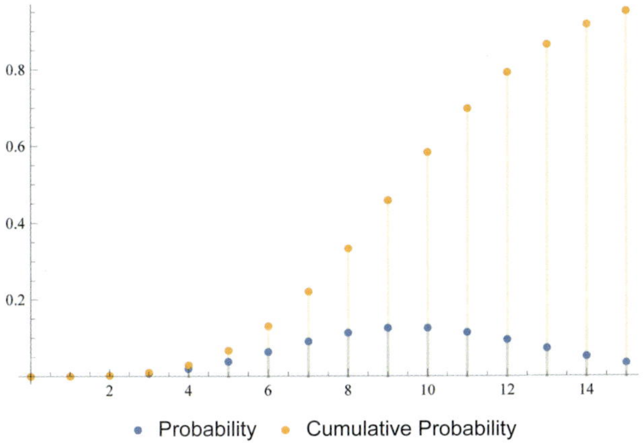

1.6 Utilities

1.6.1 Example Data

The installation of the program includes a set of files with data from different fields that can be used to test some of the functions. To download them use the command `ExampleData`.

- Let's download the original text of Charles Darwin's book *On the Origin of Species*:

```
txt = ExampleData[{"Text", "OriginOfSpecies"}];
```

- To find out how many times the word "evolution" appears compared to the word "selection" we use the function `StringCount`, specifying the chosen word or text.

```
StringCount[txt, "evolution"]
```

4

```
StringCount[txt, "selection"]
```

351

1.6.2 Accessing External Data

One of the most important features of *Mathematica* since Version 8, already mentioned, is its integration with **Wolfram|Alpha** (tutorial/DataFormatsInWolframAlpha). This enables us to access information about practically anything and once downloaded it can be used in the *Mathematica* environment to perform further calculations.

- Type ⊜ `Japan vs. Germany`. Several frames (pods) will be displayed, each of them containing information about an aspect of the question. In the right corner of each frame the symbol ⊞ appears. Click on one of them and you will see a menu with several options. In this example, we chose the **Geographic Properties** frame and clicked on the **Subpod content**. The following entry was generated afterward:

```
WolframAlpha["Japon vs. Germany",
 IncludePods → "GeographicProperties:CountryData",
 AppearanceElements → {"Pods"},
 TimeConstraint → {30, Automatic, Automatic, Automatic}]
```

Geographic properties: More | Show non-metric

	Japan	Germany
total area	377 835 km² (square kilometers) (world rank: 62nd)	357 022 km² (square kilometers) (world rank: 63rd)
land area	374 744 km² (square kilometers) (world rank: 62nd)	348 672 km² (square kilometers) (world rank: 63rd)
continent	Asia	Europe

- We repeat the same steps for "China vs USA" but in this case instead of **Subpod content** we choose **ComputableData**. The output is shown below. This has the advantage of being generated as a list allowing easy subsequent manipulation.

```
WolframAlpha["China vs USA",
 {{"GeographicProperties:CountryData", 1}, "ComputableData"}]
```

	China	United States
total area	9.597×10^6 km²	9.631×10^6 km²
land area	9.326×10^6 km²	9.162×10^6 km²
continent	Asia	North America

In the **WolframAlpha** style you can select the information you are interested in by choosing from the menu that appears after clicking on + ▶ **Subpod content** or **ComputableData** in the desired frame. An *input* will be generated to get only the desired information.

You can also access external data through Computable Data (guide/ComputableDataOverview), a group of functions specific to *Mathematica*, to obtain the desired information. We will cover them in more detail in Chapter 5.

- The graph below shows the evolution of the gross domestic product, GDP, of China from 1980 to 2015.

```
DateListPlot[CountryData["China", {{"GDP"}, {1980, 2015}}]]
```

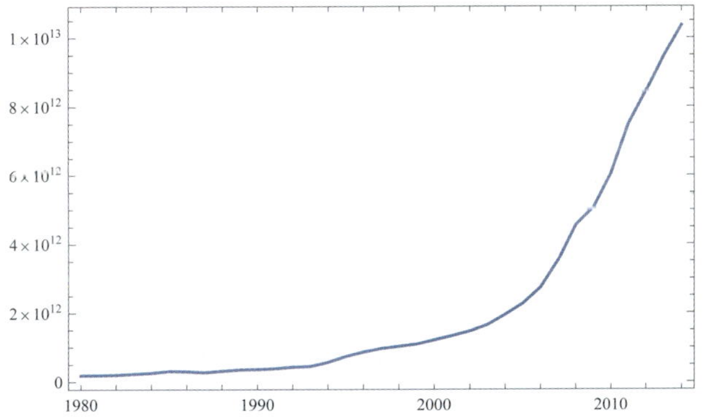

- We can even obtain information from Wikipedia using the new function (available since *Mathematica* 10.1): `WikipediaData`.

```
WikipediaData["Alhambra", "ImageList"][[5]]
```

1.6.3 Application: Everybody Out

We would like to estimate the amount of energy required to move the entire population off-planet: https://what-if.xkcd.com/7/. Without considering the weight of the rockets, the idea is to see how we can solve this problem accessing external data from within *Mathematica*. The same approach can be useful to solve other types of problems.

The data that we need to know are: the kinetic energy equation (it's been such a long time since we took high-school physics that we have forgotten it) and the values of its components. We will need to know the world population as well.

- The kinetic energy equation can be found using `FormulaData`.

```
FormulaData["KineticEnergy"]
```

- Or the free-form input format: typing **Kinetic Energy Equation**.

```
Kinetic Energy Equation
```

$$K = \frac{1}{2} m v^2$$

- From the equation we can see that we need to know: *m* (mass per person) and *v* (escape velocity from Earth). Both values can be obtained with the free-form input format typing: **average adult weight** and **earth's escape velocity** or similar statements.

```
m = average weight adult
```

82 kg

```
v = earth's escape velocity
```

1.118×10^4 m/s

- With these data we obtain the energy per person. We need to multiply them by *n,* the world population.

```
n = QuantityMagnitude[CountryData["World", "Population"]]
```

7.13001×10^9

- Now, after evaluating the kinetic energy equation, we get the necessary kinetic energy to put the entire world population in orbit. Naturally this is just a toy example that has nothing to do with reality since we would need to consider the mass of the launcher and the fact that people would have to go in spaceships that have masses much bigger than that of their passengers. Until now we have sent about 600 people into space. The trip in a Soyuz spacecraft costs about 20 million dollars per person. Multiply that figure by the number of people living on earth and the result is thousands of times bigger than the world's GDP. Let's hope that we will not need to put the entire human population into orbit!

```
UnitSimplify[1/2 m v^2 * n]
```

3.65391×10^{19} J

1.6.4 Interactive Visualizations (Demonstrations)

In http://demonstrations.wolfram.com you can find thousands of small interactive applications of great educational value. In this book we'll refer to them using the word "demonstrations". If you don't know how to build them just download and open the files; later on you'll be able to modify them to suit your needs or even make new ones. In Chapter 4 we'll learn how to create them.

- This example explores *Coulomb's law*. This law describes the interaction between two charges (based on '*Potential Field of Two Charges*' created by Herbert W. Franke: http://demonstrations.wolfram.com/PotentialFieldOfTwoCharges/).

```
Manipulate[ContourPlot[q1 / Norm[{x, y} - p[[1]]] + q2 / Norm[{x, y} - p[[2]]],
  {x, 2, -2}, {y, 2, -2}, Contours → 10],
 {{q1, -1}, -3, 3}, {{q2, 1}, 3, -3},
 {{p, {{-1, 0}, {1, 0}}}, {-1, -1}, {1, 1}, Locator}, Deployed → True]
```

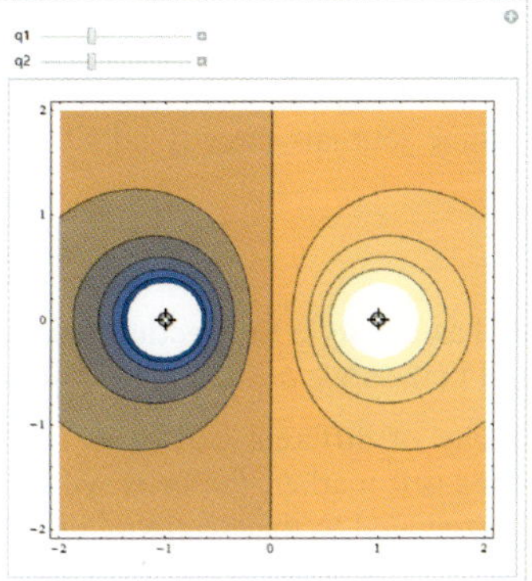

Move the locators around with the mouse to see how the potential fields change.

1.6.5 Packages

Commands can be used directly or can be put together to develop applications for specific purposes (*packages*). The packages included in *Mathematica* can be seen in **Documentation Center ► StandardExtraPackages** and **Documentation Center ► InstalledAddOns**. StandardExtraPackages are the ones included in the program installation. Many of those functions now in the packages will probably be part of the kernel of the program in the future.

Installed AddOns are packages developed by users for specific purposes. Some of them are available in http://library.wolfram.com/infocenter. This book's author has developed several packages, some of them available in http://diarium.usal.es/guillermo.

Normally these packages will be copied in Addons/Applications, located in the *Mathematica* installation directory.

To load a package you can type Needs["package name`"] or <<package name`. For example, the following package includes various functions related to the properties of the radiation emitted by a black body.

```
Needs["BlackBodyRadiation`"]
```

```
?"BlackBodyRadiation`*"
```

▼ BlackBodyRadiation`

BlackBodyProfile	MaxPower	PeakWavelength	TotalPower

- The following package function calculates the peak wavelength of a black body for a given temperature.

```
PeakWavelength[5000 Kelvin]
```

5.79554×10^{-7} Meter

In a later chapter we'll give a brief introduction to package development.

1.7 Editing Notebooks

We've seen that the interaction with *Mathematica* is done through the Front End (what we usually see when using *Mathematica*), and this generates files named "*Mathematica notebooks*". What you are currently reading is one such *notebook*.

Each *Mathematica notebook* (file.nb) is a complete and interactive document containing text, tables, graphics, calculations and other elements. The program also enables high-quality editing so notebooks can be ready for publication (some journals directly admit papers in *nb* format).

1.7.1 Notebook Structure

Each cell has a style. For example, the previous cell has the style **Subsection** associated to it. To do that we just need to select its cell marker (]) and in the *style box* choose Subsection.

Notebooks are organized automatically into a hierarchy of different cell types (title, section, input, output, figures, etc). If you click on the exterior cell marker, the cells will be grouped. For example, if you click on a section, all the cells associated to that section will collapse and you will only see the title of the section. You just need to click again to see them back.

This is an ordinary text with the color formatted Format ▶ Text Color.

Click on a text cell and modify its format.

With Insert ▶ Hyperlink we can create hyperlinks to jump from one location to another inside the same *notebook,* to another *notebook* or to an external link.

Using **Format ▶ Text Color** we can choose the font, **face**, size, **color**, etc. We can even use special characters such as ⍋.

This *notebook*, like any other, has a predefined style. The style tells *Mathematica* how to display its contents. There are many such styles. To choose the *notebook's* overall style you can use **Format ▶ Stylesheet**. With **Format ▶ Style** you can define the style of a particular cell. Additionally, inside a style, we can choose the appropriate working environment **Format ▶ Screen Environment**. Try changing the appearance of a *notebook* by modifying the working environment.

1.7.2 Choosing Your Style

To edit an article, book or manual the **Writing Assistant** palette is of great help. Load it. Note that is divided into three sections: *Writing and Formatting*, *Typesetting* and *Help and Settings*. See the help included in the palette.

If you click inside the palette on the button **Stylesheet Chooser**... (alternatively you can go to **Format ▶ Stylesheet ▶ Stylesheet Chooser**...) a separate window with icons representing different styles will appear (Figure 1.15). If you press on the upper part of any of the icons (**New**) a new sample *notebook* formatted according to that style will be generated. You will be able to use it and modify its contents to suit your needs. If you click on the bottom part of the icons (**Apply**), the style will be applied to the currently active *notebook.* For example, in the next screenshot (Figure 1.15), by clicking on the upper part of the Textbook icon, a *notebook* with a Textbook template style will be opened.

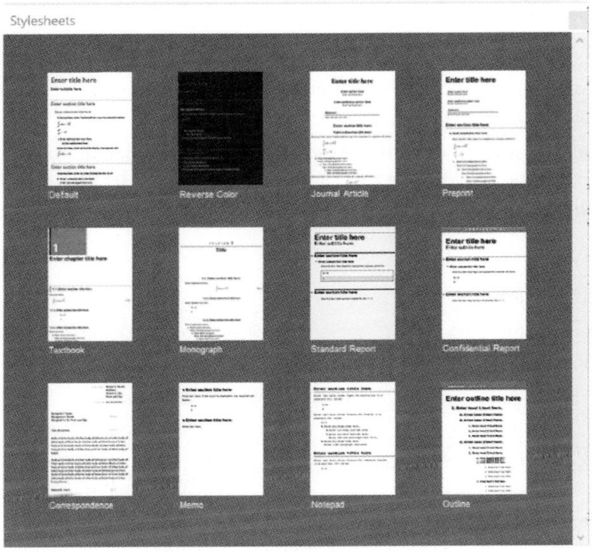

Figure 1.15 The Stylesheet Chooser.

When a new cell is created, depending on the stylesheet, by default it will be of a certain format type, for example with the *Default* stylesheet new cells are of type *Input*.

> When typing in a new *notebook* choose the stylesheet from the beginning with **Format ▶ Stylesheet ▶ Stylesheet Chooser...** . After clicking on the upper side of the icon representing the desired style (New), you will be able to use and modify the template according to your needs.

1.7.3 Typesetting Advice

If you are interested in emphasizing text containing formulas using classical mathematical notation, begin by typing in a text cell. Select the formula and define it as Inline Math Cell by pressing CTRL + 9 or, in the Writing Assistant palette, choose **Math Cells ▶ Inline Math Cell**. Finally, in the same palette in Cell Properties click on Frame and select the appropriate one . That's how the following example was done:

> ¿What is the domain of the function $f(x) = \sqrt{x^2 - 1}$?

If you'd like to type a sequence of formulas and align them at the equal signs, in Writing Assistant, choose **Math Cells ▶ Equal Symbol Aligned Math Cell**.

$$\begin{aligned} a + b &= c \\ a &= c - b \end{aligned}$$

Figure 1.16 Aligning equations at the equal sign.

Sometimes it may be useful to type both, the formula and its result, as part of the text. This can be done by writing and executing the function in the same text cell. For example: using a palette write $\int x^3 \, dx = \int x^3 \, dx$. Now select the second term of the equation and press

SHIFT+CTRL+ENTER. Or, in the menu bar, choose **Evaluation ▸ Evaluate in place**, and the previous expression will become $\int x^3\, dx = \frac{x^4}{4}$.

When creating a document, you may be interested in seeing only the result. In that case you can hide the Input cell.

- To hide the input just click directly on the output marker cell. The following example was created by executing `Plot[Sin[x], {x,-2 Pi, 2 Pi}]` and then hiding the command to display only the graph.

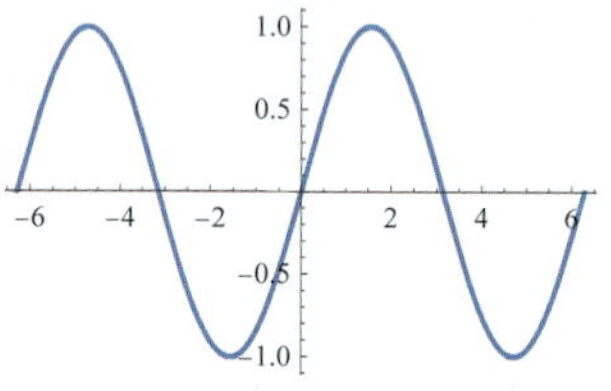

All *Mathematica* commands are in English but the menus and palettes can also be displayed in any of the additional 12 languages offered since version 11. The alternatives include Chinese, French, German, Italian, Japanese, Korean and Spanish among others. Users can take advantage of this functionality by going to **Edit ▸ Preferences ▸ Interfaces**. Once an option different from English is selected, every function will be automatically annotated with a "code caption" in the chosen language. *Mathematica* 11 also includes a real-time spell checker that works with any of the supported languages. Alternatively, users of previous *Mathematica* versions can review the English contents of a notebook with **Edit ▸ Check Spelling...** .

With **Edit ▸ Preferences** and with **Format ▸ Option Inspector** (see tutorial/OptionInspector) you will be able to customize many features of *Mathematica* that can be applied globally, to a *notebook* or to a particular cell. For example: with **Format ▸ Option Inspector ▸ Global Options ▸ File Locations** you can change *Mathematica* default directories.

1.7.4 The Traditional Style

We've seen that *Mathematica* by default uses the style: `StandardForm` in the input cells (*Input*) by default. However, if we'd like to create better looking documents, it's recommended to use the `TraditionalForm` style that is similar to the one used in traditional mathematical notation.

If the main purpose of a document is to be read by others, it may be convenient to present the inputs and outputs using the traditional notation. It's likely that the reader will not realize that it was created with *Mathematica*. This is specially true if the document is saved using the cdf format to which we will refer in the following section.

> You can write your inputs in **`StandardForm`**. Once you verify that they work you can convert them to the traditional style.

- Example: Let's type:

$$\texttt{Limit}\left[\frac{(x + \Delta x)^2 - x^2}{\Delta x}, \Delta x \to 0\right]$$

$2x$

- We convert the cell to the traditional form by selecting it and clicking on: **Cell ▸ Convert to ▸ TradicionalForm**. Then the cell above will be shown as follows:

$$\lim_{\Delta x \to 0} \frac{(x + \Delta x)^2 - x^2}{\Delta x}$$

$2x$

> If an entry is in the traditional style, to see how it was originally typed convert it to the standard style by selecting the cell and choosing from the menu bar **Cell ▶ Convert to ▶ StandardForm**.

We can make an output to be shown in the traditional form by adding `//TraditionalForm` at the end of each *input*. There are several ways to apply this style to all the outputs in a notebook. In **Edit ▶ Preferences ▶ Evaluation** you can specify for all the outputs to be displayed in the traditional form (`TraditionalForm`). Another easy one is to use the following command:

```
SetOptions[EvaluationNotebook[],
 CommonDefaultFormatTypes -> {"Output" -> TraditionalForm}]
```

It's recommended to include the command above in a cell at the end of the notebook and define it as an initialization cell (click on the cell marker and check **Cell ▶ Cell Properties ▶ Initialization Cell**). This way, the cell will be executed before any other cell.

In many of the styles available in **Stylesheet Chooser...** outputs are displayed using the standard form. In this chapter we have used a customized stylesheet that shows outputs in the traditional format. In the rest of the book, we will almost always use the standard form since what we are trying to highlight is how to create functions with *Mathematica* and for that purpose the classical format is not adequate.

1.7.5 Automatic Numbering of Equations and Reference Creation

Textbook and scientific articles frequently use numbered equations. This can be done as follows:

- First open a new notebook (**File ▶ New ▶ Notebook**) and select one of the stylesheets available in the format menu. For instance **Textbook: Format ▶ Stylesheet ▶ Book ▶ Textbook**.
- Then create a new cell and choose **EquationNumbered** as its style (**Format ▶ Style ▶ EquationNumbered**). The cell will be automatically numbered, then you can write the equation, for example: $a + b = 1$.

$$a + b = 1 \qquad (1.1)$$

If you have opened a new notebook using the Textbook style, you will probably see (0.1) instead of (1.1). It's possible to modify the numbering scheme but it is beyond the scope of this chapter. When you need to number equations, sections, and so on, remember that it is a good idea to select the stylesheet directly from the Writing Assistant palette: **Palettes ▶ Writing Assistant ▶ Stylesheet Chooser...** . After selecting **Textbook (New)**, a new notebook will be generated with many different numbered options such as equations, sections, etc. You can then use that notebook as a template.

If you wish to insert an automatic reference to the equation proceed as follows:

- Write an equation in a cell with the **EquationNumbered** style. Then go to **Cell ▶ Cell Tags ▶ Add/Remove...** and add a tag to the formula. Let's give it the tag: "par" (we write "par", from parabola, although we could have used any other name). A good idea would be also to go to **Cell ▶ Cell Tags** and check **▶ Show Cell Tags**, keeping it checked while creating the document, and only unchecking it after we are done. This will enable us to see the cell tags at all times. The first equation done this way is written below:

par

$$x^2 = 1 \qquad (1.2)$$

- If you wish to refer to this equation: Type the following: (1.), put the cursor after the dot and go to **Insert ▸ Automatic Numbering**. In the opening dialog (Figure 1.17) scroll down to **EquationNumbered** and choose the tag of the equation, in our case: "par" then press OK (see the screen below). After this has been done the number of the equation appears between the parentheses: (1.2). If you click on the "2" the cursor will go immediately to the actual equation.

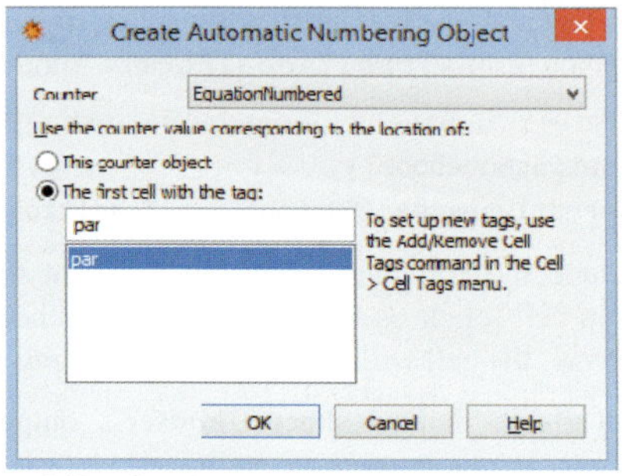

Figure 1.17 Creating an automatically numbered object.

1.7.6 Cell Labels Display

As we have mentioned previously, any time we execute an input cell, a label in the form In[*n*] is shown, but we can choose not to display it. This can be done from **Edit ▸ Preferences** or typing **SetOptions[SelectedNotebook[], ShowCellLabel → False]**. You can also reset the counter (set In[*n*] to 1) in the middle of a session by typing the following in an input cell: $Line=1.

1.8 Sharing Notebooks

1.8.1 Save as...

Documents created with *Mathematica* are saved with extension *.nb, *Mathematica's* own format, by default. If you share any of these documents with someone, that person will require access to *Mathematica.*

However, *Mathematica* can also save documents in other formats. To do that, in the menu bar, choose **File ▸ Save As ▸**. Then, from the list of options, select the desired one.

Very often we may be interested in saving a document in *pdf* format to reach a broad audience. We can also use *TeX* (widely used in professional journals in mathematics) or *HTML/XML* if we want to use it in a website. Probably it would be better to save the web files in *XML* since everything can be included in a single file in contrast to *HTML*, where there is a main file and numerous additional files associated to it. However, if we'd like to keep *Mathematica's* interactive capabilities the most appropriate format would be *cdf* (Computable Document Format).

1.8.2 The Computable Document Format (*.cdf)

As previously discussed, if you'd like other people without access to *Mathematica* to read your documents without losing their interactivity, you should save them in the cdf format: **File ▸ CDF Export ▸ Standalone** or **File ▸ Save As ▸ Computable Document** (*.*cdf*).

The main advantage of this format is that it keeps the interactivity of the objects created with `Manipulate`. In this case your documents can be read with the Wolfram *CDF Player* (available for download for free from http://www.wolfram.com/cdf-player). When sharing a cdf file, to ensure that the intended reader can open it, we recommend that when sending it the link above is included as well.

You can also create a cdf file (normally a section of a notebook) to be embedded in a web page. In the menu bar, select **File ▸ CDF Export ▸ Web Embeddable**.

For more detailed information see in the help system: "Interactivity in .cdf Files" or, even better, visit: http://www.wolfram.com/cdf.

The Wolfram *CDF Player* is not just a reader of *Mathematica* documents for users without access to the program. The player can also avoid displaying the functions used in the inputs showing only the outputs (remember the trick that we have seen previously to hide the inputs and only show the outputs). Furthermore, it can include dynamic objects created with `Manipulate` so that readers can experiment with changing certain parameters and visualizing the corresponding effects in real time. Additionally, readers will not even know that the original document was generated in *Mathematica.* To see all these capabilities in action take a look at the following cdf document:

http://www.wolfram.com/cdf/uses-examples/BriggsCochraneCalculus/BriggsCochraneCalculus.cdf

Many of the Wolfram Research websites, such as Wolfram|Alpha (http://www.wolframalpha.com), allow visitors to generate and download documents in *cdf format*. The same happens with the Wolfram Demonstrations Project (http://demonstrations.wolfram.com) where the files are available in cdf format and you can download and run them (for example, in a digital board connected to a PC with the Wolfram *CDF Player*).

1.9 The Wolfram Cloud

As stated at the beginning of the chapter, the program can be executed in the cloud, that is: in a server accessible through the Internet using a browser, in what is known the Wolfram Cloud (http://www.wolframcloud.com/). The Wolfram Cloud can also be used to store files remotely, as a matter of fact all the files generated locally are interchangeable with the cloud-generated ones. If you don't have access locally and you would like to start familiarizing yourself with *Mathematica* you can do so directly using the Wolfram Development Platform, part of the Wolfram Cloud, for free. If you are planning to use this option heavily, the cost will depend on your requirements and whether you already own a *Mathematica* license or not. In any case, you must register as a user using a Wolfram ID.

If you already registered in the user portal https://user.wolfram.com, mentioned previously, you can use the same Wolfram ID. After signing up you can access the development platform by visiting http://www.wolframcloud.com and then clicking on the **Wolfram Development Platform** icon (http://develop.wolframcloud.com/app).

After entering your Wolfram ID and password, the screenshot shown in Figure 1.18 will appear although you might see an initial screen that is somewhat different since the Wolfram Cloud is constantly evolving.

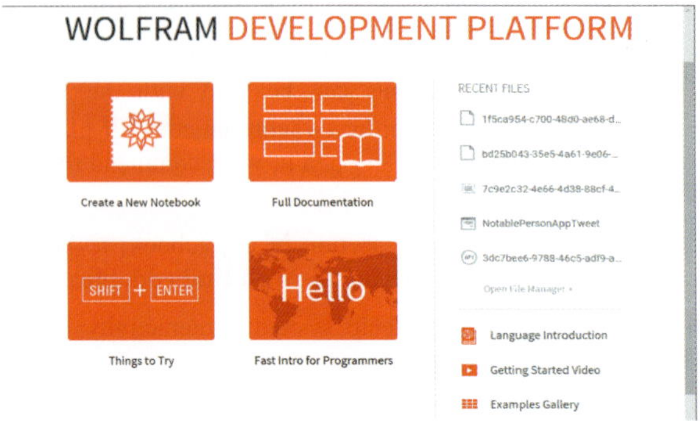

Figure 1.18 Welcome Screen in the Wolfram Development Platform.

You can now create a new notebook by clicking on: **Create a New Notebook**, and be able to reproduce most of the examples discussed in the chapter. You can also save the notebook in Wolfram Cloud, and when accessing it again, no matter from what location, you will be able to continue working on it. Additionally, you will have the possibility of accessing cloud files from a local *Mathematica* installation: **File ▸ Open from Wolfram Cloud...** or **File ▸ Save to Wolfram Cloud...** . *Mathematica* has specific functions related to the Wolfram Cloud. In later chapters we will refer to them. You can get an idea by clicking on: **Getting Started...** and watching the short video.

1.10 Additional Resources

To access the following resources, enter the web addresses in a browser:

An Elementary Introduction to the Wolfram Language by Stephen Wolfram is in **Help ▸ Wolfram Documentation ▸ Introductory Book** ≫, or free on the web:
http://www.wolfram.com/language/elementary-introduction/ (there is also a print version).

Complete program documentation: http://reference.wolfram.com/language/

Explanatory video guides: http://www.wolfram.com/broadcast

Introductory program tutorials:
http://reference.wolfram.com/language/tutorial/IntroductionOverview.html

Interactive applications: http://demonstrations.wolfram.com

Mathematics reference website: http://mathworld.wolfram.com/

Paid and free courses: http://www.wolfram.com/training/courses/

Information about the *Computable document format* (CDF): http://www.wolfram.com/cdf

News and ideas: http://blog.wolfram.com

Author's website: http://diarium.usal.es/guillermo

2

Data Analysis and Manipulation

In this chapter we are going to delve into list operations, essential for data manipulation and for understanding how Mathematica works; show how to import and export data and files in different formats; have our first contact with the statistical analysis of data; and see how to connect to big databases using ODBC.

2.1 Lists

The basic structure in *Mathematica* is the list. The elements of a list can be numbers, variables, functions, strings, other lists or any combination thereof; they are written inside braces { } and are separated by commas. If we feel comfortable with the classical terminology of mathematics we can interpret a simple list as a vector, a two-dimensional list (a list of sublists) as a matrix, and a list of three or more dimensions as a tensor, keeping in mind that the tensor elements can be sublists of any dimension.

- Below we define a list consisting of three sublists.

```
list = {{2, 3, 6}, {3, 1, 5}, {4, 3, 7}};
```

- The previous list can be shown in traditional notation as a matrix but we should not forget that internally *Mathematica* treats it as a list.

```
list // TraditionalForm
```

$$\begin{pmatrix} 2 & 3 & 6 \\ 3 & 1 & 5 \\ 4 & 3 & 7 \end{pmatrix}$$

If the notebook has been configured to display outputs using the traditional format (as in this case), it's not necessary to use //MatrixForm.

- We generate a list of lists of sublists.

```
tensor = Table[i + j + k - 1, {i, 2}, {j, 3}, {k, 4}, {l, 2}]

{{{{2, 1}, {3, 2}, {4, 3}, {5, 4}}, {{3, 2}, {4, 3}, {5, 4}, {6, 5}},
  {{4, 3}, {5, 4}, {6, 5}, {7, 6}}}, {{{3, 2}, {4, 3}, {5, 4}, {6, 5}},
  {{4, 3}, {5, 4}, {6, 5}, {7, 6}}, {{5, 4}, {6, 5}, {7, 6}, {8, 7}}}}
```

It's important to distinguish operations between lists and note that they don't behave in the same way as operations between matrices, especially with respect to multiplication.

- This is an example of list multiplication. We write * to emphasize that it's a multiplication, but if we leave an empty space instead, the result would be the same.

```
{{a, b}, {c, d}}*{x, y}
```

```
{{a x, b x}, {c y, d y}}
```

- Here we multiply matrices, or if you prefer, a matrix by a vector. To indicate that this is a matrix multiplication we use `Dot` (or "`.`" which is the same). Note that "*" and "." are different operations.

```
{{a, b}, {c, d}}.{x, y}
```

```
{a x + b y, c x + d y}
```

- In the mathematical rule for multiplying matrices a row vector is multiplied by a column vector. *Mathematica* is less strict and allows the multiplication of two lists that do not meet the mathematical multiplication rule. It chooses the most appropriate way to perform the operation.

```
{x, y, z} . {a, b, c}
```

```
a x + b y + c z
```

```
{x, y, z} . {{a}, {b}, {c}}
```

```
{a x + b y + c z}
```

- In general, the sum of lists of equal dimension is another list of the same dimension whose elements are the sum of the elements of the initial lists based on their position.

```
{1, 2, 3} - {a, b, c} + {x, y, z}
```

```
{1 - a + x, 2 - b + y, 3 - c + z}
```

- If we multiply a list by a constant, each element of the list will be multiplied by it. The same happens if we add a constant to a list.

```
{{1, 2}, {3, 4}} k
```

```
{{k, 2 k}, {3 k, 4 k}}
```

```
{{1, 2}, {3, 4}} + c
```

```
{{1 + c, 2 + c}, {3 + c, 4 + c}}
```

To extract a list element based on its position *i* in the list we use Part[*list, i*] or, in compact form [[*i*]].

- In this example we show two equivalent ways of extracting the second element of a list:

```
Part[{a, b, c}, 2]
```

```
b
```

```
{a, b, c}[[2]]
```

```
b
```

In the case of a list of sublists, first we identify the sublist and then the element within the sublist.

- In the following example we extract element 3 from sublist 2:

```
{{a1, a2}, {b1, b2, b3}, {c1}}[[2, 3]]
```

```
b3
```

- In the example below we extract elements 2 to 4. Note that the syntax `[[i;;j]]` means to extract from *i* to *j*.

```
{a, b, c, d, e, f}[[2;;4]]
```

```
{b, c, d}
```

- Here we extract elements 2 and 3 from sublists 1 and 2.

```
{{a1, a2, a3}, {b1, b2, b3}, {c1, c2, c3}}[[{1, 2}, {2, 3}]]
```

```
{{a2, a3}, {b2, b3}}
```

- In this case we use `ReplacePart` to replace a list element, specifically from sublist 2 we replace element 2 that is b2 with c2.

```
ReplacePart[{{a1, a2, a3}, {b1, b2, b3}}, {2, 2} → c2]
```

```
{{a1, a2, a3}, {b1, c2, b3}}
```

- Next, we use `[[{ }]]` to rearrange the elements in a list.

```
list = {a, b, c, d, e, f};
```

```
list[[{1, 4, 2, 5, 3, 6}]]
```

```
{a, d, b, e, c, f}
```

The solutions (*out*) are often displayed in list format. If we need a specific part of the solution, we can use the previous command to extract it.

- To extract the second solution of this second degree equation:

```
sol = Solve[1.2 x^2 + 0.27 x - 0.3 == 0, x]
```

```
{{x → -0.625}, {x → 0.4}}
```

- We can proceed as follows:

```
x /. sol[[2]]
```

```
0.4
```

Other functions very useful for list manipulation are `Flatten`, `Take` and `Drop`.

- We apply `Flatten`. With `Take` [*list, i*] we extract the first *i* elements of the list, and with $-i$ the last *i*. With `Take` [*list*, {*i,j*}] we can extract elements *i* to *j*.

```
Take[Flatten[{{a, b}, {c, d, e}, f, g}], 5]
```

```
{a, b, c, d, e}
```

- With `Drop` [*list, i*] we remove the first *i* elements of the list, and with $-i$ the last *i*. `Drop` [*list*, {*i, j*}] removes elements *i* to *j*.

```
Drop[{a, b, c, d, e, f, g}, -3]
```

```
{a, b, c, d}
```

Some functions, when applied to lists, operate on the individual list elements, e.g. f[{*a*, *b*, ...] → {f[*a*], f[*b*], ...}. In this case the function is said to be `Listable`.

- We can see below how a listable function operates. `Sin` in this case.

```
Sin[{a, b, c}]
```

```
{Sin[a], Sin[b], Sin[c]}
```

- Using Attributes, we can check that `Sin` is listable.

```
Attributes[Sin]
```

```
{Listable, NumericFunction, Protected}
```

- An alternative way to type the previous example is:

```
Sin@{a, b, c}
```

```
{Sin[a], Sin[b], Sin[c]}
```

- There are commands particularly useful when operating on lists. It's easy to identify them since they generally contain **List** as part of their names. You can check using the help system:

```
?*List*
```

We omit the output due to space constraints.

There are many more functions used to manipulate lists. We recommend you consult the documentation: guide/ListManipulation.

The most recent versions of *Mathematica,* starting with *Mathematica* 10, include datasets (`Dataset`). These constructions can be considered an extension to the concept of lists. With this new addition, the program has a very powerful way to deal with structured data.

- The example below shows a simple dataset with 2 rows and 3 columns. It consists of an association of associations. Most datasets are displayed in tabular form.

```
data = Dataset[<|"1" → <|"A" → 3, "B" → 4, "C" → 2|>,
   "2" → <|"A" → 4, "B" → 1, "C" → 3|>|>]
```

	A	B	C
1	3	4	2
2	4	1	3

- You can extract parts from datasets in a similar way to the one used with lists. To extract the element from "row 2" and "column C":

```
data["2", "C"]
```

```
3
```

In a later chapter we will see some examples using Datasets.

2.2 Importing/Exporting

- As mentioned before, we will often use as a precaution the command `Clear`["Global`*"] at the beginning of a section to avoid problems with variables previously defined:

```
Clear["Global`*"]
```

A fundamental feature of *Mathematica* is its capability for importing and exporting files in different formats. The commands we need for that are `Import` and `Export`. They enable us to import or export many different types: `XLS`, `CSV`, `TSV`, `DBF`, `MDB`, ...

In this section we are going to use, among others, the files available as examples in a subdirectory named **ExampleData** created during the program installation. Some of these examples are downloaded from www.wolfram.com (Internet connection required).

A file can be imported or exported by typing the complete source or destination path.

- In Windows, to type the file location for import or export purposes use the following syntax:
  "...*directory**file.ext*" , as shown in the following example:

```
Import["C:\\Users\\guill\\Google Drive\\Mathbeyond\\Data\\file.xls"]
```

> An easy way to type the file path is: In an *input* cell write **Import**, place your cursor inside **Import[]**. On the menu bar click: **Insert ▶ File Path...** Locate the file and click on it. The file path will automatically be copied to **Import**["path"].

If we want to use the working directory in which the files are located for importing or exporting purposes, normally the easiest way will be to create a subdirectory inside the directory where the notebook is.

- The following function checks the directory of the active notebook.

```
NotebookDirectory[]
```

```
C:\Users\guill\Google Drive\Mathbeyond\
```

In the directory above, create a subdirectory named **Data**.

- With the following command we set the subdirectory **Data** as our default working directory, the one that the `Import` and `Export` functions will use during this session.

```
SetDirectory[FileNameJoin[{NotebookDirectory[], "Data"}]]
```

```
C:\Users\guill\Google Drive\Mathbeyond\Data
```

We make the previous cell an **Initialization Cell** to ensure its evaluation before any other cell in the notebook by clicking on the cell marker and choosing from the menu bar: **Cell ▶Properties ▶Initialization Cell**). This means that all the data will be imported/exported from the **Data** subdirectory. We may even consider "hiding" the initialization cell by moving it to the end of the notebook. Remember that the order of operations is not the same as it appears on screen. It is the order in which the cells are evaluated. Therefore, even if an initialization cell is located at the end it will still be the first one to be executed.

- The total number of available formats can be seen using `$ImportFormats` and `$ExportFormats`.

```
{$ImportFormats // Length, $ExportFormats // Length}
```

```
{178, 148}
```

2.2.1 Importing

> To import files use `Import`[*"path/file.ext", "Elements"*] where *file* is the name of the file you want to import. If the file is in the **Data** directory, previously defined, you only need to type the file name. If it is located somewhere else you will need to enter the complete path (remember that you can use **Insert ▶ File Path...**). You can also import files directly from the Internet. To do that, specify the desired URL in `Import["http://`*url*`"]`. As an option you can import just part of the file, the one defined by *Elements*.

If your computer uses a non-American configuration remember that *Mathematica* uses the dot "." as a decimal separator and the comma "," to distinguish between list elements. If you have another configuration you may have problems when importing files. The program enables the configuration of the format in **Edit ▸ Preferences... ▸ Appearance** to suit the user needs. Nevertheless, it is better to modify the operating system global settings so that all programs use the same setup. In Windows that can be done by selecting in the **Control Panel** "." as decimal separator and the comma "," as list separator. In OS X, the same can be done by going to **System Preferences... ▸ Language & Region ▸ Advanced...**

- Next, we import an *xls* file containing data about the atomic elements. Note that the program can figure out the type of file from its extension. In this case it's an *xls* file, and the program identifies it as a Microsoft® Excel file.

```
Import["ExampleData/elements.xls"]
```

```
{{{AtomicNumber, Abbreviation, Name, AtomicWeight},
  {1., H, Hydrogen, 1.00793}, {2., He, Helium, 4.00259},
  {3., Li, Lithium, 6.94141}, {4., Be, Beryllium, 9.01218},
  {5., B, Boron, 10.8086}, {6., C, Carbon, 12.0107},
  {7., N, Nitrogen, 14.0067}, {8., O, Oxygen, 15.9961},
  {9., F, Fluorine, 18.9984}}}
```

- Notice that the previous output has the following structure {{{ ...}}}. The reason is that *Mathematica* assumes the file is an Excel spreadsheet that may contain several worksheets. Using **{"Data", 1}** we tell the program to import the data from worksheet 1 (do not confuse this Data with the working subdirectory Data that we have created earlier):

```
examplexls = Import["ExampleData/elements.xls", {"Data", 1}]
```

```
{{AtomicNumber, Abbreviation, Name, AtomicWeight},
 {1., H, Hydrogen, 1.00793},
 {2., He, Helium, 4.00259}, {3., Li, Lithium, 6.94141},
 {4., Be, Beryllium, 9.01218}, {5., B, Boron, 10.8086},
 {6., C, Carbon, 12.0107}, {7., N, Nitrogen, 14.0067},
 {8., O, Oxygen, 15.9961}, {9., F, Fluorine, 18.9984}}
```

- We show below two ways to extract the first 5 rows of the worksheet and display them in a table format.

```
TableForm[Take[examplexls, 5]]
```

AtomicNumber	Abbreviation	Name	AtomicWeight
1.	H	Hydrogen	1.00793
2.	He	Helium	4.00259
3.	Li	Lithium	6.94141
4.	Be	Beryllium	9.01218

```
TableForm[examplexls[[;; 5]]]
```

AtomicNumber	Abbreviation	Name	AtomicWeight
1.	H	Hydrogen	1.00793
2.	He	Helium	4.00259
3.	Li	Lithium	6.94141
4.	Be	Beryllium	9.01218

- Here we see how to import a single column. In this example, we get the fourth column containing the atomic weights.

```
Import["ExampleData/elements.xls", {"Data", 1, All, 4}]
```

```
{AtomicWeight, 1.00793, 4.00259, 6.94141,
 9.01218, 10.8086, 12.0107, 14.0067, 15.9961, 18.9984}
```

- We use `Rest` to remove the first sublist containing the headings (the same could have been done using **Drop**[data, 1]).

```
data = Rest[Import["ExampleData/elements.xls" , {"Data", 1}]];
```

- We show the imported data. `Tooltip` is included so that when the cursor is over a point in the graphic, its corresponding atomic weight can be seen.

```
ListPlot[Tooltip[data[[All, 4]]],
 Filling→ Axis, PlotLabel→ "Atomic Weight"]
```

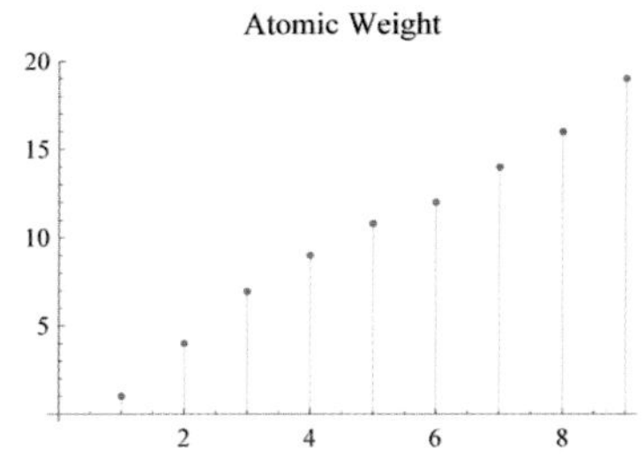

- In our next example we find out the elements available in general for this type of format. It refers to all the possible elements. It doesn't mean that this file contains all of them.

```
Import["ExampleData/elements.xls", "Elements"]
```

```
{Data, FormattedData, Formulas, Images, Sheets}
```

- We choose "Sheets" that returns the number of sheets that the file contains including the labels for each one. In this case it only contains one.

```
Import["ExampleData/elements.xls", "Sheets"]
```

```
{Spreadsheet1}
```

- To display data in a typical spreadsheet format, like Excel, we can use **TableView**, a built-in function that as of *Mathematica* 11.0 remains undocumented.

```
TableView[Import["ExampleData/elements.xls", {"Data", 1}]]
```

	1	2	3	4	5
1	AtomicNumber	Abbreviation	Name	AtomicWeight	
2	1.	H	Hydrogen	1.00793	
3	2.	He	Helium	4.00259	
4	3.	Li	Lithium	6.94141	
5	4.	Be	Beryllium	9.01218	
6	5.	B	Boron	10.8086	
7	6.	C	Carbon	12.0107	
8	7.	N	Nitrogen	14.0067	
9	8.	O	Oxygen	15.9961	
10	9.	F	Fluorine	18.9984	
11					
12					
13					
14					
15					
16					
17					
18					
19					
20					

Note: Undocumented functions in *Mathematica* are not guaranteed to be included in future releases of the program. Additionally, some users might experience software stability issues when evaluating them. For these reasons, we recommend caution when using them.

- We can use `DateListPlot`, a very useful function when the values in the x-axis are dates.

```
DateListPlot[
 Import["ExampleData/financialtimeseries.csv"], Filling→ Axis,
 PlotLabel→ "Financial Data Time Series" , ImageSize → Medium ]
```

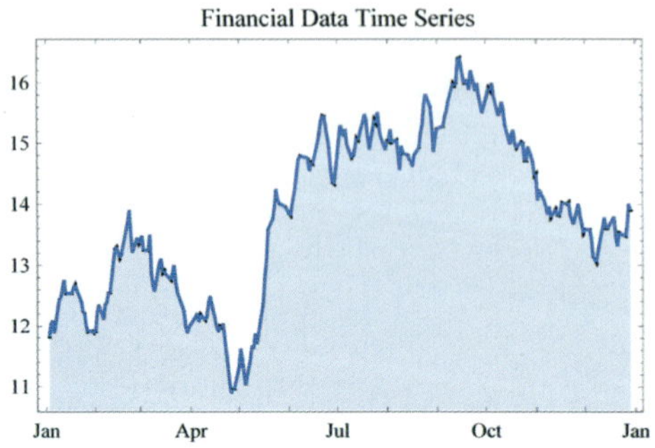

- Try to copy and paste links to any of the numerous images available in: earthobservatory.nasa.gov/Features/BlueMarble/.

```
Import[
 "http://earthobservatory.nasa.gov/Features/BlueMarble/Images/BlueMarble_
   2005_SAm_09_1024.jpg"]
```

- This function returns the size of an image.

```
Import[
 "http://earthobservatory.nasa.gov/Features/BlueMarble/Images/BlueMarble_
   2005_SAm_09_1024.jpg", "ImageSize"]
```

```
{1024, 1024}
```

Note: The examples using files imported from the Internet may not work if their web addresses have been changed or the files themselves deleted.

Importing from databases is a frequent data import operation. For example: importing from an Access *mdb* file.

- We import the complete structure of the database:

```
Import[ "ExampleData/buildings.mdb", {"Datasets", "Buildings", "Labels"} ]
```

```
{Rank, Name, City, Country, Year, Stories, Height}
```

- We use `Short` or `Shallow` to display just a few data elements after importing the entire database (this is practical if there's only one table and it's not very big).

```
Import[ "ExampleData/buildings.mdb" ] // Short
```

```
{{{1, Taipei 101, Taipei, Taiwan, 2004, 101, 508}, <<19>> }}
```

- We import the records associated to several fields and display them in a table format:

```
Text[Style[Grid[Transpose[Import[
    "ExampleData/buildings.mdb", {"Datasets", "Buildings", "LabeledData",
     {"Rank", "Name", "City", "Height"}} ]] , Frame→ All ], Small]]
```

1	Taipei 101	Taipei	508
2	Petronas Tower 1	Kuala Lumpur	452
3	Petronas Tower 2	Kuala Lumpur	452
4	Sears Tower	Chicago	442
5	Jin Mao Building	Shanghai	421
6	Two International Finance Centre	Hong Kong	415
7	CITIC Plaza	Guangzhou	391
8	Shun Hing Square	Shenzhen	384
9	Empire State Building	New York	381
10	Central Plaza	Hong Kong	374
11	Bank of China	Hong Kong	367
12	Emirates Tower One	Dubai	355
13	Tuntex Sky Tower	Kaohsiung	348
14	Aon Centre	Chicago	346
15	The Center	Hong Kong	346
16	John Hancock Center	Chicago	344
17	Shimao International Plaza	Shanghai	333
18	Minsheng Bank Building	Wuhan	331
19	Ryugyong Hotel	Pyongyang	330
20	Q1	Gold Coast	323

For big databases (ORACLE, SQLServer, etc.), with a multitude of tables and a large number of records, it's better to access them with an ODBC connection as we will show later.

From the URL address of a website we can limit the information selected to only what we are interested in. A good practice is to first open the URL in a browser and see what information suits our needs. After importing the web page you will notice that it will appear in *Mathematica* as a list of lists. Analyze how the information is structured. You need to find out what lists contain the information you want and then use `[[...]]` to extract it.

- The following command opens the specified URL using the default browser on your system.

```
SystemOpen["http://www.uxc.com/review/UxCPrices.aspx"]
```

- From the previous web page (http://www.uxc.com/review/UxCPrices.aspx) we only import the information included in the frame: Ux Month-End Spot Prices as of MM, YYYY (Month-end spot price data), in US dollars and euros. The change from the previous month is included between parentheses. The following command was created after trial and error by downloading the entire web page and trying different combinations until getting the desired results. We tell *Mathematica* to show the results using small fonts with the command Style.

```
Style[
 TableForm[Import["http://www.uxc.com/review/UxCPrices.aspx", "Data"][[
   2, 2]]], Small]
```

```
Ux Month-End Prices as of November 28, 2016 [Change from previous month]
1 US$ =                         0.94226 € ‡
U 3 O 8 Price (lb)              $18.25      ( -0.50 )
NA Conv. (kgU)                  $5.85       ( Unch. )
EU Conv. (kgU)                  $6.40       ( Unch. )
NA UF 6 Price (kgU)             $53.40      ( -1.45 )
NA UF 6 Value § (kgU)           $53.53      ( -1.31 )
EU UF 6 Value § (kgU)           $54.08      ( -1.31 )
SWU (SWU)                       $47.00      ( -2.00 )
```

The previous command will work as long as the web page keeps the same structure as when we accessed it (December 7, 2016).

If you know HTML then you might be interested in accessing the contents of a web page by using the function: `URLFetch`.

You can also import data through **`WolframAlfa`** using the free-form input.

- In this example we analyze the evolution of the temperatures in Mexico City.

Temperatures in Mexico City since 1980 to today

↳ Result

(−16 to 36) °C (average low: 9 °C | average high: 23 °C)
(1980 to Monday, August 1, 2016)

- We can expand the output by clicking on the + that appears in the upper right-hand corner. It will show us several sections available for selection. Here we have chosen the plot with the historical evolution of the average monthly temperatures by selecting from the contextual menu: **Paste input for ▸ Subpod Content**.

```
WolframAlpha["Temperatures in Mexico City since 1980 to today",
 {{"TemperatureChart:WeatherData", 1}, "Content"}]
```

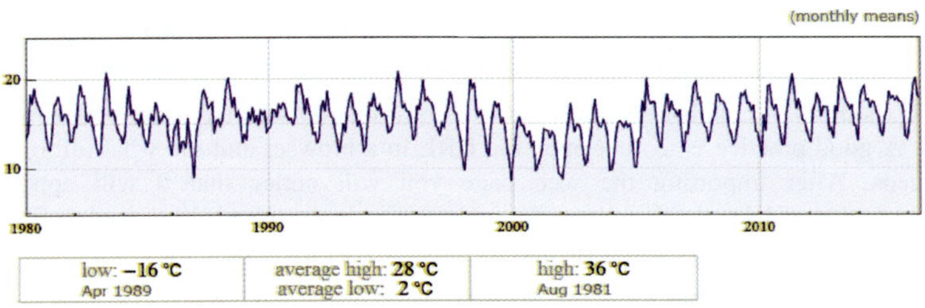

To access multiple data from several fields is better to use the computable data functions that we will discuss in Chapter 5. These functions will always return the same output structure, while WolframAlpha outputs to the same question do not always return the data in the same format.

- In this example we download and display the average temperature in a weather station (LESA) located in the province of Salamanca in Spain during a specific time period. We use the computable data function `WeatherData` combined with `DateListPlot`.

```
DateListPlot[WeatherData["LESA", "MeanTemperature",
  {{2014, 8, 1}, {2016, 7, 31}, "Day"}], Joined → True]
```

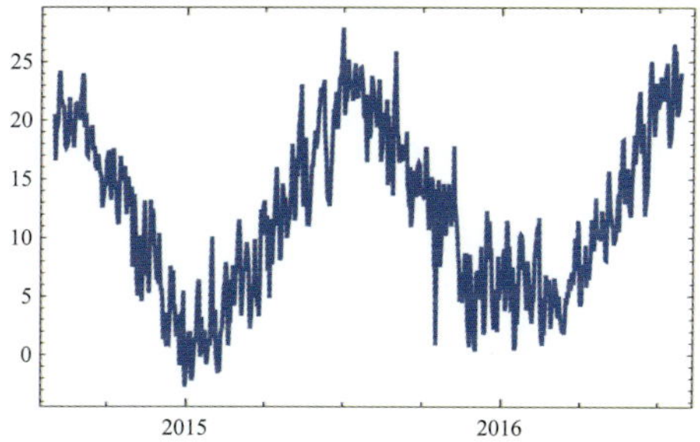

2.2.2 Semantic Import

Since *Mathematica* 10 there's a new function named `SemanticImport` that detects and interprets the type of objects imported.

- We import sales data for different cities and dates.

```
sales = SemanticImport["ExampleData/RetailSales.tsv"]
```

Date	City	Sales
1 Jan 2014	Boston	190
1 Jan 2014	New York City	220
1 Jan 2014	Paris	215
1 Jan 2014	London	225
1 Jan 2014	Shanghai	241
1 Jan 2014	Tokyo	218
2 Jan 2014	Boston	189
2 Jan 2014	New York City	232
2 Jan 2014	Paris	211
2 Jan 2014	London	228
2 Jan 2014	Shanghai	242
2 Jan 2014	Tokyo	229
3 Jan 2014	Boston	196
3 Jan 2014	New York City	235
3 Jan 2014	Paris	221
3 Jan 2014	London	229
3 Jan 2014	Shanghai	238
3 Jan 2014	Tokyo	236
4 Jan 2014	Boston	194
4 Jan 2014	New York City	237

showing 1–20 of 600

- If you move the cursor over the table above, you will see that the program identifies whether the data points are dates or cities, with the later ones displaying additional information. In *Mathematica,* objects with these properties are called entities (`Entity`) and have certain characteristics that we will explore later on. For example: If you hover the mouse over Boston, the following message will appear on the screen: **Entity**["City", "Boston", "Massachusetts", "UnitedStates"], that is, the data point refers to the city located in Massachusetts (USA) and not to other less famous Boston cities that might exist. To find the first city in the table type the following:

```
sales[1, "City"]
```

Boston

- Since it's a city you can directly obtain information about it, e.g. its population (Remember that % refers to the last *output*).

```
%["Population"]
```

655 884 people

- In this example we import rows 3 to 5:

```
SemanticImport["ExampleData/RetailSales.tsv", Automatic, "Rows"][[3 ;; 5]]
```

{{Wed 1 Jan 2014, Paris, 215}, {Wed 1 Jan 2014, London, 225},
{Wed 1 Jan 2014, Shanghai, 241}}

2.2.3 Export

> To export a file use the function Export ["*path/file.ext*", *expr*, *format*] where *file* is the name of the file that will be created from the expression *expr*. Normally the file will be exported in the format specified by *.ext, for example: file.xls will create an Excel file. Additionally, the format can also be specified with the third argument of the function (*format*). You can even export only parts of the data using the optional fourth argument (*elems*): Export ["*path/file.ext", expr, format, elems*].

The variety of export formats is quite extensive: `$ExportFormats`. We can even export in compression formats such as *zip*.

As it is the case when importing, if we don't specify the export path, the files will be copied to our working directory (Data in this case).

- Generate a 3D graph:

```
gr = BarChart3D[{RandomInteger[7, {4}], RandomInteger[{-7, -1}, {4}]},
  ChartElementFunction → "Cone"]
```

- Export the graph as a file named "cones.tif" in tiff format.

```
Export["cones.tif", gr, "TIFF"];
```

You can then import the image previously exported with: **Import["cones.tif"]**.

We are going to show how to export an animation.

- First, we create an animation using `Manipulate`. In later chapters we will see this interesting function in more detail.

```
example3DAnimation = Manipulate[
 Plot3D[{Cos[k x y], Sin[k x y]}, {x, 0, Pi},
   {y, 0, Pi}, Filling -> Automatic, PlotRange -> 1], {k, 0, 2}]
```

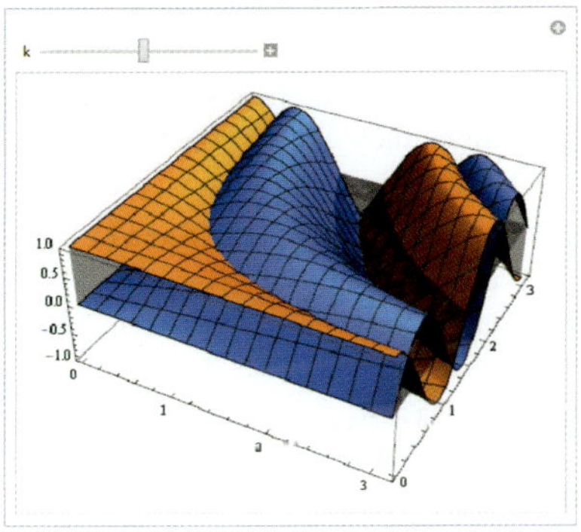

- Then, we export it to a video format, in this case *avi*. (Normally this operation may last a few minutes)

```
Export["example3DAnimation.avi", example3DAnimation]
```

```
example3DAnimation.avi
```

- Now we can either manually locate the exported file and open it or call it directly from *Mathematica* and have the system open it using the default application for *avi* files.

```
SystemOpen["example3DAnimation.avi"]
```

2.3 Descriptive Statistics

When working with big volumes of data, it's very common to perform a preliminary exploratory data analysis. In this section we are going to see how to do it in *Mathematica*: tutorial/BasicStatistics.

- The command `RandomReal[]` generates real random numbers that follow a uniform distribution (a distribution returning a fixed value when $min < x < max$, and 0 when $x < min$ or $x > max$). In the example below, after clearing the variables from memory as indicated in Section 2.2, we generate two lists with 20 numbers between 0 and 10. To ensure the replicability of the example, we have used `SeedRandom[n]`, with n representing the seed used by the algorithm to generate the random numbers.

```
Clear["Global`*"]
```

```
{data1, data2} = {SeedRandom[314];
  RandomReal[10, 20], SeedRandom[312];
  RandomReal[10, 20]}
```

```
{{9.58927, 8.7557, 5.10411, 3.65071, 3.134, 4.89832,
  2.67417, 8.43079, 9.80874, 9.33524, 4.4242, 0.633763, 4.19604,
  5.94988, 9.71835, 9.58623, 6.54041, 3.65745, 0.774322, 7.91849},
 {4.89057, 9.35402, 2.60879, 3.08132, 7.56905, 9.79594, 4.21418,
  8.11898, 9.11195, 0.125038, 4.03004, 2.59509, 7.25321,
  2.69409, 3.61778, 9.64948, 3.1916, 2.63862, 3.13724, 6.11095}}
```

Strictly speaking, we should be talking about pseudorandom numbers since they are generated by algorithms. However, for most practical applications they can be considered in all respects random numbers.

- We compute some of the most frequent statistics from the first list (**data1**). Notice the use of the syntax **{f1[#]**, ..., **fn[#]}&**[data1] equivalent to **{f1**[data1], ..., **fn**[data1]**}** to simplify writing. We will learn more about this type of functions, known as pure functions, in the next chapter.

```
{Mean[#], Variance[#], Skewness[#], Kurtosis[#],
   StandardDeviation[#], MeanDeviation[#], MedianDeviation[#],
   Quantile[#, .6], InterquartileRange[#]} &[data1]
```

```
{5.93901, 9.30229, -0.160336, 1.78156,
 3.04997, 2.6243, 2.62292, 6.54041, 5.39139}
```

- We calculate the covariance and correlation between the two lists created previously.

```
{Covariance[data1, data2], Correlation[data1, data2]}
```

```
{2.83559, 0.317937}
```

- We visualize **data1** using `Histogram`. If you position the cursor inside any of the bars, its value will be displayed.

```
Histogram[data1, 5]
```

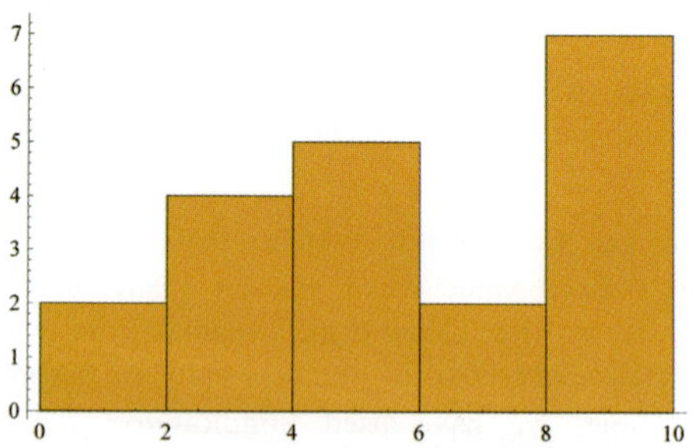

- We can also visualize the data using `BoxWhiskerChart`. When placing the cursor on the boxes you will see the statistical information associated with them.

```
BoxWhiskerChart[{data1, data2}, ChartLabels → {"a", "b"}]
```

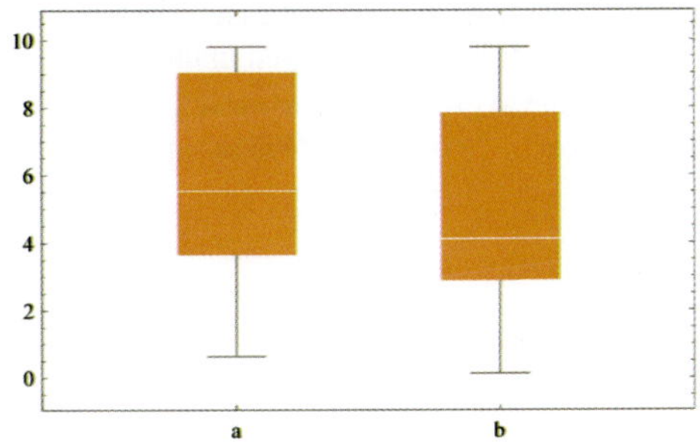

- To create stem-and-leaf plots (`StemLeafPlot`) we need to load the StatisticalPlots package.

```
Needs["StatisticalPlots`"]
```

- We create a stem-and-leaf plot to compare both lists.

```
StemLeafPlot[data1, data2]
```

Leaves	Stem	Leaves
86	0	1
7	2	6667
771	3	1126
942	4	029
91	5	
5	6	1
9	7	36
84	8	1
87663	9	1468

Stem units: 1

- Here is an example of data visualization using a bar chart and a pie chart. We display them next to each other horizontally using `GraphicsRow`.

```
data3 = {5, 2, 10, 4, 6};

GraphicsRow[{BarChart[data3], PieChart[data3]}, ImageSize → Small]
```

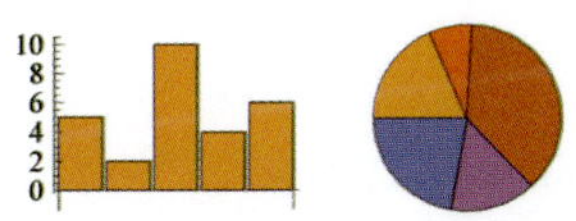

- Using the suggestions bar, located below the output, we can assign a light blue background to the previous result:

```
Show[%, Background → RGBColor[0.84, 0.92, 1.]]
```

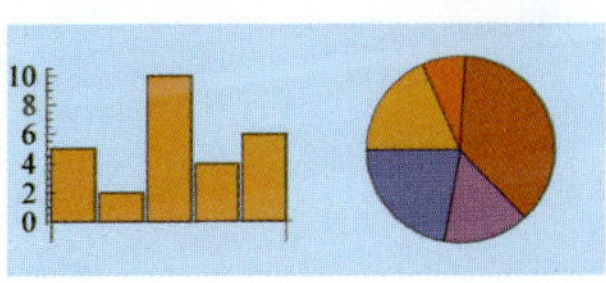

We can also create customized charts as shown next.

- We add a label to each bar and choose a color palette (in this case "DarkRainbow"). We also rotate the chart vertically so the frequencies are now displayed in the x-axis.

```
BarChart[data3, BarOrigin → Left, ChartStyle → "DarkRainbow",
 ChartLabels → {"John", "Mary", "Peter", "Michael", "Paula"}]
```

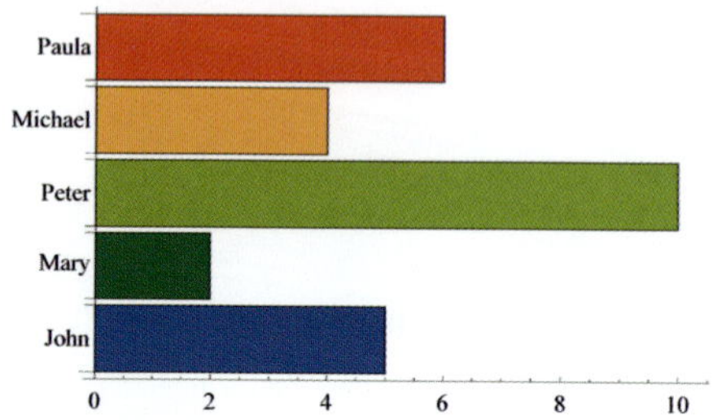

- Any image can be used to represent a bar chart.

```
BarChart[data3, ChartElements →
```

```
,
 ChartLabels → {"a", "b", "c", "d", "e"}]
```

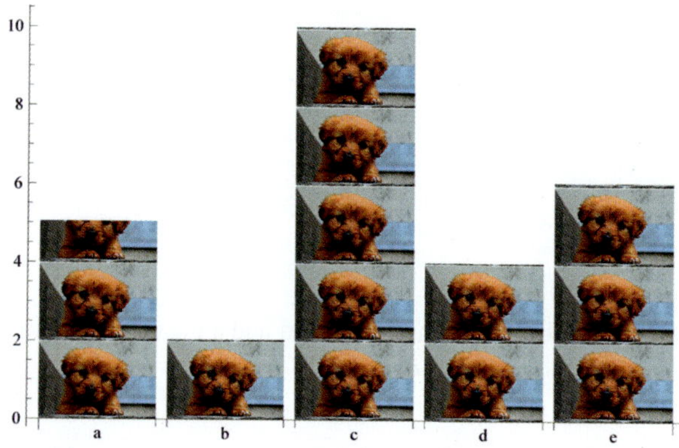

- Here we use a figure known as the "Utah Teapot", included in `ExampleData`, to make a 3D bar chart.

```
g = ExampleData[{"Geometry3D", "UtahTeapot"}];
```

```
BarChart3D[data3, ChartElements → g, BoxRatios → {4, 1, 1}]
```

2.4 Application: Analysis of the Evolution of Two Cell Populations

Let's analyze the evolution over time of two cell populations.

- After clearing all the variables from memory (see Section 2.2.), we import the data dropping the extra set of braces {} that we don't need.

```
Clear["Global`*"]

cells = Import["cells.xlsx"][[1]]

{{Time, Population 1, Population 2}, {0.5, 0.7, 0.1}, {1., 1.3, 0.2},
 {1.5, 1.9, 0.3}, {2., 2.3, 0.4}, {2.5, 2.7, 0.5}, {3., 3., 0.6},
 {3.5, 3.3, 0.7}, {4., 3.6, 0.7}, {4.5, 3.8, 0.8}, {5., 4., 0.8},
 {5.5, 4.2, 0.9}, {6., 4.3, 1.}, {6.5, 4.4, 1.}, {7., 4.5, 1.1},
 {7.5, 4.6, 1.1}, {8., 4.6, 1.2}, {8.5, 4.7, 1.2}, {9., 4.8, 1.3}}
```

- We remove the headings and reorganize the data to get two separate lists representing the evolution of the two cell populations.

```
data = Drop[cells, 1];

{ti, p1, p2} = Transpose[data];

{population1, population2} = {Transpose[{ti, p1}], Transpose[{ti, p2}]}

{{{0.5, 0.7}, {1., 1.3}, {1.5, 1.9}, {2., 2.3}, {2.5, 2.7}, {3., 3.},
  {3.5, 3.3}, {4., 3.6}, {4.5, 3.8}, {5., 4.}, {5.5, 4.2}, {6., 4.3},
  {6.5, 4.4}, {7., 4.5}, {7.5, 4.6}, {8., 4.6}, {8.5, 4.7}, {9., 4.8}},
 {{0.5, 0.1}, {1., 0.2}, {1.5, 0.3}, {2., 0.4}, {2.5, 0.5}, {3., 0.6},
  {3.5, 0.7}, {4., 0.7}, {4.5, 0.8}, {5., 0.8}, {5.5, 0.9}, {6., 1.},
  {6.5, 1.}, {7., 1.1}, {7.5, 1.1}, {8., 1.2}, {8.5, 1.2}, {9., 1.3}}}
```

- We plot the evolution of both populations.

```
ListPlot[{population1, population2}, Joined → True,
 PlotLabel -> Style["Cell Populations Evolution", Bold]]
```

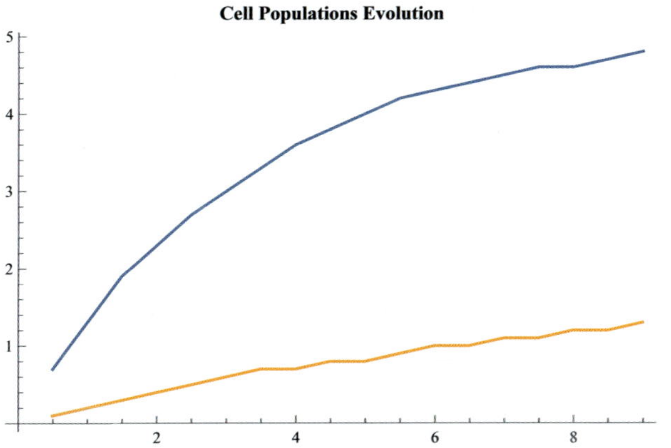

- The shape of the curves suggests that they could be approximated by functions of the form: $a + b\,\mathrm{Exp}[-k\,t]$. We use the function `FindFit` to find the best-fit parameters for the fitting functions.

```
fitpopulation1 = FindFit[population1, a1 + b1 Exp[-k1*t], {a1, b1, k1}, t]
```

```
{a1 → 5.11361, b1 → -5.14107, k1 → 0.303326}
```

```
fitpopulation2 = FindFit[population2, a2 + b2 Exp[-k2*t], {a2, b2, k2}, t]
```

```
{a2 → 1.9681, b2 → -1.9703, k2 → 0.114289}
```

The next example, showing how to customize a graph, includes options that we haven't covered yet (we will explain them later in the book). As usual, if any of the instructions is not clear, consult the help by selecting the command and pressing **F1**. You can use this code as a template for similar cases. Alternatively, you might find it easier to create the labels and plot legends with the drawing tools (CTRL+D).

- We display the fitted curves along with the original data after having replaced the symbolic parameters in the curves with the values obtained from `FindFit`. We also customize the size of the points representing the experimental data. (If you use a version prior to *Mathematica* 9 you need to load the **`PlotLegends`** package first).

```
fitsol1 = a1 + b1 Exp[-k1*t] /. fitpopulation1;
fitsol2 = a2 + b2 Exp[-k2*t] /. fitpopulation2;
Plot[{fitsol1, fitsol2}, {t, 0, 10}, AxesLabel → {"t(hours)",
   "Concentration"}, PlotLegends → Placed[{"Population 1", "Population 2"},
   Center], Epilog → {{Hue[0.3], PointSize[0.02], Map[Point,
     population1] }, {Hue[0], PointSize[0.02], Map[Point, population2]}},
  PlotLabel → Style[Framed["Evolution of 2 cell populations"],
   12, Blue, Background → Lighter[Cyan]]]
```

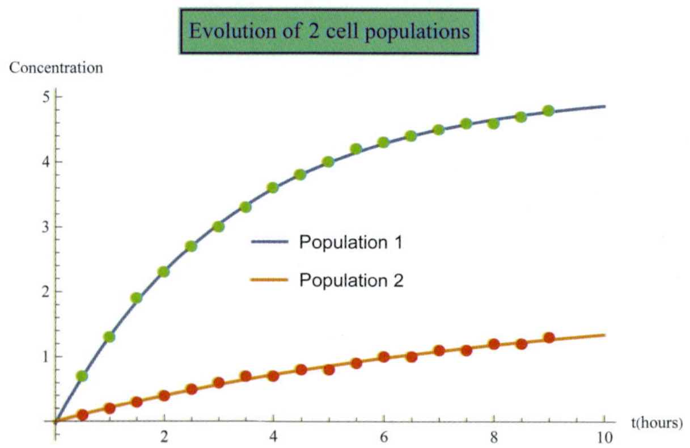

2.5 Application: Global Energy Consumption Analysis

We are going to use what we have learned so far to analyze the evolution of energy consumption globally.

- We clear all the variables from memory (see Section 2.2):

```
Clear["Global`*"]
```

- The following command will take you to a website that contains the file with historical production data for different types of energy by country and region.

```
SystemOpen[
 "http://www.bp.com/en/global/corporate/energy-economics/statistical-
   review-of-world-energy.html"]
```

- Download it by clicking on "Statistical Review - Data workbook" and save it without changing its name in the subdirectory **Data**. The file that we are going to use corresponds to the 2016 (the last one available when this was written) report containing data until the end 2015.

We are going to conduct several studies using this *xlsx* file. In principle we could directly modify the data using Microsoft® Excel and create a new workbook with only the required information for our analyses. This method would probably be the most suitable one but here, for educational purposes, we will perform all the required transformations in *Mathematica*. This will help us to have a better understanding of the file manipulation capabilities of the program, although it will require a careful attention to detail.

The file contains many sheets. Before importing it, open the spreadsheet in Excel or any other similar program and examine its contents. Note that each worksheet has a label describing its contents. This is normally very useful.

> Before importing an Excel file containing several worksheets, it's better to rename each sheet using a descriptive label replacing the one used by default: Sheet1, Sheet2, ...

With the help of the following functions we are going to import part of the file.

- We can see the labels for the worksheets by using the argument "Sheets".

```
Import[
 "bp-statistical-review-of-world-energy-2016-workbook.xlsx", {"Sheets"}]
```

```
{Contents, Oil - Proved reserves, Oil - Proved reserves history,
 Oil Production - Barrels, Oil Production - Tonnes,
 Oil Consumption - Barrels, Oil Consumption - Tonnes,
 Oil - Regional Consumption , Oil - Spot crude prices,
 Oil - Crude prices since 1861, Oil - Refinery throughput,
 Oil - Refinery capacities, Oil - Regional refining margins,
 Oil - Trade movements, Oil - Inter-area movements ,
 Oil - Trade 2014 - 2015, Gas - Proved reserves,
 Gas - Proved reserves history , Gas Production - Bcm,
 Gas Production - Bcf, Gas Production - tonnes, Gas Consumption - Bcm,
 Gas Consumption - Bcf, Gas Consumption - tonnes, Gas - Trade - pipeline,
 Gas - Trade movements LNG, Gas - Trade 2014-2015, Gas - Prices ,
 Coal - Reserves, Coal - Prices, Coal Production - Tonnes,
  Coal Production - Mtoe, Coal Consumption -  Mtoe,
 Nuclear Consumption - TWh, Nuclear Consumption - Mtoe,
 Hydro Consumption - TWh,  Hydro Consumption - Mtoe,
 Other renewables -TWh, Other renewables - Mtoe, Solar Consumption - TWh,
 Solar Consumption - Mtoe, Wind Consumption - TWh ,
 Wind Consumption - Mtoe, Geo Biomass Other - TWh,
 Geo Biomass Other - Mtoe, Biofuels Production - Kboed,
 Biofuels Production - Ktoe, Primary Energy Consumption ,
 Primary Energy - Cons by fuel, Electricity Generation ,
 Carbon Dioxide Emissions, Geothermal capacity, Solar capacity,
 Wind capacity, Approximate conversion factors, Definitions}
```

2.5.1 Global Primary Energy Consumption by Source

We analyze the global consumption of primary energy (including all types: electricity, transportation, etc.) in 2015, the last available year in the file, from different sources (oil, gas, hydropower, nuclear, etc.)

- From the sheet labels we conclude that the information we want is in the sheet "Primary Energy - Cons by fuel".

```
primaryenergyDatabyFuel =
  Import["bp-statistical-review-of-world-energy-2016-workbook.xlsx",
   {"Sheets", "Primary Energy - Cons by fuel"}];
```

- We use **TableView** to see the imported data. You can either increase the number of visible rows with ImageSize or explore the data like in a spreadsheet, using the slide bars on the right and at the bottom to move around the sheet and clicking on specific cells.

```
TableView[primaryenergyDatabyFuel]
```

	1	2	3	4	5
1	ary energy: Consumption by f				
2					
3	Million tonnes oil equivalent	Oil	Natural Gas	Coal	clear En
4					
5	US	838.074	692.681	453.787	189.872
6	Canada	103.327	93.8103	21.3568	24.1956
7	Mexico	85.242	78.1394	12.7525	2.18971
8	Total North America	1026.64	864.631	487.896	216.258
9					
10	Argentina	30.903	42.4281	1.489	1.30986
11	Brazil	143.352	35.4959	17.5511	3.47976
12	Chile	16.6842	3.40432	7.6146	0.
13	Colombia	14.6398	9.78853	5.95395	0.
14	Ecuador	11.9909	0.59378	0.	0.
15	Peru	10.2282	6.4843	0.869399	0.
16	Trinidad & Tobago	1.75293	19.7912	0.	0.
17	Venezuela	36.6564	27.6532	0.202054	0.
18	Other S. & Cent. America	63.593	6.92764	3.0238	0.
19	Total S. & Cent. America	329.8	152.567	36.7039	4.78962
20					

- Notice that the information we want is in row 3 (energy types) and row 89 (the total global consumption, in Million-ton equivalent of petroleum (MTEP) or Million tons of oil equivalent (MTOE)).

```
{energytype1, mtoe } =
  { primaryenergyDatabyFuel[[3]], primaryenergyDatabyFuel[[89]]}
```

```
{{Million tonnes oil equivalent, Oil, Natural Gas, Coal,
  Nuclear Energy, Hydro electric, Renew- ables, Total, Oil, Natural Gas,
  Coal, Nuclear Energy, Hydro electric, Renew- ables, Total},
 {Total World, 4251.59, 3081.46, 3911.18, 575.47, 884.288, 316.605,
  13020.6, 4331.34, 3135.21, 3839.85, 583.135, 892.938, 364.861, 13147.3}}
```

- We store the 2015 data, located in columns 9 to 14 and replace the words "Renew- ables" and "Hydro electric" with "Renewables" and "Hydroelectric" respectively.

```
{energytype, mtoe2015} = {energytype1[[9 ;; 14]], mtoe[[9 ;; 14]]}
```

```
{{Oil, Natural Gas, Coal, Nuclear Energy, Hydro electric, Renew- ables},
 {4331.34, 3135.21, 3839.85, 583.135, 892.938, 364.861}}
```

```
{energytype[[5]], energytype[[6]]} = {"Hydroelectric", "Renewables"};
```

- We represent the previous data in a bar chart. To see the actual consumption by source, in Mtep, place the cursor in the corresponding bar.

```
BarChart[mtoe2015, BarOrigin → Left, BarSpacing → None,
 ChartStyle → "DarkRainbow", ChartLabels → energytype,
 AxesLabel → {Style["Energy Type", Bold], Style["MTEP", Bold]} ]
```

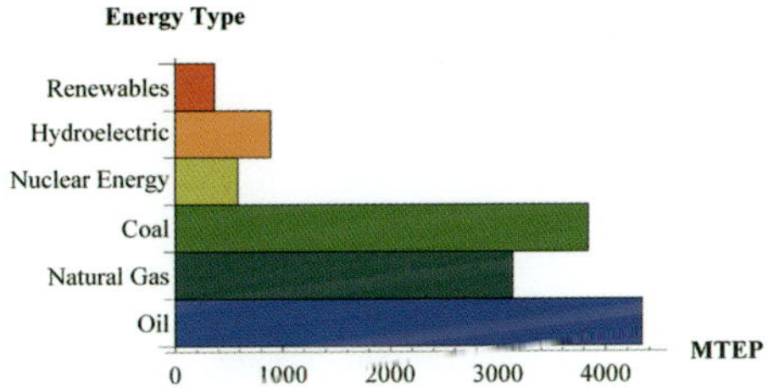

- We display the same data in a pie chart including the percentage share of the total for each energy source. Notice that fossil fuels represent the vast majority of the world primary energy consumption with renewables (excluding hydroelectric) playing a very small role (wind, sun, biomass, etc.)

```
PieChart[mtoe2015 / Total[mtoe2015], ChartStyle -> "DarkRainbow",
 ChartLabels -> energytype, LabelingFunction →
  (Placed[Row[{NumberForm[100 #, {3, 1}], "%"}], "RadialCallout"] &),
 PlotLabel → Style["Sources of Energy (%)", Bold]]
```

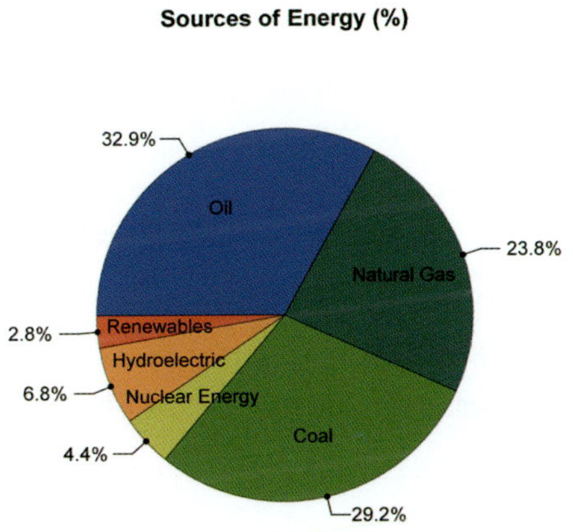

To highlight an individual source of energy just click on its wedge.

> If you'd like to customize a graph, use the information available in the help system. For example, go to the documentation page for `PieChart`, and look for an example that suits your needs.

2.5.2 Global Primary Energy Consumption Temporal Evolution

In this section we are going to analyze the evolution of the global primary energy consumption and look for a function that will enable us to extrapolate energy consumption over the coming years.

- We import the worksheet "Primary Energy - Consumption ":

```
primaryenergyDataYear =
  Import["bp-statistical-review-of-world-energy-2016-workbook.xlsx",
   {"Sheets", "Primary Energy Consumption "} ];
```

- Below we show the imported data:

```
TableView[primaryenergyDataYear]
```

	1	2	3	4	5	6	
1	ary Energy: Consumpti						
2							
3	lion tonnes oil equival	1965.	1966.	1967.	1968.	1969.	19
4							
5	US	1286.5	1359.77	1406.81	1492.34	1573.	162
6	Canada	116.316	123.486	129.559	138.368	146.184	156
7	Mexico	24.4493	25.7187	26.0014	28.3978	31.2613	33.
8	Total North America	1427.26	1508.98	1562.37	1659.1	1750.45	181
9							
10	Argentina	26.9358	28.0012	28.7237	29.5995	30.6691	28.
11	Brazil	22.0479	24.0325	24.9024	28.5187	31.1181	36.
12	Chile	6.12368	6.55866	6.671	6.64832	7.15625	7.5
13	Colombia	7.49287	8.17584	8.71642	9.15435	9.31731	10.
14	Ecuador	0.739	0.78	0.845	0.994	1.061	1.
15	Peru	4.5934	5.58155	5.67508	5.78554	5.83138	6.0
16	Trinidad & Tobago	2.95316	3.14851	3.35912	3.46428	3.57035	3.8
17	Venezuela	15.9878	16.1089	16.9442	18.293	18.4786	18.
18	ther S. & Cent. Americ	21.4213	22.6558	23.7377	24.2953	25.94	27.
19	otal S. & Cent. Americ	108.295	115.043	119.575	126.753	133.142	140
20							

- Notice that the desired information is in row 3 (year) and row 90 (Total World, in Mtep). We also eliminate the first and the last two rows since they are not relevant.

```
primaryevolution =
  Drop[Drop[Transpose[primaryenergyDataYear[[{3, 90}]]], 1], -2];
```

- We plot the data. Using Tooltip, if you click on a curve point, its value will be shown:

```
dataplot = ListLinePlot[Tooltip[primaryevolution] ,
  PlotStyle → PointSize[.01], Mesh → All, AxesLabel → {"Year", "MTEP"},
  PlotLabel → Style["Energy Consumption Evolution", Bold]]
```

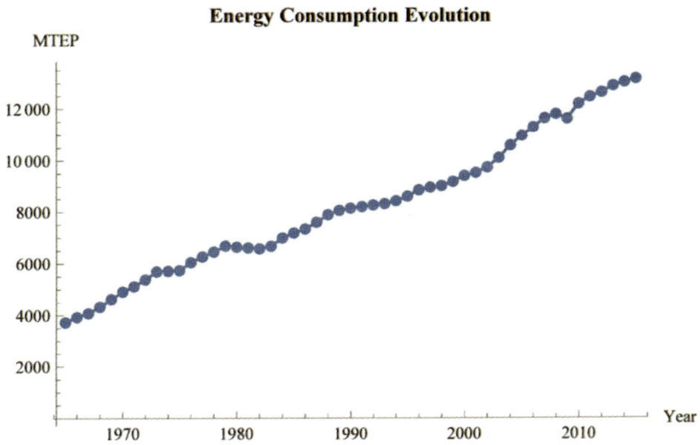

- The shape of the data suggests that the growth has been approximately linear, although a polynomial fit could probably generate better results. Here we use `Fit` (Later, in Section 6.4, we will explain more about fitting). In this case, a polynomial with degree 2 is an appropriate choice.

```
model = Fit[primaryevolution, {1, t , t^2}, t]
```

$4.20076 \times 10^6 - 4390.21\,t + 1.14738\,t^2$

- We show both the real data and the fitted curved extrapolated until the year 2020:

```
Show[
 Plot[model, {t, 1965, 2020}, PlotStyle → {Pink, Thick}, PlotRange → All],
 dataplot, AxesLabel → {"Year", "MTEP"},
 PlotLabel → Style["Energy Consumption Extrapolation", Bold]]
```

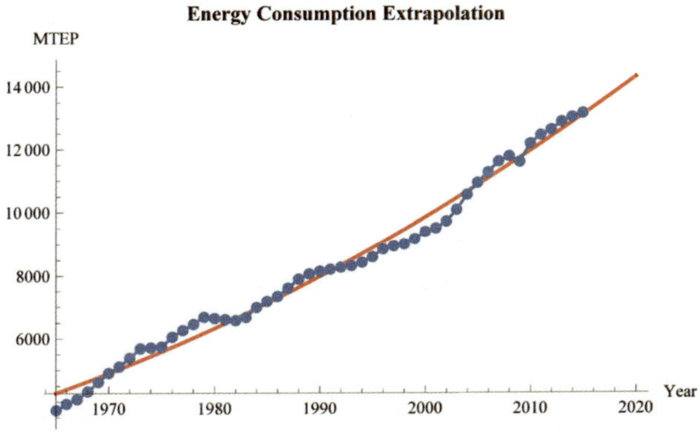

- We generate a table that includes: year (t), real consumption ($r(t)$) and fitted consumption ($f(t)$). This is done by applying the replacement rule {a, b} → {t, $r(t)$, $f(t)$} to the data from `primaryevolution`. We replace t in the object `model` *with a* obtaining the fitted value $f(t)$ for each year. We also apply `Round` to avoid the decimal separator in the last year.

```
exportData = primaryevolution /. {a_, b_} -> {Round[a], b, model /. t -> a};
```

- We include the data from 2000, that we extract with:

```
TableForm[exportData[[36 ;; 51]]
 , TableHeadings → {None, {"Year", "Consumption\n (MTEP)", "Fit\n(MTEP)"}}]
```

Year	Consumption (MTEP)	Fit (MTEP)
2000	9388.25	9840.81
2001	9501.31	10 041.2
2002	9715.28	10 244.
2003	10 081.8	10 449.
2004	10 562.6	10 656.3
2005	10 940.	10 866.
2006	11 267.8	11 077.9
2007	11 617.3	11 292.1
2008	11 780.8	11 508.6
2009	11 598.5	11 727.4
2010	12 181.4	11 948.5
2011	12 450.4	12 171.9
2012	12 622.1	12 397.6
2013	12 873.1	12 625.5
2014	13 020.6	12 855.8
2015	13 147.3	13 088.4

- We export, in Excel format, all the data included in **exportData** to a file named "primaryenergy.xlsx".

```
Export["primaryenergy.xlsx", exportData, "XLSX"];
```

- Finally, we open the file we just created (you need to have Excel or other similar application that can open xlsx files).

```
SystemOpen["primaryenergy.xlsx"]
```

2.5.3 China Population Forecast

We can use the free-form input to import data. In this example, we use it to forecast China's population.

```
China population since 1980 »
China (country) [ population | Interval[{ 1980 , Mon 1 Aug 2016 }] ]
```

```
TimeSeries[ Time: 01 Jan 1980 to 01 Jan 2016
            Data points: 37 ]
```

- Time series data can be fitted using `TimeSeriesModelFit`.

```
tsm = TimeSeriesModelFit[%]
```

```
TimeSeriesModel[ Family: ARIMA
                 Order: {3, 4, 0} ]
```

- Now, the evolution of China's population over the next 10 years can be estimated using `TimeSeriesForecast`.

```
DateListPlot[{tsm["TemporalData"], TimeSeriesForecast[tsm, {10}]},
 PlotLegends → {"Historical", "Forecast"},
 PlotLabel → Style["Chinese Population", Bold]]
```

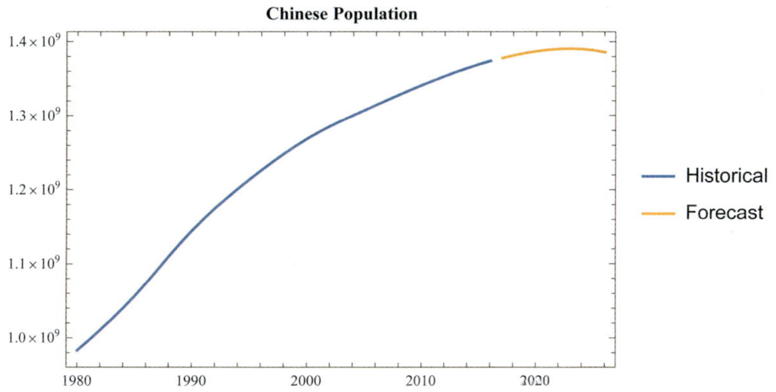

2.6 Database Access with Open Database Connectivity (ODBC)

In this section we are going to learn how to access big databases such as the ones usually found in large organizations nowadays.

- To avoid potential problems is better to start a new session by doing any of the following: Close and open *Mathematica* again; or choose in the menu bar: **Evaluation ▸ Quit Kernel ▸ Local**; or execute the command below:

```
Quit[]
```

Access to big databases (ORACLE, SQLServer, etc) installed in servers is usually done through ODBC connections. *Mathematica* offers the possibility of making such connections using the *DatabaseLink* package (see help: DatabaseLink/tutorial/Overview).

- We are going to use the examples normally found in the *Mathematica* installation directory. The directory with these examples can be found with `FileNameJoin`.

```
FileNameJoin[{$InstallationDirectory,
  "SystemFiles", "Links", "DatabaseLink", "Examples"}]
```

```
C:\Program Files\Wolfram
  Research\Mathematica\11.0\SystemFiles\Links\DatabaseLink\Examples
```

- Before using these examples for the first time they must be installed in:

```
FileNameJoin[{$UserBaseDirectory, "DatabaseResources", "Examples"}]
```

```
C:\Users\guill\AppData\Roaming\Mathematica\DatabaseResources\Examples
```

- To proceed, execute the following instructions. The first one installs them and the second one resets them in case they have been previously modified. Normally you will only have to do this once.

```
<< DatabaseLink`DatabaseExamples`;
```

```
DatabaseExamplesBuild[]
```

2.6.1 A Basic Example

- First we load the package for ODBC connections, containing the JDBC variant:

```
Needs["DatabaseLink`"]
```

Then we establish the connection to the desired database.

- In this case we are going to connect to the publisher.mdb database included in DatabaseResources /Examples
  (we write down the complete path to the database for demonstration purposes although it can also be done just by typing: **OpenSQLConnection["publisher"]**).

```
conn =
 OpenSQLConnection[ JDBC["hsqldb", FileNameJoin[ {$UserBaseDirectory,
     "DatabaseResources", "Examples", "publisher"}] ],
    "Name" -> "manualA", "Username" -> "sa"]
```

```
SQLConnection[ Name: manualA   ID: 3
               Status: Open    Catalog: PUBLIC ]
```

- Here we show the name of the tables in the database.

```
SQLTableNames[conn]
```

```
{AUTHORS, EDITORS, PUBLISHERS, ROYSCHED,
 SALES, SALESDETAILS, TITLEAUTHORS, TITLEDITORS, TITLES}
```

- We display the entries in the PUBLISHERS table with their corresponding headings.

```
SQLSelect[conn, "publishers", "ShowColumnHeadings" → True] // TableForm
```

PUB_ID	PUB_NAME	ADDRESS	CITY	STATE
0736	Second Galaxy Books	100 1st St.	Boston	MA
0877	Boskone & Helmuth	201 2nd Ave.	Washington	DC
1389	NanoSoft Book Publishers	302 3rd Dr.	Berkeley	CA

2.6.2 The Database Explorer

The Database Explorer (Figure 2.1) is a graphical interface used to interact with databases. Sometimes, it may be useful to make a query and later on copy it to a notebook using its capabilities as shown later on in the section. However, we believe it is better in general to use the SQL functions included in *DatabaseLink* as seen in the previous section.

- To launch it you need to use the command DatabaseExplorer[] and use the Connection Tool to select the database:

```
DatabaseExplorer[]
```

```
NotebookObject[ Database Explorer ]
```

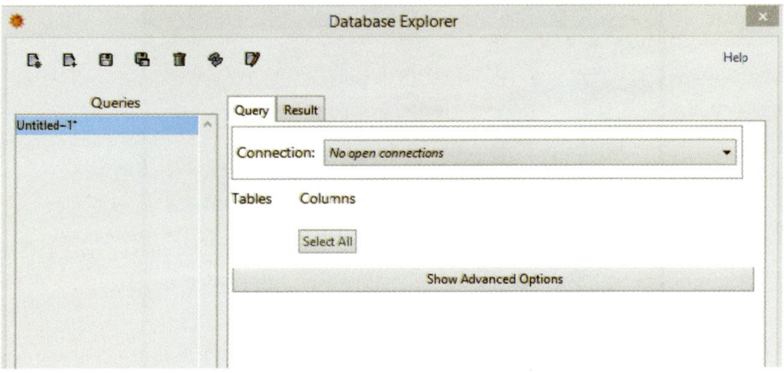

Figure 2.1 The Database Explorer.

- After clicking on the icon, the databases available in our system will be shown (Figure 2.2).

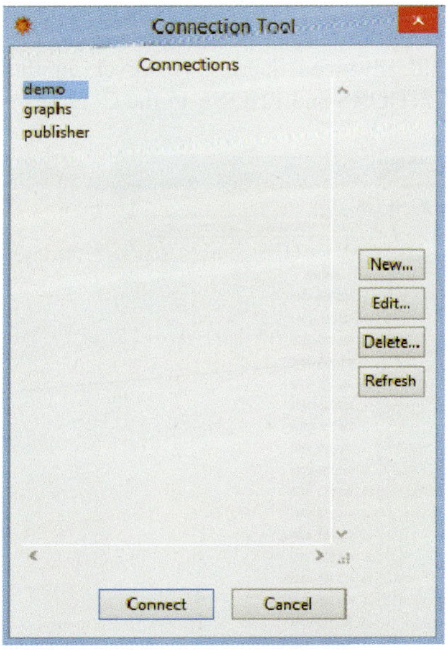

Figure 2.2 The Connection Tool displaying all the databases available in the system.

- Once we connect to a database, we will see its tables and columns. In this case we choose publisher and press **Connect** (Figure 2.3).

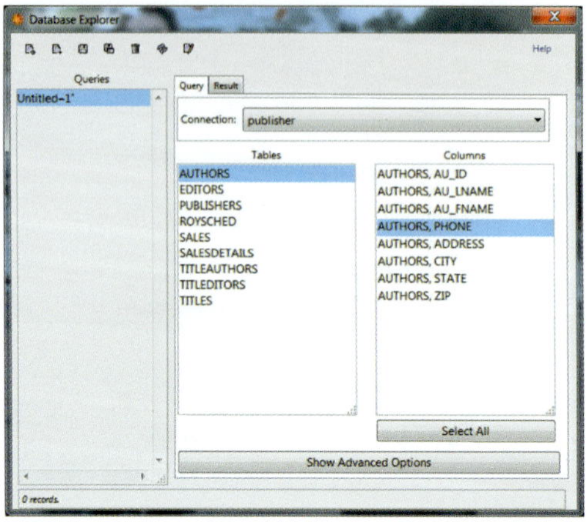

Figure 2.3 Tables and columns of the publisher database.

- With **Show Advanced Options** (the horizontal bar at the bottom in the previous screenshot) you can build advanced queries. However, in this example, to keep things simple, we just select AUTHORS and PHONE in the Columns section and click on **Result** (Figure 2.4):

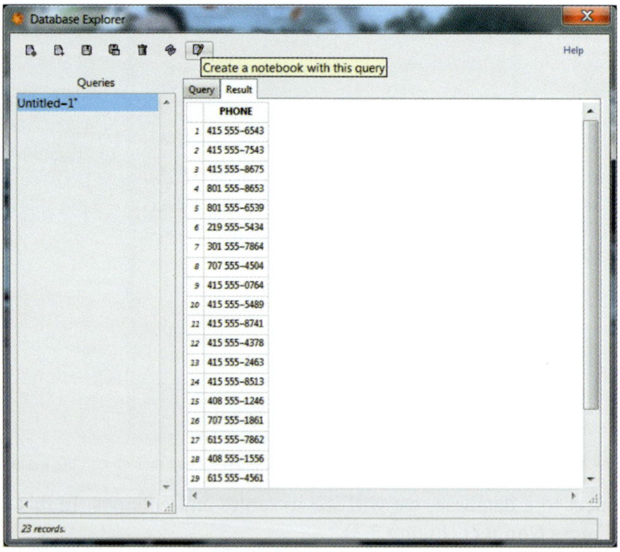

Figure 2.4 Query results.

- The query and its result can be saved by clicking on the **Create a notebook with this query** icon (the right-most icon on the image above).
- When pressed, a new *Mathematica* notebook is generated containing the SQL commands. We have copied the actual contents of the notebook in the next cell.

```
SQLExecute[SQLSelect["publisher", {"AUTHORS"},
   {SQLColumn[{"AUTHORS", "PHONE"}]}, None, "SortingColumns" → None,
   "MaxRows" → 100, "Timeout" → 10, "Distinct" → False,
   "GetAsStrings" → False, "ShowColumnHeadings" → True]] // TableForm
```

```
PHONE
415 555-6543
415 555-7543
415 555-8675
801 555-8653
801 555-6539
219 555-5434
301 555-7864
707 555-4504
415 555-0764
415 555-5489
415 555-8741
415 555-4378
415 555-2463
415 555-8513
408 555-1246
707 555-1861
615 555-7862
408 555-1556
615 555-4561
415 555-4781
415 555-1568
503 555-7865
913 555-0156
```

- Finally we need to close the connection.

```
CloseSQLConnection[conn];
```

2.6.3 Connecting to Database Servers

If you want to connect to a server, you will probably need to configure your ODBC connection tool to define the database location. This can be done in Windows XP/Vista/7/8 with: **Control Panel ► Administrative Tools ► Data Sources (ODBC)** (Figure 2.5). If you have access to your organization's database it should appear in the "User Data Sources:" window and you will be able to connect to it using this interface.

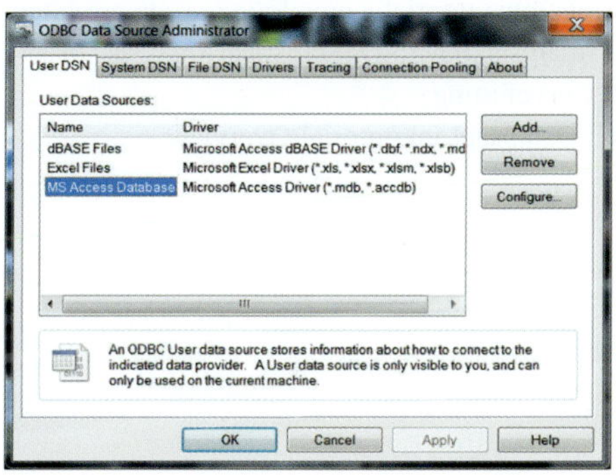

Figure 2.5 The Windows ODBC Data Source Administrator Tool.

- In this example we configure the access to a local MS Access database (Figure 2.6).

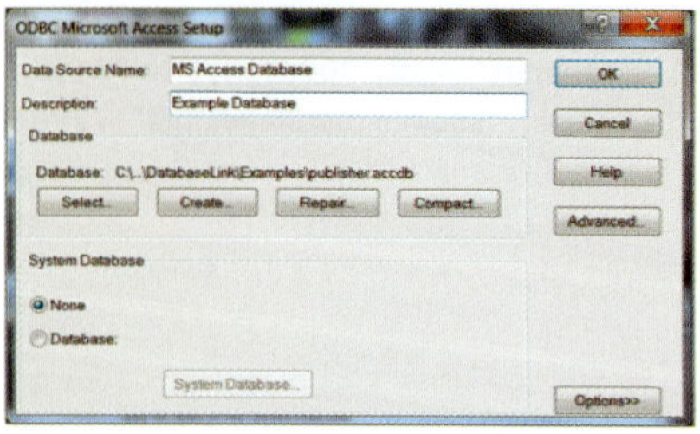

Figure 2.6 Configuring access to a local MS Access database.

- Next you will be able to connect to it using *Mathematica* following a procedure similar to this one:

```
Needs["DatabaseLink`"]
```

```
conn = OpenSQLConnection[JDBC["odbc", "Your DATABASE name"],
  "Username" → "Mylogin", "Password" → "Mypassword"]
```

- To see the tables in the database type:

```
SQLTableNames[conn]
```

Import them or make the necessary queries.

- It's recommended to close the connection at the end.

```
CloseSQLConnection[conn];
```

2.7 Additional Resources

To access the following resources, if they refer to the help files, write their locations in a notebook (e.g., guide/ListManipulation), select them and press <F1>. In the case of external links, copy the web addresses in a browser:

List Manipulation: guide/ListManipulation
Importing and Exporting Data: tutorial/ImportingAndExportingData
Basic Statistics: tutorial/BasicStatistics
Descriptive Statistics: tutorial/DescriptiveStatistics
Continuous Distributions: tutorial/ContinuousDistributions
Discrete Distributions: tutorial/DiscreteDistributions
Convolutions and Correlations: tutorial/ConvolutionsAndCorrelations
How to Do Statistical Analysis: howto/DoStatisticalAnalysis
Curve Fitting: tutorial/CurveFitting
DatabaseLink Tutorial: DatabaseLink/tutorial/Overview

3

Programming: The Beauty and Power of the Wolfram Language

In this chapter we will discuss some of the basic concepts that we need to know to program in Mathematica using the Wolfram Language. We will also cover some useful notions for generating efficient code. In particular, we will learn how to code using pure functions, very important if you want to create compact programs that take full advantage of the power of Mathematica. We will also include examples of some of the most commonly used programming functions and describe the fundamental ideas behind package development, essential to build new applications that incorporate user-defined functionality. Finally, we will provide guidance on tools available for large projects or to develop programs that can be run from a browser.

3.1 *Mathematica*'s Programming Language: The Wolfram Language

As we have seen already in previous chapters, *Mathematica* is not only a program to perform symbolic and numeric calculations. It is a complete technical system that can be used to develop anything: from traffic control applications to sophisticated image manipulation interfaces. All of this is possible thanks to the power of the Wolfram Language, the high-level general-purpose programming language used by *Mathematica*. To start using it we need to know its fundamentals and that's what we are going to learn in this chapter. This section will cover the basic components of the language.

3.1.1 Expressions

Everything in *Mathematica* is an expression, including notebooks. All expressions are written in the form: `Head[..., {}]`. Head is always capitalized and followed by a pair of enclosing brackets containing elements separated by commas. Those elements sometimes go inside braces. Here is one example:

```
Head[a + b + c]
```

```
Plus
```

- Instead of x + y + z we write:

```
Plus[x, y, z]
```

```
x + y + z
```

- With `FullForm` we can see the basic form of expressions: One head, Plus in this case, and elements separated by commas inside brackets:

```
FullForm[x + y + z]
```

```
Plus[x, y, z]
```

Cells are also expressions. Shown below is a copy of this cell displaying its contents. This was done by selecting the copied cell and clicking on: **Cell ▶ Show Expression**.

```
Cell[TextData[{
 "Cells are also expressions. Shown below is a copy of this cell displaying its contents.
 This was done by selecting the copied cell and \
clicking on: ",
 StyleBox["Cell",
  FontWeight->"Bold"],
 " \[FilledRightTriangle] ",
 StyleBox["Show Expression",
  FontWeight->"Bold"],
 ". "}], "Texto AM",
 CellChangeTimes->{{3.627622*^9, 3.627622`*^9}, {3.6396977`*^9, 3.6396977*^9},
 {3.6396978`*^9, 3.6396978*^9}, {3.6396979*^9, 3.6396979*^9}, {3.63969798*^9,
  3.63969798`*^9}}]
```

3.1.2 Atomic Objects

All objects in *Mathematica* are one of the following types:

Numbers: (i) Integers: Any number without a decimal separator, (ii) Reals: Any number with a decimal separator, (iii) Rationals: the ratio between two integers such as 3/4 o 27/8, (iv) Complex: The sum of a real number and an imaginary number: $3 + 2\,i$.

Symbols: A sequence of letters, digits or symbols not starting with a number (we will describe them later).

Strings: Any expression between double quotation marks: "this is a *string*".

SparseArrays: Matrices or lists where most of the elements have the same value:

```
MatrixForm[SparseArray[{i_, i_} :> RandomInteger[8], {4, 4}]]
```

$$\begin{pmatrix} 6 & 0 & 0 & 0 \\ 0 & 5 & 0 & 0 \\ 0 & 0 & 7 & 0 \\ 0 & 0 & 0 & 2 \end{pmatrix}$$

Graphs: A collection of objects, joined by vertices in pairs.

```
Graph[{1 ⟷ 2, 2 ⟷ 3, 3 ⟷ 1}]
```

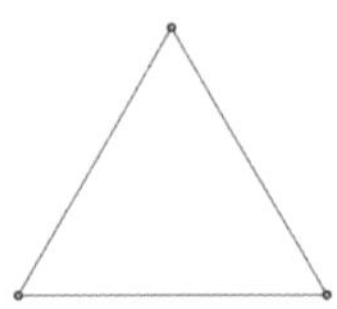

Mathematica internally always separates expressions first into their atomic components before computing them.

3.1.3 Symbols

Symbols is the name of one of the types of atomic objects in *Mathematica*. They can be: letters, digits, combinations of letters and digits not starting with a digit, and special symbols such as: $, π, ∞, e, i , j. Upper and lower-case letters are distinguished. Once a symbol has been created during a session, it will remain in memory unless explicitly removed. An alternative is to use functions that automatically delete a symbol once it's been used.

3.1.4 The *Mathematica* Paradigm

> At the core of the Wolfram Language (or *Mathematica*) is the foundational idea that everything—data, programs, formulas, graphics, documents—can be represented as symbolic expressions. And it is this unifying concept that underlies the Wolfram Language symbolic programming paradigm.

- If we have a list and we'd like to know the type of objects that it contains, we apply `Head` to each of the list elements. This can be done using `Map` (We will refer later to this important programming function).

`Map[Head, {3, 3/2, 1.3, 3 + 4 I, I, √2, Pi, {a, b}, a,`  `,`  `}]`

```
{Integer, Rational, Real, Complex, Complex,
 Power, Symbol, List, Symbol, Graphics, Image}
```

3.2 Functional vs. Procedural Programming

The power of *Mathematica* can be seen specially when we use it to create programs. Readers with experience in procedural programming using languages such as C or FORTRAN, sometimes use the same approach with *Mathematica* even though it may not be the most appropriate one. Because of this, if you are a programmer we would like to encourage you not to replicate the procedural paradigm when programming with *Mathematica*. It will be worth the effort.

Let's see an example for creating a function to calculate the factorial of a number using different programming styles without using the built-in factorial function **n!**:

- Using a procedural style, similar to C or FORTRAN, we could do as follows:

```
factorial1[n_] :=
  (temp = 1; Do[temp = temp counter, {counter, 1, n}]; temp)

factorial1[5]

120
```

- With *Mathematica* we can create a function that virtually transcribes the definition of factorial (note the use of "=" in the first definition and of ":=" in the second one; the reason will be explained later):

```
factorial2[0] = 1;
factorial2[n_] := n factorial2[n - 1]

factorial2[5]

120
```

Normally, the functional style besides being easier to understand is also more efficient.

The next example consists of building, using different programming styles, a command to add the elements of a list of consecutive integer numbers and compare execution times (we use the function `AbsoluteTiming` to show the absolute number of seconds in real time that have elapsed, together with the result obtained). We avoid directly applying the formula for calculating the sum of an arithmetic series which would be the easiest and most effective way.

- We start with a list consisting of the first one million integers.

```
list = Range[10^6];
```

- A procedural method (traditional) to add the elements of this list is:

```
AbsoluteTiming[sum = 0;
 Do[sum = sum + list[[i]], {i, Length[list]}];
 sum]
```

```
{0.819956, 500000500000}
```

The functional method (using different instructions in each case) is much more efficient.

- We perform the same operation in three different ways and display the execution time. In the first case we use `Total`, a built-in function. If available, these functions are usually the fastest way to get things done in *Mathematica*. Later on in the chapter, we will explain Apply (or @@) in more detail.

```
AbsoluteTiming[Total[list]]
```

```
{0.00111397, 500000500000}
```

```
AbsoluteTiming[Apply[Plus, list]]
```

```
{0.131744, 500000500000}
```

In this example we build a function, using different styles, to calculate the sum of the square roots of n consecutive integers : $\{\sqrt{1} + ... + \sqrt{n}\}$, and apply it with $n = 50$.

- A classical but inefficient way of doing this in *Mathematica*, is:

```
n = 50; sum = 0.0; Do[ sum = sum + N[Sqrt[i]], {i, 1, n} ];
sum
```

```
239.036
```

- In the functional style we transcribe almost literally the traditional notation in *Mathematica.* This approach is not only simpler but also more effective.

$$\text{rootsum[n\_] := } \sum_{i=1}^{n} \sqrt{i}$$

```
rootsum[50] // N
```

```
239.036
```

Why do we use "//N"? To find out, see what happens after removing it from the previous function definition.

Another way of programming where *Mathematica* displays its power and flexibility is rule-based programming.

- For example: To convert $\{\{a_1, b_1, c_1\},..., \{a_i, b_i, c_i\},...\}$ into $\{a_1/(b_1 + c_1), ..., \{a_i/(b_i + c_i)\}, ...\}$ we just create the corresponding replacement rule.

```
list = {{a1, b1, c1}, {a2, b2, c2}, {a3, b3, c3}};
```

```
list /. {a_, b_, c_} → a / (b + c)
```

$\left\{\frac{a1}{a2 + a3}, \frac{b1}{b2 + b3}, \frac{c1}{c2 + c3}\right\}$

The previous approach is very useful to perform operations, such as reordering, on lists.

- The following function uses rules-based programming to calculate the distance between two points in a two-dimensional space.

```
distance[list_] := list /. {a_, b_} → √(a^2 + b^2)
```

There's a specific function, EuclideanDistance, to calculate distances. However, *Mathematica* has thousands of functions and sometimes it is not easy to find the most appropriate one. Often the idea is to create a program to perform an operation. It doesn't matter if that approach is not the most efficient one. Later on you can always optimize the program if it's going to be used frequently or if its computing time is excessive.

3.3 Set vs. SetDelayed

A frequently asked question is how to differentiate between "=" (Set) and ":=" (SetDelayed).

- Examine the following example carefully, after clearing the variables from memory:

```
Clear["Global`*"]

x = RandomReal[]
y := RandomReal[]

0.486034

x - x

0.

y - y

0.00130115
```

What has happened? When "=" is used in **f = exp,** the result of evaluating **exp** is assigned to **f** immediately. From then on, whenever **f** is called, the program will always replace **f** with the value of **exp**. In the previous example, the value of **x** was stored right away, while in the case of **y**, what gets stored is the function that will be executed every time it's called. Because of that, when calculating **y-y**, RandomReal [] is executed twice, returning each time a different result.

- The following expression generates a list of 5 random integers between 0 and 5. We call it **data1**. These numbers have been stored and they will always be used anytime **data1** is called.

```
data1 = RandomInteger[{5}, 5]

{5, 3, 0, 4, 2}
```

- Now we type the same above expression but using ":=". We name this new list **data2**. From now on, whenever **data2** is called, the expression that generates the 5 random numbers will be executed anew. What is stored is the function **RandomInteger[{5},5]** not the actual numbers it generates.

```
data2 := RandomInteger[{5}, 5]
```

- The difference can be seen by calling and executing the above defined expressions.

```
data1
```

```
{5, 3, 0, 4, 2}
```

```
data2
```

```
{0, 1, 0, 2, 1}
```

- In the following example we combine ":=" with "=". What happens?

```
data3a := data3b = RandomInteger[{5}, 5]
```

```
data3a
```

```
{5, 2, 4, 2, 3}
```

```
data3b
```

```
{5, 2, 4, 2, 3}
```

```
data3b
```

```
{5, 2, 4, 2, 3}
```

```
data3a
```

```
{3, 3, 3, 2, 0}
```

```
data3b
```

```
{3, 3, 3, 2, 0}
```

- Notice that in the example that we saw previously, we combined "=" and "=:" to calculate the factorial. The use of "=" takes precedence over ":=".

```
factorial[0] = 1;
factorial[n_] := n factorial[n-1]
```

```
factorial[5]
```

```
120
```

It's normal to use ":="when defining functions. However, in many cases it is also possible to use " =" or ":=" without affecting the final result.

> Remember: When using ":=" (SetDelayed) in **f:=exp** the expression **exp** is not evaluated until **f** is used in a posterior calculation and it will be executed each time **f** is called. If you use "=" the function **f=exp** will only be evaluated the first time it is called and after that, every time f is evaluated it will be replaced by **exp**.

The same role that " =" y ":=" play when defining functions, "→" and ":→ (or :>)" play when making replacements:

- Immediate replacement (Rule).

```
{x, x, x} /. x -> RandomReal[]
```

```
{0.587342, 0.587342, 0.587342}
```

- Delayed replacement (RuleDelayed).

```
{x, x, x} /. x :→ RandomReal[]
```

```
{0.255118, 0.107796, 0.380331}
```

3.4 Matrices and Lists Operations

```
Clear["Global`*"]
```

In this section we are going to see some functions that are very useful when performing operations on lists, matrices and vectors. We already mentioned some of them in Chapter 2.

3.4.1 A Random Walk

Any time we want to do something in *Mathematica*, it's possible that the program already has a built-in function to perform the operation.

- In this example we sum consecutively all the elements in a list.

```
Accumulate[{a, b, c, d, e}]
```

```
{a, a + b, a + b + c, a + b + c + d, a + b + c + d + e}
```

As mentioned previously, we will not always find the specific function for what we want to do. However, there are versatile functions useful to know that we can use to perform a specific computation. When operating with lists `FoldList` is such a function.

- Here we use `FoldList` as an alternative to `Accumulate`.

```
Rest[FoldList[Plus, 0, {a, b, c, d, e}]]
```

```
{a, a + b, a + b + c, a + b + c + d, a + b + c + d + e}
```

Using any of the two previous functions we can build a simple expression to simulate a random walk, also known as the drunk man's walk. Let's suppose that a drunk man tries to cross the street and each step he takes can randomly go in any direction. Many physical phenomena, such as Brownian motion, are variations of random walks (The *Mathematica* built-in function is `RandomWalkProcess`).

- The function below is a very simple case of a random walk along a line. Each step has a maximum length of 1 and can go toward both ends of the line with equal probability. To simulate it, we generate random numbers between −1 and 1 (if it goes in one direction we use + and in the opposite direction −) What will be the distance covered after n steps?

```
RandomWalk[steps_] := Accumulate[RandomReal[{-1, 1}, {steps}]]
```

- We create a function to graphically represent the distance covered after n steps. In this case we assume $n = 500$.

```
ListPlot[RandomWalk[500], Filling → Axis]
```

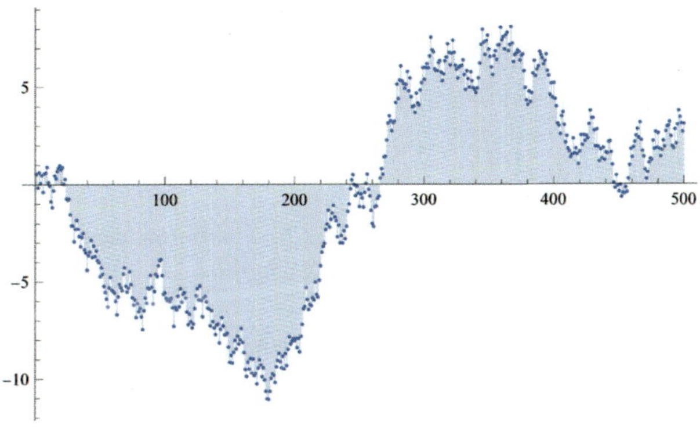

3.4.2 Matrix Operations

Matrix operations appear very often when performing algebraic calculations. The use of `SparseArray` speeds up computations significantly.

- Generate a random matrix:

```
mat = RandomInteger[10, {5, 5}]
```

```
{{9, 10, 6, 2, 7}, {7, 4, 2, 10, 9},
 {6, 10, 10, 3, 1}, {1, 0, 5, 10, 9}, {7, 10, 9, 1, 1}}
```

- The next command displays the previous output in matrix format and evaluates its transpose, inverse, determinant and eigenvalues. We use `TabView` to place the outputs under different labels visible after clicking on the corresponding tab.

```
TabView[
 {"Matrix A" → MatrixForm[mat], "Transpose" → MatrixForm[Transpose[mat]],
  "Inverse" → MatrixForm[Inverse[mat]], "Determinant" → Det[mat],
  "Eigenvalues" → N[Eigenvalues[mat]]}]
```

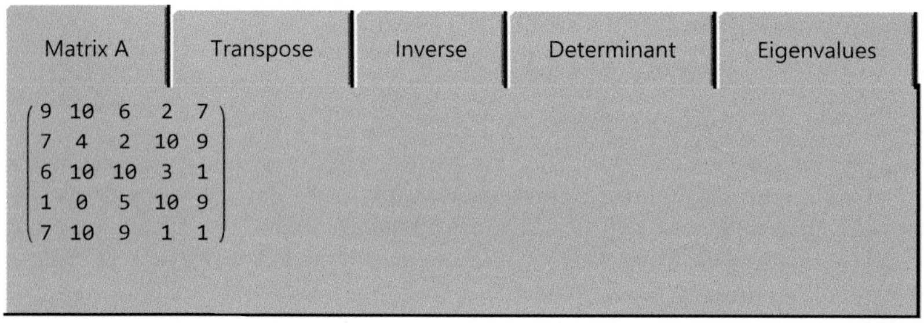

Next, we create an equation of the form **m x** = **b**, where **m** corresponds to the matrix **`mat`**. We will solve it and check that the solution is correct.

- We generate matrix **b** randomly.

```
b = RandomInteger[10, {5}]
```

```
{2, 10, 1, 4, 0}
```

- We solve the equation using `LinearSolve`.

```
xvec = LinearSolve[mat, b]
```

$$\left\{\frac{691}{1893}, \frac{234}{631}, -\frac{1468}{1893}, \frac{2026}{1893}, -\frac{671}{1893}\right\}$$

- We verify that the solution is correct ("." is used for matrix multiplication).

```
mat.xvec
```

```
{2, 10, 1, 4, 0}
```

In matrix operations it is very frequent to have matrices where only few elements have non-zero values. These kinds of matrices are called sparse arrays. In these cases is recommended to use `SparseArray`.

- We build a tridiagonal matrix 5×5 whose main diagonal elements are 9s and its adjacent 1s.

```
s = SparseArray[{{i_, i_} → 9, {i_, j_} /; Abs[i - j] == 1 → 1}, {5, 5}]
```

```
SparseArray[ Specified elements: 13
             Dimensions: {5, 5} ]
```

- We show it numerically and represent it graphically using `MatrixPlot`.

```
MatrixForm[Normal[s]]
```

$$\begin{pmatrix} 9 & 1 & 0 & 0 & 0 \\ 1 & 9 & 1 & 0 & 0 \\ 0 & 1 & 9 & 1 & 0 \\ 0 & 0 & 1 & 9 & 1 \\ 0 & 0 & 0 & 1 & 9 \end{pmatrix}$$

```
MatrixPlot[s]
```

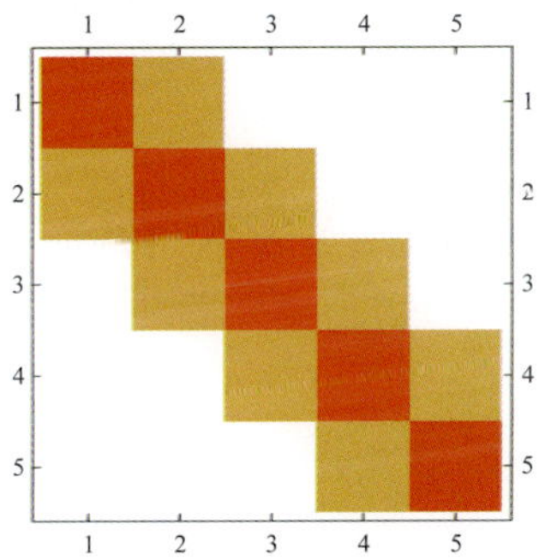

We build a linear system of 5 equations of the form **s x** = **b**, with **s** and **b** the matrices previously defined.

- We solve the equation and check the calculation time:

```
Timing[LinearSolve[s, b]]
```

$$\left\{0.,\left\{\frac{1393}{14040},\frac{1727}{1560},-\frac{22}{351},\frac{713}{1560},-\frac{713}{14040}\right\}\right\}$$

- We repeat the same process but for a system of 50000 variables and 50000 equations, of the form **s x** = **b** where **s** and **b** are called **`sLarge`** and **`bLarge`** respectively. We use ";" to avoid displaying the output since it would take too much space.

```
Timing[bLarge = RandomReal[1, {50000}];]
```

```
{0., Null}
```

```
Timing[sLarge = SparseArray[
    {{i_, i_} → 9, {i_, j_} /; Abs[i - j] == 1 → 1}, {50000, 50000}];]
```

```
{0.453125, Null}
```

```
MatrixPlot[sLarge, ColorFunction → "Rainbow"]
```

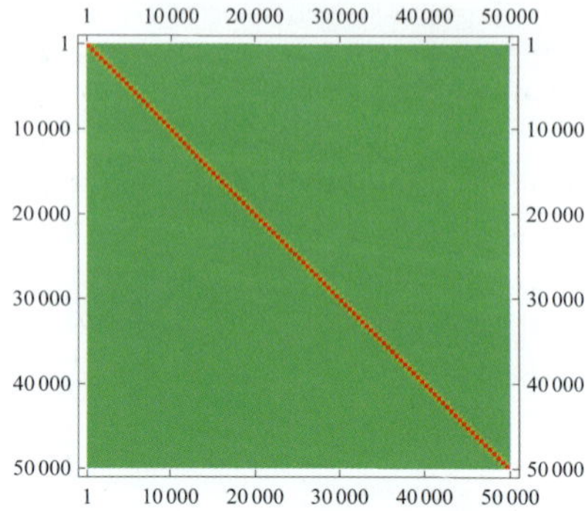

- We solve the system **sLarge x** = **bLarge** and see that the calculation time is extremely short.

```
Timing[xvec = LinearSolve[sLarge, bLarge];]
```

```
{0., Null}
```

- We don't show the result but we can check that the answer is correct by verifying that $m x - b = 0$. With Chop we eliminate the insignificant terms, unavoidable in numerical calculations involving decimal numbers.

```
Norm[sLarge.xvec - bLarge] // Chop
```

```
0
```

3.5 How *Mathematica* Works Internally

We've seen that *Mathematica* manipulates basic objects located at the head of each expression (`Head`).

```
Clear["Global`*"]
```

- Remember that with `FullForm` we can see the basic form of expressions. The example below shows that a/b internally is a multiplied by b^{-1}.

```
FullForm[ a / b]
```

```
Times[a, Power[b, -1]]
```

- Here we apply `FullForm` to a more complicated expression.

```
FullForm[ 1.5 Sin[2 Pi / 4]]
```

```
1.5`
```

- We get the result of the operation but not the structure of the initial expression as we wanted. This is because the calculation is done first and then `FullForm` is applied to the result, just a number. To avoid the execution of the calculation we use `HoldForm`.

```
FullForm[ HoldForm[1.5 Sin[2 Pi / 4]]]
```

```
HoldForm[Times[1.5`, Sin[Times[2, Times[Pi, Power[4, -1]]]]]]
```

- The structure of any expression adopts a tree-like format. Here we represent the previous expression in such a format using `TreeForm`.

```
TreeForm[HoldForm[ 1.5 Sin[2 Pi / 4]]]
```

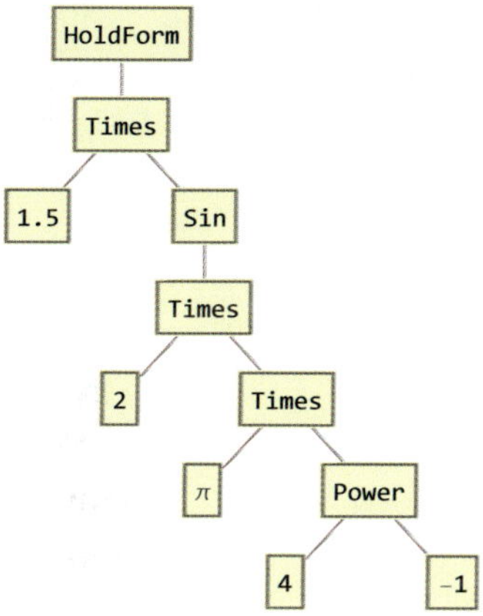

If we analyze the previous tree we can see that the first expression is 1/4. Internally it is represented as 4^{-1}. Next, that expression is multiplied by Pi, the result multiplied by 2, and so on until reaching the top of the tree.

- `Trace` enables us to see the order in which operations are executed:

```
Trace[ 1.5 Sin[2 Pi / 4]]
```

$\{\{\{\{\{\frac{1}{4}, \frac{1}{4}\}, \frac{\pi}{4}, \frac{\pi}{4}\}, \frac{2\pi}{4}, \frac{2\pi}{4}, \frac{2\pi}{4}, \frac{\pi}{2}\}, \text{Sin}[\frac{\pi}{2}], 1\}, 1.5 \times 1, 1.5\}$

- The previous output seems to duplicate certain steps. However, with `FullForm` we can see that it's actually the basic structure that is changing. So when we see $\{\frac{1}{4}, \frac{1}{4}\}$, internally there has been a structural change: the expression has changed from (4^{-1}) to $(\frac{1}{4})$.

```
FullForm[%]
```

```
List[
 List[List[List[List[HoldForm[Power[4, -1]], HoldForm[Rational[1, 4]]],
    HoldForm[Times[Pi, Rational[1, 4]]],
    HoldForm[Times[Rational[1, 4], Pi]]],
   HoldForm[Times[2, Times[Rational[1, 4], Pi]]],
   HoldForm[Times[2, Rational[1, 4], Pi]],
   HoldForm[Times[Rational[1, 4], 2, Pi]],
   HoldForm[Times[Rational[1, 2], Pi]]],
  HoldForm[Sin[Times[Rational[1, 2], Pi]]], HoldForm[1]],
 HoldForm[Times[1.5`, 1]], HoldForm[1.5`]]
```

- With a simple transformation rule we can modify the previous expression so that multiplication becomes addition.

```
HoldForm[1.5 Sin[2 Pi / 4]] /. {Times → Plus}
```

$1.5 + \text{Sin}[2 + (\pi + \frac{1}{4})]$

- The next function shows a red circumference of radius 1.

```
gr = Graphics[{Red, Circle[{0, 0}, 1]}]
```

- Its internal form is as follows:

```
FullForm[gr]
```

```
Graphics[List[RGBColor[1, 0, 0], Circle[List[0, 0], 1]]]
```

```
TreeForm[gr]
```

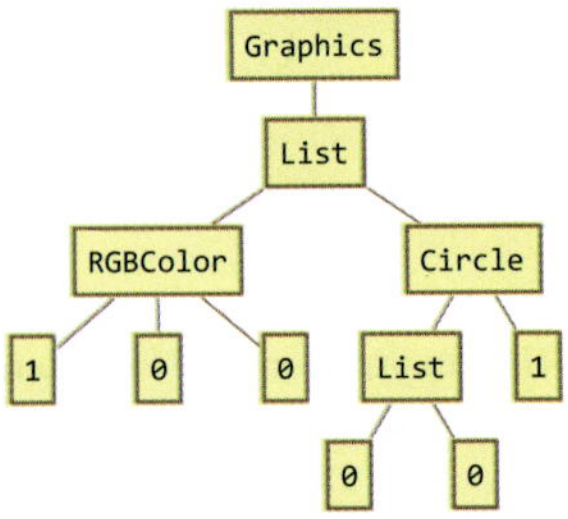

If you move the cursor over an individual frame you will notice that it displays the result of the operations until that frame.

- A small transformation will allow us to modify the expression so that instead of a red circumference (Circle), we get a red circle (Disk).

```
gr /. Circle → Disk
```

Once we know the structure of an expression, we will be able to modify it. Additionally, when encountering outputs from operations that we don't understand, their internal form can give us clues about how the program works.

3.6 Apply, Map, and Other Related Functions

```
Clear["Global`*"]
```

In this section we have selected some of the most commonly used functions for programming.

- `Apply`[*f*, *list*] or f@@list applies f to list.

```
Apply[f, {a, b, c}]
```

```
f[a, b, c]
```

```
f @@ {a, b, c}
```

```
f[a, b, c]
```

```
Apply[f, {{a, b}, c}]
```

```
f[{a, b}, c]
```

- We use FullForm to see how Apply works:

```
FullForm[f @@ {x, y, z}]
```

```
f[x, y, z]
```

- Here Apply is used to multiply the elements of a list:

```
Times @@ {a, b, c, d}
```

```
a b c d
```

- Have you realized that the previous pattern can be used to compute the factorial of a number? For example 5!

```
Apply[Times, Range[5]]
```

```
120
```

- Apply (@@@) at level 1:

```
Apply[f, {{a, b, c}, {d, e}}, {1}]
```

```
{f[a, b, c], f[d, e]}
```

```
f @@@ {{a, b, c}, {d, e}}
```

```
{f[a, b, c], f[d, e]}
```

```
Plus @@@ {{a, b, c}, {d, e}}
```

```
{a + b + c, d + e}
```

- Using @ in succession enables the creation of composite functions as shown in this example:

```
f@g@h@x
```

```
f[g[h[x]]]
```

```
f[x_] := d x^2 ; g[y_] := c Exp[y]; h[z_] := a Cos[b z];
```

```
f@g@h@x
```

$c^2\, d\, e^{2\, a\, \mathrm{Cos}[b\, x]}$

```
Clear[f, g, h]
```

- Map[*f, list*] or f/@list applies f to each of the elements of *list*. This is probably the most frequently used function in programming.

```
Map[g, {a, b, c}]
```

```
{g[a], g[b], g[c]}
```

```
g /@ {a, b, c}
```

```
{g[a], g[b], g[c]}
```

```
Map[Sin, func[x, y, 0, π/2, 90 °]]
```

```
func[Sin[x], Sin[y], 0, 1, 1]
```

- Notice that f with Map is applied to each element of the expression while with Apply actually changes the entire expression by replacing its head.

```
{Map[f, {a, b, c}], Apply[f, {a, b, c}]}
```

```
{{f[a], f[b], f[c]}, f[a, b, c]}
```

- Next we apply Map at level 1:

```
Map[f, {{a, b, c}, {d, e}}, 1]
```

```
{f[{a, b, c}], f[{d, e}]}
```

- Map at level 2:

```
Map[f, {{a, b, c}, {d, e}}, {2}]
```

```
{{f[a], f[b], f[c]}, {f[d], f[e]}}
```

- Map at levels 1 and 2:

```
Map[f, {{a, b, c}, {d, e}}, 2]
```

```
{f[{f[a], f[b], f[c]}], f[{f[d], f[e]}]}
```

- `MapAll` or f//@ applies f to each of the expressions.

```
MapAll[f, {{a, b}, {c}, {{d}}}]
```

```
f[{f[{f[a], f[b]}], f[{f[c]}], f[{f[{f[d]}]}]}]
```

```
f //@ {{a, b}, {c}, {{d}}}
```

```
f[{f[{f[a], f[b]}], f[{f[c]}], f[{f[{f[d]}]}]}]
```

- In the case of listable functions, `Map` is implicitly included in them as this example shows.

```
Map[Sin, {a, b, c}]
```

```
{Sin[a], Sin[b], Sin[c]}
```

```
Sin[{a, b, c}]
```

```
{Sin[a], Sin[b], Sin[c]}
```

- See also: `MapAt`, `MapThread` and `Outer`.

```
MapAt[f, {a, b, c, d}, 2]
```

```
{a, f[b], c, d}
```

```
MapThread[f, {{a, b, c}, {x, y, z}}]
```

```
{f[a, x], f[b, y], f[c, z]}
```

```
Outer[f, {a, b}, {x, y, z}]
```

```
{{f[a, x], f[a, y], f[a, z]}, {f[b, x], f[b, y], f[b, z]}}
```

- In the next example we apply `MapIndexed` together with `Labeled` to identify the position of each term.

```
MapIndexed[Labeled, D[Sin[x]^4, {x, 4}]]
```

$$\underset{\{1\}}{24\,\mathrm{Cos}[x]^4} + \underset{\{2\}}{-192\,\mathrm{Cos}[x]^2\,\mathrm{Sin}[x]^2} + \underset{\{3\}}{40\,\mathrm{Sin}[x]^4}$$

- A related function is `Thread`.

```
Thread[ {a, b, c} -> {x, y, z}]
```

```
{a → x, b → y, c → z}
```

3.7 Iterative Functions

The most commonly used iterative functions in *Mathematica* are `Nest`, `FixedPoint`, `Fold` and `Catch` along with their extensions `NestList`, `NestWhile`, `FoldList` and `FixedPointList`.

People with experience in other programming languages tend to make iterations using `For` and `Do` (also available in *Mathematica*) but normally we should avoid them and use the commands in the previous paragraph. In the examples that follow and throughout the rest of the book we will do just that.

- `NestList` is specially useful for iterative calculations.

```
f[x_] := 1 + x^2

NestList[f, x, 4]
```

$$\{x, 1+x^2, 1+(1+x^2)^2, 1+(1+(1+x^2)^2)^2, 1+(1+(1+(1+x^2)^2)^2)^2\}$$

- In this example we use `NestList` to build an iterative formula for computing the future value of an initial investment of *n* at an annual interest rate *i* for *j* years and show its growth year by year. We assume an initial capital of €100 and an annual interest rate of 3% for 10 years.

```
interest[n_] := (1 + 0.03) n

NestList[interest, 100, 10]

{100, 103., 106.09, 109.273, 112.551,
 115.927, 119.405, 122.987, 126.677, 130.477, 134.392}
```

3.8 Pure Functions

The concept of a pure function is absolutely fundamental to program properly in *Mathematica*. Pure functions behave like operators telling arguments what to do. Initially they may appear cryptic but their use will end up showing us how powerful they are. Therefore, it's convenient to start using them as soon as the opportunity arises.

- After removing all variables from memory, we are going to demonstrate several ways of defining the same pure function:

```
Clear["Global`*"]

Function[u, 3 + u][x]

3 + x

Function[3 + #][x]

3 + x

(3 + #) &[x]

3 + x
```

In this last case, instead of Function we have used the # symbol and to finish defining the function we have added & at the end. Since the combination of # and & represent dummy variables we don't need to store those variables in memory, reducing the computation time. Next we are going to see several examples using this notation.

- This function squares its argument and adds 3 to it.

```
#^2 + 3&[x]
```

```
3 + x^2
```

- An example of a pure function with two arguments (note the use of ***#1*** and ***#2***):

```
(#1^2 + #2^4) &[x, y]
```

```
x^2 + y^4
```

- We define an operator, name it f, and see its behavior.

```
f = (3 + #) &
```

```
3 + #1 &
```

```
{f[a], f[b]}
```

```
{3 + a, 3 + b}
```

- The same result can be obtained with `Map` (/@):

```
f /@ {a, b}
```

```
{3 + a, 3 + b}
```

```
(#1^2 + 1 &) /@ func[x, y, 1, 4]
```

```
func[1 + x^2, 1 + y^2, 2, 17]
```

- There are many ways to define a function in *Mathematica* as shown below for the function $f(x) = x^2 + \sin(x) - 3$.

```
f1[x_] := x^2 + Sin[x] - 3; f1[1]
```

```
-2 + Sin[1]
```

```
f2 = Function[{x}, x^2 + Sin[x] - 3]; f2[1]
```

```
-2 + Sin[1]
```

```
f3 = Function[#^2 + Sin[#] - 3]; f3[1]
```

```
-2 + Sin[1]
```

```
f4 = (#^2 + Sin[#] - 3) &; f4[1]
```

```
-2 + Sin[1]
```

```
(#^2 + Sin[#] - 3) &[1]
```

```
-2 + Sin[1]
```

- In practice, although we can often use several methods to solve problems, certain approaches may return incorrect results. In the example below, given a list of pair of elements: $\{\{a_1, b_1\}, \{a_2, b_2\}, ...\}$ we'd like the second element of each pair, b_i, to be multiplied by a constant k. To do this we can apply a rule:

```
list1 = {{a1, b1}, {a2, b2}, {a3, b3}};
```

```
list1 /. {a_, b_ } → {a, k b}
```

```
{{a1, b1 k}, {a2, b2 k}, {a3, b3 k}}
```

- The rule works fine except if the list has exactly two sublists. In that case, b1 and b2 are interpreted as sublist 1 and sublist 2 respectively.

```
list2 = {{a1, b1}, {a2, b2}};
```

```
list2 /. {a_, b_ } → {a, k b}
```

```
{{a1, b1}, {a2 k, b2 k}}
```

- To avoid this unexpected behavior, we use a combination of a pure function and `Apply` at level 1 (the short form for `Apply`[*f*, *expr*,{1}] is @@@).

```
({#1, k #2}) & @@@ list2
```

```
{{a1, b1 k}, {a2, b2 k}}
```

- In this example we want to solve the equation below step-by-step expressing the result as a function of x.

```
y + x == 2;
```

- Let's see the internal form of the previous expression.

```
FullForm[y + x == 2]
```

```
Equal[Plus[x, y], 2]
```

- We want to solve for *y* by moving *x* to the right side. That is equivalent to subtracting *x* from both sides of the equal sign (note that is necessary to use parenthesis to apply the pure function to the entire equation).

```
# - x & /@ (y + x == 2)
```

```
y == 2 - x
```

- Another way of doing it would be as follows (remember that % always refers to the last `Out`[t]).

```
y + x == 2;
# - x & /@ %
```

```
y == 2 - x
```

- Next, we generate a list of random values first.

```
data = RandomReal[10, 20];
```

- Then, we draw three lines, red, green and blue, each one representing the median, the mean and the geometric mean of the data respectively. We use `MapThread` to define the beginning and the end of each line.

```
lines =
  MapThread[{#1, Line[{{0, #2}, {20, #2}}]} &, {{Red, Green, Blue},
    {Median[data], Mean[data], GeometricMean[data]}}];
```

- Finally, we combine the data with the generated lines.

```
ListPlot[data, Epilog → lines]
```

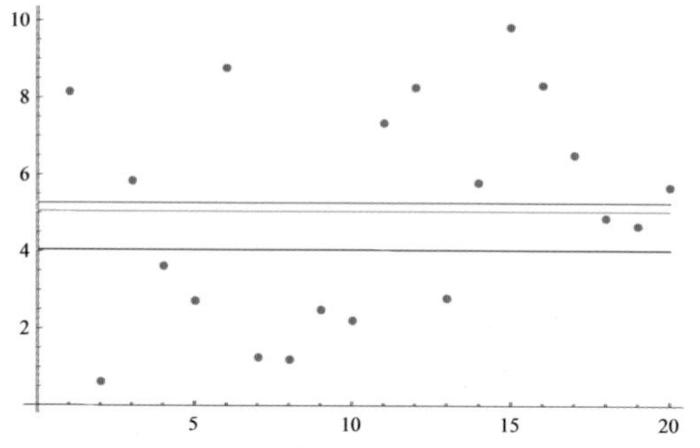

- Here we calculate $\sum_{i=1}^{n} \sqrt{i}$ for $n = 50000$ using two methods: a combination of existing *Mathematica* commands and a pure function. We compare calculation times and see that in this case they are similar.

```
Timing[Total[Sum[Sqrt[i], {i, 1, 50000}]]] // N
```

$\{0.328125, 7.45367 \times 10^6\}$

```
rootsum2[n_] := Total[Sqrt[#] & /@ Range[n]] // N
Timing[rootsum2[50000]]
```

$\{0.328125, 7.45367 \times 10^6\}$

- In the next example we use `NestList` to compute a continuous fraction.

```
NestList[1 + 1/ (1 + #) &, a, 4]
```

$$\left\{a,\ 1+\frac{1}{1+a},\ 1+\frac{1}{2+\frac{1}{1+a}},\ 1+\frac{1}{2+\frac{1}{2+\frac{1}{1+a}}},\ 1+\frac{1}{2+\frac{1}{2+\frac{1}{2+\frac{1}{1+a}}}}\right\}$$

- The previous method, with $a = 1$, can be used to calculate $\sqrt{2}$. With `Nest` only the last term is shown.

```
Nest[1 + 1/ (1 + #) &, 1, 10] // N
```

1.41421

- We use `NestList` to simulate a random walk like the one described earlier. The individual steps can be forward or backward with equal probability. We randomly generate 1 *s* and -1 *s*, 1 meaning a step forward and -1 indicating a step backward. Starting at an arbitrary point, we want to know the position after n steps.

```
RandomWalk[initial_, nsteps_] :=
    NestList[# + If [RandomReal[1] < 0.5, -1, 1] &, initial, nsteps]
```

```
ListPlot[RandomWalk[3, 100], Filling→ Axis]
```

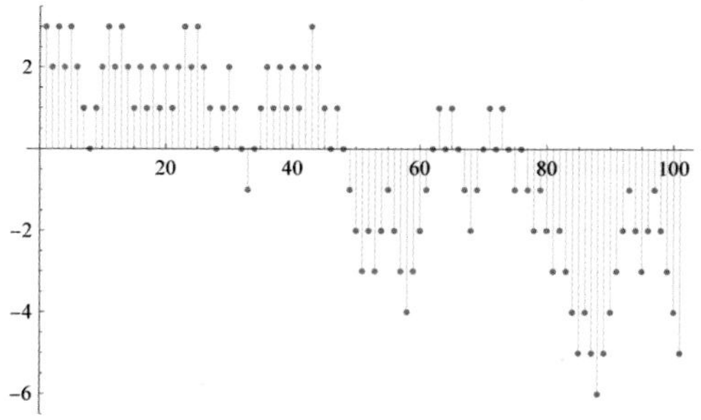

- `FixedPointList` is similar to `NestList` except that it will execute until the solution converges.

```
FixedPointList[1 + 1 / (1 + #) &, 0.1]
```

```
{0.1, 1.90909, 1.34375, 1.42667, 1.41209, 1.41458, 1.41415, 1.41422,
 1.41421, 1.41421, 1.41421, 1.41421, 1.41421, 1.41421, 1.41421, 1.41421,
 1.41421, 1.41421, 1.41421, 1.41421, 1.41421, 1.41421, 1.41421}
```

- In this example we use `FixedPoint` to implement the famous Newton's method (http://mathworld.wolfram.com/NewtonsMethod.html), where x0 is the initial point.

```
newton[f_, x0_] :=
    FixedPoint[# - f[#] / f'[#] &, N[x0], 10]
```

- We apply the previous function to solve Log[x] − $5/x = 0$ (remember that Log[x] represents the natural logarithm of x).

```
newton[(Log[#] - 5 / #) &, 0.2]
```

```
3.76868
```

3.9 Global and Local Variables

A common problem in programming languages is how to avoid conflicts among variables. Be careful and do not confuse the mathematical concept of variable with its concept in programming, the one we use here. In this case, when we mention variables we refer to assignments or definitions made for later use. If we create a variable and later on use the same name to define a new one, conflicts may arise between them.

- If you have executed all the cells until here, you'll have the following list of used names:

```
Names["Global`*"]
```

```
{a, a1, a2, a3, b, b1, b2, b3, bLarge, c, c1, c2, c3, counter, d, data,
 data1, data2, data3a, data3b, distance, e, f, f1, f2, f3, f4, factorial,
 factorial1, factorial2, func, g, gr, h, i, initial, interest, j,
 k, lines, list, list1, list2, mat, n, newton, nsteps, RandomWalk,
 rootsum, rootsum2, s, sLarge, steps, sum, temp, u, x, x0, xvec, y, z}
```

- Symbols starting with $ that you haven't defined may appear. They correspond to internal assignments done by the program. In addition, if you used Clear["Global`*"] earlier during the session, you will have deleted the previous assignments but not their names. For example: **a** will not have any value assigned to it but **RandomWalk** will. You still have not cleared the assignment, but in both cases the symbols still exist in the global context.

```
?a
```

```
Global`a
```

```
?RandomWalk
```

```
Global`RandomWalk

RandomWalk[initial_, nsteps_] :=
 NestList[#1 + If[RandomReal[1] < 0.5, -1, 1] &, initial, nsteps]
```

- With Clear["Global`*"] all the assignments made will be deleted but not the variable names themselves. They can be removed as follows:

```
Remove["Global`*"]
```

```
Names["Global`*"]
```

```
{}
```

```
?a
```

```
Information: Symbol a not found.
```

```
?RandomWalk
```

```
Information: Symbol RandomWalk not found.
```

However, a more logical way to proceed is to avoid global variables and use local variables instead, that is, variables that only have meaning inside an expression.

- Let's define the following variable in a global context.

```
y = 3
```

```
3
```

- Let's use **y** within a local context. To create local variables we will use `Module` or `With`. For example, in the next expression, **y** appears but in this case it's a local variable, defined using `With`, and its use is restricted to operations inside the defined expression.

```
f1[x_] := With[{y = x + 1}, 1 + y + y^2]
```

- If we apply the previous function when $x = 3$, since $y = x + 1$ then $y = 4$ (local variable), that is replaced by $1 + y + y$^2.

```
f1[3]
```

```
21
```

```
f2[x_] := Module[{y}, y = x + 1; 1 + y + y^2]
```

```
f2[3]
```

```
21
```

- Nevertheless, that doesn't affect the global variable y that will still be $y = 3$.

```
y
```

```
3
```

The main difference between `Module` and `With` is that with `Module` we explicitly indicate what variables are local while when using `With`, replacement rules are assigned to the local symbols, that is: `With[{x = x0, y = y0, ...}, expr]`. In many cases you can use them interchangeably.

Another related command is `Block`, very similar to `Module`. The main difference is that while `Module` creates temporary assignments for the specified variables during the evaluation, `Block` suspends any existing assignments for those variables and restores the assignments once the evaluation is done.

```
Module[{y}, Expand[(1 + y)^2]]
```

$1 + 2\,y\$1313 + y\$1313^2$

```
Block[{y}, Expand[(1 + y)^2]]
```

```
16
```

You can find an interesting comparison between them, using the new function `Inactive`, in: http://www.wolfram.com/mathematica/new-in-10/inactive-objects/optimize-code.html.

- If we want to remove the definition assigned to **y** from the global context we can use `Clear`.

```
Clear[y]
```

- Now **y** doesn't have any assignments:

```
y
```

```
y
```

- As we've been doing, if we want to eliminate all the previously defined assignments we can use:

```
Clear["Global`*"]
```

Remember that unless you exit the session, even though you create a new notebook the functions previously defined during the session will still be active.

There's another way to limit the use of variables to a specific context by choosing in the menu bar: **Notebook's Default Context**. With this method you will be able to restrict the variables to a notebook or a group of cells. For example:

- Select all the cells corresponding to this section and in the main menu choose: **Evaluation ▶ Notebook's Default Context ▶ Unique to Each Cell Group**.
- Define the function below:

```
f[x_] := 3 x
```

```
f[3]
```

```
9
```

- If you execute the same function in other section of the notebook you'll see that is not defined.

3.10 Conditional Expressions

```
Clear["Global`*"]
```

A predicate is a function that can take two values: {True, False}. This type of function is called Boolean. Figure 3.1 shows all the built-in symbols in *Mathematica* used to represent Boolean functions and their corresponding Wolfram Language command:

x && y, x ∧ y	And, Y
x || y, x ∨ y	Or, O
! x, ¬ x	Not
x ⊻ y	Xor
x == y	Equal
x != y, x ≠ y	Unequal
x > y	Greater
x < y	Less
x ≥ y	GreaterEqual
x ≤ y	LessEqual

Figure 3.1 Boolean Functions in *Mathematica.*

Predicates are used when defining conditions. These can appear under very different situations.

- Let's use f to define a function whose behavior will be different depending on whether we are using 1 or 2 arguments:

```
f[x_] := 0.2 x^2

f[x_, y_] := 0.2 x^2 y^3

f[3]

1.8

f[3, 5]

225.
```

- A frequently used format when defining conditions is "_h" to indicate that the replacement will happen for any expression whose head is h as the following examples show:

```
Clear[f, g]

f[x] + g[x] /. x_f -> 1/x

1/f[x] + g[x]

Map[Head, {a, 3, 1/3, 2^(1/3.0), 2^(1/3), 3 I, e}]

{Symbol, Integer, Rational, Real, Power, Complex, Symbol}

{a, 3, 1/3, 2^0.33, 2^(1/3), 3 I, e} /. x_Integer → x^2

{a, 9, 1/3, 1.25701, 2^(2/3), 3 i, e}
```

Notice that $2^{1/3}$ is being interpreted as an integer. The reason is that when the head is Power the check is done over the base, that in this case is an integer.

In a function, the easiest way to limit a parameter **n** so that the function will not be executed unless the parameter meets the condition *cond,* is *_cond.*

- In this example **n** is required to be an integer.

```
Clear[f]

f[x_, n_Integer] := x^n

f[3.2, -2]

0.0976562
```

```
f[3.2, 2.1]
```

```
f[3.2, 2.1]
```

Another way to establish a condition is with _?Test (guide/TestingExpressions). As a matter of fact, this is best way to do it since it's the fastest one.

- In this example, **n** must be a non-negative integer ($n \geq 0$).

```
g[x_, (n_Integer) ?NonNegative] := x^n
```

```
g[3.2, 2]
```

```
10.24
```

```
g[3.2, -2]
```

```
g[3.2, -2]
```

A condition can also be defined using "`/;`", the function **`f:=exp/;cond`** is evaluated only if the condition is met. Other equivalent forms are: **`f/;cond:=exp`** and **`f:>exp/;cond`**.

- In the next example the function takes different forms depending on the selected region (the same can be done with `Piecewise`, as we will see later on):

```
Clear[f]
```

```
f[x_] := x^2 /; x > 0
```

```
f[x_] := x /; -2 ≤ x <= 0
```

```
f[x_] := "x must be bigger than or equal to -2" /; x < -2
```

- We apply the function to different values of x.

```
{f[0.2], f[3], f[-3]}
```

```
{0.04, 9, x must be bigger than or equal to -2}
```

In the next example we combine several conditions that must be met to apply the binomial coefficients formula $\binom{n}{r}$. There is a specific function for this calculation: `Binomial`, but here we want to create it ourselves.

- The function below calculates the binomial coefficients if n is an integer, r is an integer ≥ 0, and $n >= r$ (implying that $n \geq 0$).

```
binomial[n_Integer, (r_Integer) ?NonNegative] /; n >= r :=
  n! / (r! * (n - r)!)
```

- We add a message that will be displayed if the conditions are not met.

```
binomial[n_, r_] :=
  "Check that n is an integer, r is a non-negative integer and n >= r."
```

- We test the function for several cases.

```
binomial[3, 2]
```

```
3
```

```
binomial[2, 3]
```

```
Check that n is an integer, r is a non-negative integer and n >= r.
```

```
binomial[-3, 2]
```

```
Check that n is an integer, r is a non-negative integer and n >= r.
```

```
binomial[3, -0.5]
```

```
Check that n is an integer, r is a non-negative integer and n >= r.
```

- In the following case we establish the condition that n has to be either integer, rational or complex to return n^2. The function will be returned unevaluated in any other case.

```
squareofnumber[n_Integer | n_Rational | n_Complex] := n^2
```

```
Map[squareofnumber, {3, 3/2, 1.3, 3 + 4 I, I, √2, Pi, {a, b}, a}]
```

```
{9, 9/4, squareofnumber[1.3], -7 + 24 i, -1, squareofnumber[√2],
 squareofnumber[π], squareofnumber[{a, b}], squareofnumber[a]}
```

We can also define conditions with the following functions (see the program's help): `If`, `Which`, `Switch` and `Piecewise`.

- In this example the function is defined using intervals.

```
Plot[
 Piecewise[{{x^3 + 6, x < -2}, {x, -2 ≤ x ≤ 0}, {x^2, x > 0}}], {x, -5, 5}]
```

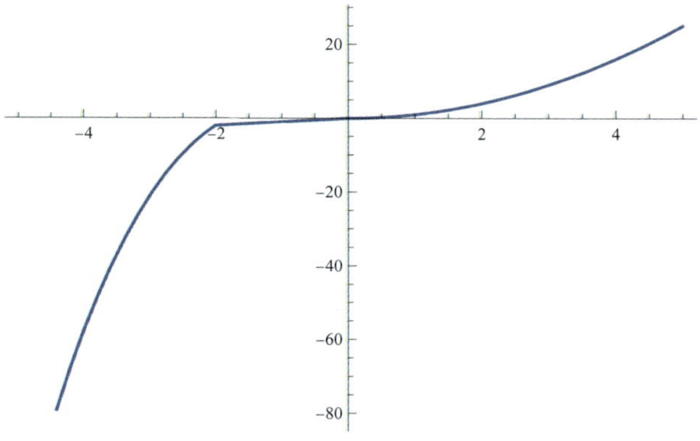

- The use of /; combined with `FreeQ` is particularly useful. For example we create a function that will consider everything that is not x as a constant.

```
f[c_ x_, x_] := c f[x, x] /; FreeQ[c, x]
```

```
{f[3 x, x], f[a x, x], f[(1 + x) x, x]}
```

```
{3 f[x, x], a f[x, x], f[x (1 + x), x]}
```

```
Clear[f, g]
```

- A double underscore __ or `BlankSequence[]` indicates a sequence of one or more expressions as shown in this example:

```
f[x_Symbol, p__Integer] := Apply[Plus, x^{p}]
```

```
f[y, 1, 2, 3]
```

$y + y^2 + y^3$

```
f[y]
```

```
f[y]
```

- A triple underscore ___ or BlankNullSequence[] enables the use of any number of arguments. This is especially useful when defining optional arguments as we'll see later on.

```
g[x_Symbol, p___Integer] := Apply[Plus, x^{p}]
```

```
g[y, 1, 2, 3]
```

$y + y^2 + y^3$

```
g[y]
```

0

- Here it is an elegant example (from *Power Programming with Mathematica: The Kernel* by D.B. Wagner) of sorting the elements of a list using the triple underscore pattern.

```
sorter[{a___, b_, c_, d___} /; b > c] := sorter[{a, c, b, d}]
```

```
sorter[x_] := x
```

```
sorter[{5, 2, 7, 1, 4}]
```

{1, 2, 4, 5, 7}

- Certainly, we could have used Sort instead of sorter.

```
Sort[{5, 2, 7, 1, 4}]
```

{1, 2, 4, 5, 7}

- In this example we use Cases to select the list elements of the form x^y.

```
Cases[{1, x, √x, x^y}, x^y_]
```

$\{\sqrt{x}, x^y\}$

- The pattern "_." makes it possible to define an argument that can be omitted. By using **x^y_.** (note the dot after the underscore), **x** is also selected since the argument of **y** can be omitted.

```
Cases[{1, x, √x, x^y}, x^y_.]
```

$\{x, \sqrt{x}, x^y\}$

In the examples that follow we use DeleteCases to eliminate those elements of a list that meet certain criteria.

- We remove the elements that are decimals (we use Real, different from Reals that refers to the real numbers).

```
DeleteCases[{1, 1, x, 2, 3, y, 3/5, 2.7, Pi, 9, y}, _Real]
```

$\{1, 1, x, 2, 3, y, \frac{3}{5}, \pi, 9, y\}$

- We eliminate from the list those sublists containing 0 as their last term.

```
DeleteCases[
 {{{1, 1}, 0}, {{2, 3}, 1}, {{y, 3/5}, .7}, {{Pi, 9}, z}}, {a_, 0}]
```

$\{\{\{2, 3\}, 1\}, \{\{y, \frac{3}{5}\}, 0.7\}, \{\{\pi, 9\}, z\}\}$

3.11 Accuracy and Precision

The number of significant digits of a number x used in calculations is called precision. To know the precision used in a calculation used `Precision[x]`. Accuracy (`Accuracy[x]`) is the number of significant digits to the right of the decimal point. Precision is a relative error measure while accuracy measures the absolute error.

- By default N uses machine precision.

```
Clear["Global`*"]
```

```
Precision[N[π]]
```

```
MachinePrecision
```

```
Precision[N[π, 40]]
```

```
40.
```

- However, if we don't use a decimal approximation, the precision of Pi is infinite.

```
Precision[π]
```

```
∞
```

`WorkingPrecision` controls the precision level during internal calculations while `PrecisionGoal` determines the precision in the final result. Other options related to accuracy and precision are: `Accuracy`, `AccuracyGoal` and `Tolerance`.

Besides knowing how to use the options mentioned above, there's an important issue that you should know: *Mathematica* has different rules for working with accuracy and precision. An approximate quantity is identified by the presence of a decimal point. Therefore, 0.14 and 14.0 are approximate numbers and we will refer to them as decimals. If we rationalize 0.14 we get 14/100, the ratio of two integers. When doing operations with integers or combinations of them (rationals, integer roots, etc.) *Mathematica* generates exact results.

- For example:

$$2 + \left(3 \times \frac{5}{12}\right)^7$$

$$\frac{110893}{16384}$$

- is different from:

$$2 + \left(3 \times \frac{5.0}{12}\right)^7$$

```
6.76837
```

There are several functions to convert a decimal number into an exact one.

- For example: `Round[x]` returns the closest integer to x, `Floor[x]` the closest integer to x that is less than or equal to x, and `Ceiling[x]` the closest integer to x that is greater than or equal to x.

```
{Round[3.3], Floor[3.3], Ceiling[3.3]}
```

```
{3, 3, 4}
```

In the next example we are going to see the difference between operating with integers (or expressions composed of integers, such as fractions) and operating with decimal approximations.

- We create the following function:

```
f[x_] := 11 x - a
```

- We iterate it and check that we always get the same result:

```
NestList[f, a / 10, 30] // Short
```

$$\left\{\frac{a}{10}, \frac{a}{10}, \frac{a}{10}, \frac{a}{10}, \ll 23 \gg, \frac{a}{10}, \frac{a}{10}, \frac{a}{10}, \frac{a}{10}\right\}$$

- However, when we replace 1/10 with its decimal approximation 0.1, we can clearly see the errors associated with machine precision.

```
NestList[f, 0.1 a, 20]
```

```
{0.1 a, 0.1 a, 0.1 a, 0.1 a, 0.1 a, 0.1 a, 0.1 a, 0.1 a, 0.1 a,
 0.1 a, 0.1 a, 0.100002 a, 0.100025 a, 0.100279 a, 0.103066 a,
 0.133729 a, 0.471014 a, 4.18116 a, 44.9927 a, 493.92 a, 5432.12 a}
```

- This effect is not associated to the way *Mathematica* operates. It will happen with any program performing the same operation using machine precision. For example, try to replicate the previous calculation, replacing **a** with a positive integer (for example 3), in a program such as Microsoft® Excel and you will see that the errors multiply.

```
f[x_] := 11 x - 3

NestList[f, 0.3, 20]
```

```
{0.3, 0.3, 0.3, 0.3, 0.3, 0.3, 0.3, 0.3, 0.3, 0.3,
 0.3, 0.299995, 0.299949, 0.299443, 0.293868, 0.232543,
 -0.442028, -7.86231, -89.4854, -987.34, -10863.7}
```

- `Rationalize` converts a decimal number to the nearest rational with the smallest denominator.

```
Rationalize[0.3]
```

$$\frac{3}{10}$$

- When used in the previous iteration we eliminate the numerical errors since *Mathematica* uses infinite precision for calculations without decimal numbers.

```
NestList[f, Rationalize[0.3], 30] // Short
```

$$\left\{\frac{3}{10}, \frac{3}{10}, \frac{3}{10}, \frac{3}{10}, \ll 23 \gg, \frac{3}{10}, \frac{3}{10}, \frac{3}{10}, \frac{3}{10}\right\}$$

Let's show another more complicated example where we will see as well the effect of using decimal approximations instead of exact calculations.

Consider the following system of differential equations:

$x_1'(t) = 1 - 2.5\, x_1(t) + x_2(t)$

$x_2'(t) = 2\, x_1(t) - x_2(t)$

Initial Conditions : $x_1(0) = x_1(0) = 0$

- We solve it using `DSolve`. Depending on the machine precision of your computer, you may find numerical problems when visualizing the solutions.

```
Clear[x1, x2]
```

```
{{x1[t_], x2[t_]}} = {x1[t], x2[t]} /. DSolve[
   {x1'[t] == 1 - 2.5 x1[t] + x2[t],
    x2'[t] == 2 x1[t] -  x2[t], x1[0] == 0, x2[0] == 0},
   {x1[t], x2[t]}, t]
```

$$\{\{e^{-3.5t}(-0.219131\, e^{0.149219t} + 4.44089\times10^{-16}\, e^{0.298438t} - 1.78087\, e^{3.35078t} + 2.\, e^{3.5t} - 6.93889\times10^{-18}\, e^{6.70156t}),\ e^{-3.5t}(0.186433\, e^{0.149219t} - 4.44089\times10^{-16}\, e^{0.298438t} - 4.18643\, e^{3.35078t} + 4.\, e^{3.5t})\}\}$$

```
Plot[{x1[t], x2[t]}, {t, 0, 100}]
```

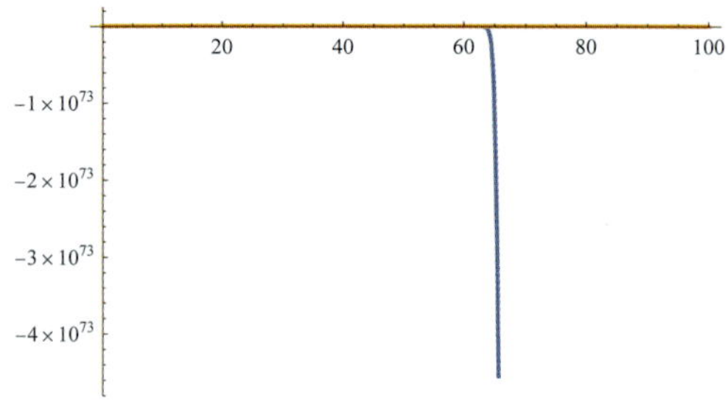

- Now we solve it rationalizing the decimal coefficients since *Mathematica* computes with integers and rationals using infinite precision. The numerical problem is fixed.

```
Clear[x1, x2]
```

```
{{x1[t_], x2[t_]}} = {x1[t], x2[t]} /. DSolve[
    {x1'[t] == 1 - Rationalize[2.5] x1[t] + x2[t],
     x2'[t] == 2 x1[t] -  x2[t], x1[0] == 0, x2[0] == 0},
    {x1[t], x2[t]}, t] // ExpandAll
```

$$\left\{\left\{2 - e^{-\frac{7t}{4}-\frac{\sqrt{41}\,t}{4}} + \frac{5\,e^{-\frac{7t}{4}-\frac{\sqrt{41}\,t}{4}}}{\sqrt{41}} - e^{-\frac{7t}{4}+\frac{\sqrt{41}\,t}{4}} - \frac{5\,e^{-\frac{7t}{4}+\frac{\sqrt{41}\,t}{4}}}{\sqrt{41}},\ 4 - 2\,e^{-\frac{7t}{4}-\frac{\sqrt{41}\,t}{4}} + \frac{14\,e^{-\frac{7t}{4}-\frac{\sqrt{41}\,t}{4}}}{\sqrt{41}} - 2\,e^{-\frac{7t}{4}+\frac{\sqrt{41}\,t}{4}} - \frac{14\,e^{-\frac{7t}{4}+\frac{\sqrt{41}\,t}{4}}}{\sqrt{41}}\right\}\right\}$$

```
Plot[{x1[t], x2[t]}, {t, 0, 100}]
```

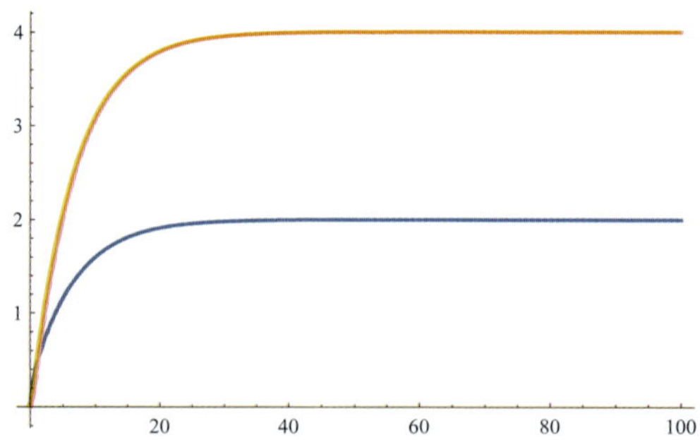

Another situation where it is common to experience difficulties similar to the previous ones is when performing matrix operations (matrix inversion, resolution of systems of differential equations with decimal coefficients, etc.) Therefore, it is often desirable to rationalize the decimal matrix elements; although you may have to consider the increase in computation time.

> When problems associated to operations with decimals arise, is always a good idea to rationalize the terms containing those decimals.

3.12 Choosing the Method of Computation

The power of *Mathematica* becomes particularly evident in the precision with which you can do calculations and the ability to choose the most appropriate method or algorithm given the required computation.

- `NIntegrate` is used for numerical integrations and like almost all other calculation-related functions, unless indicated otherwise, will choose by default a computation method. However, you can select a different method or, for a given method, choose the most appropriate parameters. In the example that follows we compare three methods after showing graphically the method of quadrature. Note the use of `Defer`.

```
f[x_] := Exp[-x]
```

```
Show[Plot[f[x], {x, -3, 3}],
 DiscretePlot[f[x], {x, -3, 3}, ExtentSize → Left] ,
 PlotLabel → Style[Defer[∫_{-3}^{3} Exp[-x] ⅆx], Small]]
```

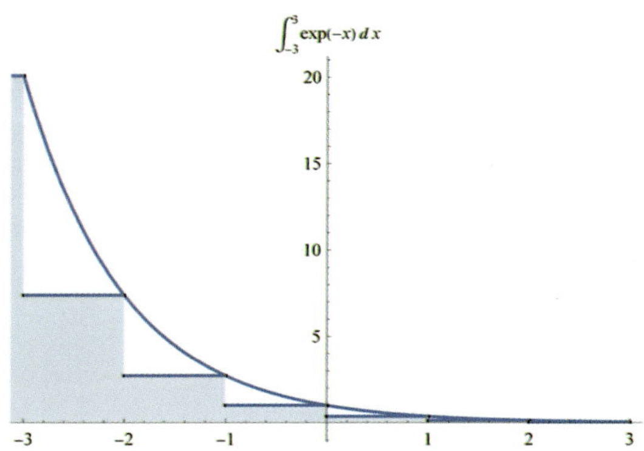


```
NIntegrate[f[x], {x, -3, 3}, Method → #, PrecisionGoal → 2] & /@
 {Automatic, {"RiemannRule", "Type" → "Left"}, "TrapezoidalRule"}
```

```
{20.0357, 20.2215, 20.036}
```

- Here we use `EvaluationMonitor` to display the value of x used in each iteration.

```
Print["x = ", x]
```

```
x = x
```

```
NIntegrate[f[x], {x, -1, 1}, Method → "TrapezoidalRule",
 PrecisionGoal → 2, EvaluationMonitor :→ Print["x = ", x]]
```

```
x = -1.
x = -0.75
x = -0.5
x = -0.25
x = 0.
x = 0.25
x = 0.5
x = 0.75
x = 1.
2.35045
```

- The next function is a piecewise function, that is, it takes different values in each of the intervals.

```
Clear["Global`*"]
```

$$\text{fun} = \begin{cases} \frac{\text{Sin}[10\,x]}{\sqrt{-x}} & -\infty < x < 0 \\ \frac{1}{\sqrt{x}} & 0 < x < 1 \\ \text{Sin}[2000\,x]\,x^2 & 1 < x < 2 \\ \frac{\text{Cos}[x]}{x} & 2 < x < \infty \end{cases};$$

- To represent it graphically *Mathematica* uses different calculation methods depending on the interval.

```
Plot[fun, {x, -10, 10}, ImageSize→ 300]
```

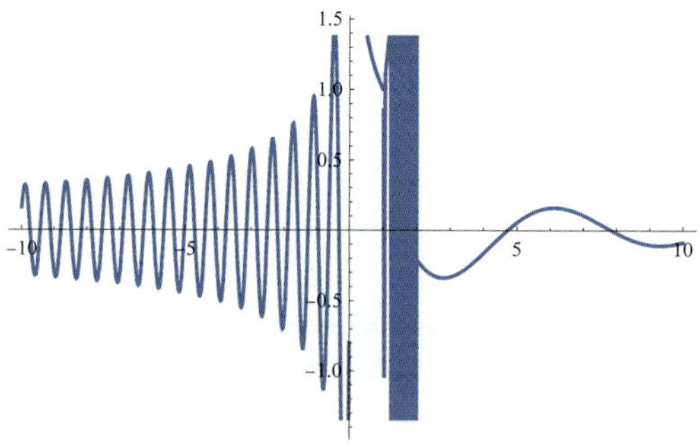

- We can control the level of precision as follows:

```
N[Integrate[Sin[2000 x] x^2, {x, 1, 2}], 10]
0.001275015530
N[NIntegrate[Sin[2000 x] x^2, {x, 1, 2}], 10]
0.00127502
NIntegrate[Sin[2000 x] x^2, {x, 1, 2}, WorkingPrecision -> 10]
0.001275015530
```

- Next, we present an example of a differential equation in which the solving method (tutorial/NDSolveOverview) depends on the value of δ. `NDSolve` chooses the best algorithm given δ. We use `Manipulate`, a function that we will refer to extensively in the next chapter.

```
Manipulate[
 res = NDSolve[{y'[t] == y[t]^2 - y[t]^3, y[0] == δ}, y, {t, 0, 2/δ}];
 p1 = Plot[(y /. First[res]) [t], {t, 0, 2/δ}, Mesh → All],
 {{δ, 0.01}, 0.0001, 0.01}]
```

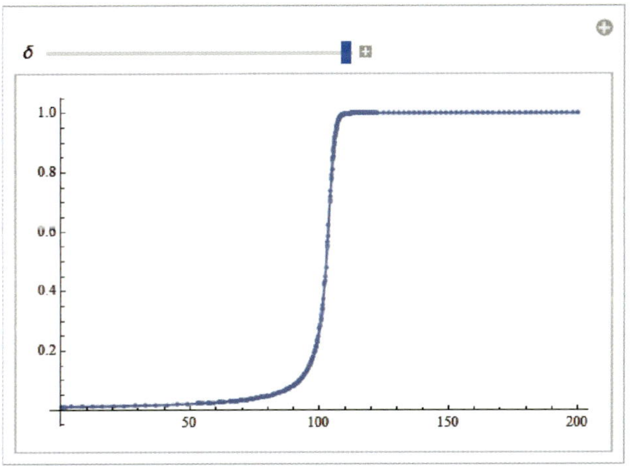

3.13 Optimizing the Computation Time

Another interesting aspect of *Mathematica* is its use of previously calculated results. Before showing it with an example, it would be better to quit the session to remove any previous information stored in memory.

```
Quit[]
```

- In the next example we show the calculation time of a complicated definite integral for different integration limits.

```
Timing[(f1[x1_] := Integrate[Sin[x^3], {x, 0, x1}];
   Table[f1[i], {i, 1, 10}]);]

{3.67188, Null}
```

- We repeat the same exercise using "=" instead of ":="

```
Timing[(f2[x1_] = Integrate[Sin[x^3], {x, 0, x1}];
   Table[f2[i], {i, 1, 10}]);]

{1.07813, Null}
```

The calculation time is faster in this last case. It seems logical, since in the first example the integral **f1** is calculated 10 times, each time applying a different integration limit, while in the second case **f2** provides the solution to the integral as a function of the integration limit x1, and then replaces the value of x1 directly in the solution 10 times.

- The surprise comes when we repeat the calculation of f1. The calculation time has been reduced from the first time we used the function.

```
Timing[(f1[x1_] := Integrate[Sin[x^3], {x, 0, x1}];
   Table[f1[i], {i, 1, 10}]);]
```

```
{1.03125, Null}
```

- And even more surprising: if instead of **f1** we define a new function identical to **f1** named **f3**, the calculation time is still faster than the first time we computed **f1**.

```
Timing[(f3[x1_] := Integrate[Sin[x^3], {x, 0, x1}];
   Table[f3[i], {i, 1, 10}]);]
```

```
{0.90625, Null}
```

The reason behind this behavior is that for functions such as: Integrate, Simplify and others, *Mathematica* keeps a cache of previous results that it will use later if necessary to make calculations more efficient (http://library.wolfram.com/infocenter/Conferences/5832/).

- If we now delete the cache, the calculation takes longer:

```
ClearSystemCache[]; Timing[(f3[x1_] := Integrate[Sin[x^3], {x, 0, x1}];
   Table[f3[i], {i, 1, 10}]); ]
```

```
{3.5625, Null}
```

In many situations we can significantly reduce the computation time by using Compile. Normally it's used with functions that only need to evaluate numerical arguments. This approach enables the program to save time by avoiding checks related to symbolic operations. The command generates a compiled function that applies specific algorithms to numerical calculations.

- Example: We use Compile to calculate the following function that only accepts real values as arguments:

```
cf = Compile[{{x, _Real}}, x^2 - 1/(1+x) Sin[x]]
```

```
CompiledFunction[ WVM  Argument count: 1
                       Argument types: {_Real} ]
```

- CompiledFunction evaluates the expression using the compiled code:

```
cf[Pi]
```

```
9.8696
```

- We can now represent graphically the compiled version of the function:

```
Plot[cf[x], {x, -1, 3}]
```

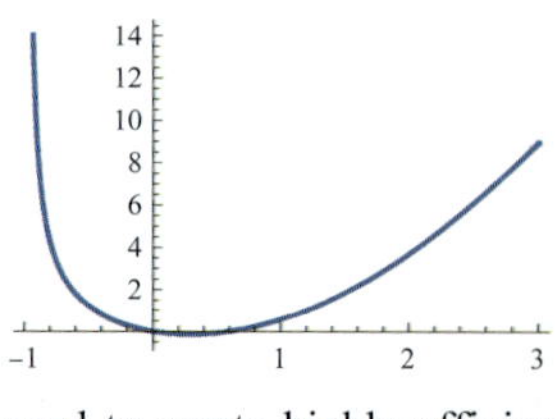

If you need to create highly efficient programs for calculations that could take hours or even days, it's recommended to try several alternative approaches first. It's not unusual for two commands generating the same result to have very different computation times (a carefully crafted programming instruction can be tens or sometimes even thousands of times faster than one that is not).

3.14 Cloud Deployment

The function CloudDeploy deploys objects to the Wolfram Cloud. Let's take a look at some examples.

- The command below creates a web page with a 100-point font message:

```
CloudDeploy[Style["My first cloud program", 100]]
```

```
CloudObject[https://www.wolframcloud.com/objects/e4f106f2-25fc-4cb8-ad22-14
```

- The first time we execute it, *Mathematica* will ask for our Wolfram ID and password as shown in Figure 3.2:

Figure 3.2 Sign-in Screen for WolframCloud.

- Clicking on the hyperlink just created will automatically take us to the web page shown in Figure 3.3:

Figure 3.3 Web page created with the CloudDeploy command.

- We can deploy any *Mathematica* function to the cloud. Figure 3.4 shows the deployment of a dynamic interface:

```
CloudDeploy[Manipulate[Plot[Sin[x (1 + k x)], {x, 0, 6}], {k, 0, 2}]]
```

```
CloudObject[https://www.wolframcloud.com/objects/7c9e2c32-4e66-4d38-88cf-40
```

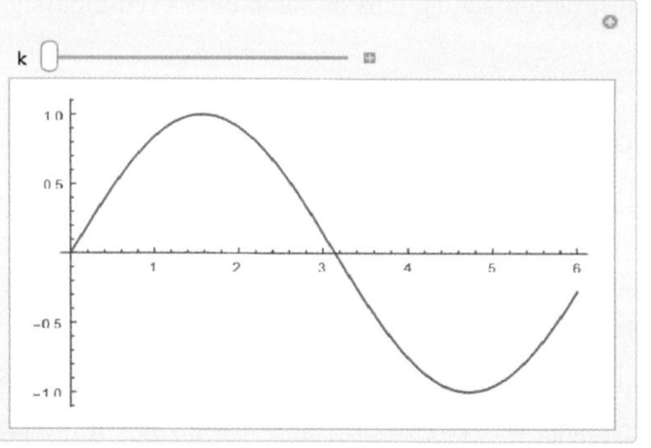

Figure 3.4 Manipulate in the cloud.

For further information:

http://reference.wolfram.com/language/guide/CloudFunctionsAndDeployment.html

3.15 Package Development

Mathematica's capabilities can be extended substantially with the use of packages. Packages are conceptually similar to what other programming languages call subroutines or subprograms. They basically involve the creation of functions, often complex, that can be called when needed without having to define them again.

3.15.1 The Structure of a Package

```
Clear["Global`*"]
```

A package has the following structure :

`BeginPackage`["name of the package`"]

f`::usage = "f brief description" ,` … (here we would include the headings of the functions that we are going to use with a brief description, that will be seen by the user later on).

`Begin`["`Private`"]

f [*args*] = *value* , … (here we would have the option to place functions that will be used by f but that will remain private. That is, the user will not be able to access them).

f [*args*] = *value* , … (here we would program the package functions accessible to the user).

`End[]`

`EndPackage[]`

A simple package

We'd like to build the function below and make sure that it's going to be available for later use in any notebook.

$$f(n, r, p) = \binom{n}{r} p^k (1-p)^{n-r}, \quad n \geq 1, \; n \geq r, \; 0 \leq p \leq 1$$

- Write the following in a code cell (**Format ▸ Style ▸ Code**)

```
BeginPackage["Binomialpackage`"]

Binomialdistribution::usage =
"Binomialdistribution[n, r, p] calculates the binomial probability
distribution under the conditions that n must be an integer, n ≥ 1,
n >= r and 0 <= p <= 1.";

Begin["`Private`"]

Binomialdistribution[n_Integer, (r_Integer)?NonNegative,
     (p_)?NonNegative]/; n >=r && 0<=p<=1 :=
   (n!/(r!*(n - r)!))*p^r*(1 - p)^(n - r);

Binomialdistribution[n_, r_, p_]:="Check that n is an integer,
r is a non-negative integer,n >= r and 0 ≤ p ≤ 1."

End[ ]

EndPackage[ ]
```

- Next, from the main menu bar select **File ▶ New ▶ Package**. A new window in the package environment will appear (Figure 3.5). After that, copy the previous cell to this new notebook, turn it into an initialization cell (**Cell ▶ Cell Properties ▶ Initialization**) and save it as a package (**File ▶ Save as ▶ Wolfram Mathematica Package**) in the Data folder located in our working directory. Use as the name the one defined in BeginPackage: "Binomialpackage.m" (remember to choose the extension .m).

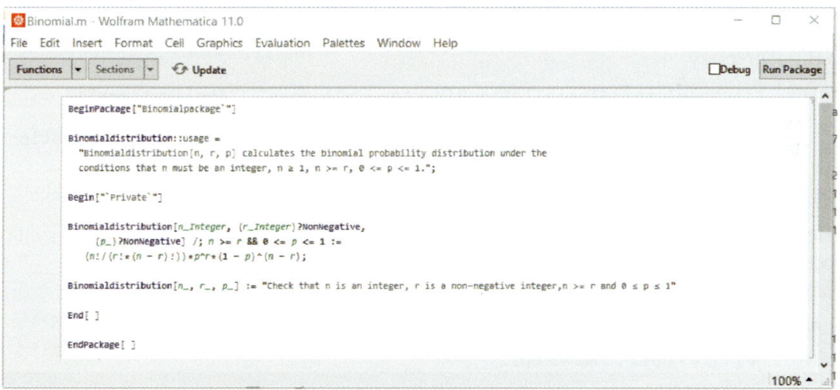

Figure 3.5 Developing packages in *Mathematica.*

- In this environment we can test, modify and even debug the package. Once we see it is working correctly we will save it. A package is an ASCII file. It can also be modified using a text editor such as Notebook (in Windows).
- As the last step, we exit the session (to be precise: we exit the kernel) to start a new one. This can be done without closing the notebook by choosing in the menu bar **Evaluation ▶ Quit Kernel** or with the command:

```
Quit[]
```

- Let's make the folder Data our working directory.

```
SetDirectory[FileNameJoin[{NotebookDirectory[], "Data"}]]
```

```
C:\Users\guill\Documents\MathematicaBeyond\Data
```

- To use the package we need to load it first (or, for testing purposes, execute it inside a notebook). We can use either << or `Needs` but in this case we choose the latter since it has the advantage of only loading the package if it hasn't been loaded previously.

```
Needs["Binomialpackage`"]
```

- One way to know what functions are available inside the package is as follows:

```
?Binomialpackage`*
```

```
Binomialdistribution[n, r, p] calculates the binomial probability distribution
    under the conditions that n must be an integer, n ≥ 1, n >= r, 0 <= p <= 1.
```

- At the time of loading the package, all its functions are executed, in this case just one: Binomialdistribution.

```
Binomialdistribution[5, 3, 0.1]
```

```
0.0081
```

```
Binomialdistribution[5, 3, 1.4]
```

```
Check that n is an integer, r
   is a non-negative integer,n >= r and 0 ≤ p ≤ 1.
```

3.15.2 Package Use

If you want the package to be available to all users, copy it to a *Mathematica* subdirectory named Applications that can be found executing the command $InstallationDirectory or $BaseDirectory.

```
FileNameJoin[{$InstallationDirectory, "Addons", "Applications"}]
```

```
C:\Program Files\Wolfram Research\Mathematica\11.0\Addons\Applications
```

3.15.3 Adding Options to Packages

Here we show a way to build a package with options. For that purpose we use the syntax: f[x_, opts___]. Note the use of the triple underscore symbol (BlankNullSequence) "___" that we have mentioned before.

- Before building a function with options remember the replacement rules:

```
m /. m -> opt1 /. m1 -> opt2
```

```
opt1
```

```
m1 /. m -> opt1 /. m1 -> opt2
```

```
opt2
```

- These rules are applied when defining a function where users can choose different options. One option will be used by default ("Method1"):

```
Opt1[function1] = Method → "Method1";
```

```
f1[x1_, opts___] := {x1, Method /. Flatten[{opts}] /. Opt1[function1]}
```

- If we don't write any option or we write Method→ "Method1" then Method 1 is applied:

```
f1[v1]
```

```
{v1, Method1}
```

- The method written in opts will be used in other cases. We have used Flatten[{opts}] to ensure that the function works even if the user writes the option between nested brackets.

```
f1[v2, {{Method → "Method2"}}]
```

```
{v2, Method2}
```

- Now we will develop a package with options. Let's suppose that we would like to do an operation between two lists (a = list1, b = list2). The operation f will be different depending on the chosen method: If the method is "Automatic" then f = a * b; if it's "Method1" then f = a / b; and if it's "Method2" then f = a + b. If the user doesn't choose a method then the default method, "Automatic", will be applied. We will also include a message in case there are problems. For example, if we mistype the name of the method we will get the message "Something is wrong".

```
BeginPackage["PackageWithOptions`"]
(*This is a comment. It doesn't affect the package functionality
 and cannot be seen by the user. Comments can be useful to
 understand the package if we'd like to modify it later on.*)
(*We associate a message to the function name to be
 displayed when we request help about the function.*)

listOperation::usage =
  "listOperation[listA, listB, options]. This function
    will multiply,divide  or add the elements of
    two lists depending on the chosen method.";

Options[listOperation] = {Method → "Automatic"};


Begin["`Private`"]

listOperation[listA_List, listB_List, opts___] :=
  With[{m = Method /. Flatten[{opts}] /. Options[listOperation]},
   Module[{a, b, r}, a = listA;
     b = listB;
     r = a b] /; m === "Automatic"];

listOperation[listA_List, listB_List, opts___] :=
  With[{m = Method /. Flatten[{opts}] /. Options[listOperation]},
   Module[{a, b, r}, a = listA;
     b = listB;
     r = a / b] /; m === "Method1"];

listOperation[listA_List, listB_List, opts___] :=
  With[{m = Method /. Flatten[{opts}] /. Options[listOperation]},
   Module[{a, b, r}, a = listA;
     b = listB;
     r = a + b] /; m === "Method2"];
```

```
listOperation[listA_, listB_, opts___] := "Something is wrong";

End[]
EndPackage[]
```

Now, we proceed to save the package under the name "PackageWithOptions". This can be done once again by copying the cell into a new notebook, changing the cell style to **Code**, initializing the cell and finally saving the notebook as a Mathematica package with the ".m" extension. It's recommended to copy the file in the subdirectory Applications.

- As mentioned before, if you want the package to be available to all users, copy it to Addons\Applications in:

```
FileNameJoin[{$InstallationDirectory, "Addons", "Applications"}]
```

```
C:\Program Files\Wolfram Research\Mathematica\11.0\Addons\Applications
```

- If you'd like only the current user to access it, place it in the following Applications subdirectory:

```
$UserBaseDirectory
```

```
C:\Users\guill\AppData\Roaming\Mathematica
```

```
Needs["PackageWithOptions`"]
```

```
?PackageWithOptions`*
```

listOperation[listA, listB, options]. This function will multiply,
divide or add the elements of two lists depending on the chosen method.

```
listOperation[{1, 2, 3}, {1, 2, 3}]
```

```
{1, 4, 9}
```

```
listOperation[{1, 2, 3}, {1, 2, 3}, Method → "Method1"]
```

```
{1, 1, 1}
```

```
listOperation[{1, 2, 3}, {1, 2, 3}, Method → "Method2"]
```

```
{2, 4, 6}
```

```
listOperation[1, 2]
```

```
Something is wrong
```

In later chapters we will see other examples of real applications of packages.

3.15.4 Error and Warning Messages

In the previous section we have learned how to display a very simple error message. If we want to create more detailed ones we can proceed as follows: A message has the form `Message[`*symbol*`::`*tag*`]`, where *symbol* is the function associated to the message, and *tag* is a message identifier.

- Let's create a message associated with a function **fu** using **generalmsg** as the tag (as a precaution we remove any previous reference to **fu**.)

```
Clear[fu]
```

```
fu::generalmsg = "This a message about fu.";
```

- We can see the output type that **Message** uses.

```
Message[fu::generalmsg]
```

fu: This a message about fu.

- Let's define **fu** so that **Message** will be generated when **fu** is called with just one argument.

```
fu[_] := Message[fu::generalmsg]
```

```
fu[0]
```

fu: This a message about fu.

- We define **fac** to calculate the factorial of a number and we limit the argument to positive integers.

```
fac[n_Integer?Positive] := n fac[n - 1];
```

```
fac[0] = 1;
```

- The following message is generated if we use **fac** with the wrong argument type.

```
fac::wrongargs:=
  "the argument of fac must be a single positive integer."
```

```
fac[___] := Message[fac::wrongargs]
```

- Let's see some examples with incorrect inputs.

```
fac[2, 3]
```

fac: the argument of fac must be a single positive integer.

```
fac[2.3]
```

fac: the argument of fac must be a single positive integer.

- We can include more complex messages to display specific information related to the type of error. For example, we can include a particular message in the position defined by a double backquote punctuation mark (``) inside a general message.

```
fac::error :=
 "A positive integer is expected in fac[``]."
```

```
fac[wrong_?NumericQ] := Message[fac::error, wrong]
```

```
fac[-6]
```

fac: A positive integer is expected in fac[−6].

You can find additional information about the use of messages in: Message.

3.15.5 Integrating Package Help Files with *Mathematica's* Help System

If you are going to develop an application that includes multiple notebooks, you may find the AuthorTools package useful: AuthorTools/tutorial/MakeProject.

Another alternative way of creating packages would be to use Wolfram's IDE, Wolfram Workbench (http://www.wolfram.com/products/workbench), which we will cover in Chapter 12.

3.16 Advanced Programming Tools

In addition to Workbench, there's a complementary set of *Mathematica* tools that we will cover as well in Chapter 12 such as: WLGM (Wolfram Lightweight Grid Manager) to connect multiple computers and use them in parallel computations and webMathematica, to develop applications that run on a web server and can be accessed from any browser. The latest *Mathematica* versions also include the possibility of CUDA programming (to take advantage of the GPUs in NVidia graphics cards) and incorporate several CUDA functions optimized for list processing, image processing, and linear algebra and Fourier transforms applications among others.

3.17 Additional Resources

As in most cases, the best available information comes from *Mathematica*'s own documentation or Wolfram's website:

Fast Introduction for Programmers, a short introduction to the Wolfram Language for programmers, is available from **Help ▸ Wolfram Documentation ▸ Intro for programmers** ≫, or on the web: (http://www.wolfram.com/language/fast-introduction-for-programmers/)

Stephen Wolfram's book: *An Elementary Introduction to the Wolfram Language* covers the basic principles of the Wolfram Language. It's available directly from within the Wolfram documentation: **Help ▸ Wolfram Documentation ▸ Introductory Book** ≫ or on the web: http://www.wolfram.com/language/elementary-introduction

Wolfram's reference page for the Wolfram Language: http://www.wolfram.com/language/

Functions and references:
http://reference.wolfram.com/mathematica/tutorial/FunctionsAndProgramsOverview.html

The following sources contain links to *Mathematica* programming documents available on the Internet. Even though they were written for previous versions, most of their contents are still relevant.

Ted Ersek's collection of *Mathematica* Tricks:
http://www.verbeia.com/mathematica/tips/Tricks.html

Roman E. Maeder's *Mathematica*'s Programming Language:
http://library.wolfram.com/infocenter/Conferences/183/

Michael A. Morrison's *Mathematica*'s Tips, Tricks and Techniques: Syntax
http://www.nhn.ou.edu/~morrison/*Mathematica*/TipSheets/Syntax.pdf

Leonid Shifrin's *Mathematica* programming - an advanced introduction:
http://www.mathprogramming-intro.org

There are many books for learning how to program with *Mathematica.* One of the best ones is *Power Programming with Mathematica: The Kernel* by David B. Wagner (available as a free download); though written for *Mathematica* 3, its ideas are still useful. Another one is *The Computer Science with Mathematica* by Roman E. Maeder. The bible of programming with *Mathematica* is probably *The Mathematica GuideBooks* collection written by Michael Trott (in *Mathematica* 5.) A great reference for those who aspire to become *Mathematica* programming gurus.

http://packagedata.net/ is a global repository of *Mathematica* packages that makes it easy for users to find additional functionality in a wide variety of scientific and technical fields.

4

Interactive Applications, Image Processing and More

In this chapter we explore the use of dynamic functions for creating interactive simulations (demonstrations). We also introduce image processing, graphs and networks and show some of the commands added to the most recent Mathematica versions for advanced calculus. We will use examples throughout the chapter to illustrate the functionality of all the commands.

4.1 Manipulate

One of the most interesting functions in *Mathematica* is `Manipulate`. We can easily build dynamic applications with it. This function has numerous options, and in this section we show some of them. Normally we will use more than one option at a time; if you prefer you can try them separately and then combine them. Remember that if you don't know the syntax of a function you can always use the free-form input first and later modify the *Mathematica* output to suit your needs.

- We type in an *input* cell: **manipulate sin x from 0 to 1**. An input and an output are generated.

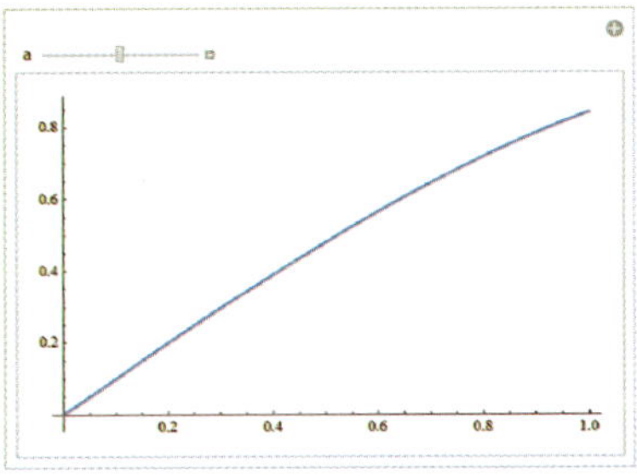

- In the input cell above we click on the floating label "Replace cell with this input" that appears when placing the mouse below "Result". This way we obtain the function below that we can use as a basic `Manipulate` template. Notice that **a** is a parameter that can take values between 0 and 2 with a default value of 1. Try moving the slide bar with the cursor.

```
Manipulate[Plot[Sin[x*a], {x, 0, 1}], {{a, 1}, 0, 2}]
```

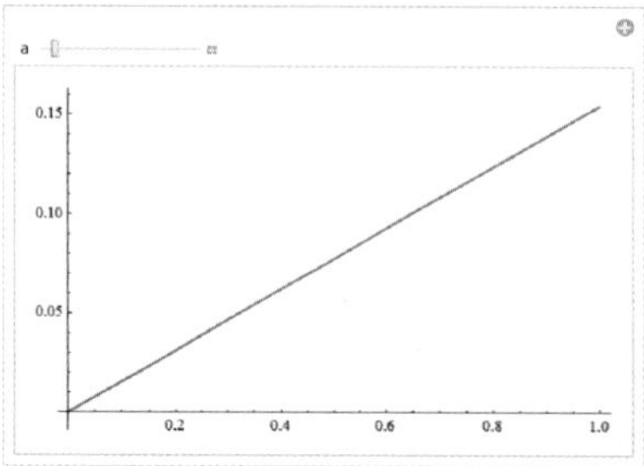

The output area containing `Manipulate` changes dynamically depending on the input. However, sometimes we may be interested in giving it a fixed size.

- To set the size of that area we use the option `ContentSize`. Note that if the result does not fit, a sliding bar is displayed. We also use `BaseStyle` to choose the output format.

```
Manipulate[D[Sin[x] / Cos[x], {x, n}],
 {n, 1, 10, 1}, ContentSize → {200, 50},
 BaseStyle → {Blue, FontFamily → "Times", 14}]
```

In some of the examples that come next we use `ContentSize` to reduce the print size. To replicate them in a computer you can remove the option.

- In the following example we use `Manipulate` to dynamically represent the famous Lissajous curves: $x(t) = \cos(a\,t)$, $y(t) = \sin(b\,t)$, for integers a and b. By adding the option **`Appearance → "Labeled"`** we can display to the right of the sliding bars the actual values of the parameters. We can also label the parameters using the syntax: *{{a, initial a value, "a label"}, minimum a value, maximum a value}*:

```
Manipulate[
 ParametricPlot[{Cos[a t], Sin[b t]}, {t, 0, 2 π} ],
 {{a, 3, "Amplitude a"}, 1, 8, Appearance → "Labeled"},
 {{b, 3, "Amplitude b"}, 1, 13, Appearance → "Labeled"},
 ContentSize → {150, 150} ]
```

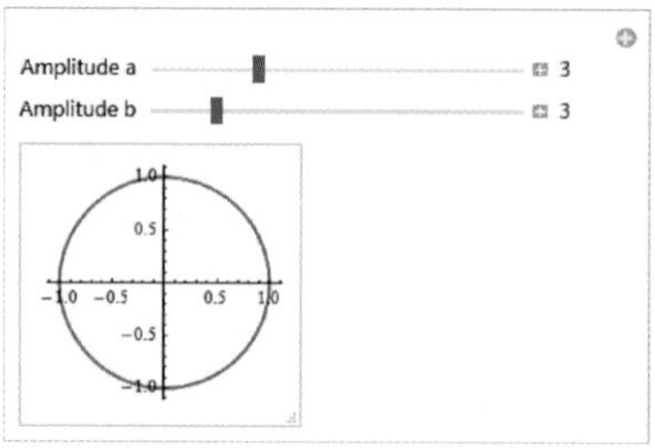

- To display in the previous output the controls associated to the parameters a and b you can click on ⊞. If you'd like them to be shown by default, you should add the option Appearance → "Open".

```
Manipulate[Plot[Sin[n x], {x, 0, 2 π}],
 {n, 1, 20, Appearance → "Open"}, ContentSize → {200, 150} ]
```

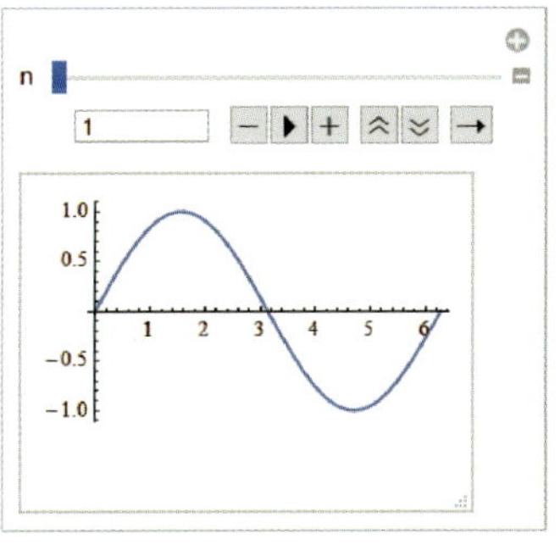

- An alternative way to display the controls is AppearanceElements → All.

```
Manipulate[Plot[Sin[n x], {x, 0, 2 π}], {n, 1, 20},
 AppearanceElements → All, ContentSize → {200, 150}]
```

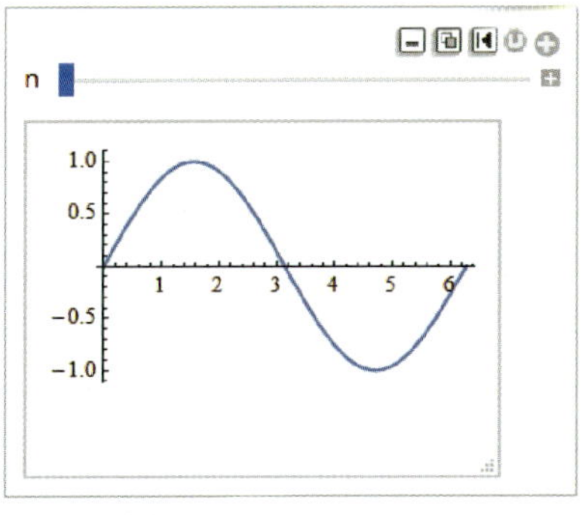

- You can also include a drop-down button. This will automatically be added when a parameter contains five or more choices as shown in the next example. In this case we have the possibility to choose several filling alternatives for displaying a graph. We use the `Plot` option: `Filling`. We also use `PlotLabel` to show the parameter value when labeling the graph. Notice that the combination of `Plot` options with `Manipulate` greatly increases the number of possible combinations for dynamic interactions.

```
Manipulate[Plot[Sin[n x], {x, 0, 2 π}, Filling → choice, PlotRange → 2,
  PlotLabel → Style[StringForm["Sin(n t) representation for  n = `1`",
     PaddedForm[n, {4, 2}]], 10]],
 {n, 1, 20}, {{choice, None, "Choice"}, {None, Axis, Top, Bottom,
   Automatic, 1, 0.5, 0, -0.5, -1}}, ContentSize → {200, 150}]
```

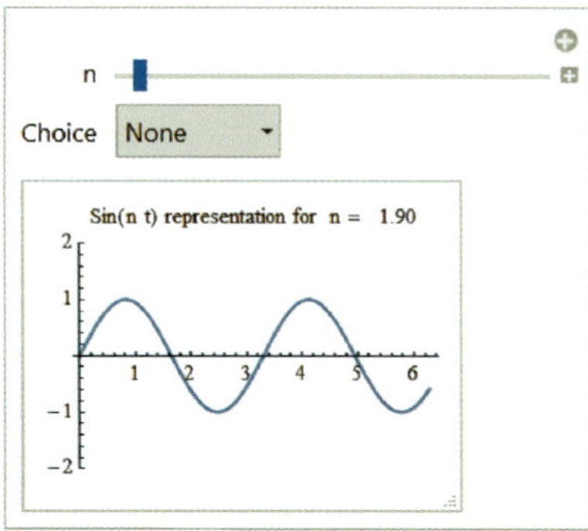

- Here is a slightly more complicated example showing how we can format the output of `Manipulate`. In this case we take advantage of the flexibility provided by the options of `Grid`.

```
Manipulate[
 Grid[Table[{i, i^m}, {i, 1, n}],
  Alignment → {{Left, Right}, Automatic}, Frame → All,
  ItemStyle → {Blue, FontFamily → "Helvetica"}, Background → LightGray],
 {n, 5, 20, 1}, {m, 5, 100, 1}]
```

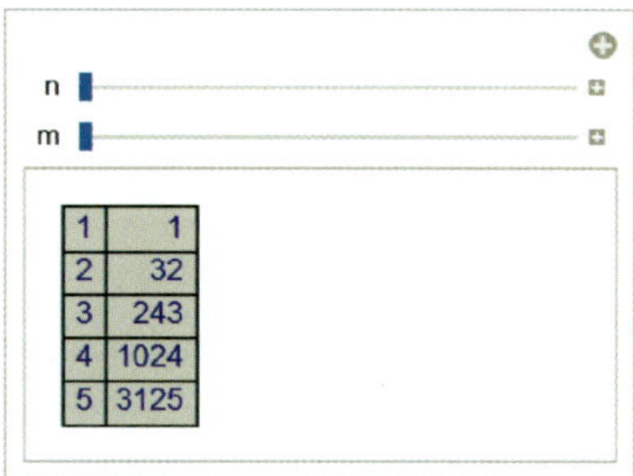

Another interesting option is `ControlType`. For example, we can use `Checkbox`, an option not only available in `Manipulate`, to include check boxes. The syntax is: $\{a, \{a1, a2\}\}$, where $\{a1, a2\}$ refers to the alternatives available, $a1$ is the default option. When there are more than two choices you can use `RadioButtonBar`.

- In the example below there are options to display the axes and to choose the filling color of the disk. The filling color by default is red.

```
Manipulate[
 Graphics[{EdgeForm[Thick], color, Disk[]}, Axes → axes, ImageSize → Tiny],
 {{axes, True, "Axes"}, {True, False}, ControlType → Checkbox},
 {{color, Red, "Color"}, {Red → "Red", Green → "Green", Blue → "Blue"},
  ControlType → RadioButtonBar}]
```

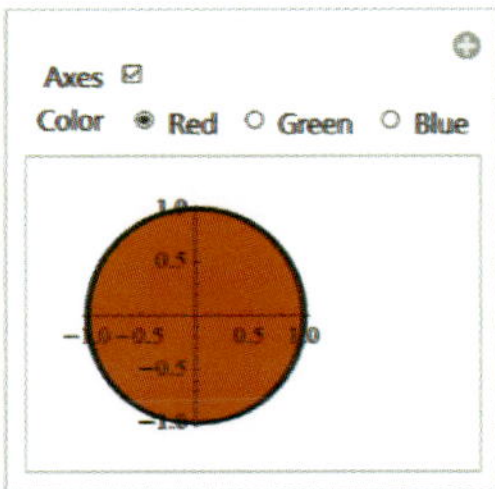

Once we execute the previous command we can change the color and/or choose not to display the axes.

- The next example uses ColorSlider to dynamically change the background color as we move along the color bar starting with LightBlue. We use the label "Disk color". In this case we don't need to type **ColorSlider** because it's implicitly stated.

```
Manipulate[Graphics[{EdgeForm[Thick], col, Disk[]}, ImageSize → Tiny],
 {{col, LightBlue, "Disk Color"}, Red}]
```

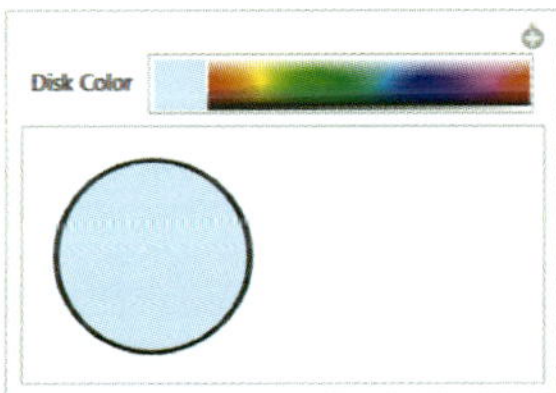

- We can also do something similar using ColorSetter. Clicking on the colored square will launch the operating system color palette to enable us to dynamically modify the disk background.

```
Manipulate[Graphics[{EdgeForm[Thick], color, Disk[]}, ImageSize → Tiny],
 {{color, LightBlue, "Color"}, LightBlue, ControlType → ColorSetter}]
```

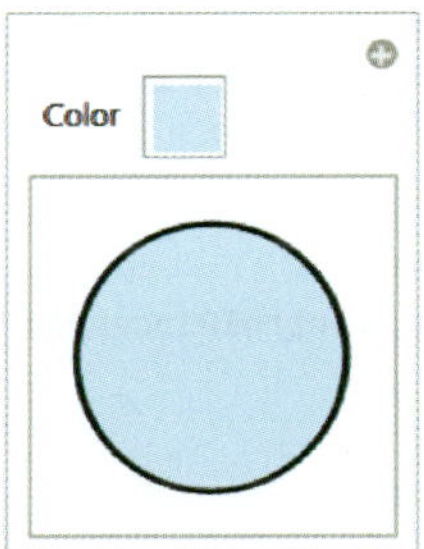

- In the command below we use **ControlType**→Slider2D to move a point of a certain size around a square.

```
Manipulate[Graphics[{PointSize[0.1], Point[pt]},
  PlotRange → 1, ImageSize → Tiny], {pt, {-1, -1}, {1, 1}}]
```

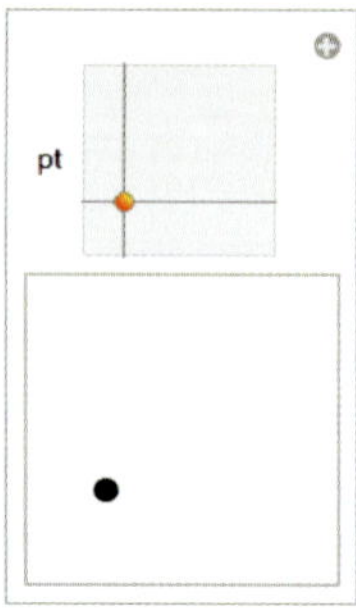

- This example is more advanced.

```
Manipulate[
 Graphics[{Line[Table[{{Cos[t], Sin[t]}, pt}, {t, 2π/30, 2π, 2π/30}]]},
  PlotRange → 1, ImageSize → Small],
 {pt, {-1, -1}, {1, 1}}]
```

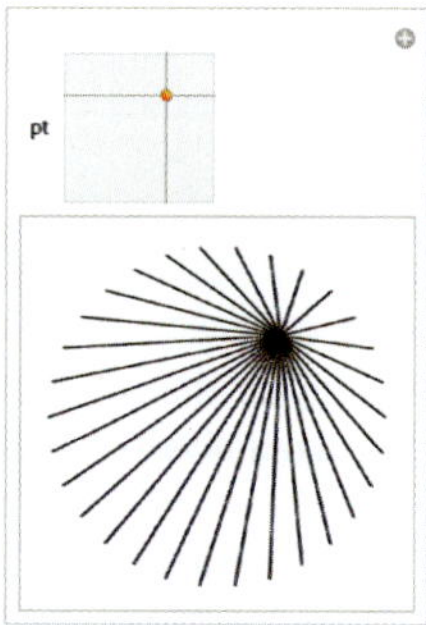

- Here we combine `Manipulate` with a gauge (`HorizontalGauge`). Click on it and move it around.

```
Manipulate[Plot[Sin[p t], {t, 0, 2π},
  PlotRange → {{0, 2π}, {-1, 1}}, ImageSize → Small],
 {{p, 1, "Period"}, 0, 5, HorizontalGauge[##] &}, ControlPlacement → Bottom]
```

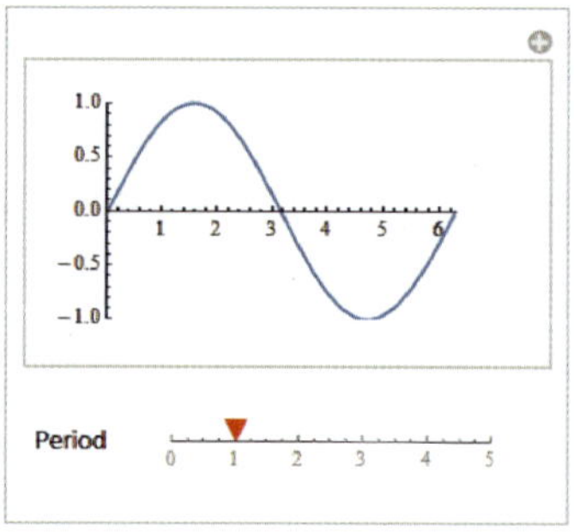

If we use Locator with `Manipulate` we will be able to move objects in the output with the mouse. Additionally, all the information associated to those objects will be dynamically updated. The syntax of `Locator` is: `{{`*parameter*`, {`*x*`,` *y*`}},` *Locator*`}`.

In some of the examples that follow we have added the option **`ImageSize → Tiny`** or **`ImageSize → Medium`** to reduce the print size. To replicate them in a computer remove the option.

- In this example we can move the three vertices of a polygon and see how the triangle area is filled automatically.

```
Manipulate[
 Graphics[{Pink, Polygon[pts]}, PlotRange → 1.1, ImageSize → Tiny],
 {{pts, {{0, 0}, {1, 0}, {0, 1}}}, Locator}]
```

We are going to extend the functionality of the previous command to calculate and display the area of the triangle, but first let's introduce a way to call a function from within `Manipulate`.

- Use the option SaveDefinitions.

```
g[x_] := x^2
```

```
Manipulate[g[y], {y, 0, 100, 1}, SaveDefinitions → True]
```

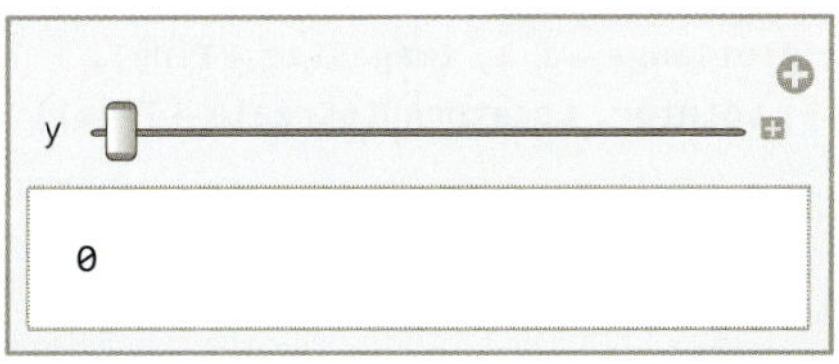

Alternatively, we can use Initialization with the function that we would like to evaluate first when `Manipulate` is executed:

```
Manipulate[h[y], {y, 0, 100, 1}, Initialization :→ (h[x_] := x^2)]
```

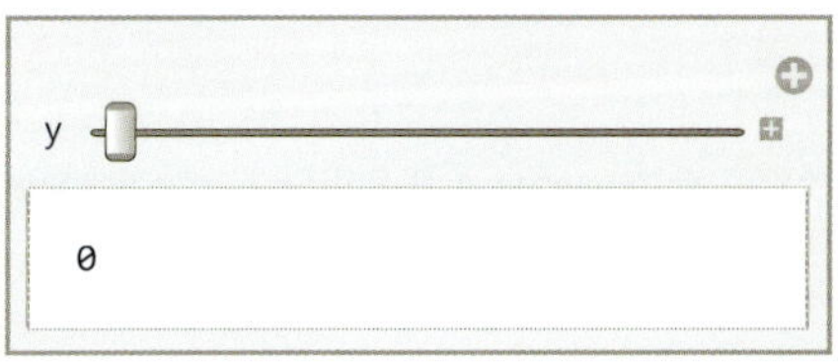

- Going back to the example of the triangle, to explicitly include the value of its area we have to define a function that calculates it from the Cartesian coordinates of the its three vertices: v_1, v_2, and v_3.

```
areaTriang[{v1_, v2_, v3_}] :=
 Abs[Det[Join[Transpose[{v1, v2, v3}], {{1, 1, 1}}]]] / 2
```

```
areaTriang[{{3, 2}, {4, 1}, {7, -3}}]
```

$\frac{1}{2}$

- Now we include **areaTriang** in the previously typed `Manipulate`. Move the vertices with the mouse to see how the area is dynamically updated.

```
Manipulate[Graphics[{Pink, Polygon[pts]}, PlotRange → 1,
  PlotLabel → StringForm["Triangle Area  = `1`", areaTriang[pts]],
  ImageSize → Tiny],
 {{pts, {{0, 0}, {1, 0}, {0, 1/2}}}, Locator},
 Initialization :→ (areaTriang[{v1_, v2_, v3_}] :=
    Abs[Det[Join[Transpose[{v1, v2, v3}], {{1, 1, 1}}]]] / 2) ]
```

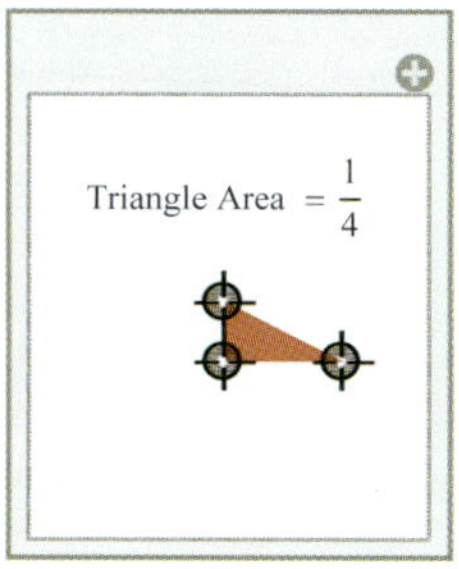

- If we use `LocatorAutoCreate` → **True** we will be able to add or remove points using ALT+click (CMD+click in OS X).

```
Manipulate[
 Graphics[{Pink, Polygon[pts]}, PlotRange → 1.1, ImageSize → Tiny],
 {{pts, {{0, 0}, {1, 0}, {0, 1}}}, Locator, LocatorAutoCreate → True}]
```

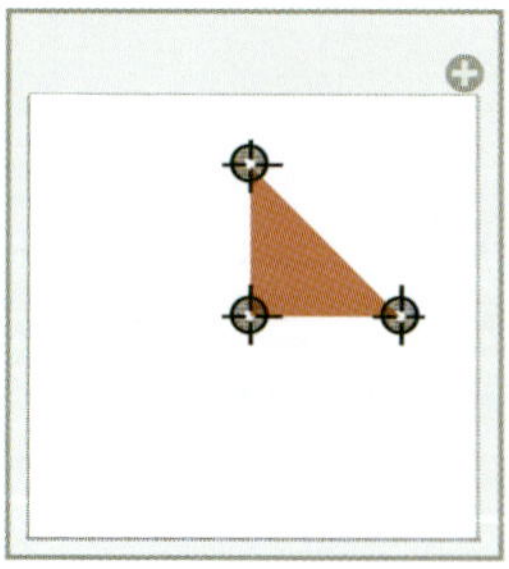

- In this example we use `BSplineFunction` to generate a B-Spline curve associated with all the displayed points.

```
Manipulate[
 Show[Graphics[{{Point[p], Green, Line[p]}}, ImageSize → Small],
  ParametricPlot[BSplineFunction[p][t], {t, 0, 1}]],
 {{p, pts}, Locator},
 Initialization :→ (pts = {{1, 5}, {3, 2}, {6, 4}})]
```

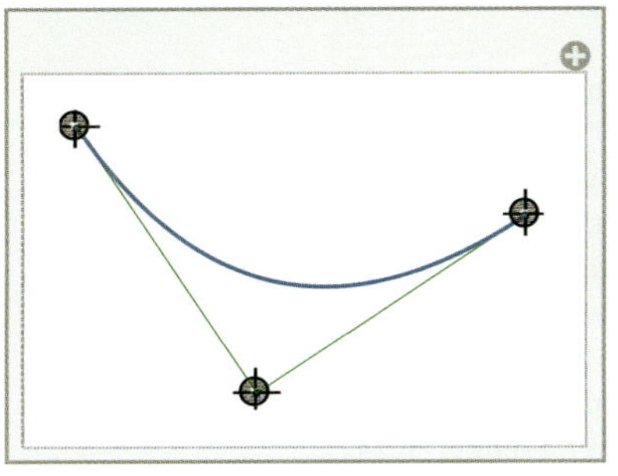

If we are going to use PlotPoints with a large number of points, we can save computation time by using ControlActive.

- While moving the slider button, PlotPoints performs the calculations using a preset value. In this example the first argument in `ControlActive` is 8; once we stop, PlotPoints will use another value, normally significantly bigger than the preset one (in this example 40). If we replace 8 with 40 in `ControlActive`, the update of the plot will slow down. We also include `Defer`, a function that displays the unevaluated form of its input to label the graph.

```
Manipulate[Plot[Sin[1/(x - a)], {x, -1, 1}, PlotPoints → ControlActive[8, 40],
  PlotRange → 1, PlotLabel → Style[Defer[Sin[1/(x - "a")]], Medium]], {a, 0, 1}]
```

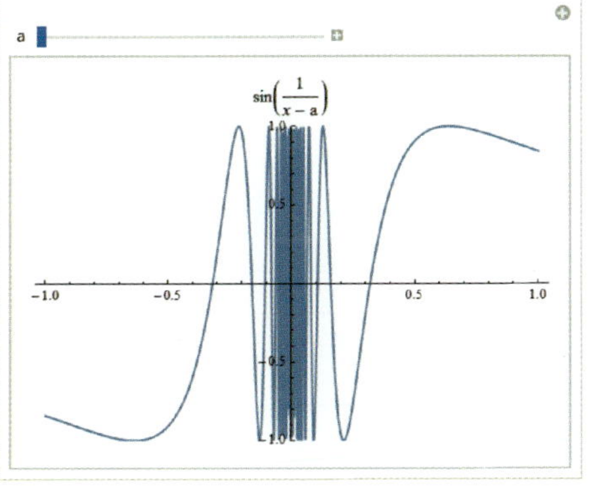

- `DynamicModule` is another function used for dynamic applications. The example below shows how to use it to display some of the different control types available in *Mathematica*.

```
DynamicModule[{pua, rbba, cbba, tgba}, Grid[{
   {Style["Control Type", FontWeight → "Bold"],
    Style["Example", FontWeight → "Bold"]},
   {"Slider", Manipulate[x, {x, 0, 1, ControlType → Slider},
     LabelStyle → 16]},
   {"2DSlider", Manipulate[x, {x, 0, 1, ControlType → Slider2D},
     LabelStyle → 16]},
   {"PopupMenu", Row[{PopupMenu[Dynamic[pua], {a, b, c, d}],
      "        ", InputField[Dynamic[pua], FieldSize → 5]}]},
   {"RadioButtonBar", Row[{RadioButtonBar[Dynamic[rbba], Range[5]],
      "        ", InputField[Dynamic[rbba], FieldSize → 5]}]},
   {"CheckBoxBar", Row[{CheckboxBar[Dynamic[cbba], {1, 2, 3}],
      "        ", InputField[Dynamic[cbba], FieldSize → 5]}]},
   {"TogglerBar", Row[{TogglerBar[Dynamic[tgba], Range[5]],
      "        ", InputField[Dynamic[tgba], FieldSize → 7]}]},
   {"InputField", Manipulate[input, {input, 3.14},
     ControlType → InputField, LabelStyle → 16]},
   {"ColorSlider", ColorSlider[Pink]}
  }, BaseStyle → {"TraditionalForm", FontFamily -> "Helvetica"},
  Alignment → {Left}, Background → LightYellow,
  Frame → All, Spacings → {2, 0.62}]]
```

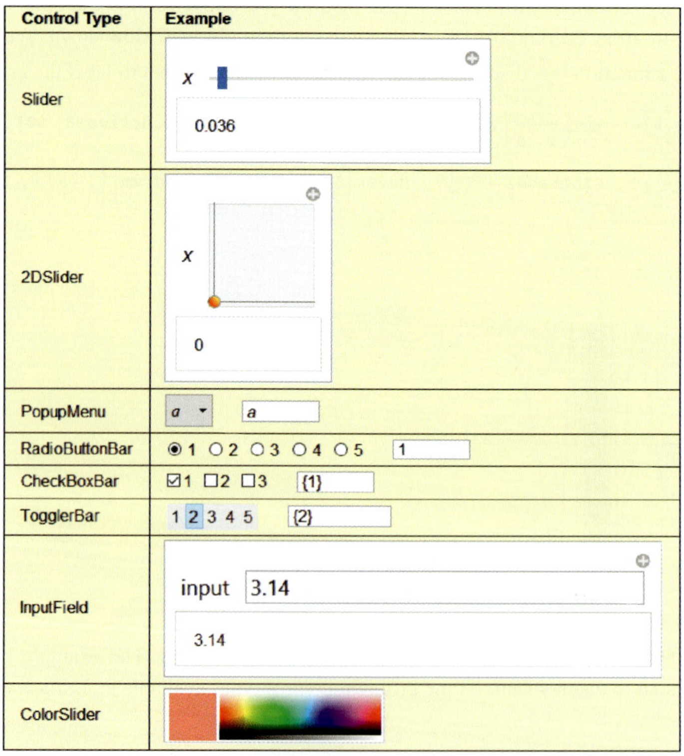


```
Quit[]
```

4.2 Creating Demonstrations

The main use of `Manipulate` is to create eye-catching interactive visualizations with just a few lines of code. In the Demonstrations site (http://demonstrations.wolfram.com) you can find thousands of them. Some of the demonstrations that we will be showing here are adaptations of existing ones available on the website.

4.2.1 A Random Walk

- We use the function `RandomWalkProcess` to represent a random walk on a line with the probability of a positive unit step p and the probability of a negative unit step $1 - p$. By default we assume n steps = 500 and set the interval from 100 to 1000 and $p = 0.5$. Each time we execute the command, even though the number of steps is the same, the distance covered will be different since we are assuming random walks.

```
Manipulate[
 ListPlot[RandomFunction[RandomWalkProcess[p], {1, n}], Filling→ Axis],
 {{n, 500, "Steps"}, 100, 1000 , 1, Appearance→ "Labeled"},
 {{p, 0.5, "Probability"}, 0, 1, Appearance→ "Labeled"}]
```

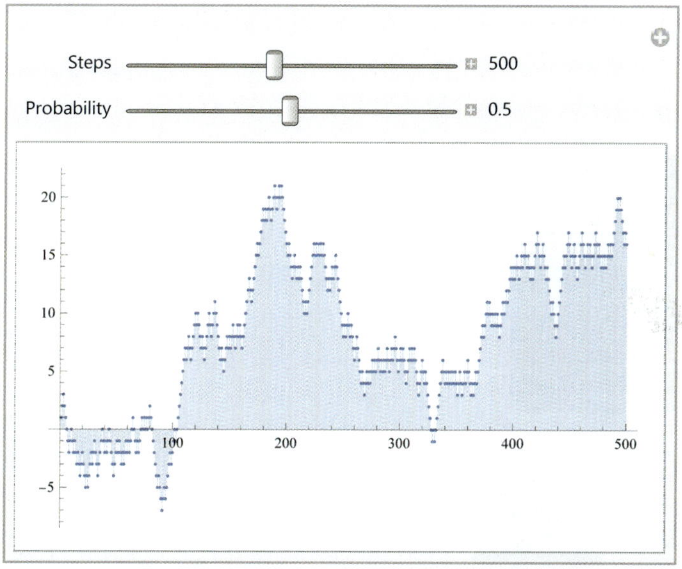

4.2.2 Zooming and the Rosetta Stone

In this example we show how we can zoom on an image and scroll through it. In this case the image is a picture of the Rosetta Stone but any other picture would do as well.

- We can import it directly with:

```
Clear["Global`*"] (* Remove all the variables from memory *)

rosettaimage =
  Import["http://upload.wikimedia.org/wikipedia/commons/thumb/c/ca/
     Rosetta_Stone_BW.jpeg/800px-Rosetta_Stone_BW.jpeg"];

rosetta = Image[rosettaimage, ImageSize→ 300]
```

- Now we can copy it, as shown below, or type **rosetta** instead.

`Manipulate[`

`ImageResize[ImageCrop[` 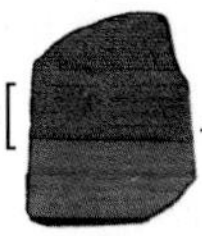 `, {300, 300} / zoom, -pts],`

`{300, 300}, Resampling → ControlActive["Nearest", "Lanczos"]],`

`{{pts, {0, 0}, "pts"}, {-1, -1}, {1, 1}},`

`{{zoom, 1, "Zoom"}, 1, 10}]`

- We can generate a similar effect using `DynamicImage` (New in *Mathematica* 11.0).

`DynamicImage[`  `, ZoomFactor → 2, ImageSize → {300, 300}]`

4.2.3 Zooming and Google Maps

- In this example we access https://maps.google.com using `URLFetch` and we move through the chosen area by zooming on it. We use functions not yet covered in the chapter that can be found in the help system. `InputField` can be especially useful. It enables us to open a window for inputting the data interactively, in this case the city that we are interested in.

```
Manipulate[ImportString[
  ExportString[URLFetch["http://maps.googleapis.com/maps/api/staticmap",
    "ContentData", "Parameters" → {"center" → c, "zoom" → ToString[z],
      "size" → "512x256", "maptype" → ToLowerCase["Map"], "sensor" → "false",
      "style" → "inverse_lightness:true", "format" → "png"}], "Byte",
   "DataFormat" → {"Bytes"}], "PNG", "DataFormat" → {"Byte"}],
 {{c, "Salamanca", "Location"}, "",
  InputField[#1, String, FieldSize → 35] &},
 {{z, 11, "Zoom"}, 1, 20, 1}, ContinuousAction → False]
```

- An easier way of doing it is with `DynamicGeoGraphics` (available since *Mathematica* 11.0).

```
DynamicGeoGraphics[Salamanca (city), GeoRange → Quantity[10, "km"]]
```

4.2.4 Derivatives in Action

```
Clear["Global`*"]
```

■ Here is a derivative evaluation using the definition:

```
Panel[DynamicModule[{f = Sin[x]}, lhs = DifferenceQuotient[f, {x, h}];
  Column[{InputField[Dynamic[f]], Dynamic[Block[{e = Inactivate[
       Limit[lhs, h -> 0]]}, e == Activate[e]]]}]] // TraditionalForm]
```

Sin[x]

$$\lim_{h\to 0} \frac{2\sin\left(\frac{h}{2}\right)\cos\left(\frac{h}{2}+x\right)}{h} = \cos(x)$$

■ In this example we use derivatives to show the tangent at the selected point for different curves. This type of visualization is useful to represent the tangent to a curve and introduce the concept of the derivative.

```
Manipulate[Plot[{f[x], f[x1] + f'[x1] * (x - x1)}, {x, -2 Pi, 2 Pi},
  PlotRange -> {-2, 2}, Epilog -> {PointSize[0.025], Point[{x1, f[x1]}]}],
 {x1, -2 Pi, 2 Pi}, { f, {Sin, Cos, Tan, Sec, Csc, Cot}}]
```

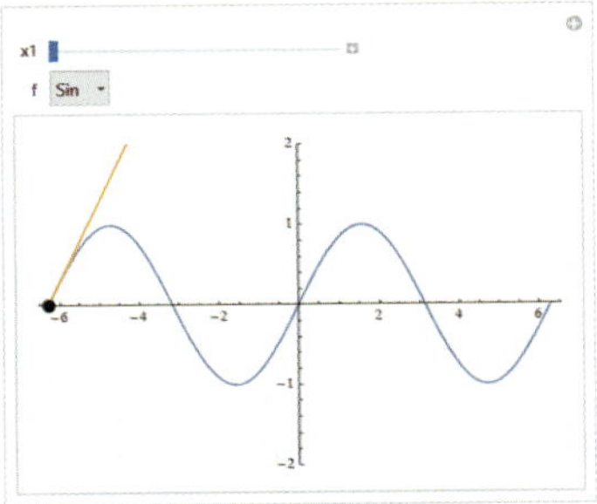

4.2.5 Tsunamis

We can model very complex natural phenomena. The following demonstration simulates a tsunami using the following system of partial differential equations:

$$\frac{\partial^2 u(t, x)}{\partial t^2} = \frac{\partial^2 u(t, x)}{\partial x^2} + a\,u(t, x)^3 + b\,u(t, x)^2 + c\,u(t, x) + d,$$

$$\text{with } u(0, x) = e^{-x^2}, \quad \frac{\partial u(t, x)}{\partial t} = 0 \;\; u(t, -x_0) = u(t, x_0)$$

(http://mathworld.wolfram.com/news/2005-01-14/tsunamis)

- We solve it with `NDSolve` using a, b, c, d, x_0 as parameters. We can show dynamically the behavior of the system by modifying the parameter values.

```
Manipulate[
 Plot3D[Evaluate[u[t, x] /. Quiet[NDSolve[Evaluate[{D[u[t, x], t, t] ==
        D[u[t, x], x, x] + a u[t, x]^3 + b u[t, x]^2 + c u[t, x] + d,
       u[0, x] == e^-x^2, D[u[t, x], t] == 0 /. t → 0, u[t, -x0] == u[t, x0]}],
     u, {t, 0, x0}, {x, -x0, x0}, Method → {"MethodOfLines",
       "SpatialDiscretization" → {"TensorProductGrid",
         "DifferenceOrder" → "Pseudospectral", "MinStepSize" → 0.2}}]]],
  {x, -x0, x0}, {t, 0, x0}, MeshFunctions → {#3 &},
  Mesh → None,
  ColorFunction → "Rainbow",
  PlotPoints → 20,
  MaxRecursion → 1],
 {{a, -0.3}, -4, 0}, {{b, -0.6},
  -4, 0},
 {{c, 0.8}, -4, 1}, {{d, 0}, -1, 1},
 {{x0, 16, "Solution Range"}, 5, 20},
 ControlPlacement → Left,
 ContentSize → {400, 300}]
```

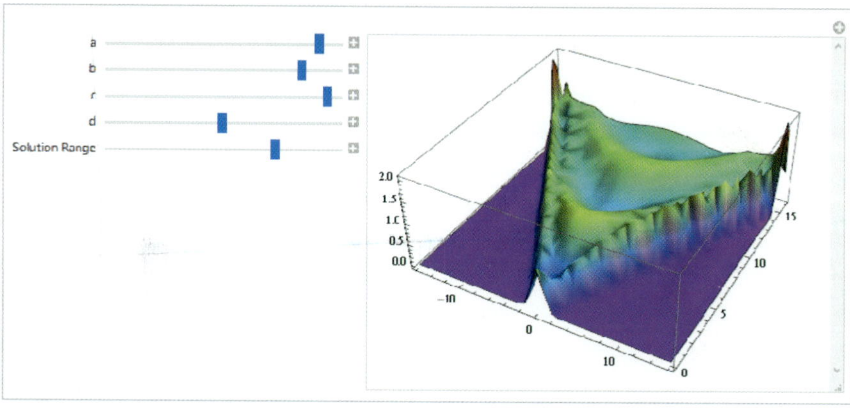

■ A more realistic simulation is available at:

http://demonstrations.wolfram.com/TsunamiStrikingALandscape

4.2.6 Bouncing Balls

■ This function combines `Manipulate` with `DynamicModule` to show bouncing balls. Each time you click inside the area, a new ball will be generated.

```
Manipulate[
 DynamicModule[{contents = {}, lastpt = {0, 1}}, Graphics[{PointSize[0.1],
    Point[Dynamic[If[pt =!= lastpt, AppendTo[contents, {pt, {0, 0}}];
       lastpt = pt];
      contents = Map[If[#[[1, 2]] ≥ 0, {#[[1]] - #[[2]], #[[2]] +
             {0, 0.001}}, {{#[[1, 1]], 0}, {1, -0.8} #[[2]]}] &, contents];
      Map[First, contents]]], Line[{{0, -.05}, {1, -.05}}]},
   PlotRange → {{0, 1}, {-.1, 1}}]], {{pt, {0, 1}}, {0, 0},
  {1, 1}, Locator, Appearance → None}, SynchronousUpdating → True]
```

■ In this example we simulate the bounce of a ball falling down stairs. You can modify the impulse given to the ball, its rebound and the step size. *Mathematica* can solve the problem both analytically and numerically, by replacing DSolve with NDSolve (numeric solution).

```
Manipulate[
 sol = DSolve[{y''[t] == -9.8, y[0] == 13.5, y'[0] == 5, a[0] == 13,
    WhenEvent[y[t] - a[t] == 0, y'[t] → - c y'[t]],
    WhenEvent[Mod[t, n], a[t] → a[t] - 1]}, {y, a}, {t, 0, 8},
   DiscreteVariables → {a}];
 Plot[Evaluate[{y[t], a[t]} /. sol], {t, 0, 8}, Filling → {2 → 0}],
 {c, 0.6, 1},
 {n, 0.6, 1.75}]
```

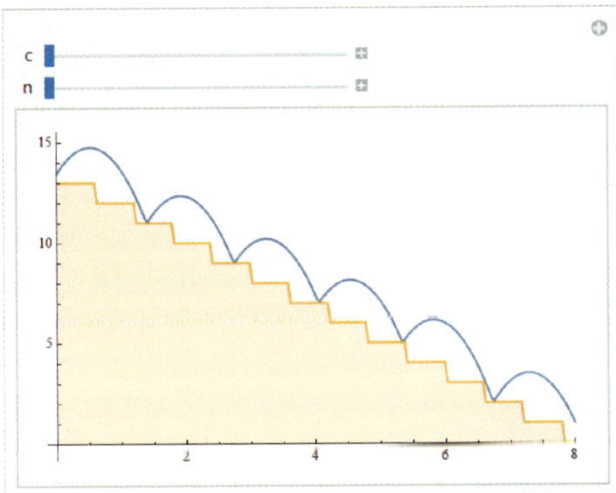

4.2.7 Language Detection

- One of *Mathematica*'s latest features as mentioned earlier is `Classify`. Here we use it to identify the language of a sentence.

```
Classify["Language", {"the house is blue",
  "la maison est bleu", "la casa es azul", "das Haus ist blau",
  "房子是蓝色的", "المنزل باللون الأزرق", "будинок синій"}]
```

{English, French, Spanish, German, Chinese, Arabic, Ukrainian}

```
Manipulate[Classify["Language", sentence],
 {{sentence, "la casa es azul", "Sentence"}, {"the house is blue",
   "la maison est bleu", "la casa es azul", "das Haus ist blau",
   "房子是蓝色的", "المنزل باللون الأزرق", "будинок синій"}}]
```

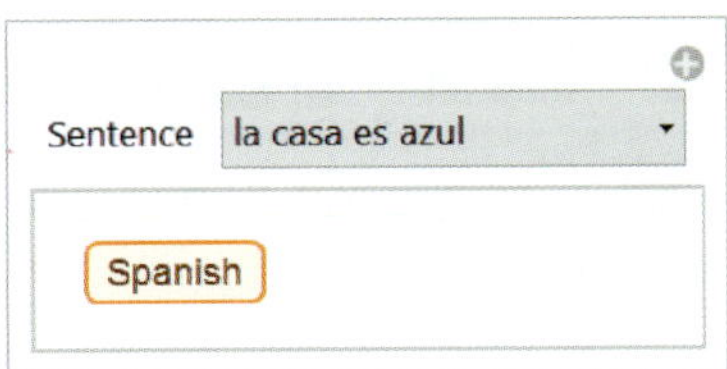

4.3 Image Processing

The image processing functionality of *Mathematica* has greatly improved in the latest versions of the program. We are going to give a brief overview of it. You can extend your knowledge by going through this tutorial: tutorial/ImageProcessing. At the end of this chapter there are several links in case you may be interested in learning more about the topic.

In the first chapter we saw that when you click on an image, the **Image Assistant**, a toolbar providing several options to modify the image will appear right below it. As a matter of fact, many of the operations that we will show here can be executed directly using this toolbar. However, for clarity purposes we will show the *Mathematica* commands instead of describing how to use the toolbar. Besides, if we were interested in creating a program we would have to type the required instructions.

4.3.1 Playing with The Giralda

- We import a JPG image (in this example, The Giralda of Seville, although with a search engine you can find many images similar to the one below or even use other images as well). The resolution of the image was preserved when downloaded. You can also cut and paste from a website.

```
Clear["Global`*"] (*Remove all the variables from memory*)
```

-

```
giralda =
```

```
;
```

- We can check the image size in pixels. It consists of a $n \times m$ grid of triplets, each one representing one pixel.

```
ImageDimensions[giralda]
```

```
{635, 635}
```

- We can see the RGB value of the individual pixels:

```
ImageData[giralda]
```

```
{{{0.607843, 0.733333, 0.886275}, {0.607843, 0.733333, 0.886275},
  {0.607843, 0.733333, 0.886275}, ... 629 ...,
  {0.458824, 0.572549, 0.792157}, {0.458824, 0.572549, 0.792157},
  {0.458824, 0.572549, 0.792157}}, ... 633 ..., {... 1 ...}}
large output   show less   show more   show all   set size limit...
```

- We adjust the contrast and the brightness (if you prefer, instead of copying and pasting the image, just type **giralda**).

```
Manipulate[ImageAdjust[
```

```
, {a, b}],
 {{a, -1.5, "Contrast"}, -2, 2}, {{b, 0.5, "Brightness"}, 0, 1}]
```

- The following command detects the edges of an image.

```
edges = EdgeDetect[giralda]
```

- Here we compare both images.

```
GraphicsRow[{giralda, edges}]
```

- `ImagePartition` splits an image into a collection of subimages of a certain size. (image downloaded from http://noticias-24.net/wp-content/uploads/2011/07/Machu-Picchu.jpg).

`Grid[ImagePartition[`  `, 100], Frame → All]`

Execute the function to see the output.

4.3.2 Background Removal and Image Composition

- In this example we remove the background of an image and create a new one by combining the result with an image of the Andromeda Galaxy.

`foreground = RemoveBackground[`  `, {"Background", Green}];`

`ImageCompose[`  `, foreground]`

- In the example below, available in the documentation, we eliminate small image components to extract a flock of flamingos.

`DeleteSmallComponents[`
`RemoveBackground[`  `, {"Background", {"Uniform", 0.08}}], 2]`

4.3.3 3D Images

Since *Mathematica* 9 there are many more functions for 3D image processing as this example shows:

```
ctHead = ExampleData[{"TestImage3D", "CThead"}];
head = Image3D[ctHead, BoxRatios → {1, 1, 1}];
Manipulate[
 Image3D[head, ClipRange → {{x, 256}, {0, y}, {z, 100}}],
 {{x, 100}, 0, 256, 1}, {{y, 150}, 0, 256, 1}, {{z, 50}, 0, 100, 1}]
```

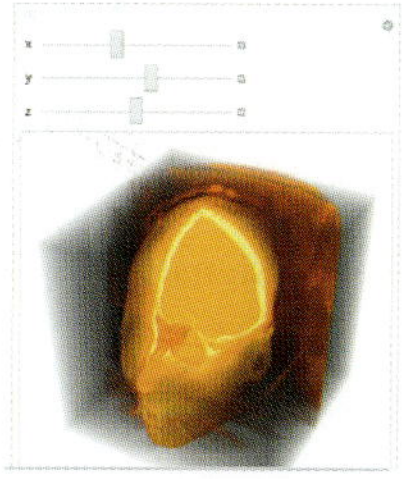

http://www.wolfram.com/mathematica/new-in-10/enhanced-3d-image-processing/index.html contains examples that may be useful for those readers interested in 3D image processing.

4.3.4 Eyesight Recovery

Next we show how to improve image resolution.

- Here's an example of an image with fuzzy text (you can find the picture in the `ImageDeconvolve` help file):

fuzzy = 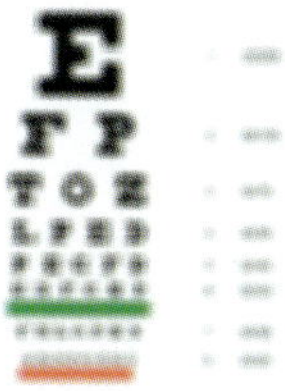 ;

- With a deconvolution we can substantially improve the image quality.

```
ImageDeconvolve[fuzzy, BoxMatrix[3] / 49, Method → "TotalVariation"]
```

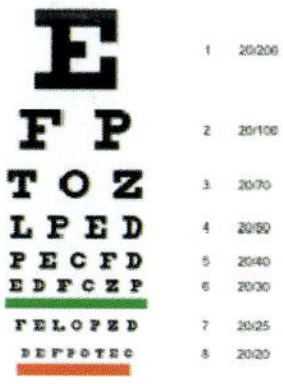

- Here is simulate a signal with noise.

```
data = Table[Sin[i^2 + i] + RandomReal[{-.3, .3}], {i, 0, Pi, 0.001}];
ListLinePlot[data]
```

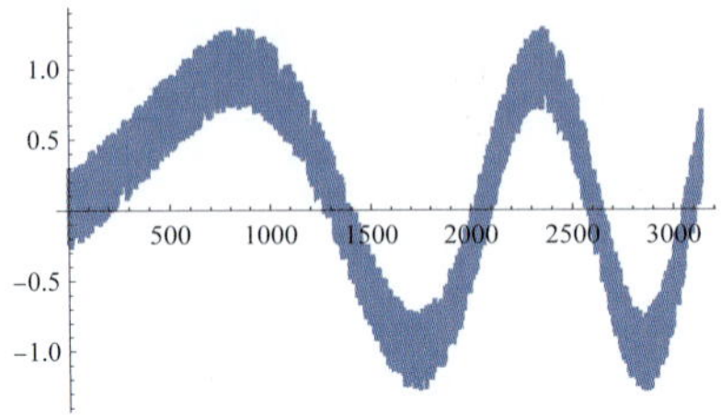

- The next function uses a filter to remove the noise component of the signal.

```
Manipulate[With[{res = LowpassFilter[data, ω]},
  GraphicsColumn[{Spectrogram[res], ListLinePlot[res]}]],
 {ω, Pi, .01, Appearance → "Labeled"}]
```

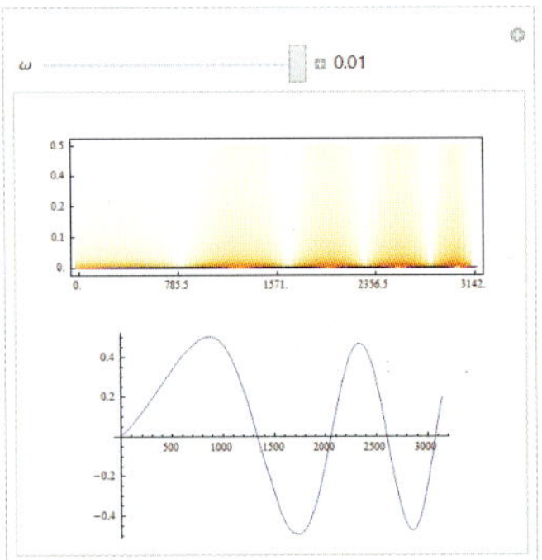

4.3.5 Edge Detection

Mathematica has several functions for detecting edges and borders that are very useful for image processing.

- In the next example we capture the image of the author while writing this book and apply the `FindFaces` command to detect his face.

```
picture = CurrentImage[]
```

```
faces = FindFaces[picture, {100, ∞}]
```

```
{{{145.5, 36.5}, {258.5, 149.5}}}
```

```
Show[picture,
 Graphics[{EdgeForm[{Red, Thick}], Opacity[0], Rectangle @@@ faces}]]
```

4.3.6 Barcode Recognition

- This example shows the use of `BarcodeRecognize` to identify barcodes.

```
boardingPass =
```

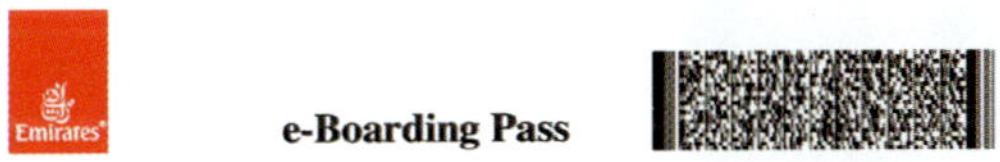


```
;
```

```
BarcodeRecognize[boardingPass, "Data", "PDF417"]
```

```
M2VANALMSICK/MARKUS    ECSEZLJ HNDDXBEK 0313 271Y018K0254
  136>20B1WW3271BEK 251762174509010 1    EK 116054035        CSEZLJ
  DXBDUSEK 0055 271Y019K0249 127251762174509011 1    EK 116054035
```

4.3.7 A Painting with a Surprise

The next example shows how to visualize an image from a certain perspective and how to select a part of it. The image was imported from a Wolfram blog post (http://blog.wolfram.com/2010/10/27/the-ambassadors).

```
ambassadors = Image[Import[
 "http://blog.wolfram.com/data/uploads/2010/10/TheAmbassadorsBlogPost01
   .jpg"], ImageSize -> 150]
```

The image above corresponds to the painting "The Ambassadors" by Holbein. There's a skull in it. Can you see it?

- With `ImagePerspectiveTransformation` we can apply a transformation to see the image from a different perspective and, combined with `TransformationFunction`, display the part of the image that we want to visualize.

```
ImagePerspectiveTransformation[ambassadors,
 TransformationFunction[( 0.5   -0.87 -0.05
                          -1.34  2.7   0.43
                          -0.62  0.2   0.525 )]]
```

4.3.8 Where Are the Matching Points?

Can you identify the location of the common points in these two images? (Images available in `ImageCorrespondingPoints`).

`imgs = {`  `,`  `};`

- The output displays the matching positions between the two images.

```
matches = ImageCorrespondingPoints @@ imgs;
MapThread[
 Show[#1, Graphics[{Yellow, MapIndexed[Inset[#2[[1]], #1] & , #2]}]] &,
 {imgs, matches}]
```

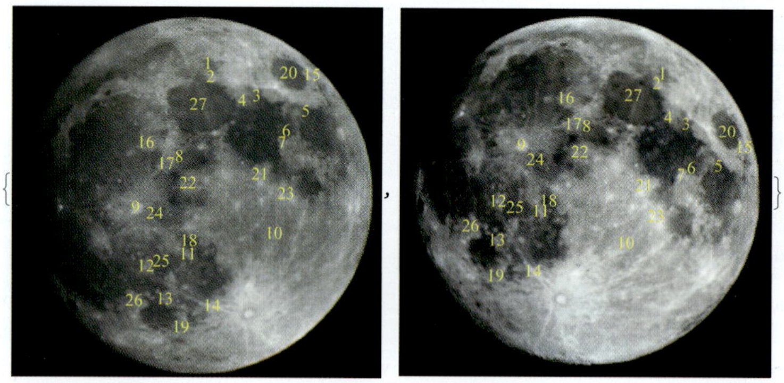

4.3.9 Image Identify

A new area of functionality available since *Mathematica* 10.1 is image identification: https://www.imageidentify.com/

- The function `ImageIdentify` is used to identify the image in a picture. If you have a camera connected to your computer, you can use the command below to identify the captured image in real time.

```
{#, ImageIdentify[#]} &[CurrentImage[]]
```

4.3.10 Metadata

As we have previously seen, data from image files usually carry additional information about the images themselves. In this case we download a file in the ARC Grid format, used in topography for representing terrain elevation.

- We import the following ARCGrid file.

```
Import[ "http://exampledata.wolfram.com/ArcGRID.zip", "ARCGrid"]
```

- Notice that this is a compressed file. We can decompress it to find out what files are included.

```
Import[ "http://exampledata.wolfram.com/ArcGRID.zip", "ZIP"]
```

```
{NED_93751926\.DS_Store, NED_93751926\info\arc.dir,
 NED_93751926\info\arc0000.dat, NED_93751926\info\arc0000.nit,
 NED_93751926\info\arc0001.dat, NED_93751926\info\arc0001.nit,
 NED_93751926\meta1.html, NED_93751926\Metadata.xml,
 NED_93751926\ned_93751926\dblbnd.adf, NED_93751926\ned_93751926\hdr.adf,
 NED_93751926\ned_93751926\prj.adf, NED_93751926\ned_93751926\sta.adf,
 NED_93751926\ned_93751926\w001001.adf,
 NED_93751926\ned 93751926\w001001x.adf, NED_93751926\NED_93751926.aux}
```

- Next, we represent the terrain elevations in 3D (we use MaxPlotPoints→100 to speed up the computations).

```
ListPlot3D[Import["http://exampledata.wolfram.com/ArcGRID.zip",
  {"ArcGRID", "Data"}], MaxPlotPoints → 100,
 ColorFunction → "SouthwestColors", Mesh → None]
```

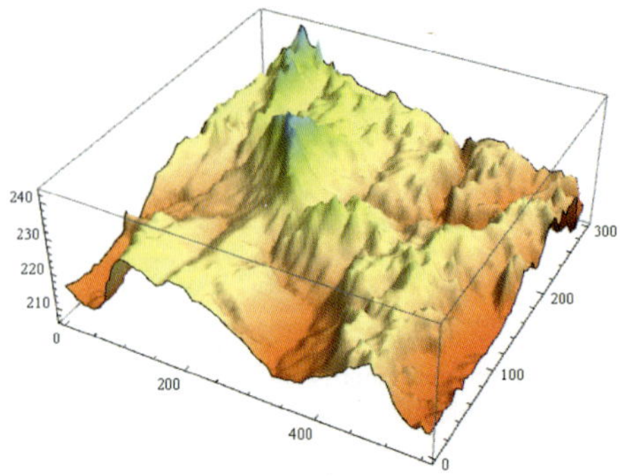

- Let's take a look at the metadata inside.

```
Import[ "http://exampledata.wolfram.com/ArcGRID.zip",
 {"ArcGRID", "Elements"}]
```

```
{Centering, CentralScaleFactor, CoordinateSystem,
 CoordinateSystemInformation, Data, DataFormat, Datum, ElevationRange,
 Graphics, GridOrigin, Image, InverseFlattening, LinearUnits, Projection,
 ProjectionName, RasterSize, ReferenceModel, ReliefImage, SemimajorAxis,
 SemiminorAxis, SpatialRange, SpatialResolution, StandardParallels}
```

- Now, we extract the information related to the coordinates system.

```
Import[ "http://exampledata.wolfram.com/ArcGRID.zip",
 {"ArcGRID", "CoordinateSystemInformation"}]
```

```
GEOGCS → {NAD83, DATUM → {North_American_Datum_1983,
    SPHEROID → {GRS 1980, 6378137, 298.257, AUTHORITY → {EPSG, 7019}},
    TOWGS84 → {0, 0, 0, 0, 0, 0, 0}, AUTHORITY → {EPSG, 6269}},
  PRIMEM → {Greenwich, 0, AUTHORITY → {EPSG, 8901}},
  UNIT → {degree, 0.0174533, AUTHORITY → {EPSG, 9108}},
  AXIS → {Lat, NORTH}, AXIS → {Long, EAST}, AUTHORITY → {EPSG, 4269}}
```

- The following command enables us to create a link to http://maps.google.com using the latitude and the longitude.

```
CoordinatesToMapLink[{lat_, long_}] := Hyperlink[
  StringJoin["http://maps.google.com/maps?q=+",
   ToString@lat, "+", ToString@long, "&z=12&t=h"]
 ]
```

- We can now get the coordinates of the previous file and set up a link to Google Maps.

```
coords = Import[ "http://exampledata.wolfram.com/ArcGRID.zip",
  {"ArcGRID", "SpatialRange"}]
```

```
{{40.0617, 40.1447}, {-88.3158, -88.1619}}
```

- Since we are establishing the link with the center of the imported grid, we need to calculate the average of the latitude and longitude.

```
CoordinatesToMapLink[Map[Mean, coords]]
```

```
http://maps.google.com/maps?q=+40.1032+-88.2389&z=12&t=h
```

4.4 Graphs and Networks

Graphs and networks is another area where the latest *Mathematica* versions incorporate new functionality: guide/GraphsAndNetworks. A graph can be represented using `Graph`, a function with many options. Let's see a basic example.

- We begin by defining the connections between vertices: i ⟷ j or i ⟶j.

```
graph1 = {"a1" ⟷ "a2", "a2" ⟷ "a3", "a2" ⟷ "a1", "a3" ⟷ "a4"};
```

- Next we define the capacity of each edge and vertex. Finally we add labels to identify the vertices.

```
Graph[graph1, EdgeCapacity → {2, 3, 4, 5},
 VertexCapacity → {2, 3, 4, 6}, VertexLabels → "Name"]
```

- We can use special symbols as labels.

g = Graph[graph1, EdgeCapacity → {2, 3, 4, 5}, VertexCapacity → {2, 3, 4, 6},

VertexLabels → "Name", EdgeLabels → {"a1" ⟷ "a2" → ,

"a2" ⟷ "a3" → , "a2" ⟷ "a1" → , "a3" ⟷ "a4" → }]

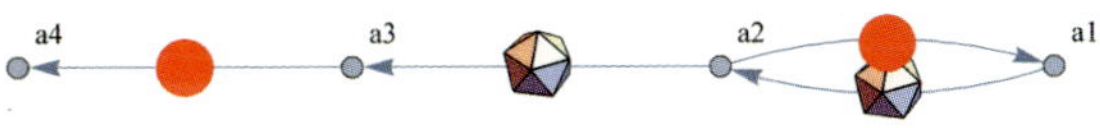

- We may have a graph representation in a certain format from which we would like to extract information. We can do it as follows:

```
EdgeRules[g]
```

{a1 → a2, a2 → a3, a2 → a1, a3 → a4}

```
PropertyValue[g, EdgeCapacity]
```

{2, 3, 4, 5}

```
PropertyValue[g, VertexCapacity]
```

{2, 3, 4, 6}

```
PropertyValue[g, EdgeLabels]
```

{a2 ⟷ a3 → , a3 ⟷ a4 → , a2 ⟷ a1 → , a1 ⟷ a2 → }

- In the next example we find the shortest way to link all the locations in Figure 4.1 with fiber optic cable:

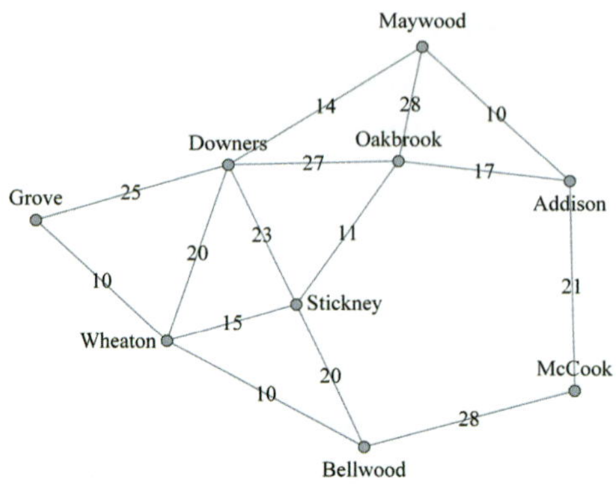

Figure 4.1 Network graph of villages in Illinois.

- After creating the network graph with *Mathematica* (code not shown) and naming it "g", we find the minimum spanning tree containing all the villages (Figure 4.2):

```
fibernetwork = FindSpanningTree[g];
```

```
HighlightGraph[g, fibernetwork, GraphHighlightStyle → "Thick"]
```

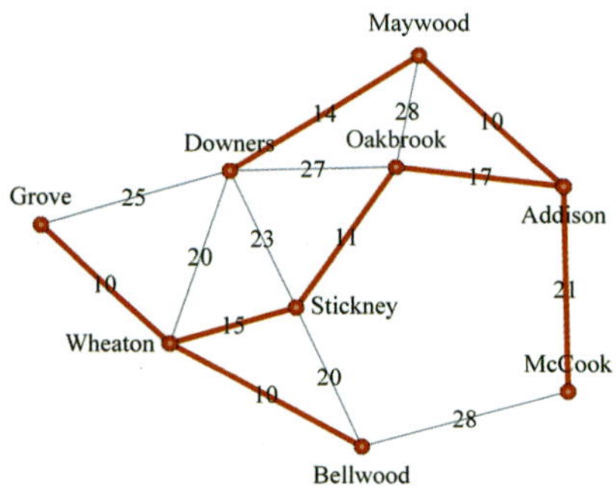

Figure 4.2 Minimum Spanning Tree.

- In this example we generate random graphs and search for cliques (highlighted in red):

```
Grid@Table[HighlightGraph[g = RandomGraph[{8, 16}],
   Subgraph[g, First[FindClique[g]]]], {2}, {4}]
```

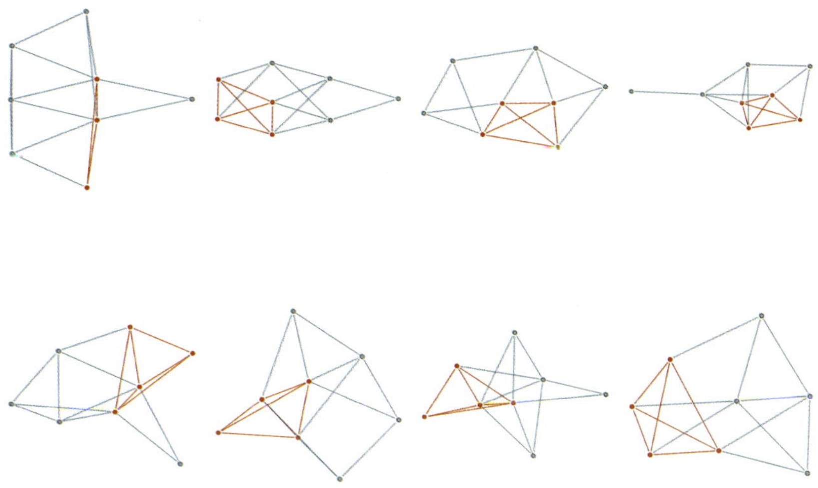

- The command below returns the edges and vertices of a dodecahedron in graph format:

```
{Graphics3D[{Opacity[.8], PolyhedronData["Dodecahedron", "Faces"]}],
 g = PolyhedronData["Dodecahedron", "SkeletonGraph"]}
```

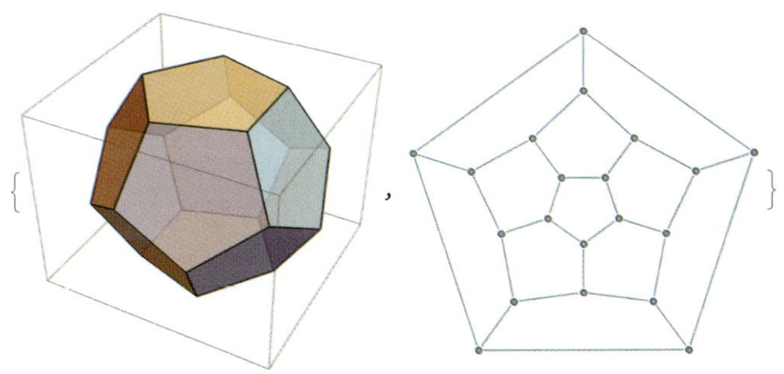

- Here we are trying to find the shortest path passing once through each vertex (this is the famous Hamilton's Icosian game):

```
h = PathGraph@First[FindHamiltonianCycle[g]];
```

```
{HighlightGraph[g, h, GraphHighlightStyle → "Thick"],
 Graphics3D[{Opacity[.8], PolyhedronData["Dodecahedron", "Faces"],
   Red, Tube[PolyhedronData["Dodecahedron", "VertexCoordinates"][[
     Append[VertexList[h], VertexList[h][[1]]]]], 0.1]}]}
```

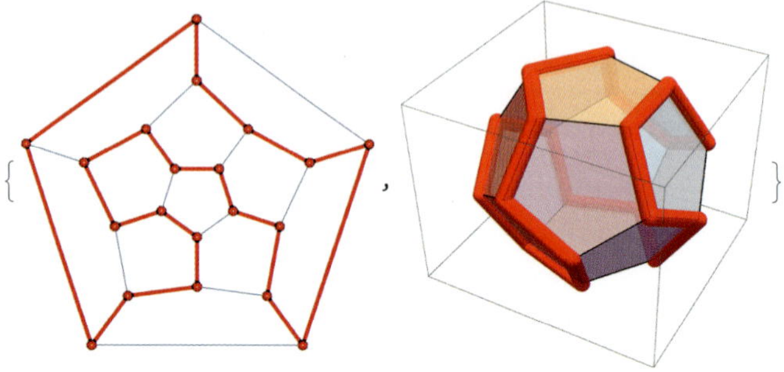

4.5 Mazes

The function `Thinning` looks for the skeleton of an image through morphological analysis.

- In this example the image is a labyrinth. **`Thinning`** replaces the background with lines showing possible paths.

```
labyrinth =
  Pruning[Thinning[
```

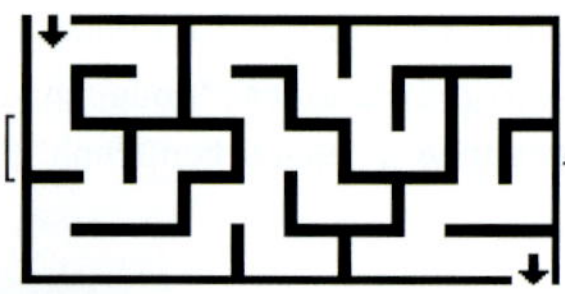

```
, Padding → 1], Padding → 1]
```

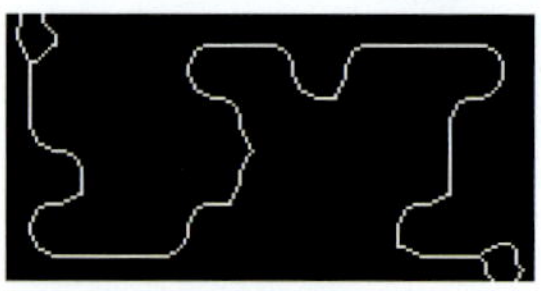

- Now we create the graph of the image:

```
MorphologicalGraph[labyrinth]
```

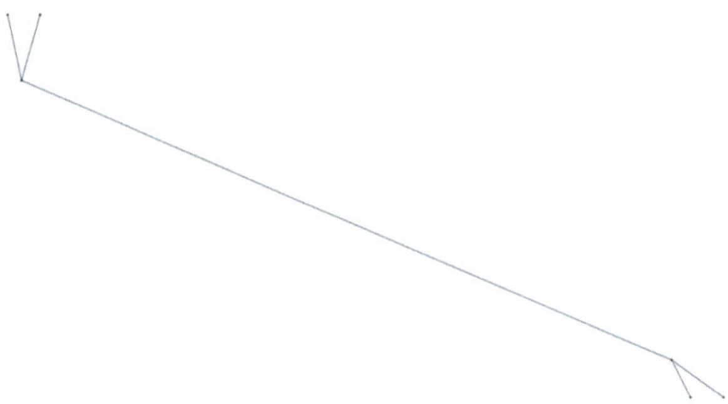

4.6 Application: Finding the Period of a Pendulum

```
Clear["Global`*"]
```

The example that follows is significantly more difficult than the previous ones (obtained from http://www.wolfram.com/services/education/seminars/s51.html).

Our objective is to find out the period of a pendulum by analyzing a sequence of video images. The images were obtained with `CurrentImage` (if your computer has a webcam you can use this function). **`CurrentImage`** adds a time stamp to the image, information that we will use later on in the example:

```
ims = {
```

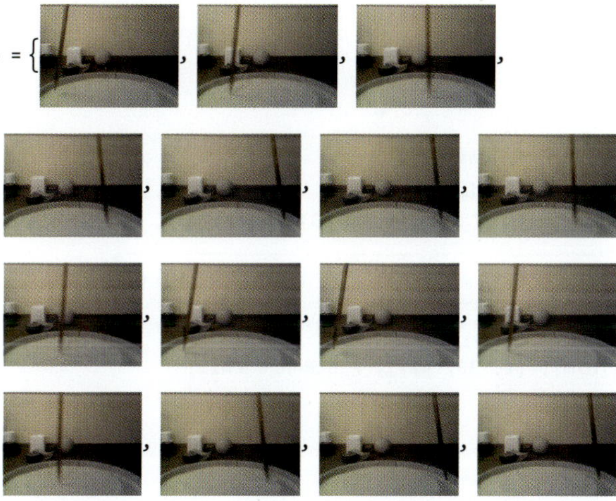

```
};
```

- We begin by highlighting the most prominent lines to identify the ones corresponding to the pendulum.

```
edges = EdgeDetect /@ ims;
```

- In the images above, the pendulum lines are the ones close to the vertical. We can identify them using Radon.

```
radons = ImageAdjust /@ (Radon[#, Automatic, {-Pi/4, Pi/4}] & /@ edges);
```

- `Binarize` detects the brightest points and extracts their x-coordinates to obtain the angle.

```
xs = ComponentMeasurements[#, "Centroid"][[1, 2, 1]] & /@
  (Binarize[#, .9] & /@ radons)

{89.5, 96., 111.5, 124., 125.5, 122.5, 112.5,
 100.5, 87.5, 89., 95.5, 106., 121.5, 127.5, 124.}
```

- In the next step we obtain the relationship between the positions and the time by converting the x-coordinates to radians and combining them with the images time stamps.

```
thetas = ((xs/215) - .5) * Pi/4;

times = Mod[Options[#, "MetaInformation"][[1, 2, 1]], 100] & /@ ims

{36.0593128, 36.3133382, 36.6473716, 36.9003969, 37.1534222,
 37.5664635, 37.8204889, 38.1605229, 38.4995568, 38.9105979,
 39.1666235, 39.4196488, 39.8376906, 40.1757244, 40.4267495}
```

```
datapoints = Transpose[{times, thetas}];
ListPlot[datapoints]
```

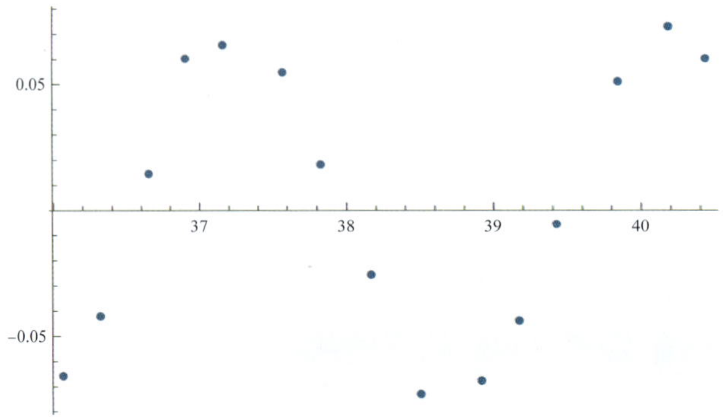

- We fit the data to a sinusoid curve.

```
nlm = NonlinearModelFit[datapoints,
  a Sin[b (t - c)] + d, {{a, .07}, {b, 3}, {c, 2}, {d, 0}}, t]
```

```
FittedModel[ -0.00305312 + 0.0746392 Sin[2.16036 (10.0063 + t)] ]
```

```
Show[ListPlot[datapoints],
 Plot[Normal[nlm], {t, 36, 41}], PlotRange → All]
```

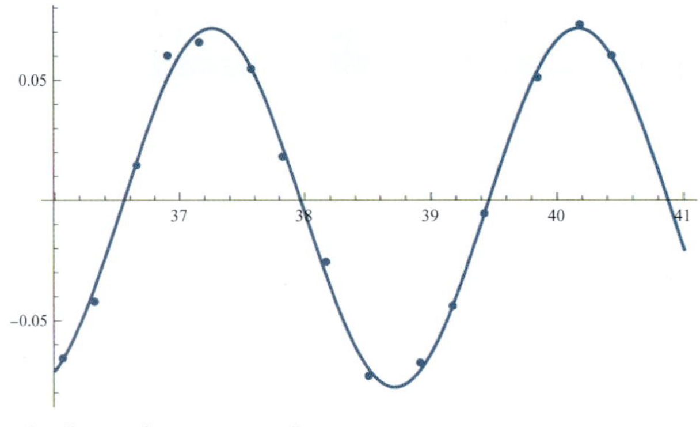


```
nlm["BestFitParameters"]
```

```
{a → 0.0746392, b → 2.16036, c → -10.0063, d → -0.00305312}
```

- We estimate the period, in seconds:

```
period = 2 Pi / (b /. %)
```

```
2.90839
```

- Here is the video sequence using the calculated period:

```
ListAnimate[ims, DefaultDuration -> 2.9]
```

- Next, we find the formula to calculate the length of the pendulum:

Pendulum formula »

```
{FormulaData[{"PendulumSmallOscillations", "Standard"}],
 FormulaData[{"Pendulum", "Standard"}]}
```

$$\left\{\left\{T == 2\pi\sqrt{\frac{l}{g}},\ f == \frac{1}{T},\ v_{max} == \sqrt{2}\sqrt{(1-\text{Cos}[\theta_0])\, g\, l}\right\},\right.$$

$$\left.\left\{T == 4\,\text{EllipticK}\left[\text{Sin}\left[\frac{\theta_0}{2}\right]^2\right]\sqrt{\frac{l}{g}},\ f == \frac{1}{T},\ v_{max} == \sqrt{2}\sqrt{(1-\text{Cos}[\theta_0])\, g\, l}\right\}\right\}$$

```
pendulumeq = FormulaData[{"PendulumSmallOscillations", "Standard"}][[1]]
```

$$T == 2\pi\sqrt{\frac{l}{g}}$$

```
pendulumeq // InputForm
```

```
QuantityVariable["T", "Period"] ==
 2*Pi*Sqrt[QuantityVariable["l",
     "Length"]/QuantityVariable["g",
     "GravitationalAcceleration"]]
```

- Finally, we apply the formula to compute the length, in meters, of our pendulum:

```
WolframAlpha["Gravitational Acceleration value",
 {{"Value", 3}, "QuantityData"}]
```

$9.807\ \text{m}/\text{s}^2$

```
Solve[pendulumeq /. {QuantityVariable["T", "Period"] → period,
   QuantityVariable["l", "Length"] → l,
   QuantityVariable["g", "GravitationalAcceleration"] → 9.807}, l]
```

```
{{l → 2.10127}}
```

With `ImageDisplacements`, you can use an optical flow to interpolate between existing frames (https://www.wolfram.com/language/11/image-and-signal-processing/slow-motion-using-optical-flow.html). In the following example we generate a video sequence twice as long as the original.

- The interpolated images are constructed by warping the original frames according to the optical flow.

```
flows = Map[vecField ↦ Transpose@Reverse[vecField],
   ImageDisplacements[ims, MaxIterations -> 1] / 2];
```

```
bg = First@DominantColors[First@ims, 1]
```

- Next, we construct the interpolated frames and insert them into the final sequence:

```
vecTrafo[xy_, motion_] := xy + Extract[motion, Ceiling[xy]];
```

```
oflow =
  Table[ ImageForwardTransformation[ims[[i]], vecTrafo[#, flows[[i]]] &,
    DataRange -> Full, Background -> bg], {i, Length@ims - 1}];
```

```
slowmo = Riffle[ims, oflow];
```

- Finally, we animate the result with a simple blending-based reference using a period twice as long as the original:

```
compresults = MapThread[ ImageAssemble[{{Show[#2, Graphics@
          Text[Style["blended frames", Larger, Red], {60, 180}]]}}] &,
   {slowmo, Riffle[ims, Blend /@ Partition[ims, 2, 1]]}];
```

```
ListAnimate[compresults, DefaultDuration -> 2*2.9]
```

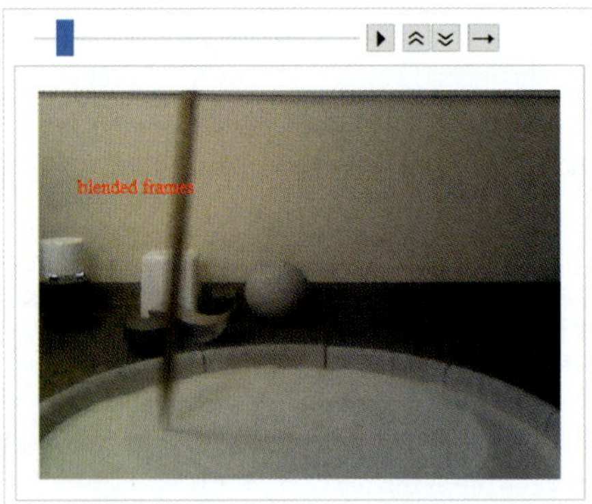

4.7 Advanced Calculus

```
Clear["Global`*"]
```

So far in this chapter we have seen some of the new functionality included in the latest versions of *Mathematica.* In this section we continue with our exploration of some of the latest features and their applications for solving advanced calculus problems. It's important to remember that some of the options that we use are common to other functions.

- We use CoordinateTransform to represent a curve with origin at the center that moves toward the surface of a sphere of radius 1 and calculate its length with ArcLength.

```
cSpherical[t_] := {t, (1 + 2 Sin[2 t + 1]), 1 + 2 t^2}
```

```
cCartesian[t_] =
  CoordinateTransform["Spherical" → "Cartesian", cSpherical[t]];
```

```
Show[Graphics3D[{Opacity[.25], Sphere[]}],
 ParametricPlot3D[cCartesian[t], {t, 0, 1}, PlotStyle → Thick,
  ColorFunction → (Hue[.8 #4] &)], ImageSize → Small]
```

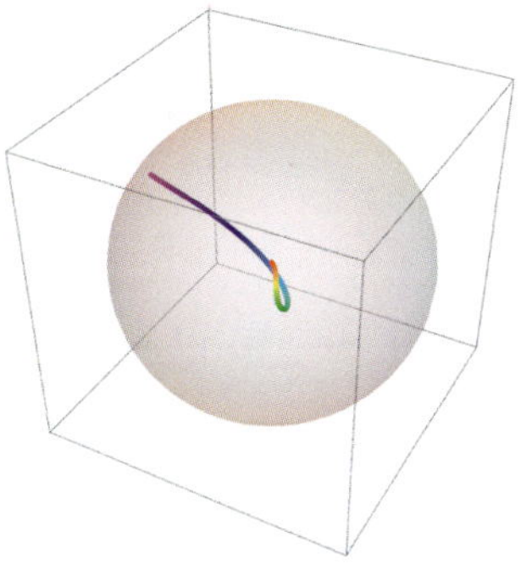

```
ArcLength[cCartesian[t], {t, 0., 1}]
```

```
2.17417
```

- In the example below, we compute the period of a function and plot it.

```
p = FunctionPeriod[Cos[2 x] + Sin[3 x], x]
```

2π

```
Plot[Cos[2 x] + Sin[3 x], {x, 0, 2 p }, Mesh → 1,
 MeshShading → {Green, Blue}, Ticks → {Range[0, 2 p, p / 2], {-1, 1}}]
```

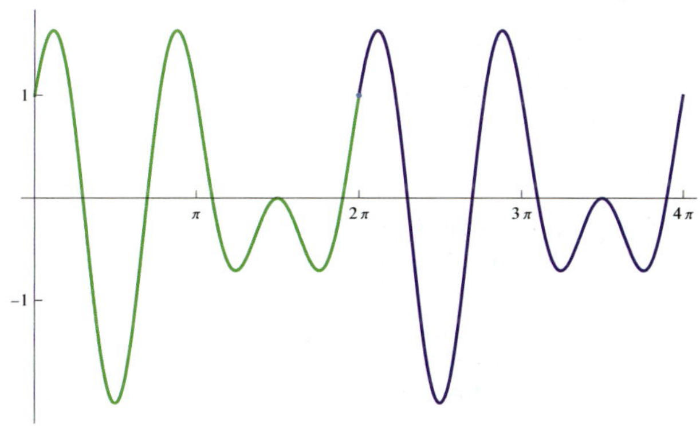

- Here, we define three spheres and calculate the intersection point from their graphical representation.

```
{s1, s2, s3} =
  Table[Sphere[{1 / 3 Cos[k 2 π / 3], 1 / 3 Sin[k 2 π / 3], 0}], {k, 0, 2}];
```

```
sol = Solve[p ∈ s1 ∧ p ∈ s2 ∧ p ∈ s3, p]
```

$$\left\{\left\{p \to \left\{0, 0, -\frac{2\sqrt{2}}{3}\right\}\right\}, \left\{p \to \left\{0, 0, \frac{2\sqrt{2}}{3}\right\}\right\}\right\}$$

```
Graphics3D[
 {{Opacity[0.5], s1, s2, s3}, {Red, PointSize[Large], Point[p /. sol]}}]
```

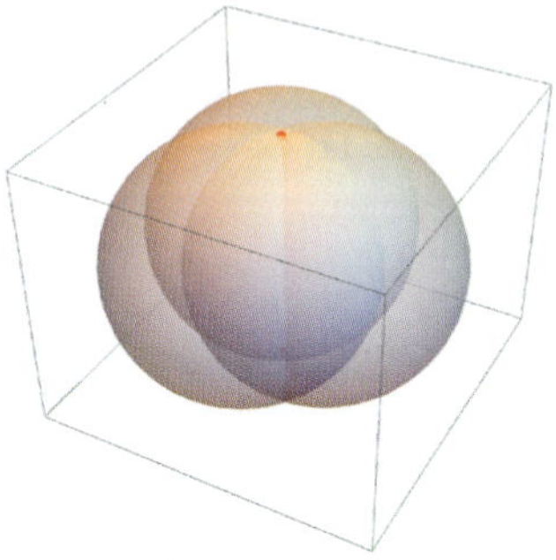

- The following command plots two concentric spheres:

```
ℛ = ImplicitRegion[4 ≤ x^2 + y^2 + z^2 ≤ 9, {x, y, z}];
RegionPlot3D[ℛ, ColorFunction → "BlueGreenYellow", PlotStyle → Opacity[0.5]]
```

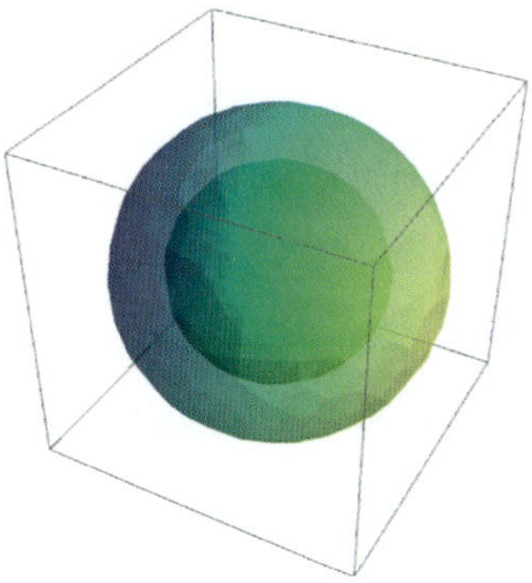

```
Integrate[1, x ∈ ℛ]
```

$$\frac{76\pi}{3}$$

- In this example we download the 3D figure "StanfordBunny", discretize it using `DiscretizeGraphics`, calculate its centroid with `RegionCentroid` and finally visualize it.

```
bunny =
 Show[ExampleData[{"Geometry3D", "StanfordBunny"}], ImageSize → Small]
```

```
ℛ = HighlightMesh[DiscretizeGraphics[bunny],
   {Style[1, None], Style[2, Opacity[0.5]]}];
c = RegionCentroid[ℛ]
```

```
{-0.0267934, -0.00829883, 0.0941362}
```

```
Show[Graphics3D[Prepend[First[bunny], Opacity[0.5]]],
 Graphics3D[{Red, Point@c}], ImageSize → Small]
```

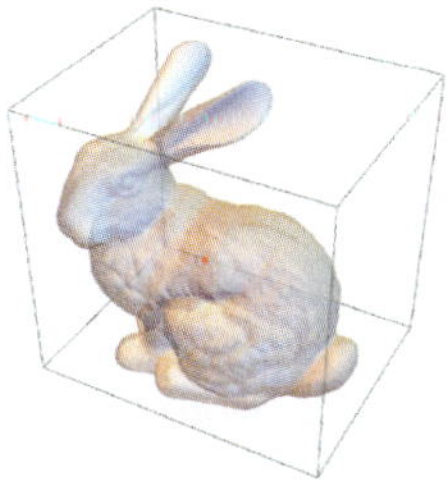

- We can also calculate the centroid using the definition (notice how `Integrate` is able to integrate over a region defined by the figure of the bunny):

```
m = RegionMeasure[ℛ]; Integrate[p, p ∈ ℛ] / m
```

```
{-0.0267934, -0.00829883, 0.0941362}
```

- The command below defines a region based on the union of two disks of radius 2 centered at {0, 0} and {1, 0}.

```
ℛ = RegionUnion[Disk[{0, 0}, 2], Disk[{1, 0}, 2]];
```

- Now, we represent the difference between the region ℛ previously defined and a disk of radius 1 centered at {1/2, 0}.

```
ℛ1 = RegionDifference[ℛ, Disk[{1/2, 0}, 1]];
```

- To visualize ℛ1 we discretize the region with `MeshRegion`.

```
DiscretizeRegion[ℛ1]
```

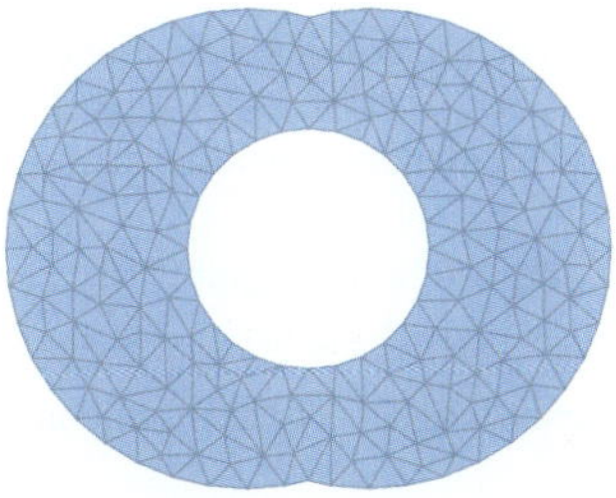

- Finally, we solve the partial differential equation over ℛ1.

```
sol = NDSolveValue[{∇²{x,y} u[x, y] == 0,
    DirichletCondition[u[x, y] == Sin[x y], True]}, u, {x, y} ∈ ℛ1];
Plot3D[sol[x, y], {x, y} ∈ ℛ1] // Quiet
```

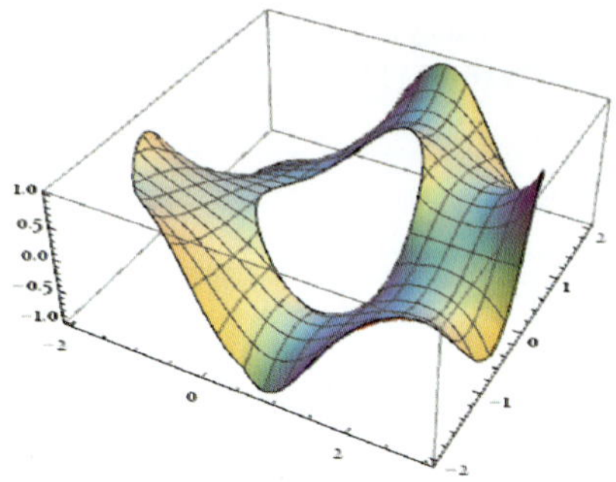

4.8 Additional Resources

To access the following resources, if they refer to the help files, write them down in a notebook, select them and press <F1>. In the case of external links, copy them in a browser.

About demonstrations

Manipulate: tutorial/IntroductionToManipulate

Advanced Manipulate functionality tutorial: tutorial/AdvancedManipulateFunctionality

Introduction to Dynamic tutorial: tutorial/IntroductionToDynamic

The Wolfram Demonstrations Project: http://demonstrations.wolfram.com

About images

Image processing tutorial: tutorial/ImageProcessing

Image Processing & Analysis guide page: guide/ImageProcessing;

Image and Signal Processing: http://www.wolfram.com/language/11/image-and-signal-processing/

Computational Photography: http://www.wolfram.com/language/11/computational-photography/

A collection of video presentations and seminars available in: http://www.wolfram.com/training/courses/image-processing/

An interesting demonstration about image processing: The Incredible Convenience of *Mathematica* Image Processing (http://blog.wolfram.com/2008/12/01/the-incredible-convenience-of-mathematica-image-processing)

The Wolfram Language Image identification Project: https://www.imageidentify.com/

About Advanced Calculus

Symbolic & Numeric Calculus: http://www.wolfram.com/language/11/symbolic-and-numeric-calculus/

Differential Eigensystems: http://www.wolfram.com/language/11/differential-eigensystems/

Algebra and Number Theory: http://www.wolfram.com/language/11/algebra-and-number-theory/

Partial Differential Equations: http://www.wolfram.com/language/11/partial-differential-equations/

5

Accessing Scientific and Technical Information

In this chapter we introduce commands to access databases containing up-to-date information from many different fields such as: astronomy, chemistry, economy, genetics, geography, physics and many more. These commands can be part of Mathematica-based programs. In later chapters we will make extensive use of some them.

5.1 Computable Data: Doing Computations with Data from Different Fields

Mathematica has functions to access databases from many different fields such as:

Physics and Chemistry: ElementData, ChemicalData, IsotopeData, ParticleData ...
Earth Sciences: WeatherData, GeodesyData, GeoDistance ...
Astronomy: StarData , PlanetData ...
Life Sciences and Medicine: GenomeData, GenomeLookup, ProteinData ...
Economics and Finance: FinancialData, CountryData ...
Mathematics: FiniteGroupData, GraphData, KnotData, LatticeData, PolyhedronData ...
Linguistics: DictionaryLookup, WordData, ExampleData ...
Engineering: AircraftData, BridgeData, NuclearReactorData ...
People and History: PersonData, SocialMediaData, MovieData, Historical PeriodData ...
Geography: GeoGraphics, GeoListPlot ...

The number of fields covered by *Mathematica* keeps on growing with each new release of the program. The big advantage of these functions is that they import the data – internally known as computable data or collections – in a format that makes their manipulation easy. When using them, *Mathematica* will connect to a server at Wolfram Research which in turn will access a specialized server containing the relevant data. There are three different types of data collections: (i) **Pre-computed**: most of the mathematical data is precomputed in a Wolfram server: e.g., GraphData, KnotData, LatticeData, PolyhedronData; (ii) **Aggregated from publicly available sources**: e.g., AstronomicalData, CityData, CountryData, ChemicalData, ElementData, IsotopeData, ParticleData, WordData; **Direct connection to web services-based provider**: e.g., FinancialData, WeatherData. The execution time may vary depending on the speed of your Internet connection and that of the Wolfram Research server itself.

- In many cases, the word "Data" is included in this type of functions. You can use **`?*Data*`** to explore some of them (output not shown).

```
?*Data*
```

- One of the functions included in the data collections is `FinancialData` for manipulating finance-related data. This command is complemented by some others that we will cover in Chapter 11. For example: we can get the euro/dollar exchange rate at the time of the query.

```
FinancialData["EUR/USD"]
```

```
1.1087
```

If you get an error message when evaluating the previous expression it may be due to problems with your firewall or proxy configuration for accessing the Internet. Check your Mathematica settings in: **Edit ► Preferences ► Proxy settings**. If the issue is proxy-related, access the help system for information about how to troubleshoot it: tutorial/TroubleshootingInternetConnectivity.

- *Mathematica* has an extensive selection of scientific and technical formulas. Use `FormulaLookup` to find their precise names. For instance: If you need the formula for relativistic kinetic energy, you can type:

```
FormulaLookup["Relativistic Kinetic Energy"]
```

```
{KineticEnergyRelativistic}
```

- Now use `FormulaData` data to obtain the actual formula:

```
FormulaData["KineticEnergyRelativistic"]
```

$$\left\{K == \left(1\, c^2\right) m \left(-1+\gamma\right),\ \gamma == \frac{1}{\sqrt{1+\left(-1/c^2\right) v^2}}\right\}$$

- Of course, you can use free form input as usual: typing "=" at the beginning of an input cell followed by the relevant words in plain English.

Kinetic Energy Relativistic formula »

```
FormulaData["KineticEnergyRelativistic"]
```

$$\left\{K == \left(1\, c^2\right) m \left(-1+\gamma\right),\ \gamma == \frac{1}{\sqrt{1+\left(-1/c^2\right) v^2}}\right\}$$

This alternative way can give us the information that we are looking for using computable data. As a matter of fact, it will not only display the results but also the appropriate *Mathematica* input function. Nevertheless, in this chapter we will type the computable function directly, the best approach if we want to write programs. However, we recommend the reader to take advantage of the free-form input and use it as the starting point for more sophisticated queries.

The following pages contain illustrations about the use of some the functions for computable data. The examples shown may be more difficult than the average ones included in this book. If that's the case, you may want to use them as templates and make the necessary modifications to suit your needs. As usual, we strongly recommend seeking further information about the functions in the help files to find out more about all of their potential uses.

In general, computable data functions contain a list of properties related to the entity (`Entity`) to which they refer. Typing **Function**[*"Properties"*] will tell us all the available properties of **Function**. Alternatively, we can always go to the help page for the function.

- For example, the properties available for WordData are:

```
WordData["Properties"]
```

```
{AmericanSpelling, Antonyms, BaseAdjective, BaseForm, BaseNoun,
 BritishSpelling, BroaderTerms, CausesTerms, CompositeTerms, ConceptWeight,
 ConsequencesTerms, Definitions, DerivedAdjectives, DerivedAdverbs,
 EntailedTerms, Examples, GeographicDomain, Hyphenation, InflectedForms,
 MaterialTerms, MorphologicalDerivatives, MorphologicalSource,
 NarrowerTerms, PartsOfSpeech, PartTerms, PhoneticForm, PorterStem,
 SentenceFrames, SimilarAdjectives, SimilarVerbs, SubsetTerms,
 SupersetTerms, Synonyms, UsageField, UsageType, WholeTerms, WordNetID}
```

- We can ask for specific word properties using the command: WordData[*"word", "property"*]:

```
WordData["confused", "SimilarAdjectives"]
```

```
{{confused, Adjective, Dazzled} →
  {addlebrained, addled, addlepated, befogged, befuddled, clouded,
   dazed, dazzled, muddled, muddleheaded, muzzy, puddingheaded,
   punch-drunk, silly, slaphappy, spaced-out, stunned, stupefied,
   stupid, trancelike, woolly, woolly-headed, wooly, wooly-minded},
 {confused, Adjective, Disjointed} → {incoherent},
 {confused, Adjective, Broken} → {disorganised, disorganized},
 {confused, Adjective, Disoriented} → {unoriented},
 {confused, Adjective, Mazed} → {perplexed}}
```

- In the next example we use DictionaryLookup to search for English (other languages are also available) words starting with "sh" and ending in "y":

```
DictionaryLookup[{"English", "sh" ~~ __ ~~ "y"}]
```

```
{shabbily, shabby, shadily, shadowy, shady, shaggy, shakily,
 shaky, shallowly, shamefacedly, shamefully, shamelessly, shandy,
 shanty, shapelessly, shapely, sharply, shatteringly, shay, sheeny,
 sheepishly, sherry, shiftily, shiftlessly, shifty, shimmery,
 shimmy, shinny, shiny, shirty, shiveringly, shivery, shockingly,
 shoddily, shoddy, shortly, shortsightedly, shorty, showery, showily,
 showy, shrewdly, shrilly, shrinkingly, shrubbery, shrubby, shyly}
```

- SpokenString [*expr*] transcribes the indicated expression:

```
transcribe = SpokenString[Sqrt[x / (y + z)]]
```

```
square root of the quantity x over the quantity y plus z
```

We can create a button that when pressed will return a spoken representation of the argument:

```
Button["Press", Speak[transcribe]]
```

```
Press
```

- Next we are going to obtain the phonetic transcription (in American English) of the previous expression. We use SpokenString to split the phrase into words and the property **"PhoneticForm"**, from WordData, to get the phonetic transcription of each word. Given that the output is a list containing the words' phonetic transcriptions, we apply **Row**[list, " "] to join the elements substituting the commas with blank spaces.

```
Row[WordData[#, "PhoneticForm"] & /@ StringSplit[transcribe], " "]
```

```
skw'ɛr r'ut 'ʌv ðə kw'ɒntəti 'ɛks 'oʊvɝ ðə kw'ɒntəti w'aɪ pl'ʌs z'i
```

In general, any computable data function is of the form: **Function**[*"name or Entity", "property"*], or **Function**[*"name" or "Entity"*, {*"property"*, ...}] where Entity is an entity identified by its name: Country, Element, Isotope ... and *property* is information related to such entity.

- To see all the properties available for a specific function we can access the documentation or type:

```
CityData["Properties"]
```

```
{AlternateNames, Coordinates, Country, Elevation,
 FullName, Latitude, LocationLink, Longitude, Name,
 Population, Region, RegionName, StandardName, TimeZone}
```

- If you type the entity, in this case "Salamanca", *Mathematica* will return all the cities in the world with that name:

```
CityData["Salamanca"]
```

```
{Salamanca, Salamanca, Salamanca,
 Salamanca, Salamanca, Salamanca}
```

- By hovering the mouse over the list elements, you will see the details for each of the cities. If you prefer to see them explicitly, just type the following:

```
salamanca =
 SemanticInterpretation["Salamanca",
  Entity["City", _], AmbiguityFunction → All]
```

```
AmbiguityList[{Salamanca, Salamanca, Salamanca,
  Salamanca, Salamanca, Salamanca}, Salamanca]
```

```
Take[First[salamanca], 6] // InputForm
```

```
{Entity["City", {"Salamanca", "Coquimbo", "Chile"}],
 Entity["City", {"Salamanca", "Guanajuato", "Mexico"}],
 Entity["City", {"Salamanca", "Salamanca", "Spain"}],
 Entity["City", {"Salamanca", "NewYork", "UnitedStates"}],
 Entity["City", {"Salamanca", "NegrosOccidental", "Philippines"}],
 Entity["City", {"Salamanca", "Colon", "Panama"}]}
```

- If we intend to refer to a particular one, we need to be more precise and tell *Mathematica* exactly what we want (if we let the program select the city for us it may not make the right choice). For example: if we'd like to know the population of the Spanish city of Salamanca, we can use Entity with the output returned from the previous command.

```
CityData[Entity["City", {"Salamanca", "Salamanca", "Spain"}], "Population"]
```

```
159 754 people
```

- Alternatively we can use `Interpreter` (this function can be used any time there is ambiguity risk).

```
Interpreter["City"]["Salamanca, Spain"]
```

```
Salamanca
```

- Copy the output and paste it below:

```
CityData[ Salamanca (city) , "Population"]
```

```
159754 people
```

- Notice that the previous output displays the units explicitly, in this case *people*. This is the default behavior since *Mathematica* 10. To see the output without units use `QuantityMagnitude`. You can also type the entity as follows, a more convenient way if you are writing a program.

```
QuantityMagnitude[CityData[
  Entity["City", {"Salamanca", "Salamanca", "Spain"}] , "Population"]]
```

```
159754
```

- Another way is setting the system options as follows:

```
SetSystemOptions["DataOptions" → "ReturnQuantities" → False];
```

```
CityData[ Beijing (city) , "Population"]
```

```
12458000
```

After executing the previous function, the units won't be displayed for the rest of the *Mathematica* session. This can be useful if you have notebooks created with versions of the program prior to Version 10.

System options specify parameters relevant to the internal operations of the program. If we make changes to those options, we should always reset the parameters as soon as we are done with our calculations to avoid potential problems.

- In this case we can re-enable the display of units by setting the value of the system option to true again.

```
SetSystemOptions["DataOptions" → "ReturnQuantities" → True];
```

```
CityData[ New York City (city) , "Population"]
```

```
8491079 people
```

> Since *Mathematica* 10 the output of the computable data functions returns the units explicitly. In previous versions that was not the default behavior. This change may cause potential problems with notebooks and packages developed in *Mathematica* 9 or earlier. To avoid them use `QuantityMagnitude`.

- If you wish to know all the information (properties) of a given entity of a computable data function (**DF**) you can use the syntax:

`Transpose`[{**DF**["Properties"], **DF**[entities, #]& /@ **DF**["Properties"]}]

- In this case we find all the information available in `CityData` for the city of Salamanca, Salamanca province, Spain: {"Salamanca","Salamanca","Spain"}.

```
Transpose[{CityData["Properties"],
  CityData[Entity["City", {"Salamanca", "Salamanca", "Spain"} ], #] & /@
   CityData["Properties"]}]
```

```
{{AlternateNames, {Salamanque}}, {Coordinates, {40.97, -5.67}},
 {Country, Spain}, {Elevation, 818 m},
 {FullName, Salamanca, Salamanca, Spain}, {Latitude, 40.97°},
 {LocationLink, http://maps.google.com/maps?q=+40.97,-5.67&z=12&t=h},
 {Longitude, -5.67°}, {Name, Salamanca}, {Population, 159 754 people},
 {Region, Salamanca}, {RegionName, Salamanca},
 {StandardName, {Salamanca, Salamanca, Spain}}, {TimeZone, 1 h}}
```

If in the previous output we select "http://maps.google.com/..." and press <F1> we will see the location of Salamanca in GoogleMaps.

- Most of the computable functions establish groupings classes (entities that share some common characteristics) as displayed in the example below.We add `Shallow` to show only the first few lines of the result. Remove it to see the complete output.

```
StarData["Classes"] // Shallow
```

```
{stars, Algol variable stars, α 2 Canum Venaticorum variable stars,
 α Centauri, AM Herculis variable stars, Am stars, Ap stars,
 AR Lacertae variable stars, barium stars, Bayer stars, <<75>>}
```

```
StarData[ α Centauri (stars) ]
```

```
{Proxima Centauri, Rigel Kentaurus A, Rigel Kentaurus B}
```

- The following function displays a table with the properties and classes available for the specified collection:

```
DataTable[fun_Symbol] := Module[{data},
  data = {
    {Style[fun, "Section",
      FontFamily → "Lucida Grande", FontSize → 14], SpanFromLeft},
    {"", "Length", "Examples"},
    {"Entities", Length[fun[]], Append[RandomChoice[fun[], 2], "..."]},
    {"Properties", Length[fun["Properties"]],
     Append[RandomChoice[fun["Properties"], 2], "..."]},
    {"Classes", Length[fun["Classes"]],
     Append[RandomChoice[fun["Classes"], 2], "..."]}
   };
  Grid[data, Frame → All]]
```

- Here's the function applied to CountryData:

```
DataTable[CountryData]
```

CountryData		
	Length	*Examples*
Entities	240	{Somalia, Turkey, ...}
Properties	223	{CoastlineLength, FemaleElderlyPopulation, ...}
Classes	335	{Latin America and the Caribbean, Agency of Cultural and Technical Cooperation, ...}

5.2 Astronomy

- Very often, when starting a new section, we execute this command to remove functions that we may have previously defined to avoid potential name conflicts when defining new ones.

```
Clear["Global`*"]
```

The latest *Mathematica* version contains several computable functions for astronomical calculations as we will see in Chapter 8. Until *Mathematica* 9 the only available function was: `AstronomicalData` (still available) but since *Mathematica* 10 this function has been replaced by others containing new functionality: `PlanetData`, `StarData` ....

The curve describing the sun and the visible planets in the sky every day of the year always observed at the same location and time of the day is called analemma.

A solar analemma can be obtained with a camera and patience, taking pictures of the horizon every day or almost every day of the year from the same position, with the same exposure and at the same solar time. The result will be a composition similar to the one below. It shows an analemma taken by Jack Fishburn between 1998-1999 from Bell Laboratories in Murray Hill, New Jersey:

```
WikipediaData["Analemma", "ImageList"][[3]]
```

You can also make analemmas for other celestial objects such as Venus or Mercury.

With less patience and without having to wait for a year, we can simulate an analemma with *Mathematica* from our own computer.

- We need to define the location (latitude and longitude).

```
salamanca = GeoPosition[Salamanca (city)]
GeoPosition[{40.97, -5.67}]
```

SunPosition returns the sun position from our position (azimuth and latitude) at 9:00 AM in GMT (choosing TimeZone→0) on January 1, 2016.

```
SunPosition[salamanca, DateObject[{2016, 1, 1, 9}, TimeZone → 0]]
```

```
{132.67°, 9.91°}
```

We apply the same function to get the position every 5 days, at the same time, to complete a year:

```
Graphics[{Orange, PointSize[Large],
  Point@Map[QuantityMagnitude, SunPosition[salamanca,
     DateRange[DateObject[{2016, 1, 1, 9, 0}, TimeZone → 0],
      DateObject[{2016, 12, 31, 9, 0}, TimeZone → 0], 5]][
    "Values"], {2}]}, Frame → True]
```

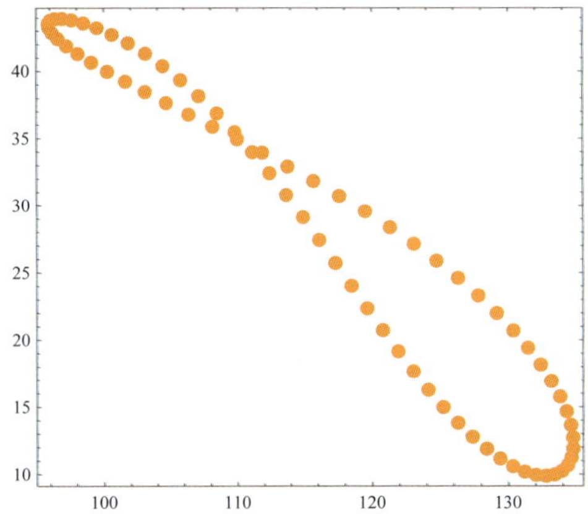

5.3 Nuclear and Particle Physics

The IsotopeData and ParticleData functions deal with nuclear and particle physics applications. We will make extensive use of them in Chapter 9.

- IsotopeData["*isotope*", "*property*"] returns the value of the specified property for a given isotope.

  You can use the name of an isotope (i.e. "Uranium235") or refer to it as {Z, A} with Z being its atomic number and A its atomic mass (i.e. instead of "Uranium235" we can write {92, 235}).

```
Clear["Global`*"]
```

```
IsotopeData["Properties"] // Shallow
```

```
{AtomicMass, AtomicNumber, BindingEnergy,
 BranchingRatios, DaughterNuclides, DecayEnergies, DecayModes,
 DecayModeSymbols, DecayProducts, ExcitedStateEnergies, <<23>>}
```

- The next example shows the properties of Iodine–131.

```
IsotopeData["Iodine131", #] & /@
 {"FullSymbol", "AtomicMass", "BranchingRatios", "DecayModes",
  "DaughterNuclides", "HalfLife", "IsotopeAbundance"}
```

```
{ 131 I   , 130.906124609 u , {1.00},
   53  78
 {BetaDecay}, { xenon-131 }, 6.9338×10^5 s , 0.}
```

The output tells us that this isotope, whose complete symbol is ${}^{131}_{53}I_{78}$, has an atomic mass of 130.09061 atomic mass units (amu), with a single type of radioactive decay, a beta decay, that disintegrates into Xenon–131 with a disintegration period of 6.9338×10^5 and that it doesn't exist in natural form (0.).

5.4 Engineering

Mathematica has a group of functions to access data related to engineering structures.

- For example, to explore the properties of a bridge:

```
BridgeData[ London Bridge , {"Length", "Crosses", "Image"}]
```

```
{262 m , {River Thames},
```

```
}
```

- In the command below we use NuclearReactorData to find out the number of nuclear reactors under construction per country (notice the use of the free-form input).

```
BarChart[#[[All, 2]], ChartLabels → #[[All, 1]],
   BarOrigin → Left, ImageSize → Medium] &@
 Tally@Flatten@NuclearReactorData[
    nuclear power reactors under construction , "Countries"]
```

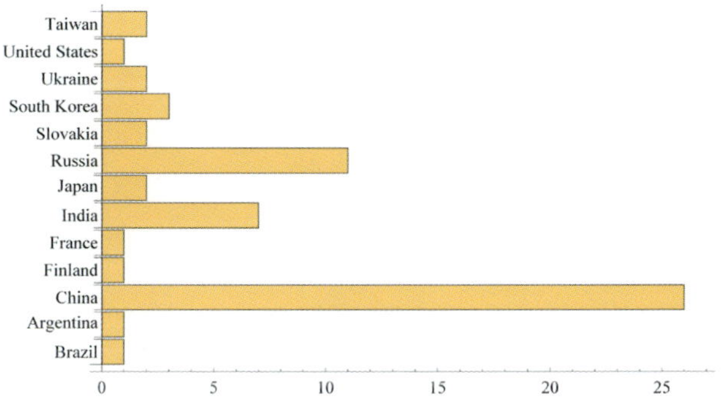

5.5 Chemical and Physical Properties of Elements and Compounds

With `ElementData` and `ChemicalData` we can access a wide variety of information about chemical elements and compounds.

■ Both functions use the following color scheme when representing atoms and molecules:

```
Clear["Global`*"]
```

```
ColorData["Atoms", "Panel"]
```

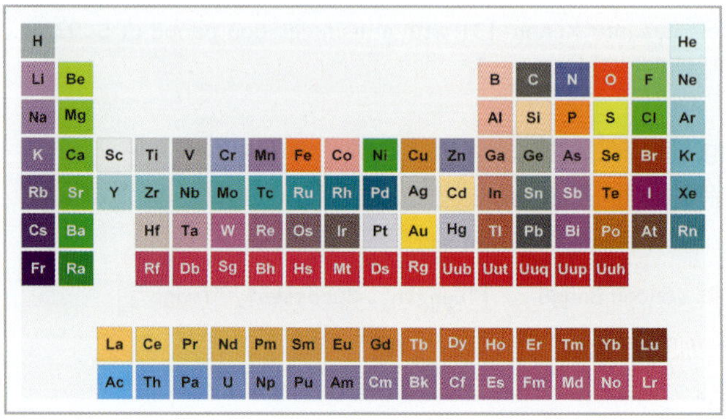

If you click on an element symbol, you will see the color code assigned by the function.

■ Let's take a look at the properties available in `ElementData`:

```
ElementData["Properties"] // Shallow
```

```
{Abbreviation, AdiabaticIndex, AllotropeNames,
 AllotropicMultiplicities, AlternateNames, AlternateStandardNames,
 AtomicMass, AtomicNumber, AtomicRadius, Block, <<75>> }
```

`ElementData["`*element*`", "`*property*`"]` displays the requested property information for a given element.

■ The next function displays the crystalline structure of an element in its solid phase:

```
ElementData["Uranium", "CrystalStructureImage"]
```

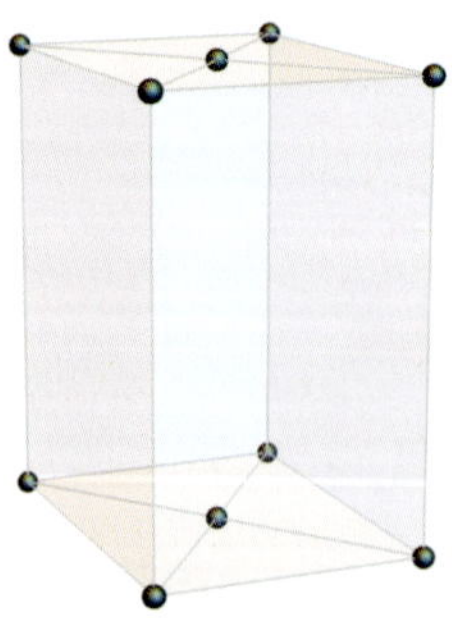

■ We choose several carbon properties and present them in a table format using `Grid`.

```
carbonproperties =
  {"AbsoluteBoilingPoint", "AbsoluteMeltingPoint", "AtomicRadius",
   "AtomicWeight", "Density", "FusionHeat", "IsotopeAbundances",
   "LatticeAngles", "LatticeConstants", "MagneticType", "MolarVolume",
   "Name", "Phase", "QuantumNumbers", "Resistivity", "SolarAbundance",
   "SpecificHeat", "ThermalConductivity", "ThermalExpansion"};
```

```
Grid[{#, ElementData["C", #]} & /@ carbonproperties, Frame → True,
 Alignment → {Left}, Background → {None, {{LightRed, Pink}}}]
```

AbsoluteBoilingPoint	4300. K
AbsoluteMeltingPoint	3823. K
AtomicRadius	67. pm
AtomicWeight	12.0107 u
Density	2260. kg/m^3
FusionHeat	105. kJ/mol
IsotopeAbundances	{{12, 0.9893}, {13, 0.0107}}
LatticeAngles	$\{\frac{\pi}{2}$ rad , $\frac{\pi}{2}$ rad , $\frac{2\pi}{3}$ rad$\}$
LatticeConstants	{246.4 pm , 246.4 pm , 671.1 pm}
MagneticType	diamagnetic elements
MolarVolume	5.314×10^{-6} m^3/mol
Name	carbon
Phase	solid elements
QuantumNumbers	$^3P_0$
Resistivity	0.000010 mΩ
SolarAbundance	0.0030 g/g
SpecificHeat	710. J/(kg K)
ThermalConductivity	140. W/(m K)
ThermalExpansion	7.1×10^{-6}/K

- We can know the units, in the international system (SI), of the corresponding property.

```
units = {#, ElementData["C", #, "UnitsNotation"]} & /@ carbonproperties
```

```
{{AbsoluteBoilingPoint, K}, {AbsoluteMeltingPoint, K},
 {AtomicRadius, pm}, {AtomicWeight, Missing[NotAvailable]},
 {Density, kg/m^3}, {FusionHeat, Missing[NotAvailable]},
 {IsotopeAbundances, Missing[NotApplicable]},
 {LatticeAngles, rad}, {LatticeConstants, pm},
 {MagneticType, Missing[NotApplicable]}, {MolarVolume, m^3/mol},
 {Name, Missing[NotApplicable]}, {Phase, Missing[NotApplicable]},
 {QuantumNumbers, Missing[NotApplicable]}, {Resistivity, Ωm},
 {SolarAbundance, Missing[NotAvailable]}, {SpecificHeat, J/(kg K)},
 {ThermalConductivity, W/(m K)}, {ThermalExpansion, 1/K}}
```

Before showing them, we eliminate non-dimensional properties or those without SI units. In such cases the output generated indicates ElementData or Missing (you can check it by removing the ";" from the previous command). To perform the operation we use `DeleteCases` in combination with "_ElementData" and "_Missing.

```
DeleteCases[DeleteCases[units, {_, _ElementData}], {_, _Missing}]
```

```
{{AbsoluteBoilingPoint, K}, {AbsoluteMeltingPoint, K}, {AtomicRadius, pm},
 {Density, kg/m^3}, {LatticeAngles, rad}, {LatticeConstants, pm},
 {MolarVolume, m^3/mol}, {Resistivity, Ωm}, {SpecificHeat, J/(kg K)},
 {ThermalConductivity, W/(m K)}, {ThermalExpansion, 1/K}}
```

- The following command tells us the abundance (as a fraction of the mass) of the different elements in the human body.

```
humanabundance =
  Table[{QuantityMagnitude[ElementData[z, "HumanAbundance"]],
    ElementData[z]}, {z, 92}];
```

We keep those whose presence in the body is greater than 0.001. We order them by comparing the first element of each sublist using **`Sort`** and a pure function.

```
Sort[Cases[humanabundance, {x_, y_} /; x > 0.001], #1[[1]] > #2[[1]] &]
```

```
{{0.61, oxygen}, {0.23, carbon},
 {0.10, hydrogen}, {0.026, nitrogen}, {0.014, calcium},
 {0.011, phosphorus}, {0.0020, potassium},
 {0.0020, sulfur}, {0.0014, sodium}, {0.0012, chlorine}}
```

ChemicalData["compound", "property"] shows the information specified in "property" for a given compound.

- In the example that follows we show the caffeine molecular plot. Once the command has been executed we can move the plot with the mouse to visualize the molecule from different perspectives.

```
ChemicalData["Caffeine", "MoleculePlot"]
```

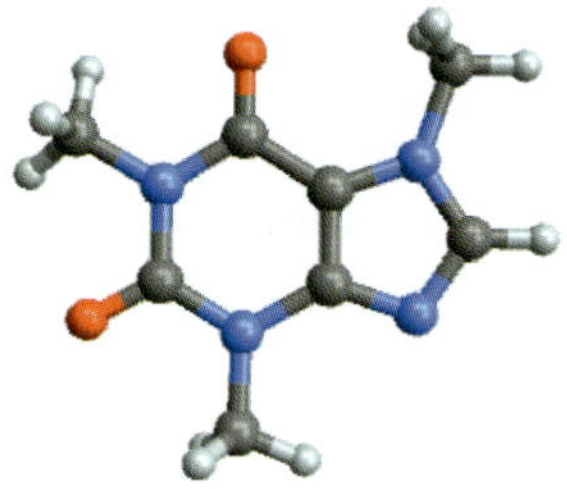

- We can compare two biomolecules based on their chemical formulas and molecular plots:

```
Grid[Outer[ChemicalData, {"Glycine", "L-Tyrosine"},
  {"Name", "ColorStructureDiagram", "MoleculePlot"}]]
```

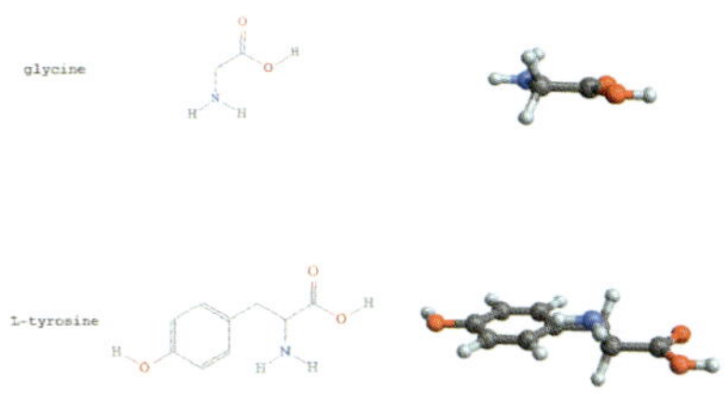

- Given a compound, the following function returns its name, structure diagram, formula, molecular weight, dielectric constant, density and boiling point:

```
ChemicalTable1[chemicals_List] :=
 Grid[Transpose[(Table[ChemicalData[#, property],
       {property, {"Name", "ColorStructureDiagram", "FormulaDisplay",
         "MolecularWeight", "DielectricConstant", "DensityGramsPerCC",
         "BoilingPoint"}}] & /@ chemicals)], Frame → All]

ChemicalTable1[{"Cyclohexanol", "Acetone"}]
```

cyclohexanol	acetone
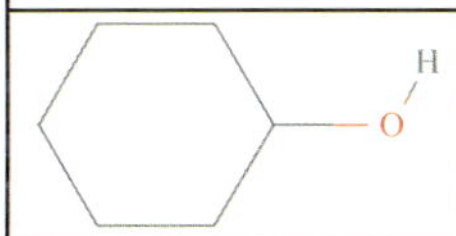	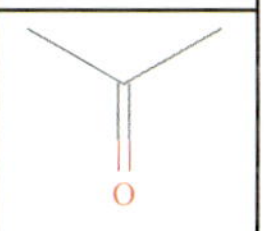
$C_6H_{12}O$	CH_3COCH_3
100.159 u	58.0791 u
15	20.56
0.9624 g/mL	0.791 g/mL
160.84 °C	56. °C

- The function below is similar to the previous one in terms of its contents but the information displayed is more elaborated. You can use it as a template.

```
ChemicalTable[chemicals_List] := Module[{labels, style, data, img},
  labels = {"Name", "Color Structure Diagram", "Formula",
    "Molecular Weight", "Density (g cm^-3)", "Dielectric Constant",
    "Boiling Point (°C)", "Flash Point (°C)", "NFPA Lavel"};
  style := Style[#, FontFamily → "Helvetica", FontWeight → Bold] &;
  data = (Table[ChemicalData[#, property],
       {property, {"Name", "ColorStructureDiagram", "FormulaDisplay",
         "MolecularWeight", "DensityGramsPerCC", "DielectricConstant",
         "BoilingPoint", "FlashPoint", "NFPALabel"}}] & /@ chemicals);
  img = Text[Style[Grid[Transpose[Join[{style /@ labels}, data]],
      Frame → All,
      Background → {LightYellow, {{GrayLevel[0.9], White}}},
      Alignment → {{Left, {Center}}, Center}],
     FontFamily → "Helvetica", Bold, 13]
   ];
  img /. {(ImageSize → _?NumericQ) → (ImageSize → 110),
    (ImageSize → x_List) → (ImageSize → 1.2 * x)}
 ]
```

This function can be applied directly to a list of arbitrary length. In this case we use it with two compounds.

```
ChemicalTable[{"Cyclohexanol", "Acetone"}]
```

Name	cyclohexanol	acetone
Color Structure Diagram	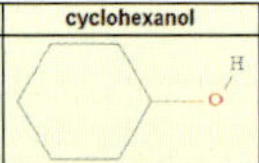	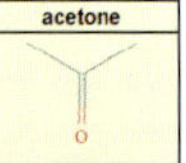
Formula	$C_6H_{12}O$	CH_3COCH_3
Molecular Weight	100.159 u	58.0791 u
Density (g cm^{-3})	0.9624 g/mL	0.791 g/mL
Dielectric Constant	15	20.56
Boiling Point (°C)	160.84 °C	56. °C
Flash Point (°C)	63. °C	−17.2222 °C
NFPA Lavel		

5.6 Genomics and Proteomics

The following examples refer to typical functions applicable to genetic and protein data related to humans.

- GenomeLookup shows the chromosomes in which a specified DNA sequence appears along with its position inside of them.

```
Clear["Global`*"] (*Clears values and
 definitions for all the previously defined variables*)
```

```
GenomeLookup["AATTTCCGTTAAAT"]
```

```
{{{Chromosome2, 1}, {213319362, 213319375}},
 {{Chromosome3, -1}, {83249225, 83249238}},
 {{Chromosome7, -1}, {78235739, 78235752}},
 {{Chromosome9, -1}, {5746733, 5746746}},
 {{Chromosome11, 1}, {41489916, 41489929}},
 {{Chromosome11, -1}, {124909525, 124909538}},
 {{Chromosome22, -1}, {11132978, 11132991}},
 {{ChromosomeX, 1}, {144568117, 144568130}}}
```

- `GenomeData["`*gene*`", "`*property*`"]` gives the DNA sequence for a specified gene on the human genome. With "*property*" we can specify a characteristic of the chosen gene. In this example we show the biological processes where the gene "ACAA2" plays a role.

```
GenomeData["ACAA2", "BiologicalProcesses"]
```

```
{CholesterolBiosyntheticProcess, FattyAcidMetabolicProcess,
 LipidMetabolicProcess, MetabolicProcess}
```

- `ProteinData["`*protein*`"]` returns the amino acids of a given protein. In this case we display the sequence of the protein A2M. We use `Short` to avoid displaying the complete output.

```
ProteinData["A2M"] // Short
```

```
MGKNKLLHPSLVLLLLVLLPTDASVSGKPQYMVL ... RDLKPAIVKVYDYYETDEFAIAEYNAPCSKDLGNA
```

- `ProteinData["`*protein*`", "`*property*`"]` gives us the option to choose the property of the protein that we want to know about. As an illustration, to know the graphical representation of the A2M protein, we can use "MoleculePlot" (this property is not yet available for all proteins at the time of writing).

```
ProteinData["A2M", "MoleculePlot"]
```

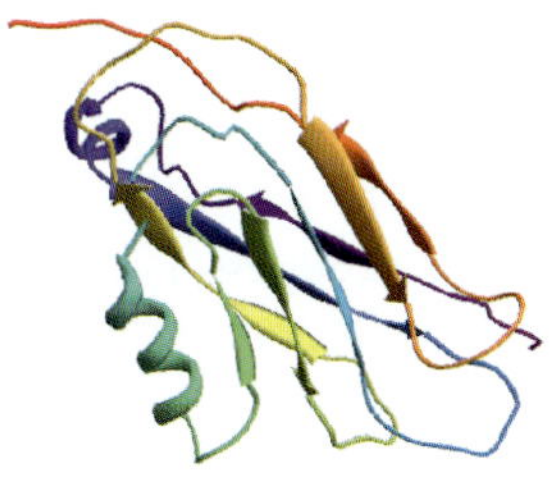

The following functions are considerably more complicated (some of them have been adapted from examples and documentation provided by Wolfram Research). You may find them useful as they are or you could use them as templates and make the necessary adjustments to suit your needs.

- This first example represents each atom of the A2M protein. The atoms are scaled. Their atomic radii have been obtained from `ElementData` and their colors follow the scheme in `ColorData`, as previously mentioned.

```
With[{p = "A2M", cr = Dispatch@ColorData["Atoms", "ColorRules"]},
 Graphics3D[MapThread[{#2 /. cr, Sphere[#1,
      QuantityMagnitude[ElementData[#2, "VanDerWaalsRadius"]]]} &,
   {QuantityMagnitude[ProteinData[p, "AtomPositions"]],
    Flatten[ProteinData[p, "AtomTypes"]]}], Boxed → False]]
```

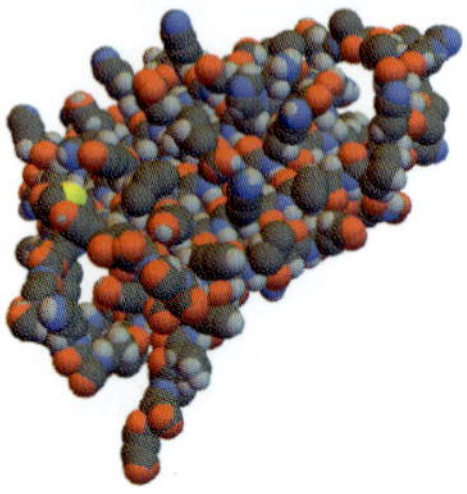

- The function below identifies the chromosomes and the position inside of them where a DNA sequence is located.

```
FindSequence[sequence_] :=
  chrpos =
   SortBy[
    Module[{chr = #, len, scale = N[GenomeData[#, "SequenceLength"] /
           GenomeData["Chromosome1", "SequenceLength"]],
       rls},
      rls = Flatten[{#[[1, 1, 1]] → (#[[All, 2, 1]] * scale / len) & /@
           SplitBy[Select[GenomeLookup[sequence],
             #[[1, 2]] == 1 &], #[[1, 1]] &], _ → {}}];
      len = GenomeData[chr, "SequenceLength"];
      {chr,
       Extract[GenomeData[chr, "GBandScaledPositions"],
           Position[GenomeData[chr, "GBandStainingCodes"],
            "acen", 1, 1]][[1, 1]] * scale, scale,
       GenomeData[#, "AlternateStandardNames"][[1]],
       GenomeData[chr, "AlternateStandardNames"][[2]], # /. rls}] & /@
    Most@GenomeData["Chromosomes"], #[[5]] &];
```

- Here we use the previously defined function (**FindSequence[]**) to find, for a given sequence, the chromosomes and the position in which they appear.

```
FindSequence["AACCTTGGGAAATT"] // Shallow
```

```
{{Chromosome1, 0.471457, 1., 1, 1, {<<3>>}},
 {Chromosome2, 0.362134, 0.982614, 2, 2, {<<5>>}},
 {Chromosome3, 0.354132, 0.806884, 3, 3, {<<3>>}},
 {Chromosome4, 0.199253, 0.773603, 4, 4, {<<8>>}},
 {Chromosome5, 0.166101, 0.731479, 5, 5, {<<3>>}},
 {Chromosome6, 0.23412, 0.691204, 6, 6, {<<1>>}},
 {Chromosome7, 0.227125, 0.642352, 7, 7, {<<3>>}},
 {Chromosome8, 0.168017, 0.591608, 8, 8, {<<1>>}},
 {Chromosome9, 0.180048, 0.567334, 9, 9, {}},
 {Chromosome10, 0.163645, 0.547522, 10, 10, {<<1>>}}, <<14>>}
```

- The next function is equivalent to the last one but its output is shown as a graphical representation.

```
PlotSequence :=
 Graphics@
  MapIndexed[
   With[{i = #2[[1]]},
     {White, EdgeForm[{Black, Thickness[0.005]}], Rectangle[
       {0, i/15.}, {#[[2]], i/15. + 1/30.}, RoundingRadius → 0.5/30],
      Rectangle[{#[[2]], i/15.}, {#[[3]], i/15. + 1/30.},
       RoundingRadius → 0.5/30], Red,
      Disk[{#, i/15. + 1/60.}, 0.01] & /@ #[[6]], Black,
      Text[#[[4]], {-0.1, i/15. + 1/60.}]}] &, Reverse@chrpos]
PlotSequence
```

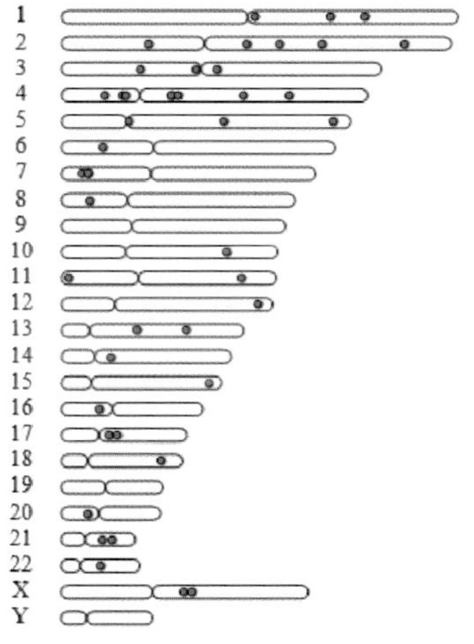

5.7 Meteorology

```
Clear["Global`*"]
```

With computable data we can have access to historical as well as real-time weather information: `WeatherData`, `WeatherForecastData`, `AirTemperatureData`, `AirPressureData`, `WindSpeedData`, `WindDirectionData`, `WindVectorData`.

- The command `Function["`*location*`", "`*property*`", {`*start*`, `*end*`, `*step*`}]` returns the value of the variable specified in "*property*" for a given location and period (by default it returns the most recent value).

  In this example we get the average daily temperature in Salamanca (Spain) from January 1, 1995 to December 31, 2015 and represent it graphically. Notice the seasonality.

```
DateListPlot[
 AirTemperatureData[Entity["City", {"Salamanca", "Salamanca", "Spain"}],
  {{2015, 1, 1}, {2015, 12, 31}, "Day"}]]
```

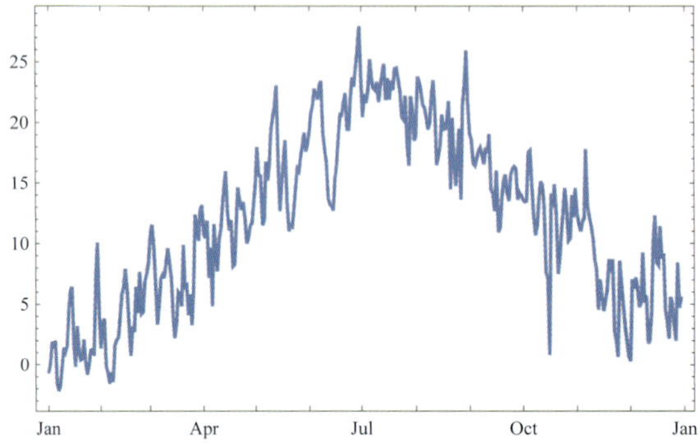

- The commands below show how to obtain a weather station code and visualize the range of daily temperatures over the past 180 days that corresponds to that station code.

```
WeatherData[
 Entity["City", {"Salamanca", "Salamanca", "Spain"}], "StationName"]
```

```
LESA
```

```
DateListPlot[(WeatherData["LESA", #1,
     {DateString[DatePlus[-180]], DateString[DatePlus[0]], "Day"}] &) /@
  {"MeanTemperature", "MaxTemperature", "MinTemperature"},
 Joined → {True, False, False}, Filling → {2 → {3}}, Mesh → All]
```

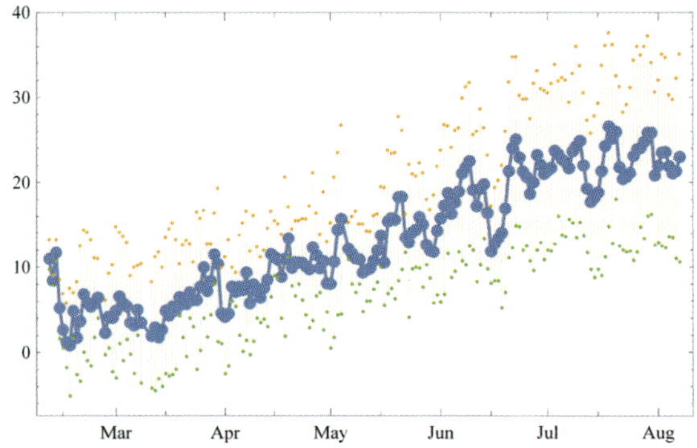

- Here we display the forecast for the following week:

```
Manipulate[data = WeatherForecastData[Salamanca];
 DateListPlot[data[g] /. Interval[x_] :→ Mean[x], FrameLabel → Automatic],
 {g, {"Temperature", "WindSpeed", "WindDirection", "PrecipitationAmount"}},
 SaveDefinitions → True]
```

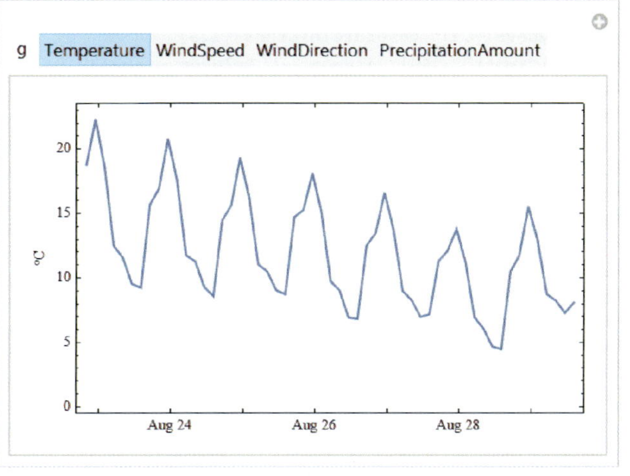

- The function that follows (obtained from the `WindVectorData` examples) returns wind data (speed and direction), at the moment of the query, for the coordinates 35ºN and 45ºN, in latitude and -10º W and 5º W in longitude (includes the Iberian peninsula and the Balearic islands). To do that we are going to use `CountryData` and `WindVectorData` in combination with `ListStreamPlot`.

```
stationnames =
  DeleteDuplicates[Flatten[Table[WindVectorData[{y, x}, "StationName"],
     {x, -10.0, 5.0, 1}, {y, 35.0, 45.0, 1}]]];
windVelocity[station_] :=
 Module[{coords = First@WindVectorData[station, "Coordinates"]},
  {Reverse@coords, -Reverse@QuantityMagnitude[WindVectorData[station]]
    {1/Cos[coords[[2]] °], 1}}]
windData = windVelocity[#] & /@ stationnames //
   Cases[#, u_ /; FreeQ[u, "NotAvailable"]] &;
```

```
Show[ListStreamPlot[windData, VectorStyle → Black, AspectRatio → 1],
 Graphics[{GrayLevel[.2], AbsoluteThickness[1.5],
   CountryData["Spain", "FullPolygon"] /. Polygon → Line}]]
```

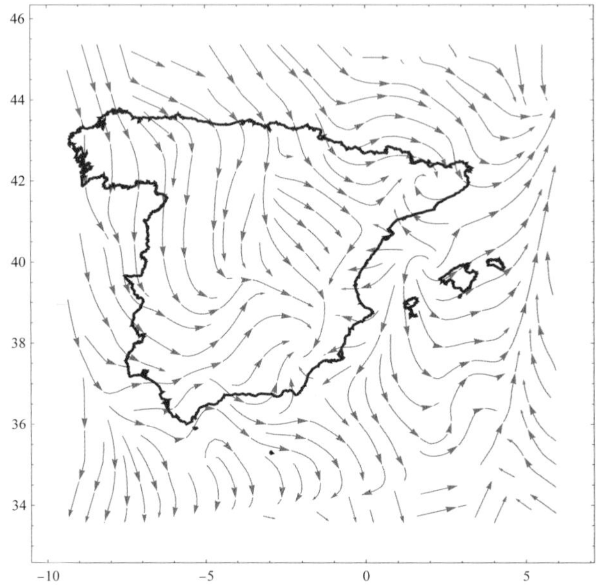

5.8 Combining Data and Graphics

```
Clear["Global`*"]
```

- Use Outer to get several properties for a set of entities.

```
Outer[f, {a, b}, {x, y, z}]
```

```
{{f[a, x], f[a, y], f[a, z]}, {f[b, x], f[b, y], f[b, z]}}
```

- Here we obtain the life expectancy in Spain, USA, France and Mexico using CountryData. The result is displayed in which each country is represented by its flag.

```
data = Outer[CountryData, {"Spain", "UnitedStates", "France", "Mexico"},
   {"LifeExpectancy", "FemaleLifeExpectancy", "MaleLifeExpectancy"}];
```

```
flags = {
   ImageResize[CountryData["Spain", "Flag"], 50],
   ImageResize[CountryData["UnitedStates", "Flag"], 50],
   ImageResize[CountryData["France", "Flag"], 50],
   ImageResize[CountryData["Mexico", "Flag"], 50]};
```

```
BarChart3D[data,
 ChartElementFunction → "Cylinder", ChartStyle → "Pastel",
 ChartLegends → {"Life Expectancy(\!\(\*Cell[\"♀ + ♂\"]\))",
   "Female (♀)", "Male (♂) "},
 ChartLabels → {Placed[flags, {.5, 0, 1.15}], Placed[
    {"\!\(\*Cell[\"♀+♂\"]\)", "♀", "♂"}, Center]}, BarSpacing → {.1, .6},
 PlotLabel → Style["Life expectancy", 16, FontFamily → "Times"],
 ImageSize → Large]
```

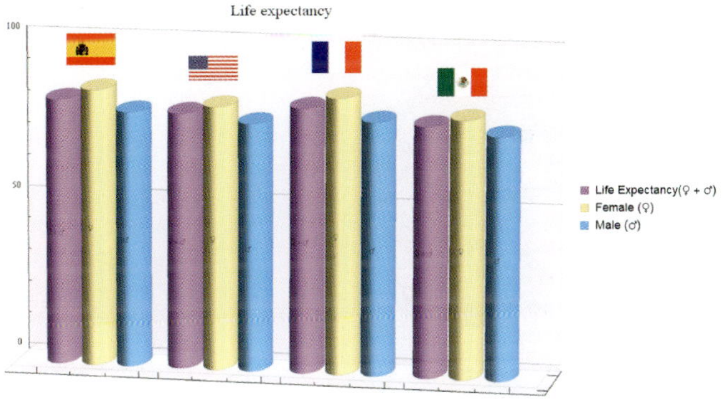

5.9 Geodata

```
Clear["Global`*"]
```

There are many functions available for the computation of data related to the earth sciences.

- We can get an idea of them by asking *Mathematica* to show us all the functions that start with the letter **Geo** (output omitted).

```
?Geo*
```

The names of the functions shown enable us to identify those related to geography, almost all of them except the ones starting with Geometric. We can always access the help documentation to see their syntax. Let's take a look at some of them.

- Using the below functions you can find your current geolocation and the name of the city of your computer.

```
{FindGeoLocation[], GeoIdentify[]}
```

```
{GeoPosition[{40.97, -5.66}], {world, Salamanca, Castile and Leon, Spain,
  Europe, Europe/Madrid, Spain, GMT+1, Castile and Leon, Spain}}
```

`FindGeoLocation[]` may use built-in GPS or other capabilities. Alternatively, it may also use geoIP lookup through the Internet. Different methods may give results with different accuracy.

- Next we use `GeoPosition` and `GeoGraphics` to first find the coordinates (latitude and longitude) of that location and then visualize its geographical image in 2D (the easiest way to do it is to copy and paste the first element of the output returned by the previous command as the argument for both commands). Remember that in many cases you will find the free-form input very useful as well (**Insert ▸ Inline Free-form Input** and type

Salamanca).

```
GeoPosition[Salamanca (city)]
```

```
GeoPosition[{40.97, -5.67}]
```

- Here we display a map of Salamanca (a City of Spain) and its surroundings.

```
GeoGraphics[GeoVisibleRegionBoundary[
  GeoPosition[{40.97, -5.67, Quantity[0.1, "Kilometers"]}]]]
```

- The commands below return the elevation of Salamanca above sea level in meters, inform us about the minimum and maximum elevations, and create a relief plot:

```
elevation =
 GeoElevationData[GeoDisk[Salamanca (city), Quantity[100, "km"]]]
```

```
QuantityArray[ Dimensions: {328, 432}  Unit: Meters ]
```

```
MinMax[elevation]
```

```
{178. m, 2359. m}
```

```
ReliefPlot[elevation, DataReversed -> True,
 ImageSize -> 500, ColorFunctionScaling -> False,
 ColorFunction -> ColorData["HypsometricTints"],
 PlotLegends -> Placed[Automatic, Below]]
```

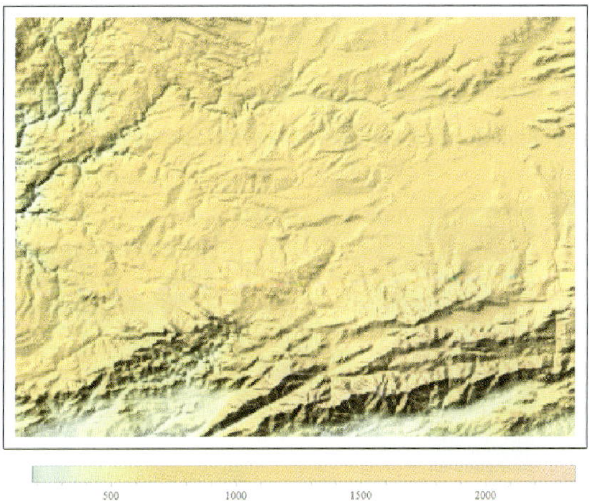

- We can also make a 3D plot:

```
ListPlot3D[Reverse[elevation],
 ImageSize -> 500, ColorFunctionScaling -> False,
 ColorFunction -> ColorData["HypsometricTints"],
 BoxRatios -> {2, 1, 1/2},
 PlotTheme -> {"FilledSurface", "Minimal"}, PlotRange → All]
```

- Here we use that approach to see the closest locations to Salamanca:

```
GeoNearest["City", Salamanca (city), {All, 10 km}]
```

```
{Salamanca, Villamayor, Santa Marta de Tormes,
 Villares de la Reina, Carbajosa de la Sagrada, Aldeatejada, Cabrerizos,
 Carrascal de Barregas, Doninos de Salamanca, Monterrubio de Armuna,
 San Cristobal de la Cuesta, Moriscos, Arapiles, Florida de Liebana,
 Castellanos de Villiquera, Castellanos de Moriscos, Pelabravo}
```

- Shown below is an example of how to zoom out of the location of Salamanca in a series of relief maps:

```
GeoGraphics[{Red, PointSize[Large], Point[ Salamanca (city) ]},
   GeoZoomLevel → #, GeoBackground → GeoStyling["ReliefMap"]] & /@ {6, 4, 2}
```

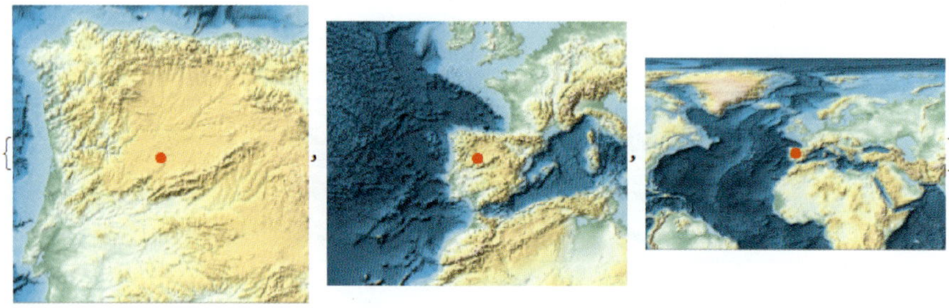

- Quite often, the data related to the United States is more complete than for other world locations. For example: New Orleans seems to have many more buildings and monuments of interest than London or Paris. We hope that the information about places of interest outside the US will be expanded in the future.

```
GeoEntities[#, Entity["Building"]] & /@
  { New Orleans (city) , London (city) , Paris (city) }
```

```
{{ Hibernia Bank Building , World Trade Center , Plaza Tower ,
   Latter & Blum Building , Medical Center of Louisiana Charity Hospital ,
   One Shell Square , Louisiana Superdome , Jackson Brewery }, {}, {}}
```

- This example returns the location of all the earthquakes that happened in Chile between 2011-1-1 and 2016-6-30 with magnitudes between 6 and 10:

```
data = EarthquakeData[Entity["Country", "Chile"],
  {6, 10}, {{2011, 1, 1}, {2016, 6, 30}}, "Position"]
```

```
TimeSeries[ Time: 02 Jan 2011 to 07 Nov 2015
            Data points: 21 ]
```

```
GeoListPlot[data]
```

- `GeoListPlot` can generate maps for locations outside the Earth. Here we visualize the landing of Apollo 11, the first manned lunar landing mission:

```
GeoListPlot[Apollo 11 (manned space mission), GeoRange → All,
 GeoProjection → "Orthographic", GeoLabels → True, LabelStyle → White]
```

```
GeoListPlot[ Apollo 11 (manned space mission) ,
 GeoRange → 10 km , GeoLabels → True, LabelStyle → White]
```

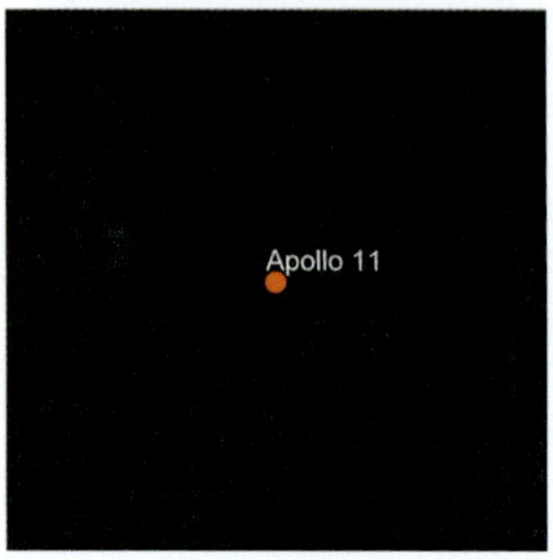

To make the visualization of the information easier, we use `GeoRegionValuePlot` to represent the values on a map. We can use **"MissingStyle" → Color** for those cases where there are no available data.

- We use `CountryData` and `GeoRegionValuePlot` to show the Literacy Rate in the world.

```
map = GeoRegionValuePlot[CountryData[] → "LiteracyRate",
  ColorFunction → ColorData["LightTemperatureMap"],
  "MissingStyle" → White, PlotLegends → None]
```

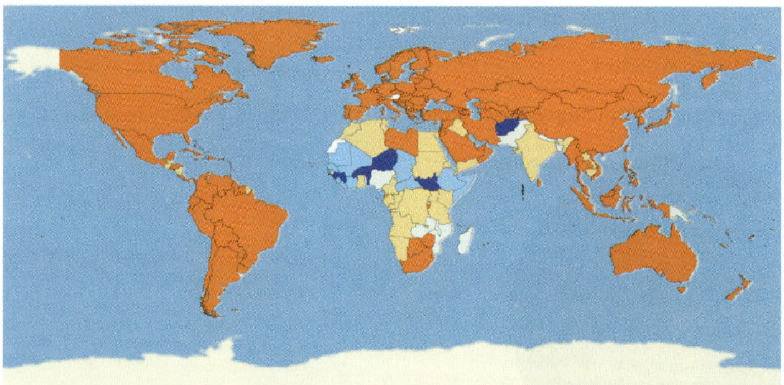

- Next, we project the previous map onto a sphere. To rotate it just click inside.

```
ParametricPlot3D[{Cos[u] Sin[v], Sin[u] Sin[v], Cos[v]},
 {u, 0, 2 Pi}, {v, 0, Pi}, Mesh → None, Boxed → False,
 TextureCoordinateFunction → ({#4, 1 - #5} &),
 PlotStyle → Texture[Show[map, ImageSize → 1000]],
 Lighting → "Neutral", Axes → False, RotationAction → "Clip"]
```

- The following example uses `EntityClass` to display the identity class that corresponds to the given name, in this case Europe. Therefore, it includes all the European nations except Russia which has most of its territory in Asia.

```
EntityClass["Country", "Europe"]
```

Europe

- Now we use `EntityList` to see the "*Entities*" included in the "Europe" class.

```
EntityList[Europe (countries)] // Shallow
```

{Albania, Andorra, Austria, Belarus, Belgium, Bosnia and Herzegovina, Bulgaria, Croatia, Cyprus, Czech Republic, <<41>>}

- With `EntityValue` we can see the properties available for a given *entity*, in this case for the European countries.

```
EntityValue[Europe (countries), "Properties"] // Shallow
```

{*adjusted net national income*, *seasonal bank borrowings from Fed, plus adjustments*, *regions*, *adult population*, *obese adults*, *number of aggravated assaults*, *rate of aggravated assault*, *aggregate home value*, *aggregate home value, householder 15 to 24 years*, *aggregate home value, householder 25 to 34 years*, <<737>>}

- The complete output, not shown here due to its length, includes emissions of greenhouse gases (GreenhouseEmissions). To see those emissions per country we can type:

```
EntityValue[ Europe (countries) ,
  greenhouse gas emissions , "EntityAssociation"] // Short
```

```
⟨| Albania → Missing[NotAvailable],
  <<49>>, Vatican City → Missing[NotAvailable] |⟩
```

- In the last example of the section we combine several functions to show some travel routes originating in Salamanca. We first define the cities that we want to travel to. When typing the city is better to add the country to avoid ambiguities in case there are several cities with the same name.

{"London, England", "Nuuk, Greenland", "Buenos Aires, Argentina", "Los Angeles", "Malmo, Sweden", "Ulan Bator, Mongolia", "Sidney, Australia"}

```
cities =
  { London (city) , Nuuk (city) ,
    Buenos Aires (city) , Los Angeles (city) ,
    Malmö (city) , Ulaanbaatar (city) , Sydney (city) }
```

```
{ London , Nuuk , Buenos Aires ,
  Los Angeles , Malmö , Ulaanbaatar , Sydney }
```

```
GeoGraphics[{{Gray, Table[GeoCircle[GeoPosition[ Salamanca (city) ],
      k*Quantity[1000, "Kilometers"]], {k, 1, 20}]},
  {Red, AbsoluteThickness[2],
   GeoPath[{#, GeoPosition[ Salamanca (city) ]}] & /@
    GeoPosition /@ cities}}, ImageSize → 800]
```

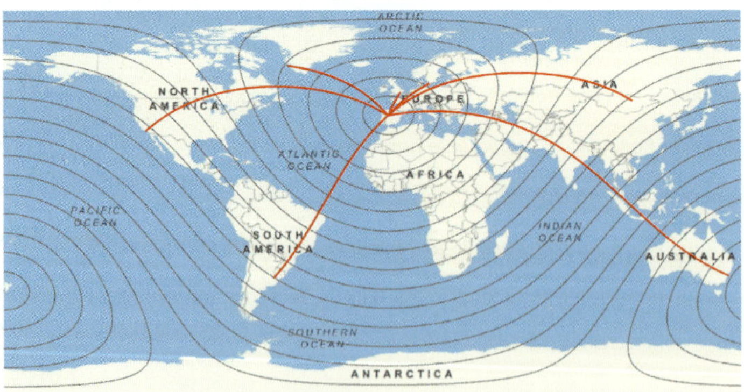

We can use `CloudDeploy` in combination with `GeoGraphics` to build a web application.

- The command below generates a map of a chosen location (we will normally enter the name of a city) and its surroundings for a given radius:

```
CloudDeploy[
  FormFunction[{"Location" -> "Location", "Radius" -> "Quantity"},
   GeoGraphics[GeoDisk[#Location, #Radius]] &, "PNG"]]
```

```
CloudObject[https://www.wolframcloud.com/objects/ed8baf92-f7b2-445f-9d60-82
```

- We can now copy this newly created link to our website to make it accessible from any device with web browsing capabilities:

  < a href = "https://www.wolframcloud.com/objects/..." > City Map (html) < /a >

- The application is now available to any user, although the first time we access it, we may have to enter our Wolfram ID and password. However, we don't need to have *Mathematica* installed in our computers. Once we enter the required data (Figure 5.1) we get the corresponding map (Figure 5.2).

Figure 5.1 Choosing the location and the radius.

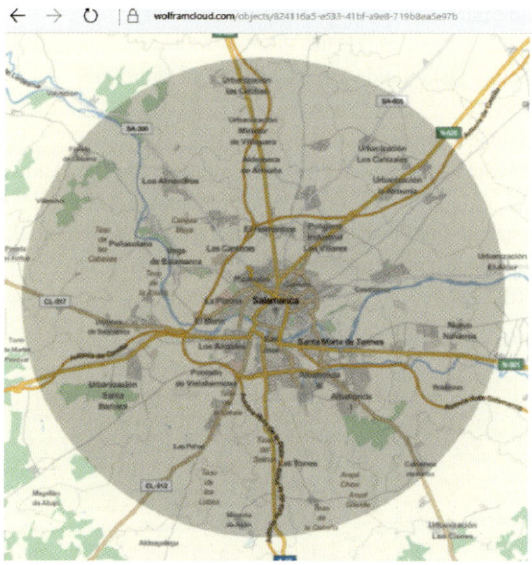

Figure 5.2 WolframCloud generated map.

5.10 Some Recommendations

- The first time you execute a data collection function, it will usually take a long time. After that, the data will be stored in your computer to speed up subsequent computations. The cached data are located in:

```
SystemOpen[FileNameJoin[{$UserBaseDirectory, "Paclets"}]]
```

- Occasionally, you might want to rebuild the cached paclet information. The following command rescans the paclet directories and rebuilds the cached indexes for each paclet.

```
RebuildPacletData[]
```

- If you want to force *Mathematica* to work from your cached directories and avoid automatic updates (useful during presentations, for example), you can turn this feature off with an internal function:

```
PacletManager`$AllowDataUpdates = False
```

5.11 Additional Resources

To access the following resources, if they refer to the help files, write their locations in a notebook (e.g., guide/EarthSciencesDataAndComputation), select them and press <F1>. In the case of external links, copy the web addresses in a browser:

Working with data collections tutorial:
http://library.wolfram.com/infocenter/Conferences/7196
Astronomical Data: guide/AstronomicalComputationAndData
Earth Sciences: guide/EarthSciencesDataAndComputation
EngineeringData: guide/EngineeringData
Financial and Economic Data: guide/FinancialAndEconomicData
Geodesy: guide/Geodesy
GeoGraphics:tutorial/GeoGraphics
Life Sciences and Medicine: guide/LifeSciencesAndMedicineDataAndComputation
Socioeconomic Data: guide/SocioeconomicAndDemographicData
Physics and Chemistry: guide/PhysicsAndChemistryDataAndComputation
Transportation Data: guide/TransportationData
WordData: ref/WordData

6

Probability and Statistics

Mathematica contains numerous functions for probability, statistics and visualization. Its most recent versions include commands for modeling nonparametric and derived statistical distributions, performing survival analysis, finding clusters and fitting data to linear, generalized linear and nonlinear models among other features. Mathematica probably contains more statistics functions than most of the specialized programs in the field. At the end of the chapter we will show an example of how to build a package for quality control.

6.1 The Latest Features

Mathematica's functionality for data analysis is quite extensive. Simply listing all the functions related to it would exceed the length of this chapter. Earlier in the book we saw some examples of basic statistical functions but if you still don't feel comfortable doing probability and statistics in *Mathematica*, you may want to read the tutorial for numerical operations on data: tutorial/NumericalOperationsOnDataOverview and from there go to the area you are interested in or simply choose one of the following:

Basic Statistics: tutorial/BasicStatistics

Descriptive Statistics: tutorial/DescriptiveStatistics

Continuous Distributions: tutorial/ContinuousDistributions

Discrete Distributions: tutorial/DiscreteDistributions

Convolutions and Correlations: tutorial/ConvolutionsAndCorrelations

Apart from the above areas, with *Mathematica* you can even explore further as some of its capabilities for statistical computing go beyond the ones available in specialized software such as R, SAS or SPSS. The list bellow contains additional details about *Mathematica*'s commands related to advanced probability and statistical analysis:

Parametric distributions: guide/ParametricStatisticalDistributions. To create discrete and continuous distributions using parameters.

Nonparametric distributions: guide/NonparametricStatisticalDistributions. To get probability distributions from empirical data.

Derived distributions: guide/DerivedDistributions. To generate new distributions from existing ones.

Statistical distribution functions: guide/StatisticalDistributionFunctions. To describe probability distributions.

Statistical model analysis: : guide/StatisticalModelAnalysis. To fit models to data.

Random variables: guide/RandomVariables. To model uncertainty.

Random processes: guide/RandomProcesses. To simulate and predict the behavior of stochastic processes.

Statistical visualization: guide/StatisticalVisualization. To understand how data is distributed.

Time series: guide/TimeSeriesProcesses. To simulate and forecast time series.

Reliability analysis: guide/Reliability. To measure the robustness of complex systems.

Survival analysis: guide/SurvivalAnalysis. To analyze the time before a certain event happens.

Text analysis: guide/TextAnalysis. To analyze and visualize text.

Machine learning: guide/MachineLearning. To classify and predict data.

6.2 Statistics Data

According to *The Oxford Dictionary of Statistical Terms*, statistics is the study of the collection, analysis, interpretation, presentation, and organization of data. Therefore, without data there can be no statistics. In statistical studies we will frequently reduce the original data, frequently called population, to a reduced number of selected representative values, also known as samples. Their visualization will be very helpful.

Mathematica gives us the possibility of accessing an extensive collection of scientific and technical data. It also gives us many tools to analyze them.

- With `AirTemperatureData` we can obtain the temperature measurement values over the course of a period for a specific city or station. They can be subsequently visualized in a histogram:

```
data =
  Flatten[Table[AirTemperatureData[ Valparaiso (city) , {{i, 6, 20},
     {i, 6, 20}}, UnitSystem → "Metric"]["ValueList"], {i, 1993, 2015}]];
Histogram[data, {1}]
```

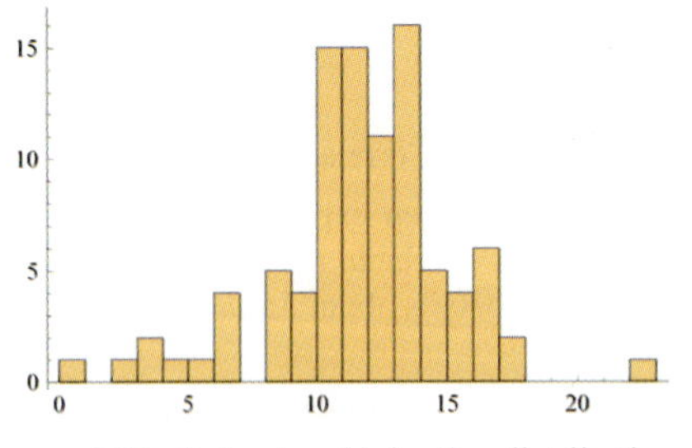

- We use `CityData` to obtain the distribution of the US population by city. We include `DeleteMissing` to eliminate those cases where the information is not complete.

```
usacities = QuantityMagnitude[DeleteMissing[
     CityData[#, "Population"] & /@ CityData[{All, "USA"}]]] // N;
```

The previous command is equivalent to **Map**[func[arg, #] &, {$city_1$, ..., $city_n$}] where the list of cities is generated with: **CityData**[{All, "USA"}]. We use `Shallow` to show a small sample of the total output.

```
usacities // Shallow
```

```
{8.49108×10^6, 3.92886×10^6, 2.72239×10^6,
 2.23956×10^6, 1.5603×10^6, 1.53706×10^6, 1.4367×10^6,
 1.38107×10^6, 1.28105×10^6, 1.01579×10^6, <<28 375>>}
```

- We can find out the number of cities in the list:

```
Length[usacities]
```

```
28 385
```

- We build a simple frequency table using `BinCounts` $[\{x_1, x_2, \ldots\}, \{\{b_1, b_2, \ldots\}\}]$ to count the number of x_i in the intervals $[b_1, b_2), [b_2, b_3), \ldots$..

```
TableForm[{{"0-10^2", "10^2-10^3", "10^3-10^4",
   "10^4-10^5", "10^5-10^6", "10^6-10^7"}, BinCounts[usacities,
   {{0, 100, 1000, 10000, 100000, 1000000, 10000000}}]}]
```

$0-10^2$	10^2-10^3	10^3-10^4	10^4-10^5	10^5-10^6	10^6-10^7
1654	10 584	11 897	3944	296	10

- We calculate the most common statistical parameters:

```
{Mean[#], StandardDeviation[#], Skewness[#], Kurtosis[#],
   Quantile[#, .6], InterquartileRange[#]} &[usacities]
```

```
{8710.85, 69482.9, 76.5378, 8346.35, 2126., 4402.5}
```

In previous chapters we have seen that *Mathematica* includes data files that can be used as examples to practice some of its commands. Let's learn how we can know the contents of the files and choose the most appropriate ones for our purposes.

- To find out the areas for which we have examples available:

```
ExampleData[]
```

```
{AerialImage, Audio, ColorTexture, Dataset, Geometry3D, LinearProgramming,
 MachineLearning, Matrix, NetworkGraph, Sound, Statistics,
 TestAnimation, TestImage, TestImage3D, TestImageSet, Text, Texture}
```

- We are interested in the ones related to statistics. The following function displays the name of the example, a brief description and the dimensions of its data matrix.

```
Shallow[
 {#[[2]], ExampleData[#, "Description"], Dimensions[ExampleData[#]]} &/@
  ExampleData["Statistics"], {5, 3}]
```

```
{{AirlinePassengerMiles,
  Revenue passenger miles flown by commercial airlines., {24}},
 {AirplaneGlass, Time to failure for airplane glass., {31}},
 {AnimalWeights, Brain and body weights for 28 animal species., {28, 3}},
 <<108>>}
```

- We're going to analyze the "OldFaithful" dataset related to geyser emissions. Let's see a detailed description of the file.

```
ExampleData[{"Statistics", "OldFaithful"}, "Properties"]
```

```
{ApplicationAreas, ColumnDescriptions, ColumnHeadings, ColumnTypes,
 DataElements, DataType, Description, Dimensions, EventData, EventSeries,
 LongDescription, Name, ObservationCount, Source, TimeSeries}
```

```
ExampleData[{"Statistics", "OldFaithful"}, "LongDescription"]
```

```
Data on eruptions of Old Faithful geyser, October 1980. Variables
  are the duration in seconds of the current eruption, and the
  time in minutes to the next eruption. Collected by volunteers,
  and supplied by the Yellowstone National Park Geologist. Data
  was not collected between approximately midnight and 6 AM.
```

```
Transpose[{ExampleData[{"Statistics", "OldFaithful"}, "ColumnHeadings"],
   ExampleData[{"Statistics", "OldFaithful"},
    "ColumnDescriptions"]}] // TableForm
```

```
Duration        Eruption time in minutes
WaitingTime     Waiting time to next eruption in minutes
```

- We assign names to the column data and create a histograms for each column:

```
OldFaithfuldata = ExampleData[{"Statistics", "OldFaithful"}];
```

```
{duration, waitingtime} = Transpose[OldFaithfuldata];
```

```
GraphicsRow[
 {Histogram[duration, 12, "PDF", ChartElementFunction → "FadingRectangle",
   PlotLabel → Style["Duration", 10, Blue]], Histogram[waitingtime,
   12, "PDF", ChartElementFunction → "GradientRectangle",
   PlotLabel → Style["Waiting Time", 10, Blue]]}, ImageSize → Medium]
```

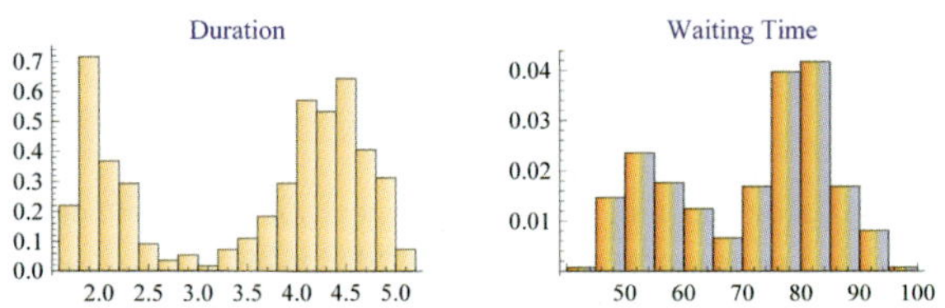

- We can measure the correlation between the two variables.

```
Correlation[duration, waitingtime]
```

```
0.900811
```

- They are highly correlated. To visualize it we can use **PairedHistogram** after rescaling the data.

```
sclData1 =
  With[{dat1 = MinMax[#] &[duration], dat2 = MinMax[#] &[waitingtime]},
   Rescale[#, dat1, dat2] & /@ duration];
```

```
PairedHistogram[sclData1, waitingtime, 12, "PDF",
 ChartElementFunction → "GlassRectangle", ChartLabels →
  {Style["Scaled Duration", "Text"], Style["Waiting Time", "Text"]}]
```

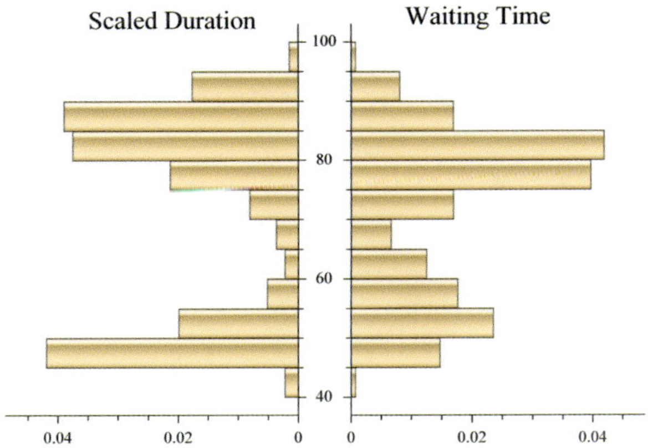

6.3 Probability Distributions

```
Clear["Global`*"]
```

6.3.1 Data and Probability Distributions

Many processes in biology, economics, engineering, finance and many other scientific and technical fields can be modeled using probability distributions.

The best known and probably the most commonly occurring distribution in nature is the Normal or Gaussian (although it was not Carl F. Gauss who discovered it).

- Using the free-form input we can easily get information about it.

```
WolframAlpha["normal distribution definition",
 {{"DefinitionPod:MathWorldData", 1}, "Content"}]
```

A normal distribution in a variate X with mean μ and variance σ^2 is a statistic distribution with probability density function

$$P(x) = \frac{1}{\sigma\sqrt{2\pi}} e^{-(x-\mu)^2/(2\sigma^2)}$$

on the domain $x \in (-\infty, \infty)$. While statisticians and mathematicians uniformly use the term "normal distribution" for this distribution, physicists sometimes call it a Gaussian distribution and, because of its curved flaring shape, social scientists refer to it as the "bell curve." Feller uses the symbol $\varphi(x)$ for $P(x)$ in the above equation, but then switches to $\mathrm{n}(x)$ in Feller.

One of the main features of probability distributions (also known as probability density functions for continuous variables) is that if certain experimental data follow a specific distribution, the difference between the theoretical distribution and the experimental one will get smaller and smaller as the number of experimental data points gets larger and larger.

- In this example we use `Histogram` to display a histogram of experimental data (in this case simulated data) following a normal probability distribution with mean 0 and standard deviation 1, commonly written as N(0,1), and compare it with its theoretical distribution. We place the command inside `Manipulate` using *n,* the number of data points, as the parameter. We can clearly see that the shape of the histogram gets closer to the probability density function (PDF) of the theoretical N(0,1) as the number of simulated data points increases.

```
Manipulate[
 Histogram[RandomVariate[NormalDistribution[0, 1], a], Automatic, "PDF",
  PlotRange → {{-4, 4}, {0, 0.45}}, AxesOrigin → {-4, 0}, Epilog → First@
    Plot[PDF[NormalDistribution[0, 1], x], {x, -4, 4}, PlotStyle → Red]],
 {{a, 1000, "Number of data points"}, 50, 10000}]
```

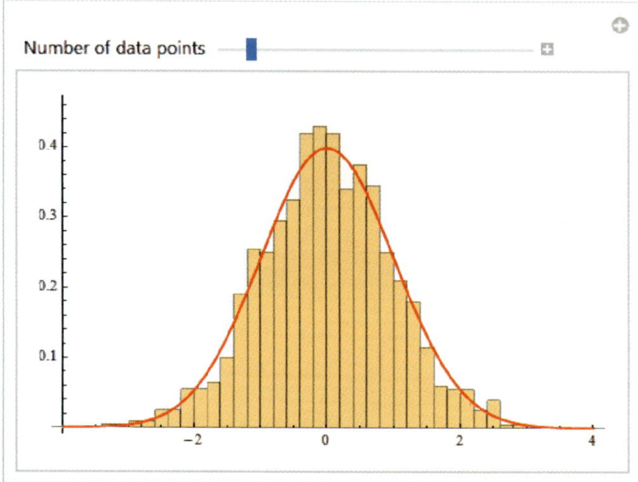

The number of probability distributions (guide/ParametricStatisticalDistributions) available in *Mathematica* is substantially higher than the one available in other well-known statistical programs. Additionally, *Mathematica*'s symbolic capabilities enable us to build our own distributions using variations of the available ones.

The new function FindDistribution can be applied to find the distribution of your data.

- Create a list of normal distributed random:

```
RandomVariate[NormalDistribution[0, 1], 100];
```

- Find the underlying distribution from the data:

```
FindDistribution[%]
```

```
NormalDistribution[0.0488666, 0.881174]
```

6.3.2 Properties

Many characteristics of distributions such as the probability density function (PDF) or the cumulative distribution function (CDF) can be computed easily.

- In this example we show the PDF and CDF functions of a *t* Student distribution for a random variable X with *ν* degrees of freedom.

```
PDF[StudentTDistribution[v], x]
```

$$\frac{\left(\frac{\nu}{x^2+\nu}\right)^{\frac{1+\nu}{2}}}{\sqrt{\nu}\ \text{Beta}\left[\frac{\nu}{2}, \frac{1}{2}\right]}$$

```
CDF[StudentTDistribution[ν], x]
```

$$\begin{cases} \frac{1}{2}\,\text{BetaRegularized}\left[\frac{\nu}{x^2+\nu}, \frac{\nu}{2}, \frac{1}{2}\right] & x \le 0 \\ \frac{1}{2}\left(1 + \text{BetaRegularized}\left[\frac{x^2}{x^2+\nu}, \frac{1}{2}, \frac{\nu}{2}\right]\right) & \text{True} \end{cases}$$

- The next command displays the PDF and CDF functions of a binomial distribution (`BinomialDistribution`). Note that the use of `DiscretePlot` is appropriate when the distribution takes discrete values as with the binomial.

```
Show[DiscretePlot[PDF[BinomialDistribution[20, .3], i], {i, 0, 20}],
 DiscretePlot[CDF[BinomialDistribution[20, .3], i],
  {i, 0, 20}, ExtentSize → Right], PlotRange → All]
```

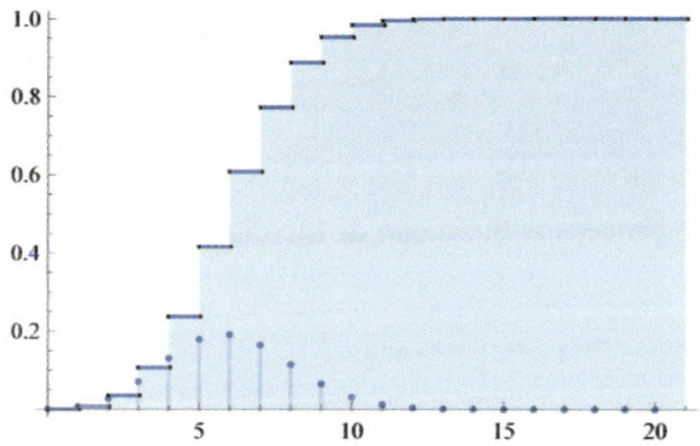

- Here several statistical parameters (mean, variance, skewness and kurtosis) for the noncentral χ^2 distribution are computed.

```
{Mean[#], Variance[#], Skewness[#], Kurtosis[#]} &[
 NoncentralChiSquareDistribution[ν, λ]]
```

$$\left\{\lambda + \nu,\ 4\lambda + 2\nu,\ \frac{2\sqrt{2}\,(3\lambda + \nu)}{(2\lambda + \nu)^{3/2}},\ 3 + \frac{12\,(4\lambda + \nu)}{(2\lambda + \nu)^2}\right\}$$

`Expectation`[*expr*, *x* ≈ *dist*] (to write ≈ you can use the palettes or the shortcut: **Esc key** "dist" **Esc key**) calculates the mathematical expectation of *expr* assuming that *x* follows a probability distribution *dist*.

- The moment of order 2 of a Poisson distribution is:

```
Expectation[x^2, x ≈ PoissonDistribution[μ]]
```

$\mu + \mu^2$

- If we want to calculate the *n-th* moment, we have to make certain assumptions about *n* (*n* must be an integer greater than zero).

```
Expectation[x^n, x ≈ PoissonDistribution[μ],
 Assumptions → n ∈ Integers && n > 0]
```

```
BellB[n, μ]
```

- The characteristic function of a distribution, such as Laplace, can be computed as follows:

```
CharacteristicFunction[LaplaceDistribution[μ, β], t]
```

$$\frac{e^{i t \mu}}{1 + t^2 \beta^2}$$

Example: A pacemaker has a redundant system consisting of independent units. Suppose that the average life of each unit is 10 years, and when a unit fails, the pacemaker is not replaced.

- The most commonly used distribution for modeling equipment failure is the Gamma distribution (`GammaDistribution`). The case in the example can be modeled with the following instruction:

```
gamma = GammaDistribution[2, 10];
```

- The graph of its density function is as follows:

```
Plot[PDF[gamma, x], {x, 0, 50}, Filling → Axis]
```

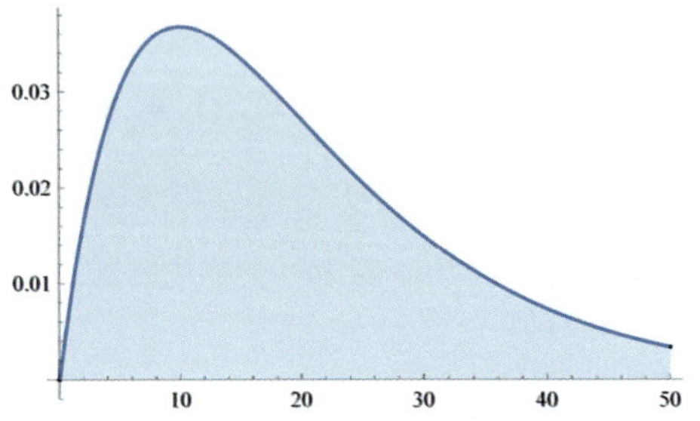

- The average time to failure of the entire process is the mean of the distribution:

```
Mean[gamma]
```

```
20
```

- The probability that the pacemaker will still be working after 20 years is:

```
Probability[x ≥ 20, x ≈ gamma] // N
```

```
0.406006
```

6.3.3 Multivariate Data

Many empirical data are represented by more than one random variable; we will refer to them as multivariate data. For the calculation of some of the properties of multivariate distributions it may be necessary to load the `` MultivariateStatistics` `` package first.

- Let's use the "Black Cherry Trees" dataset included in `ExampleData`. First, we show a description of its contents.

```
ExampleData[ {"Statistics", "BlackCherryTrees"}, "LongDescription"]
```

```
Measurements of the girth, height and
  volume of timber in 31 felled black cherry trees.
```

- We choose only the data.

```
multidata = ExampleData[ {"Statistics", "BlackCherryTrees"}];
```

- We can compute the most common statistics (mean, variance, etc.) the usual way. In this case there are three results, one for each variable.

```
{Mean[#], StandardDeviation[#], Skewness[#], Kurtosis[#],
   Quantile[#, .6], InterquartileRange[#]} &[multidata]
```

$$\left\{\{13.2484, 76, 30.171\}, \left\{3.13814, \sqrt{\frac{203}{5}}, 16.4378\right\}, \left\{0.526316, -\frac{159\sqrt{\frac{93}{406}}}{203}, 1.06436\right\}, \left\{2.44421, \frac{200353}{82418}, 3.46605\right\}, \{13.7, 79., 27.4\}, \{4.6, 8., 18.55\}\right\}$$

- The covariance and correlation are calculated to explore the relationships between the variables. The output in both cases is a 3×3 matrix from which we can infer the existence of correlation in the dataset.

```
Covariance[multidata] // MatrixForm
```

$$\begin{pmatrix} 9.84791 & 10.3833 & 49.8881 \\ 10.3833 & 40.6 & 62.66 \\ 49.8881 & 62.66 & 270.203 \end{pmatrix}$$

```
Correlation[multidata] // MatrixForm
```

$$\begin{pmatrix} 1. & 0.51928 & 0.967119 \\ 0.51928 & 1. & 0.59825 \\ 0.967119 & 0.59825 & 1. \end{pmatrix}$$

- You can have access to additional functions for multivariate analysis through the `MultivariateStatistics` package or using the new functions included in *Mathematica* 11 (Compatibility/tutorial/MultivariateStatistics).

```
Needs["MultivariateStatistics`"]
```

- We calculate the multivariate skewness (MultivariateSkewness) and kurtosis (`MultivariateKurtosis`).

```
{MultivariateSkewness[multidata], MultivariateKurtosis[multidata]}
```

```
{4.06069, 14.6777}
```

- Since *Mathematica* 11 a multivariate normality Baringhaus–Henze–Epps–Pulley (BHEP) test is available.

```
htd = BaringhausHenzeTest[multidata, "HypothesisTestData"];
```

```
htd["PValue"]
```

```
0.0375973
```

```
htd["TestDataTable"]
```

	Statistic	P-Value
Baringhaus-Henze	0.906272	0.0375973

6.3.4 Statistical Graphs

A fundamental aspect of statistics is the graphical representation of data. This section will present several types of graphs for displaying multivariate data. For additional information see: guide/StatisticalVisualization.

> To build statistical graphs load the palette : **Palettes ► Chart Elements Schemes**.

- Here we plot the PDF *Student's t* distribution for two variables.

```
Plot3D[PDF[MultivariateTDistribution[{{1, 1/2}, {1/2, 1}}, 20], {x, y}],
 {x, -3, 3}, {y, -3, 3}]
```

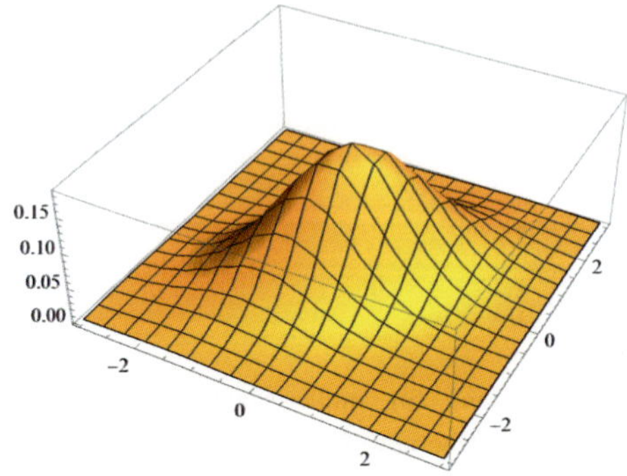

Some statistical studies may contain simulated instead of empirical data. In those cases, random number generators are used. *Mathematica* has the functions `RandomReal` and `RandomInteger` to generate random numbers. Since these functions use algorithms, strictly speaking they don't provide truly random data. However, in most situations we can assume that they do.

- We generate random numbers from a multivariate Poisson distribution. If you remove the ";" you will notice that the result is a list of sublists, that can be interpreted as a matrix with three columns:

```
multidata =
  RandomVariate[MultivariatePoissonDistribution[1, {2, 3, 4}], 125];
```

- We represent each sublist with a point. We include a red point to show the mean.

```
Show[{ListPointPlot3D[multidata],
  Graphics3D[{Red, PointSize[Large], Point[Mean[multidata]]}]}]
```

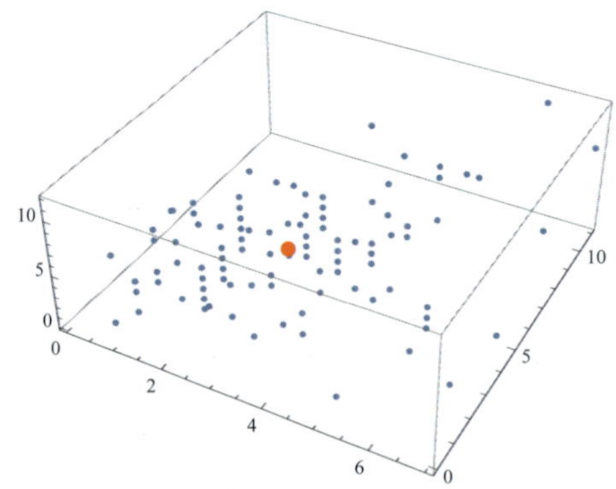

- The next graph shows a histogram of the first two dimensions of the random data previously generated (**multidata**) along with a histogram of 5,000 pairs of normally distributed random numbers (Remember that to extract list elements we use: `Part` or `[[...]]`).

```
GraphicsRow[{Histogram3D[multidata[[All, 1 ;; 2]], 5], Histogram3D[
   RandomReal[NormalDistribution[], {5000, 2}]]}, ImageSize → Medium]
```

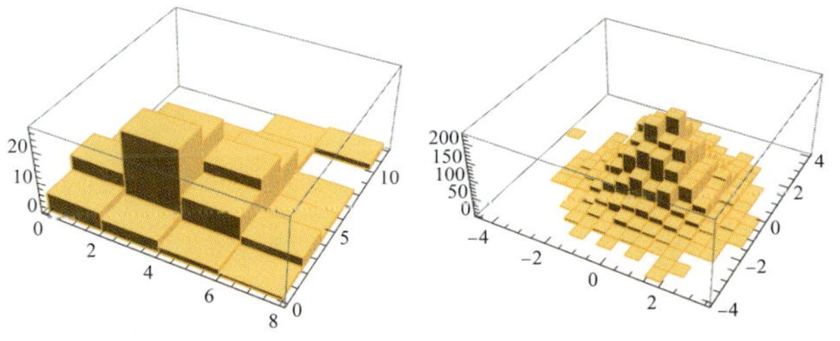

- An alternative way of representing the same data with the edges smoothed would be by using SmoothHistogram3D and SmoothDensityHistogram (the change in color is associated to the actual data value).

```
Row[{SmoothHistogram3D[multidata[[All, 1 ;; 2]], ImageSize → Medium],
  SmoothDensityHistogram[multidata[[All, 1 ;; 2]], ImageSize → Medium]}]
```

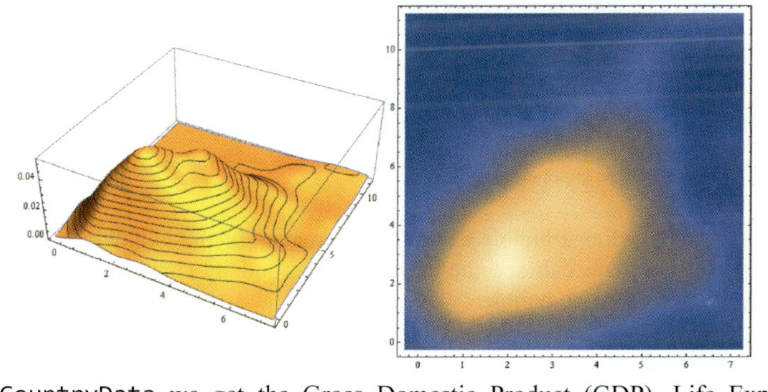

- With CountryData we get the Gross Domestic Product (GDP), Life Expectancy and Population for all the countries in South America:

```
data1 = Table[Tooltip[{(CountryData[c, #1] &) /@
      {"LifeExpectancy", "GDPPerCapita", "Population"}},
    Column[{CountryData[c, "Name"], CountryData[c, "Population"]}]],
   {c, CountryData["SouthAmerica"]}];
```

- The countries that don't have available information are deleted:

```
data2 = DeleteMissing[data1, 1, 3]
```

```
{{{ 76.305 yr , $12509.5 per person per year , 41827217 people }},
 {{ 67.26 yr , $3124.08 per person per year , 10573218 people }},
 {{ 73.937 yr , $11726.8 per person per year , 201700544 people }},
 {{ 79.955 yr , $14528.3 per person per year , 17722868 people }},
 {{ 74.038 yr , $7903.93 per person per year , 48770570 people }},
 {{ 76.471 yr , $6345.84 per person per year , 15255149 people }},
 {{ 77.121 yr , $8300. per person per year , 249408 people }},
 {{ 66.295 yr , $4053.9 per person per year , 760972 people }},
 {{ 72.259 yr , $4712.82 per person per year , 6913774 people }},
 {{ 74.826 yr , $6541.03 per person per year , 30418591 people }},
 {{ 71.019 yr , $9680.12 per person per year , 543497 people }},
 {{ 77.23 yr , $16806.8 per person per year , 3415252 people }},
 {{ 74.633 yr , $12771.6 per person per year , 30786062 people }}}
```

- The result is displayed as a bubble chart. The diameters of the bubbles are proportional to the population size. You can see the country name and its corresponding population by putting the mouse over each bubble:

```
BubbleChart[data2, FrameLabel → {"GDP/capita", "Life expectancy"}]
```

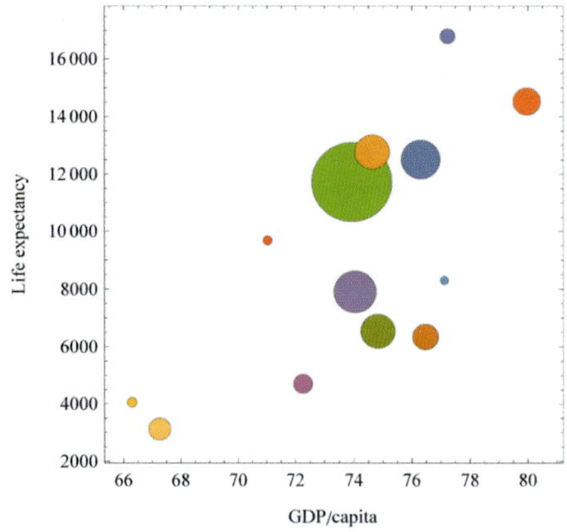

- We can also create visualizations from custom data. In the following example, the word size increases based on its frequency of appearance in a text. In this case we use the first 500 characters of the *Declaration of Independence* included in `ExampleData`.

```
Text[
 Row[With[{data = ReadList[StringToStream[StringTake[ExampleData[{"Text",
          "DeclarationOfIndependence"}], 500]], Word]},
   Table[Style[data[[i]], 10 N[Log[Count[Take[data, i], data[[i]]]]]],
    {i, Length[data]}]], " "]]
```

the to the of the the and to which the of and of to the of the which them to the to that that they are

- Create a bar chart of the frequency of characters:

```
letters = LetterCounts[
  ExampleData[{"Text", "DeclarationOfIndependence"}], IgnoreCase → True]
```

```
<|e → 928, t → 686, o → 578, a → 540, n → 537, s → 519, i → 495, r → 489, h → 393,
 l → 278, d → 266, u → 218, c → 204, f → 190, m → 178, p → 151, g → 150,
 w → 117, b → 105, y → 91, v → 76, j → 33, k → 21, x → 10, q → 6, z → 4|>
```

```
BarChart[letters, ChartLabels → Keys[letters]]
```

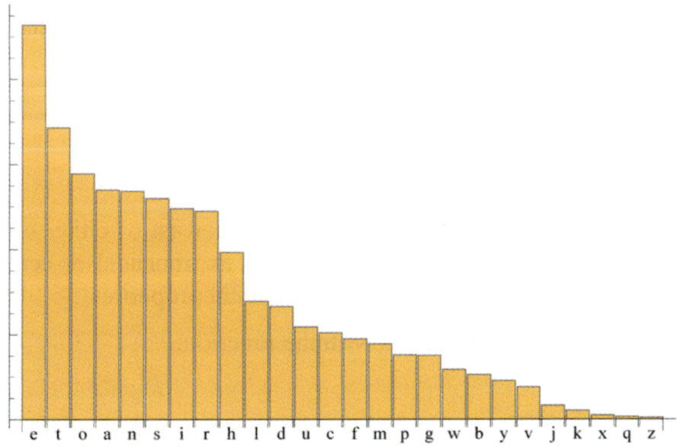

6.3.5 Mixture Distributions

This section discusses how to use mixture distributions to solve problems where it is necessary to combine several distributions.

If we have two normal distributions (or any other type), we can obtain the combined distribution with the following command:

```
MixtureDistribution[{p1, p2}, {N[m1, s1] , N[m2, s2]}]
```

where `p1, p2` represent the proportion of N[m1,s1] and N[m2,s2] respectively.

- Let's build a mixture distribution using the "OldFaithful" dataset again.

```
{duration, waitingtime} =
  Transpose[ExampleData[{"Statistics", "OldFaithful"}]];
```

```
mixDist1 = EstimatedDistribution[waitingtime, MixtureDistribution[{p1, p2},
   {NormalDistribution[m1, s1], NormalDistribution[m2, s2]}]]
```

```
MixtureDistribution[{0.639114, 0.360886},
 {NormalDistribution[80.0911, 5.86773],
  NormalDistribution[54.6149, 5.87122]}]
```

- Now let's plot together both, the original data and the mixture distribution, approximating it:

```
Show[{Histogram[waitingtime, 25, "PDF"],
  Plot[PDF[mixDist1, x], {x, 41, 100}, PlotStyle → Red]}]
```

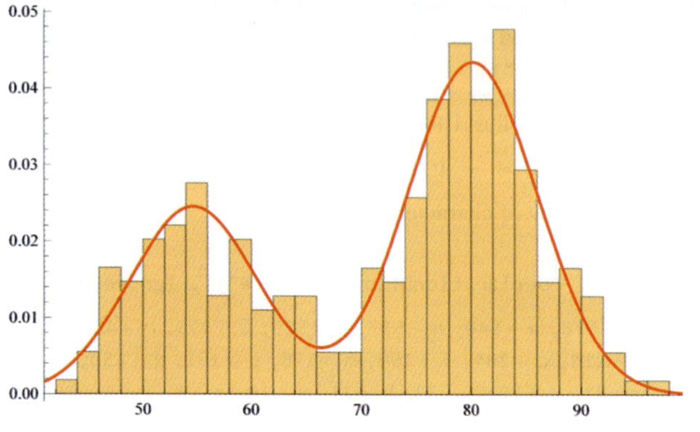

The new distribution can be used for statistical computations such as:

- The probability that the time between two emissions will be 80 minutes or more:

```
Probability[x ≥ 80, x ≈ mixDist1]
```

```
0.323517
```

6.3.6 Derived Distributions

Derived distributions are created from basic ones and they can consist of either combinations of them or special cases of individual basic distributions such as truncated or censored ones. As in the case of the original ones, we can calculate their statistical properties.

Mathematica enables us to get derived distributions with the function: `TransformedDistribution`.

- Consider the case of a normal random variable x with mean μ and standard deviation σ transformed into Exp[x]. We know that this type of transformation will generate a lognormal probability distribution and that's what *Mathematica* returns:

```
TransformedDistribution[Exp[x], x ≈ NormalDistribution[μ, σ]]
```

```
LogNormalDistribution[μ, σ]
```

- What is the derived distribution of $u + v$ with u and v being Poisson random variables with mean μ_1 and μ_2 respectively?

```
TransformedDistribution[ u + v,
 {u ≈ PoissonDistribution[μ1], v ≈ PoissonDistribution[μ2]}]
```

```
PoissonDistribution[μ1 + μ2]
```

- The same command can be used for more complex cases such as:

```
SetAttributes[fun, HoldFirst]; fun[x_] := {HoldForm[x], x};
Grid[Map[Style[#, ScriptLevel -> 0] &,
   Join[{{"Transformed Distribution", "Simplified Distribution"}},
    { fun[TransformedDistribution[ Min[Subscript[X, 1], Subscript[X, 2]],
       {Subscript[X, 1] ≈ BernoulliDistribution[Subscript[p, 1]],
        Subscript[X, 2] ≈ BernoulliDistribution[Subscript[p, 2]]}]],
     fun[TransformedDistribution[1 / X,
       X ≈ LogLogisticDistribution[γ, σ]]],
     fun[TransformedDistribution[k * X, X ≈ ChiDistribution[ν]]], fun[
      TransformedDistribution[1 / X, X ≈ BetaPrimeDistribution[a, b]]]}],
   {2}], Dividers -> All, Spacings -> {4, 2},
  Background -> {None, {{None, GrayLevel[.9]}},
    {{1, 1} -> Hue[.6, .4, 1], {1, 2} -> Hue[.6, .4, 1]}},
  BaseStyle -> {FontFamily -> Times, FontSize -> 13},
  Alignment -> {Center, Center}] // TraditionalForm
```

Transformed Distribution	Simplified Distribution
TransformedDistribution[min(X_1, X_2), {$X_1 \approx$ BernoulliDistribution[p_1], $X_2 \approx$ BernoulliDistribution[p_2]}]	BernoulliDistribution[$p_1\ p_2$]
TransformedDistribution[$\frac{1}{X}$, $X \approx$ LogLogisticDistribution[γ, σ]]	LogLogisticDistribution[γ, $\frac{1}{\sigma}$]
TransformedDistribution[$k\,X$, $X \approx$ ChiDistribution[ν]]	NakagamiDistribution[$\frac{\nu}{2}$, $k^2\,\nu$]
TransformedDistribution[$\frac{1}{X}$, $X \approx$ BetaPrimeDistribution[a, b]]	BetaPrimeDistribution[b, a]

- The following code defines a CDF that contains jump discontinuities as a result of combining continuous and discrete distributions:

```
cdf = CDF[ MixtureDistribution[{1 / 2, 1 / 2}, {LaplaceDistribution[0, 1],
     TransformedDistribution[x , x ≈ BinomialDistribution[1, 1 / 2]]}], z]
```

$$\begin{cases} \frac{e^z}{4} & z < 0 \\ \frac{1}{2} + \frac{1}{2}\left(1 - \frac{e^{-z}}{2}\right) & z \geq 1 \\ \frac{1}{2}\left(1 - \frac{e^{-z}}{2}\right) + \frac{1}{2}\,\text{BetaRegularized}\left[\frac{1}{2}, 1 - \text{Floor}[z], 1 + \text{Floor}[z]\right] & \text{True} \end{cases}$$

```
Plot[cdf, {z, -4, 4}, PlotTheme -> "Detailed",
 Filling -> Axis, ImageSize -> Medium, PlotLegends -> None]
```

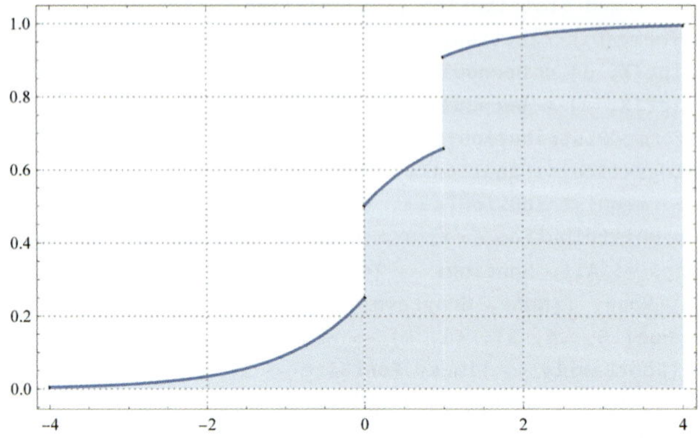

Example: A high-speed train covers 340 km at an average speed of 245 km/h. Knowing that the speed follows a normal distribution with a standard deviation of 10 km/h, what is the average time that it takes the train to cover such distance?

- In this case the idea is to transform the distribution of the speed into the distribution of the travel time. To reduce the computation time we use the `Assumptions` option to specify the condition $v > 0$:

```
travellingtime = TransformedDistribution[340 / v,
   v ≈ NormalDistribution[245, 10], Assumptions → v > 0];
```

- We calculate the average travel time. Since the speed is in km/h, we multiply the result by 60 to convert it to minutes:

```
Chop[60 Mean[travellingtime] // N] "minutes"
```

```
83.4047 minutes
```

Often in experiments, there are values that cannot be measured, as they fall outside the detection range of the measurement technique. When the observation falls below certain range, we say that the value is below the lower detection limit (LDL) and if it falls above the detection range, we state the value is above the upper detection limit (UDL).

When situations like these happen, in which we don't have the values outside the measurement range of the equipment, we find ourselves with what is known as censored distribution functions.

For calculations related to these kind of distributions we have the command
`CensoredDistribution[{{`x_{min}, x_{max}`}, {`y_{min}, y_{max}`}, ...}, `*dist*`]`

where `{{`x_{min}, x_{max}`}, {`y_{min}, y_{max}`}, ...}` refers to the limits outside of which our data are either truncated or censored.

- As an example of a left-censored distribution, we can consider the case of alpha decay measurements. Radioactive decay usually follow a Poisson distribution and are expressed in becquerels (Bq) (1 Bq representing 1 disintegration per second).
  Let's supposed that a Poisson distribution has a mean of 15 Bq and that we cannot measure values lower than 8 Bq. This process would be modeled as follows:

```
leftcensored = CensoredDistribution[{8, ∞}, PoissonDistribution[15]];
```

```
DiscretePlot[Evaluate[PDF[leftcensored, x]], {x, 0, 30}, Filling→Axis]
```

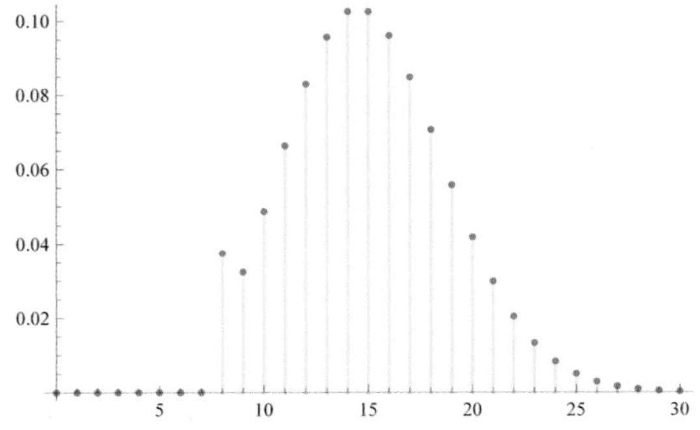

In other cases we may eliminate some data points from a measurement dataset. For example: we manufacture parts and reject those with a weight that is lower than a given value. This is a typical example of a truncated distribution: we take individual measurements and reject those that fall below a certain limit.

To deal with this situation we can use:

`TruncatedDistribution[{{`x_{min}, x_{max}`}, {`y_{min}, y_{max}`}, ...}, `*dist*`]`).

A more complicated case is when we have composite distributions consisting of combinations of independent individual distributions. In these cases we use `ProductDistribution` to get the joint distribution. We can even combine censored or truncated distributions.

- For example, let's model a joint distribution consisting of two independent Poisson distributions: one with mean 5 and an UDL of 12 and another one with mean 6 and a LDL of 2.

```
product =
  ProductDistribution[PoissonDistribution[5], PoissonDistribution[6]];
censored = CensoredDistribution[{{-∞, 12}, {2, ∞}}, product];
```

- We graphically represent the resulting function:

```
DiscretePlot3D[Evaluate[CDF[censored, {x, y}]],
 {x, 0, 12}, {y, 0, 12}, PlotRange→All, AxesLabel→Automatic]
```

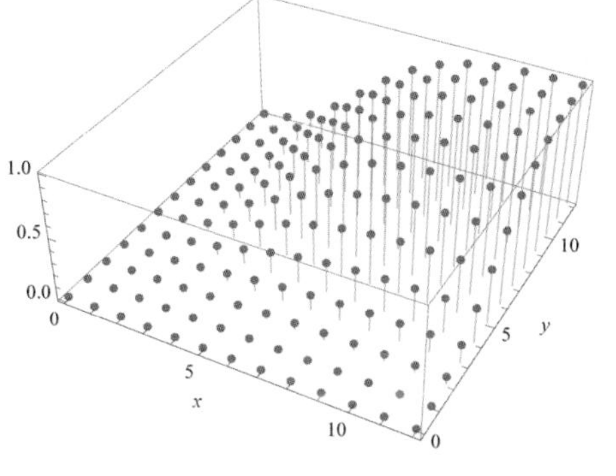

6.3.7 Empirical and Adjusted Distributions

In this example, based on a case study from a Wolfram virtual seminar, data from **WeatherData** is used to investigate the feasibility of using wind to generate electricity for

Salamanca (Spain).

- Find the name of the weather station near Salamanca:

```
wsSalamanca = WeatherData[{"Salamanca", "Spain"}, "StationName"]
```

```
LESA
```

- Get the average daily wind speed, in km/h, for the area beginning in the year 2000 and eliminate incomplete records. We use DeleteMissing[list, 1, 2] to remove any sublists containing missing data within the first 2 levels. We also use "Path" to extract Time-value pairs:

```
wsData =
  WeatherData[wsSalamanca, "MeanWindSpeed", {{2000, 1, 1}, Date[], "Day"}]
```

```
TimeSeries[ Time: 01 Jan 2000 to 29 Nov 2016
            Data points: 6148 ]
```

```
wData = DeleteMissing[wsData["Path"], 1, 2];
```

- Let's display the first five elements of our list:

```
Take[wData, 5]
```

```
{{3155673600, 2.78}, {3155760000, 2.78},
 {3155846400, 2.04}, {3155932800, 2.04}, {3156019200, 2.22}}
```

- The entries are ordered pairs containing the date and the corresponding average wind speed. The figures assigned to the dates correspond to the number of seconds elapsed since January 1, 1900. The reason behind this way of representing time is that *Mathematica* often uses `AbsoluteTime`. By default it gives us the absolute time at the moment the function is executed.

```
AbsoluteTime[]
```

$3.6896193801983815\times10^{9}$

- We can show it in a date format:

```
DateList[AbsoluteTime[]]
```

```
{2016, 12, 1, 22, 16, 20.2009}
```

- We need to extract the wind speeds from the data. We use the **Part** function to get a list of the daily average speeds, convert them to new units, and visualize the result with histograms.

```
wndData = 1000 wData[[All, 2]] / 60²;
GraphicsRow[{histplotpdf, histplotcdf} = {Histogram[wndData, 45,
    "PDF", PlotLabel → Style["Daily average speed", "Title", 12]],
   Histogram[wndData, 45, "CDF", PlotLabel → Style[
      "Accumulated daily average speed", "Title", 12]]}, ImageSize → Large]
```

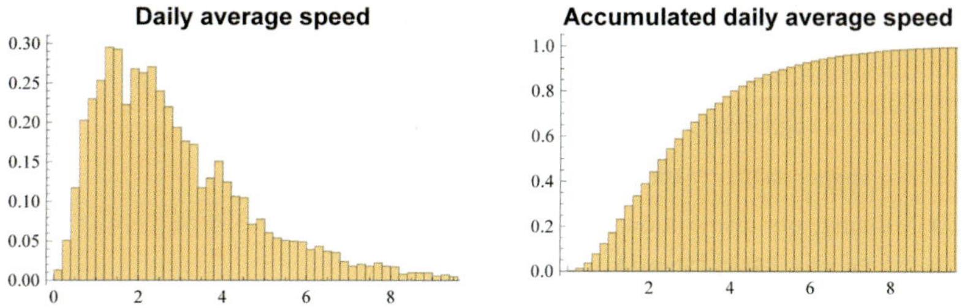

- We can fit the previous data to an existing distribution. However, we can also create a new one based on the data. This last operation can be done automatically with the function `SmoothKernelDistribution`.

```
𝒟e = SmoothKernelDistribution[wndData]
```

```
DataDistribution[ Type: SmoothKernel
                  Data points: 5989 ]
```

- While the ability to estimate a distribution is a valuable tool in statistical analysis, one must be careful when actually using it. It is possible for estimated distributions to have very small negative values in the tails. Estimated distributions of this sort behave in a similar manner to interpolating functions and can show unexpected behavior.

```
Show[{
  histplotpdf,
  estplot = Plot[PDF[𝒟e, x], {x, 0, 13}, PlotStyle → {Thick, Red}]
 }]
```

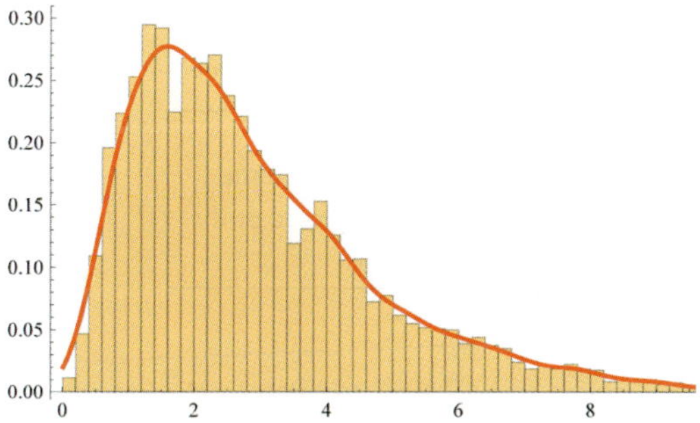

- Assuming that the generator we are using for our analysis requires at least a wind speed of 3.5 m/s to generate electricity, we can use the **Probability** function to estimate the proportion of time the generator will be contributing to the grid.

```
Probability[x > 3.5, x ≈ 𝒟e]
```

```
0.298856
```

- An alternative method would be to estimate distribution parameters and construct the distribution. The graphical representation of the data suggests that the wind speed could be approximated using a `LogNormalDistribution`. However, we need to sort the records first and if there are zeros remove them since they are not compatible with the Lognormal distribution (in this case only the first element after sorting is zero).

```
params = FindDistributionParameters[
  Drop[Sort[wndData], 1], LogNormalDistribution[μ, σ]]
```

```
{μ → 0.84165, σ → 0.696617}
```

```
𝒟p = LogNormalDistribution[μ, σ] /. params
```

```
LogNormalDistribution[0.84165, 0.696617]
```

- Visualize our distribution estimate:

```
GraphicsRow[{Show[{
    histplotpdf,
    estplot = Plot[PDF[𝒟p, x], {x, 0, 14}, PlotStyle → {Thick, Red}]
   }], Show[{
    histplotcdf,
    estplot = Plot[CDF[𝒟p, x], {x, 0, 14}, PlotStyle → {Thick, Red}]}
  ]}]
```

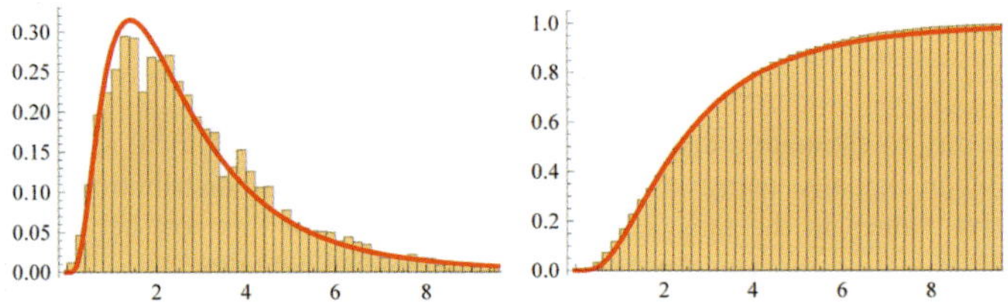

- We perform a test to assess the quality of the fit:

```
tstData =
  DistributionFitTest[Drop[Sort[wndData], 1], 𝒟p, "HypothesisTestData"];
```

```
Column[{Style[tstData["AutomaticTest"], "Text"],
  Style[tstData["TestConclusion"], "Text"]}, Frame → True]
```

CramerVonMises
The null hypothesis that the data is distributed according to the LogNormalDistribution[0.84165, 0.696617] is rejected at the 5 percent level based on the Cramér-von Mises test.

- Get the test statistic and p-value.

```
tstData["TestDataTable"]
```

	Statistic	P-Value
Cramér-von Mises	1.97958	0.0000142047

- The test tells us that the fit is not good but when there is a large number of data points, the best possible test is the graphical representation of the data alongside the density function of the estimated distribution. In that case the fit is quite reasonable. We can recalculate the probability of the wind speed being at least 3.5 m/s and see that the result is very close to the one previously obtained.

```
Probability[x > 3.5, x ≈ 𝒟p]
0.277543
```

6.3.8 Automatic Probability Calculations

In this example we use the following functions related to probability calculations: `NProbability`, `Expectation`, and `OrderDistribution`:

A person is standing on a road counting the number of cars that pass before seeing a black one. Assuming that 10% of the cars are black, find the expected number of cars that will pass before a black one appears.

- We use the geometric distribution (`GeometricDistribution`[p]) to model the number of trials until success, given a probability of success of p. The expected number of cars will be the mean of the GeometricDistribution with $p = 0.1$. It can be computed in two different ways:

```
Mean[GeometricDistribution[0.1]]
9.
Expectation[x, x ≈ GeometricDistribution[0.1]]
9.
```

Find the probability of counting 8 or fewer cars before seeing a black one.

```
NProbability[x ≤ 9, x ≈ GeometricDistribution[0.1]]
0.651322
```

Three friends are standing on a road in different places counting the number of cars until a black one appears. The winner will be the first one to spot a black car. Assuming that 10% of the cars are black, find the expected number of cars that will pass before one of the friends sees a black car.

- We use `OrderDistribution`[{*dist*, *n*}, *k*] where k represents the kth-order statistics distribution for *n* observations from the distribution *dist*.

```
Expectation[x, x ≈ OrderDistribution[{GeometricDistribution[0.1], 3}, 3]]
16.9006
```

The output indicates the average total number of cars that the friends will count before a black one appears. Since it's an average, the number doesn't have to be an integer.

What is the probability that the winner will count less than 10 cars and at least 4?

```
Probability[Conditioned[x < 10, x ≥ 4],
 x ≈ OrderDistribution[{GeometricDistribution[0.1], 3}, 3]]
0.245621
```

There are many more statistical functions available in *Mathematica* than the ones discussed in this section. You may be interested in exploring some of the latest ones such as: CopulaDistribution, MarginalDistribution and OrderDistribution.

6.3.9 Application: An Election Tie

The example in this section refers to a real scenario that happened in Spain during an extraordinary party congress discussing a highly controversial proposal. The party decided to put the proposal to a secret ballot and the participants had to either accept it or reject it. The result was a tie: 1,515 people voted Yes and 1,515 voted No. The outcome was so unexpected that the media speculated whether the result had been rigged and the blogosphere fueled a debate that included the use of statistical software about how likely a tie was in a voting process involving two options and 3,030 participants. In our analysis we make the assumption that all the participants vote for either option so there are no null or blank votes even though those kinds of votes would not affect the result.

Next, we answer the following question: What is the probability that a Yes/No vote with 3,030 people results in a tie?

These types of processes with only two possible outcomes are known as Bernoulli processes and can be modeled using a Binomial distribution: X ~ Binomial(n, p) with n in this case being the number of voters and p the probability of a Yes vote. The answer will depend on our previous knowledge regarding the value of p.

- If we don't know anything about p in advance, then the probability can take any value between 0 and 1. This is equivalent to the assumption that p follows a Uniform distribution between 0 and 1. For any value of n, the solution is:

```
sol1[n_] := Probability[x == n/2,
x ≈ ParameterMixtureDistribution[
   BinomialDistribution[n, p], p ≈ UniformDistribution[{0, 1}]]]
```

- With $n = 3030$, the probability of a tie is:

```
sol1[3030]
```

```
  1
────
3031
```

A more direct way of calculating p is to enumerate all the different possibilities: {{3030 Yes, 0 No}, {3029 Yes, 1 No},...., {1 Yes, 3029 No},{ 0 Yes, 3030 No}}. There are a total of 3,031 possible cases and since we know nothing about p, we assume that all of them have the same probability of happening. Therefore p = 1/3031 for any result and a tie is one of the potential outcomes.

- In the actual case, there was some previous knowledge. There had been earlier votes with outcomes very close to a tie. Let's assume that $\hat{p} = 0.49$ with a standard deviation of 0.02. Then we can calculate the probability assuming that p follows a Normal probability distribution: X ~ Normal(0.49, 0.02):

```
NIntegrate[PDF[NormalDistribution[0.49, 0.02], p] *
   PDF[BinomialDistribution[3030, p], 1515], {p, 0, 1}] // N
```

```
0.0054025
```

- If we'd like to perform a Bayesian analysis, it's convenient to assume that p follows a Beta distribution with parameters α and β whose actual values depend on our *prior* knowledge of p.

```
sol2[n_, α_, β_] := Probability[x == n/2,
x ≈ ParameterMixtureDistribution[
   BinomialDistribution[n, p], p ≈ BetaDistribution[α, β]]]
```

- Without previous knowledge about p, then $\alpha = 1$ and $\beta = 1$ and the result is the same as the one previously obtained.

```
sol2[3030, 1, 1]
```

$\frac{1}{3031}$

- For the secret ballot scenario, we have some prior knowledge, since the results of the previous 3 or 4 votes had been very close to a tie. We can therefore assume that p is approximately 1/2. This is equivalent to α and β taking very large values. A extreme case:

```
sol2[3030, 100000, 100000] // N

0.0143853
```

- If we knew for sure that p was exactly 1/2, this problem would be equivalent to computing the probability that after tossing a fair coin 3,030 times, we would get the same number of heads and tails.

```
PDF[BinomialDistribution[3030, 1/2], 1515] // N

0.0144938
```

In summary, the probability that a tie could happen was not that small (keeping in mind that earlier votes were close affairs). Highly unlikely events happen all the time. An extreme example occurred shortly after the vote, also in Spain. The same person won the biggest prize in the national lottery for two consecutive weeks. In this game, the probability of winning the highest prize ("el gordo") is 1/100000. This means that a priori, the probability that a person purchasing just one ticket each time will win twice in a row is $1/100000 \times 1/100000 = 10^{-10}$ quite a small likelihood. Actually, the odds of dying from lightning are higher. It's not unusual that with so much going on around us at any given moment, the small probability events are the ones drawing our attention.

6.3.10 Machine Learning Functions

One of the most exciting new capabilities of *Mathematica* is its ability to predict and classify data using functions such as: Predict, PredictorFunction, PredictorMeasurements, PredictorInformation, Classify and FeatureExtraction.

The next example, available in the documentation, shows this new functionality in action:

- Load a dataset of wine quality as a function of the wines' physical properties:

```
trainingset =
  ExampleData[{"MachineLearning", "WineQuality"}, "TrainingData"];
```

- Get a description of the variables in the dataset:

```
ExampleData[{"MachineLearning", "WineQuality"}, "VariableDescriptions"]

{fixed acidity, volatile acidity, citric acid, residual sugar,
  chlorides, free sulfur dioxide, total sulfur dioxide, density,
  pH, sulphates, alcohol} → wine quality (score between 1-10)
```

- Visualize the distribution of the "alcohol" and "pH" variables:

```
{Histogram[trainingset[[All, 1, 11]], PlotLabel → "alcohol"],
 Histogram[trainingset[[All, 1, 9]], PlotLabel → "pH"]}
```

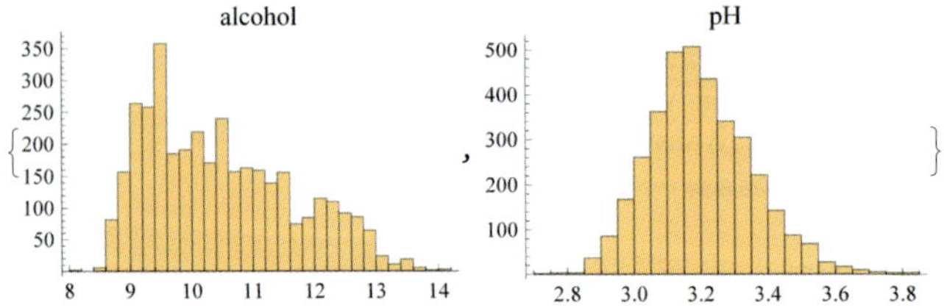

- Train a predictor on the training set:

```
p = Predict[trainingset]
```

```
PredictorFunction[ Method: NearestNeighbors
                   Feature type: NumericalVector (11) ]
```

- Predict the quality of an unknown wine:

```
unknownwine =
  {7.6, 0.48, 0.31, 9.4, 0.046, 6., 194., 0.99714, 3.07, 0.61, 9.4};
```

```
p[unknownwine]
```

```
5.4
```

- Create a function that predicts the quality of the unknown wine as a function of its pH and alcohol level:

```
quality[pH_, alcohol_] :=
  p[{7.6, 0.48, 0.31, 9.4, 0.046, 6., 194., 0.99714, pH, 0.61, alcohol}];
```

- Plot this function to have a hint on how to improve this wine:

```
Show[Plot3D[quality[pH, alcohol], {pH, 2.8, 3.8} , {alcohol, 8, 14},
  AxesLabel → Automatic, Exclusions → None], ListPointPlot3D[
  {{3.07`, 9.4`, p[unknownwine]}}, PlotStyle → {Red, PointSize[.05]}]]
```

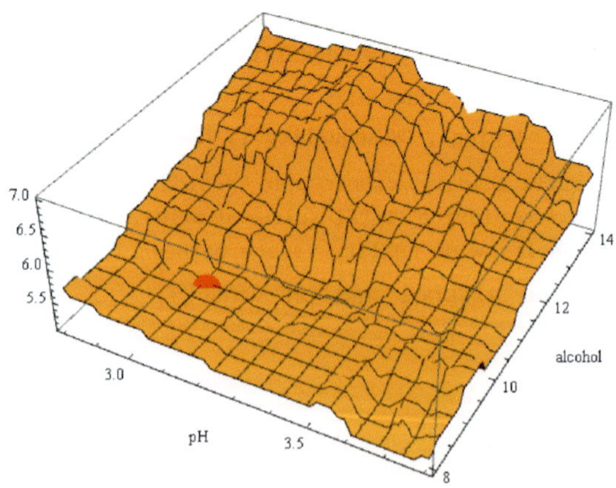

6.4 Application: Fitting Experimental Data

```
Clear["Global`*"]
```

When performing experimental studies, it's common to fit the results to a model. *Mathematica* has several functions for doing that: LinearModelFit, GeneralizedLinearModelFit, FittedModel, FindFit, Fit, NonlinearModelFit, TimeSeriesModelFit and FindFormula. For more details please consult: tutorial/StatisticalModelAnalysis.

This section will cover how to do linear and nonlinear fitting using examples. We use two files with the data we are going to fit: noisydata.xls and pulsar1257.dat. These two files must be in a subdirectory, named Data, in the directory where is notebook is located.

- We first make the subdirectory containing the data files the current working directory:

```
SetDirectory[FileNameJoin[{NotebookDirectory[], "Data"}]]
```

```
C:\Users\guill\Documents\MathematicaBeyond\Data
```

6.4.1 A Linear Regression Model

We are going to fit the data in noisydata.xls to a third-degree polynomial using `LinearModelFit`, the *Mathematica* command most commonly used when fitting linear regression models and nonlinear ones such as polynomials. This command will create a linear model of the form $\hat{y} = \beta_0 + \beta_1 f_1 + \beta_2 f_2 + \ldots$ (even though the model is linear, the f_i don't need to be linear functions; they can be, for example, polynomials.

The assumptions required for this type of fitting to be valid are that the original data y_i must be independent and normally distributed, with mean $\hat{y}_i$ and a constant standard deviation s.

- We import the data from the *xls* file.

```
data = Import["noisydata.xls", {"Data", 1}];
dataplot = ListPlot[data]
```

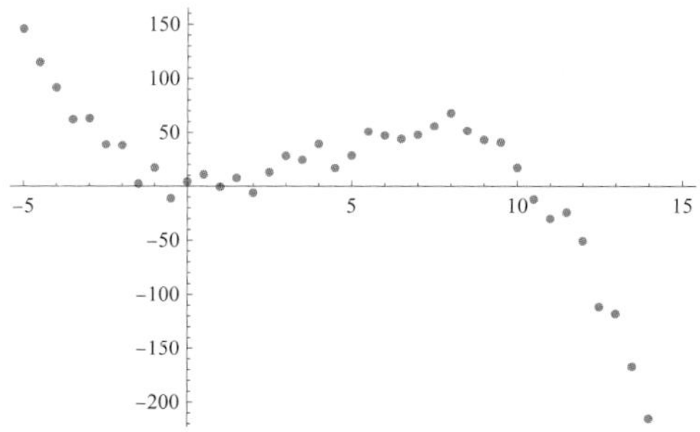

- We fit a third-degree polynomial.

```
fit = LinearModelFit[data, {1, x, x^2, x^3}, x];
```

- This is the ANOVA table:

```
fit["ANOVATable"]
```

	DF	SS	MS	F-Statistic	P-Value
x	1	165 320.	165 320.	1626.9	3.47061×10^{-32}
x^2	1	58 481.7	58 481.7	575.515	3.78263×10^{-24}
x^3	1	114 956.	114 956.	1131.27	2.41932×10^{-29}
Error	37	3759.8	101.616		
Total	40	342 517.			

- Now we graphically show both the data and the fitted function.

```
Show[dataplot,
  Plot[fit[x], {x, -5, 15}, PlotStyle → Red]]
```

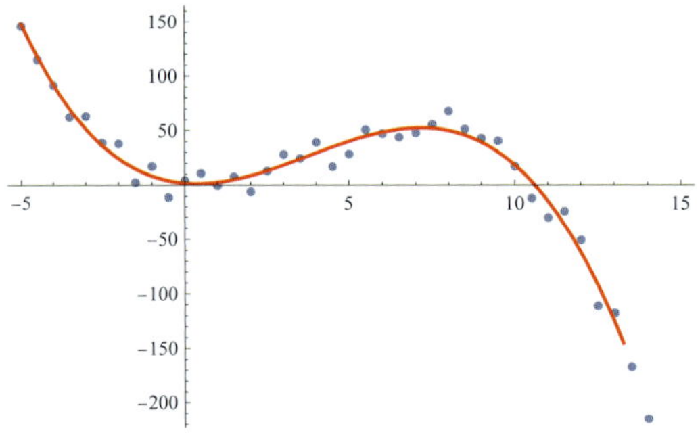

6.4.2 A Nonlinear Regression Model

In this example we fit the experimental light measurement data collected over a three-year period from pulsar PSR1257+12 (originally obtained by Alex Wolszczan and available in the file pulsar1257.dat) to the following function:

$f(t) = \epsilon + \beta \cos(t\,\theta) + \delta \cos(t\,\phi) + \alpha \sin(t\,\theta) + \gamma \sin(t\,\phi).$

For nonlinear fit we use `NonlinearModelFit`.

- We import the file:

```
data = Import["pulsar1257.dat"];
```

- Represent it graphically:

```
dataplot = ListPlot[data, AspectRatio → 1/4, PlotStyle → Red]
```

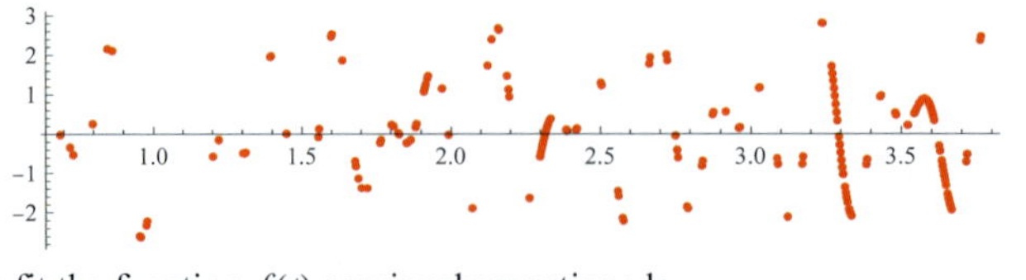

- We fit the function $f(t)$ previously mentioned:

```
pulsarfit =
 NonlinearModelFit[data, α Sin[θ t] + β Cos[θ t] + γ Sin[ϕ t] + δ Cos[ϕ t] + ϵ,
  {{α, 1}, {β, 1}, {γ, 1}, {δ, 0}, {θ, 23.31}, {ϕ, 34.64}, {ϵ, 0}}, t]
```

```
FittedModel[ 0.0769581 + 1.33261 Cos[«18» t] + «1» «1» «19» «1» - 1.29803 Sin[34.5111 t] ]
```

- We put the data and the function on the same graph to check the quality of the fit:

```
Show[Plot[pulsarfit[t], {t, 0.68, 3.76}],
 dataplot]
```

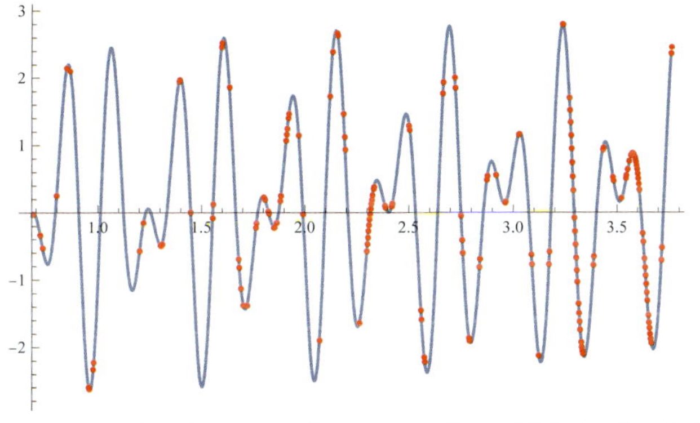

- Once the model has been created, we can perform all kinds of statistical analyses. However, it would be useful to list all the properties available for the model:

```
pulsarfit["Properties"] // Shallow
```

```
{AdjustedRSquared, AIC, AICc, ANOVATable,
 ANOVATableDegreesOfFreedom, ANOVATableEntries, ANOVATableMeanSquares,
 ANOVATableSumsOfSquares, BestFit, BestFitParameters, «40» }
```

- For example, here we display the residuals:

```
ListPlot[pulsarfit["FitResiduals"], GridLines → {None, Automatic}]
```

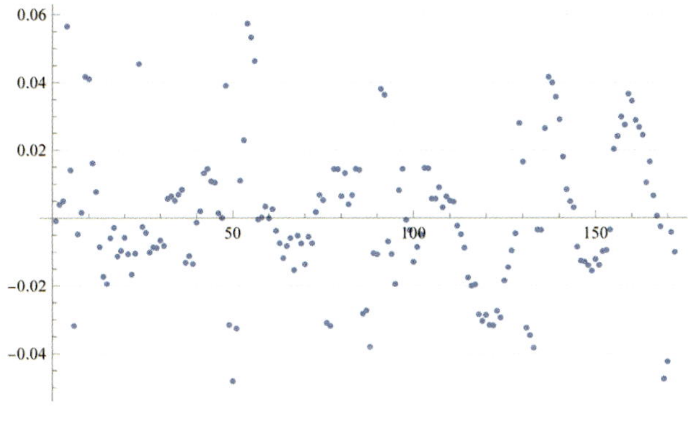

6.5 Time Series Analysis

Another example of *Mathematica*'s new functionality since Version 10 is its ability to analyze time series. *Mathematica*'s capabilities for time series analysis are so extensive that to cover them all would require a chapter on its own. In this section we are just going to see an example using weather data to estimate its future evolution.

- We start collecting the daily average temperatures in Salamanca since 1973 (the oldest record accessible with this function at the moment of writing these lines) until the end of 2014 and then display them graphically. Notice the seasonality.

```
Clear["Global`*"]
```

```
averagetempSalamanca =
  WeatherData["Salamanca", "MeanTemperature", {{1973}, {2015}, "Day"}]
```

```
TimeSeries[ Time: 01 Jan 1973 to 30 Dec 2015
            Data points: 15 683 ]
```

- *Mathematica* returns the information about temperatures in a time series format that is very useful for performing statistical calculations later on. However, we can display the actual values using `Normal`:

```
Normal[averagetempSalamanca]
```

```
{{Mon 1 Jan 1973, 0.39 °C},
 {Tue 2 Jan 1973, -2.44 °C}, {Wed 3 Jan 1973, -2.22 °C},
 {Thu 4 Jan 1973, -1.44 °C}, {Fri 5 Jan 1973, -0.39 °C},
 {Sat 6 Jan 1973, -1.72 °C}, ··· 15671 ···,
 {Fri 25 Dec 2015, 5 °C}, {Sat 26 Dec 2015, 3.17 °C},
 {Sun 27 Dec 2015, 2 °C}, {Mon 28 Dec 2015, 8.44 °C},
 {Tue 29 Dec 2015, 4.72 °C}, {Wed 30 Dec 2015, 5.5 °C}}
large output | show less | show more | show all | set size limit...
```

- The next function fits the data to a typical time series model:

```
tsm = TimeSeriesModelFit[averagetempSalamanca]
```

··· TemporalData: The data is not uniformly spaced and will be automatically resampled to the resolution of the minimum time increment.

```
TimeSeriesModel[ Family: ARMA
                 Order: {2, 7} ]
```

- We can use the fitted model to forecast the average daily temperature at the time of the query:

```
today = DateValue[Today, {"Year", "Month"}]
```

```
{2016, 8}
```

```
tsm[DatePlus[today, {1, "Month"}]]
```

```
11.4576
```

- Next, we estimate the average monthly temperatures for 12 months using the data from the last 10 years:

```
start = DateValue[DatePlus[Today, {-10, "Year"}], {"Year", "Month"}];
tspec = {start, today, "Month"};
```

```
temp = TimeSeries[WeatherData["Salamanca", "Temperature", tspec, "Value"],
   {start, Automatic, "Month"}];
```

```
DateListPlot[temp, Joined → True]
```

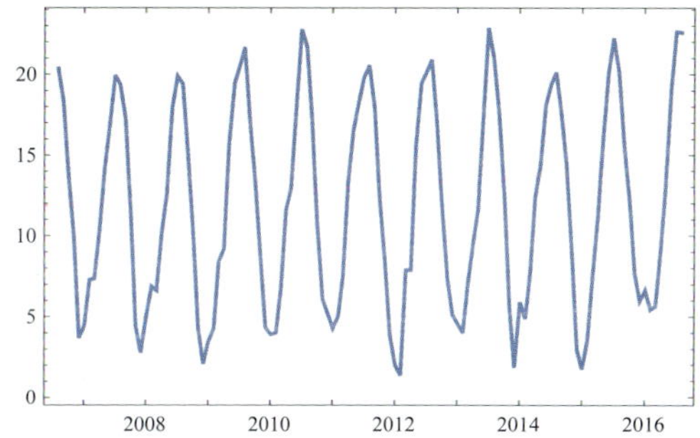

- We fit the data to a time series model:

```
tsm = TimeSeriesModelFit[temp]
```

```
TimeSeriesModel[ Family: SARIMA
                 Order: {{0, 0, 1}, {2, 1, 0}12} ]
```

- We visualize the next year's forecast for the average monthly temperature using 95% confidence bands:

```
fdates = DateRange[today, DatePlus[today, {1, "Year"}], "Month"];
```

```
forecast = tsm /@ fdates;
bands = tsm["PredictionLimits"][#] & /@ fdates;
```

```
DateListPlot[{bands⟦All, 1⟧, forecast, bands⟦All, 2⟧},
 {today, Automatic, "Month"}, Joined → True, PlotStyle → {Gray, Automatic}]
```

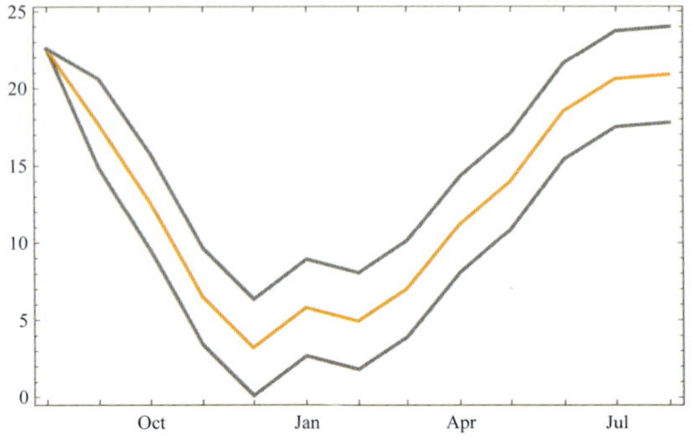

- The data records are highly correlated:

```
ListPlot[Table[AutocorrelationTest[temp, i], {i, 1, 6}], Filling → Axis]
```

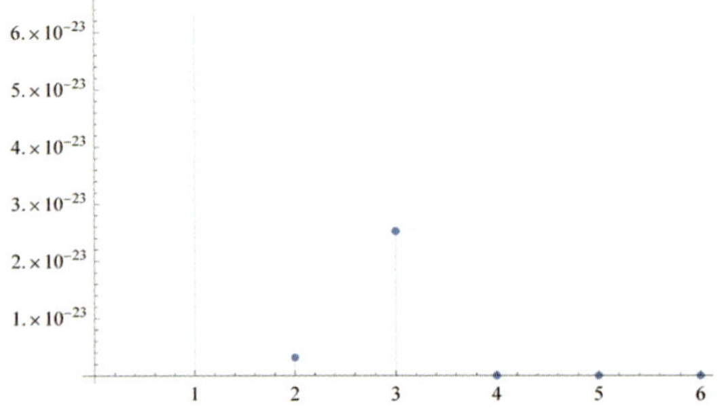

6.6 Cluster Analysis

6.6.1 Identifying Clusters

The task to identify potential data groupings or to sort things according to their similarity is known as cluster analysis.

The function `FindClusters` is very useful for this kind of analysis.

Mathematica also has several other functions for clustering based on similarity or distance measures depending on the type of data being analyzed: numerical, Boolean, string or images and colors. You can find additional details in: guide/DistanceAndSimilarityMeasures.

- The next command uses three different algorithms to measure the distance between two vectors.

```
Through[{EuclideanDistance, ManhattanDistance, ChessboardDistance}[
   {1, 2, 3, 4, 5}, {4, 7, 8, 7, 10}]] // Column
```

$\sqrt{93}$
21
5

- In this example we want to know the Boolean dissimilarity between two binary vectors.

```
DiceDissimilarity[{1, 0, 1, 0, 0, 1, 0}, {0, 0, 1, 1, 0, 0, 0}]
```

$\frac{3}{5}$

- This example calculates the Levenshtein distance (or *Edit distance*) between two text strings. This function is especially useful when analyzing genomes, for example when comparing genes from different species. This topic will be covered in more detail in Chapter 12.

```
EditDistance["This is a text string.", "This is a different text string."]
```

10

- The function `Dendrogram` plots a dendrogram from the hierarchical clustering of the elements. Alternatively, we can also use `DendrogramPlot`, a function included in the package `HierarchicalClustering`.

```
Show[Dendrogram[{1, 2, 10, 4, 8}, Left,
  DistanceFunction -> ManhattanDistance], ImageSize → Small]
```

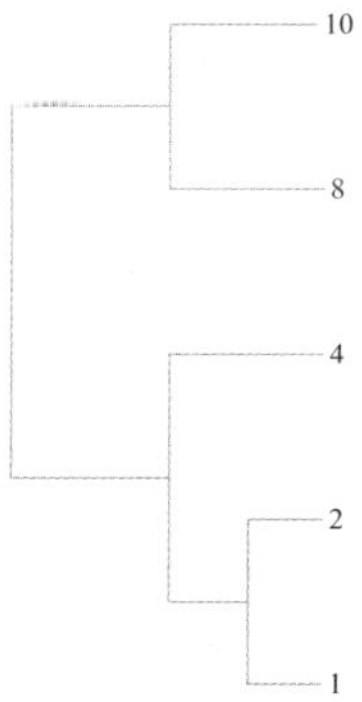

6.6.2 Finding Clusters

To find clusters we use `FindClusters`[*data*]. If we know the number *n* of clusters, we use the syntax `FindClusters`[*data, n*]. Let's see an example:

- We first simulate 10 samples from a bivariate normal distribution. Notice the use of the command `SeedRandom`[*n*]. This function generates identical random numbers for the same value of *n*. It's commonly used when we want to reproduce the results.

```
Needs["MultivariateStatistics`"]

origclusters = Round[BlockRandom[SeedRandom[123 456 789];
muvals = RandomReal[{-10, 10}, {10, 2}];
    sigmavals = Table[Block[{s1 = RandomReal[], s2 = RandomReal[], s12},
       s12 = RandomReal[] * Sqrt[s1 * s2];
       {{s1, s12}, {s12, s2}}], {10}];
    Join[MapThread[RandomReal[MultinormalDistribution[#1, #2],
         RandomInteger[{20, 75}]] &, {muvals, sigmavals}]]], 0.001];
```

- Next, we visualize the simulated clusters.

```
ListPlot[origclusters]
```

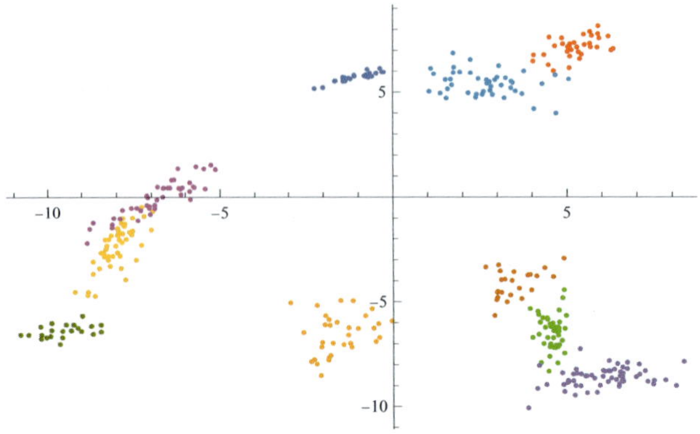

- Now we add random values to the original data. We use `FindClusters` to identify the cluster data.

```
newclusters = BlockRandom[SeedRandom[1];
   RandomSample[Flatten[origclusters, 1]]];
```

```
clusters = FindClusters[newclusters];
```

- `FindClusters` identifies 3 distinct clusters.

```
Length[clusters]
```

```
5
```

6.6.3 Visualizing Clusters

- To visualize both the original clusters and the new ones, we can proceed as follows:

```
GraphicsRow[{ListPlot[origclusters], ListPlot[clusters]}, Frame → All]
```

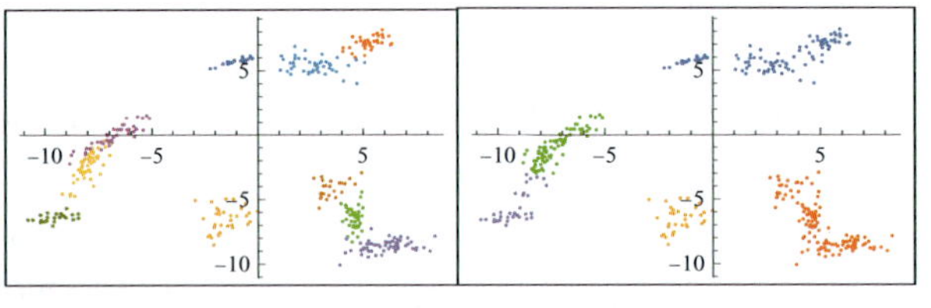

- If you know the number of clusters you can include it in the command:

```
ListPlot[FindClusters[newclusters, 4]]
```

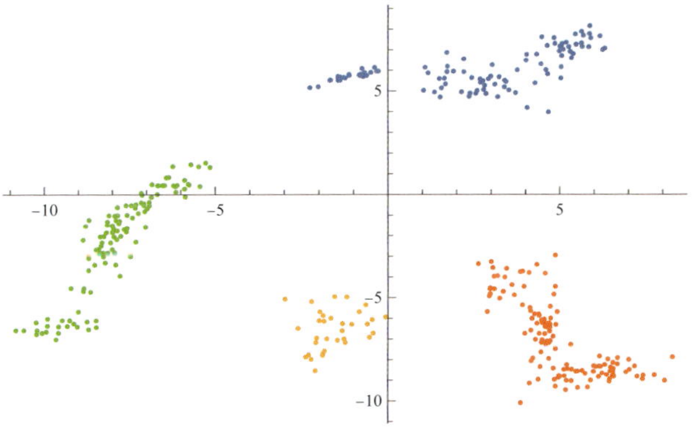

6.6.4 The Human Development Index (HDI)

The Human Development Index (HDI) is a measure of the welfare of a country used in the United Nations' Human Development Reports. It takes into account: quality of life, access to knowledge, and standard of living. The HDI is the geometric mean of the normalized indices for each of those three dimensions: Health, Education, and Income. The technical notes included in http://hdr.undp.org/sites/default/files/hdr2015_technical_notes.pdf describe the data sources for the HDI, steps to calculating the HDI and the methodology used to estimate missing values. Countries are ranked according to their HDI, which generates a classification comprising four categories: low, medium, high, and very high development. In the 2015 UN Human Development Report, the HDI intervals (based on 2014 data) corresponding to these four categories were: 0–0.55, 0.55–0.7, 0.7–0.8, and 0.8–1.

Instead of this criterion, we apply clustering techniques to determine how similar countries are with respect to the HDI. Similar approaches can be found in Julio Abad-González and Ricardo Martínez's article titled "Endogenous categorization of the human development" and published in *Applied Economics Letters* in May 2016.

- We start our analysis by downloading "Table 1: Human Development Index and its components" from http://hdr.undp.org/en/composite/HDI in XLSX format and saving the file in the Data directory as "2015StatisticalHDI.xlsx".
- Next, we import the worksheet named "HDI" containing five columns labeled: COUNTRY, Health, Education, Income, and HDI.

```
SetDirectory[FileNameJoin[{NotebookDirectory[], "Data"}]];
```

```
data = Import["2015StatisticalHDI.xlsx", {"Sheets", "HDI"}];
```

- Now, we can use the undocumented **TableView** function to see the imported data:

```
TableView[data]
```

	1	2	3	4	5	6	7
1	COUNTRY	Health	:ducation	Income	HDI		
2	Norway	0.948	0.907	0.978	0.9439		
3	ιustralia	0.96	0.932	0.913	0.9348		
4	itzerlan	0.969	0.866	0.957	0.9295		
5	Denmark	0.926	0.924	0.92	0.9233		
6	therland	0.948	0.894	0.924	0.9217		
7	Germany	0.937	0.894	0.919	0.9165		
8	Ireland	0.937	0.907	0.903	0.9155		
9	ited Sta	0.909	0.89	0.947	0.915		
10	Canada	0.954	0.875	0.913	0.9134		
11	w Zealan	0.951	0.917	0.875	0.9138		
12	ingapore	0.969	0.782	1.	0.9117		
13	ng Kong,	0.985	0.805	0.95	0.9099		
14	echtenst	0.923	0.811	1.	0.908		
15	Sweden	0.957	0.842	0.925	0.9067		
16	ited King	0.934	0.885	0.902	0.9068		
17	Iceland	0.963	0.853	0.886	0.8995		
18	rea (Rep	0.952	0.866	0.88	0.8986		

- After dropping the first row containing the column headers, we assign the remaining data to 5 different variables:

```
{countries, healthIndex, EducationIndex, IncomeIndex, hdiIndex} =
  Transpose[Drop[data, 1]];
```

- The next step is to transform hdiIndex into an `Association`: <|Country1→HDI1 ,Country1→HDI2,... |>.

```
hdi = Association[Thread[countries → hdiIndex]];
```

```
Dataset[hdi]
```

Norway	0.944
Australia	0.935
Switzerland	0.93
Denmark	0.923
Netherlands	0.922
Germany	0.916
Ireland	0.916
United States	0.915
Canada	0.913
New Zealand	0.914
Singapore	0.912
Hong Kong, China (SAR)	0.91
Liechtenstein	0.908
Sweden	0.907
United Kingdom	0.907
Iceland	0.9
Korea (Republic of)	0.899
Israel	0.895
Luxembourg	0.892
Japan	0.89

showing 1–20 of 188

- We group the countries into 4 clusters using the "KMeans" method. We use ";" to avoid displaying the output, which will be shown later.

```
hdiclusters = FindClusters[hdi, 4, Method → "KMeans" ];
```

- Figure 6.1 shows the part of the dendrogram plot obtained with Mathematica (**Dendrogram**[hdi,Left]) that corresponds to the 4th cluster, the group of the least developed countries measured by their HDIs:

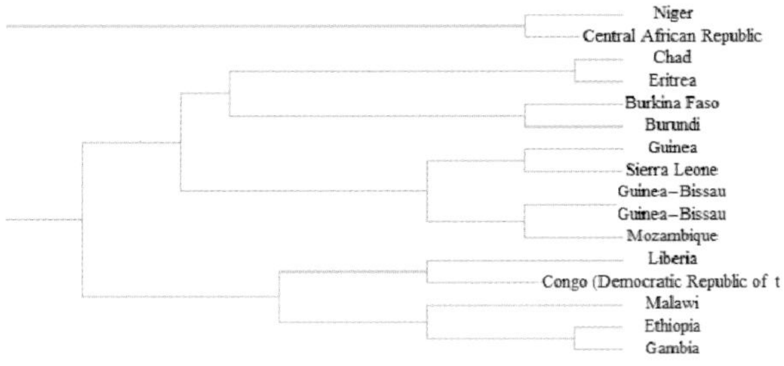

Figure 6.1 Dendrogram extract showing the cluster of the least developed countries measured by the HDI.

- If instead of the standard HDI index used in the previous example, the geometric mean of the three indices, we want to create four clusters based on the similarities of the values of the Health, Education and Income dimensions among countries, we would proceed as follows:
- First, we create a new association between the countries and the values of their indices for those three variables:

```
newhdi = Association[Thread[
   countries -> Transpose[{healthIndex, EducationIndex, IncomeIndex}]]];
newhdi // Short
```

```
<|Norway → {0.948, 0.907, 0.978}, Australia → {0.96, 0.932, 0.913},
 <<184>>, Central African Republic → {0.472, 0.341, 0.266},
 Niger → {0.637, 0.198, 0.333} |>
```

- In this case we choose the Canberra distance to find the clusters. This metric is useful for measuring distance when the data are continuous or when comparing ranked lists. To find out more, we just ask *Mathematica*:

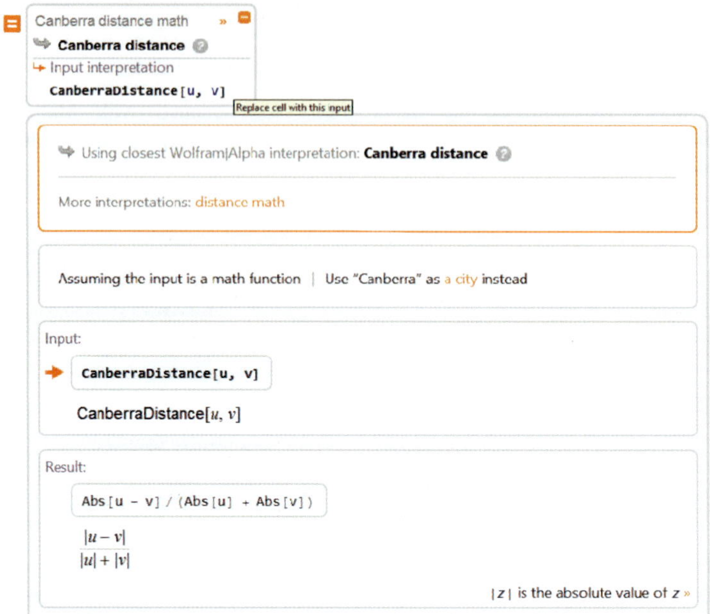

- This function is the same as in the previous example but we explicitly tell *Mathematica* what distance function we want to use:

```
canberraclusters = FindClusters[newhdi, 4,
   DistanceFunction → CanberraDistance, Method → "KMeans" ];
```

- Finally we compare both approaches:

```
Text@Grid[Transpose[Prepend[{hdiclusters, canberraclusters },
    {"Group 1", "Group 2", "Group 3", "Group 4"}]],
  Background → {{LightYellow, {White, LightBlue}}}, Dividers →
   {{Black, {Gray}, Gray}, {Black, LightGray, {LightGray}, Black}},
  Alignment → Left, ItemSize → {{4, 20, 20}},
  ItemStyle → 9, Spacings → {Automatic}]
```

Group 1	{Norway, Australia, Switzerland, Denmark, Netherlands, Germany, Ireland, United States, Canada, New Zealand, Singapore, Hong Kong, China (SAR), Liechtenstein, Sweden, United Kingdom, Iceland, Korea (Republic of), Israel, Luxembourg, Japan, Belgium, France, Austria, Finland, Slovenia, Spain, Italy, Czech Republic, Greece, Estonia, Brunei Darussalam, Cyprus, Qatar, Andorra, Slovakia, Poland, Lithuania, Malta, Saudi Arabia, Argentina, United Arab Emirates, Chile, Portugal, Hungary, Bahrain, Latvia, Croatia, Kuwait}	{Norway, Australia, Switzerland, Denmark, Netherlands, Germany, Ireland, United States, Canada, New Zealand, Singapore, Hong Kong, China (SAR), Liechtenstein, Sweden, United Kingdom, Iceland, Korea (Republic of), Israel, Luxembourg, Japan, Belgium, France, Austria, Finland, Slovenia, Spain, Italy, Czech Republic, Greece, Estonia, Brunei Darussalam, Cyprus, Qatar, Andorra, Slovakia, Poland, Lithuania, Malta, Saudi Arabia, Argentina, United Arab Emirates, Chile, Portugal, Hungary, Bahrain, Latvia, Croatia, Montenegro, Belarus, Russian Federation, Oman, Kazakhstan}
Group 2	{Montenegro, Belarus, Russian Federation, Oman, Romania, Uruguay, Bahamas, Kazakhstan, Barbados, Antigua and Barbuda, Bulgaria, Palau, Panama, Malaysia, Mauritius, Seychelles, Trinidad and Tobago, Serbia, Cuba, Lebanon, Costa Rica, Iran (Islamic Republic of), Venezuela (Bolivarian Republic of), Turkey, Sri Lanka, Mexico, Brazil, Georgia, Saint Kitts and Nevis, Azerbaijan, Grenada, Jordan, The former Yugoslav Republic of Macedonia, Ukraine, Algeria, Peru, Albania, Armenia, Bosnia and Herzegovina, Ecuador, Saint Lucia, China, Fiji, Mongolia, Thailand, Dominica, Libya, Tunisia, Colombia, Saint Vincent and the Grenadines, Jamaica, Tonga, Belize, Dominican Republic, Suriname, Maldives, Samoa, Botswana, Moldova (Republic of), Egypt, Turkmenistan, Gabon, Indonesia}	{Kuwait, Romania, Uruguay, Bahamas, Barbados, Antigua and Barbuda, Bulgaria, Palau, Panama, Malaysia, Mauritius, Seychelles, Trinidad and Tobago, Serbia, Cuba, Lebanon, Costa Rica, Iran (Islamic Republic of), Venezuela (Bolivarian Republic of), Turkey, Sri Lanka, Mexico, Brazil, Georgia, Saint Kitts and Nevis, Azerbaijan, Grenada, Jordan, The former Yugoslav Republic of Macedonia, Ukraine, Algeria, Peru, Albania, Armenia, Bosnia and Herzegovina, Ecuador, Saint Lucia, China, Fiji, Mongolia, Thailand, Dominica, Libya, Tunisia, Colombia, Saint Vincent and the Grenadines, Jamaica, Tonga, Belize, Dominican Republic, Suriname, Maldives, Samoa, Botswana, Moldova (Republic of), Egypt, Turkmenistan, Gabon, Indonesia, Paraguay, Palestine, State of, Uzbekistan, Philippines, El Salvador, South Africa, Viet Nam, Bolivia (Plurinational State of), Kyrgyzstan, Micronesia (Federated States of), Tajikistan}
Group 3	{Paraguay, Palestine, State of, Uzbekistan, Philippines, El Salvador, South Africa, Viet Nam, Bolivia (Plurinational State of), Kyrgyzstan, Iraq, Cabo Verde, Micronesia (Federated States of), Guyana, Nicaragua, Morocco, Namibia, Guatemala, Tajikistan, India, Honduras, Bhutan, Timor-Leste, Syrian Arab Republic, Vanuatu, Congo, Kiribati, Equatorial Guinea, Zambia, Ghana, Lao People's Democratic Republic, Bangladesh, Cambodia, Sao Tome and Principe, Kenya, Nepal}	{Iraq, Cabo Verde, Guyana, Nicaragua, Morocco, Namibia, Guatemala, India, Honduras, Bhutan, Timor-Leste, Syrian Arab Republic, Vanuatu, Congo, Kiribati, Equatorial Guinea, Zambia, Ghana, Lao People's Democratic Republic, Bangladesh, Cambodia, Sao Tome and Principe, Kenya, Nepal, Pakistan, Myanmar, Angola, Swaziland, Tanzania (United Republic of), Nigeria, Cameroon, Madagascar, Mauritania, Yemen, Lesotho, Sudan}
Group 4	{Pakistan, Myanmar, Angola, Swaziland, Tanzania (United Republic of), Nigeria, Cameroon, Madagascar, Zimbabwe, Mauritania, Solomon Islands, Papua New Guinea, Comoros, Yemen, Lesotho, Togo, Haiti, Rwanda, Uganda, Benin, Sudan, Djibouti, South Sudan, Senegal, Afghanistan, Côte d'Ivoire, Malawi, Ethiopia, Gambia, Congo (Democratic Republic of the), Liberia, Guinea-Bissau, Mali, Mozambique, Sierra Leone, Guinea, Burkina Faso, Burundi, Chad, Eritrea, Central African Republic, Niger}	{Zimbabwe, Solomon Islands, Papua New Guinea, Comoros, Togo, Haiti, Rwanda, Uganda, Benin, Djibouti, South Sudan, Senegal, Afghanistan, Côte d'Ivoire, Malawi, Ethiopia, Gambia, Congo (Democratic Republic of the), Liberia, Guinea-Bissau, Mali, Mozambique, Sierra Leone, Guinea, Burkina Faso, Burundi, Chad, Eritrea, Central African Republic, Niger}

6.7 Stochastic Processes

The latest versions of *Mathematica* can simulate random processes, estimate their parameters from data and compute state probabilities at different times. They also contain additional functionality for special classes of random processes such as Markov chains and stochastic differential equations.

- The following command simulates a random process given by the stochastic equation $d\,x[t] == -x[t]\,d\,t + \sigma\,d\,w[t]$:

```
Manipulate[
 proc = ItoProcess[dx[t] == -x[t] dt + σ dw[t],
   x[t], {x, 1}, t, w ≈ WienerProcess[]];
 ListLinePlot[RandomFunction[proc, {0., 10, 0.05}, 10],
  PlotRange → {-0.5, 1.5}],
 {{σ, 0.05}, 0., 0.5}, SaveDefinitions → True]
```

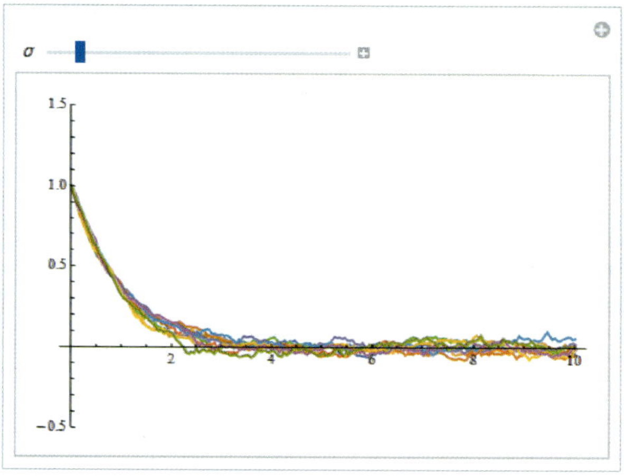

6.8 Reliability and Survival Analysis

In this section we'll show with examples how to use functions related to reliability and survival analyses.

6.8.1 Reliability Analysis

The analysis of the reliability of systems is very important for safety as well as for economic reasons. In the example, we estimate the reliability of a car.

- The safety of a car depends on its wheels, engine, transmission and brakes (we assume two independent types). Let's compute the average time between failures using Figure 6.2:

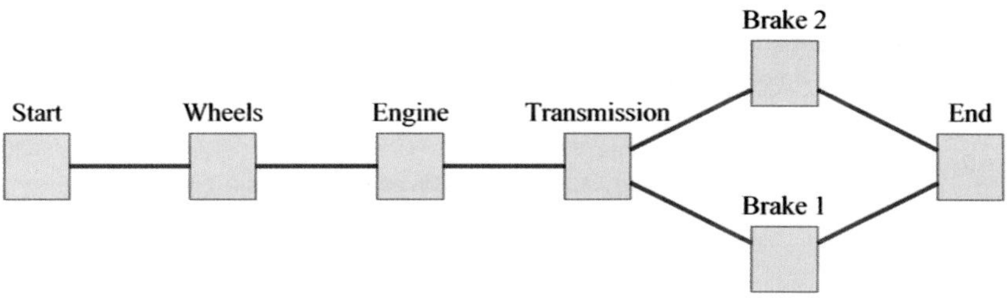

Figure 6.2 Basic Car Reliability Diagram.

- We build the reliability function of the car using the previous diagram:

```
ℛcar = ReliabilityDistribution[wheels ∧ engine ∧ transmission ∧
    (brake1 ∨ brake2), {{wheels, ExponentialDistribution[λ1]},
    {engine, WeibullDistribution[a1, b1]},
    {transmission, ExponentialDistribution[λ2]},
    {brake1, ExponentialDistribution[λ3]},
    {brake2, ExponentialDistribution[λ3]}}];
```

- Let's assume the following values for λ_i, a1 and b1:

```
vals = {λ1 → 1/5, λ2 → 1/10, λ3 → 1/15, a1 → 5, b1 → 12};
mttf = NExpectation[t, t ≈ ℛcar /. vals]
3.03247
```

6.8.2 Survival Analysis

The time that goes by until an event happens plays a crucial role when analyzing technical or biological systems. In the next example we examine the survival rate of cancer patients based on their age and the severity of the disease.

- We first collect and classify the data for 90 patients with larynx cancer. The patients are classified using four severity levels (from 1 to 4 in increasing order of severity):

```
larynx = ExampleData[{"Statistics", "LarynxCancer"}];
cov = larynx[[All, {1, 3}]];
surv = EventData[larynx[[All, 2]], larynx[[All, 5]]];
cox = CoxModelFit[{cov, surv}, {stage, age}, {stage, age},
   NominalVariables → stage];
```

- The parameters table shows the relationship between severity and risk:

```
cox["ParameterTable"]
```

	Estimate	Standard Error	Relative Risk	Wald-χ^2	DF	P-Value
stage[2.]	0.138564	0.462306	1.14862	0.0898343	1	0.764388
stage[3.]	0.63835	0.35608	1.89335	3.21382	1	0.0730189
stage[4.]	1.69306	0.422208	5.43607	16.0801	1	0.0000607172
age	0.0189018	0.014251	1.01908	1.7592	1	0.184724

- In the next graph we visualize the survival rate based on age and severity:

```
Manipulate[
 Plot[Evaluate[Table[cox["SF"][{i, aged}][t], {i, 4}]], {t, 0, 12},
  PlotLabel → Row[{"Edad: ", aged}],
  PlotStyle → Thick,
  PlotLegends → Thread[stage[Range[4]]],
  Exclusions → None], {aged, 45, 85, 1}, SaveDefinitions → True]
```

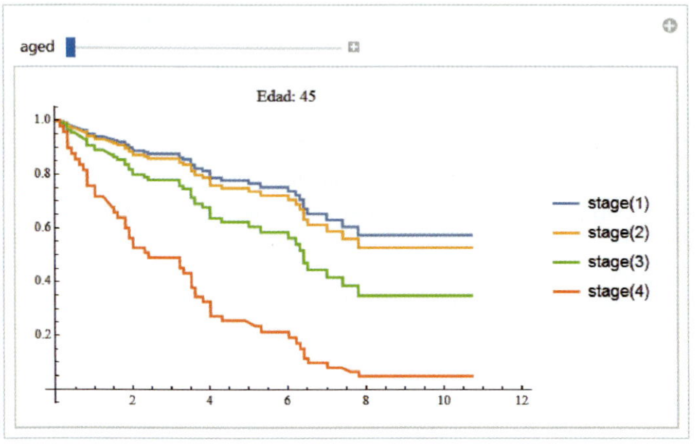

6.9 R Integration with *RLink*

Version 9 and above enables the integration of R in *Mathematica*. With RLink (paclet:RLink/tutorial/UsingRLink), we can perform operations such as exchange data between R and *Mathematica* and execute R commands and code inside the program.

6.10 Application: Predicting Outputs Using Machine Learning Methods

Mathematica 11 enhances its machine leaning capabilities with some new and improved functions (http://www.wolfram.com/language/11/improved-machine-learning/).

`Predict` and `PredictorFunction` are two of those new additions to the Wolfram Language. In the next example we use them to fit a curve using a Gaussian process and then proceed to visualize the quality of the fit.

- First, we enter the data and ask *Mathematica* to generate a prediction function based on the input-output pairs using the "GaussianProcess" method:

```
data = {1.2 -> 1.5, 2.1 -> 1.7, 5.1 -> 2.8, 3.5 -> 2.7};
p = Predict[data, Method -> "GaussianProcess"]
```

```
PredictorFunction[ Method: GaussianProcess
                   Feature type: Numerical ]
```

- Then, we display the actual data and the prediction with a one-standard-deviation confidence interval:

```
Show[Plot[{ p[x], p[x] + StandardDeviation[p[x, "Distribution"]],
    p[x] - StandardDeviation[p[x, "Distribution"]] }, {x, 0, 6},
   PlotStyle -> {Blue, Gray, Gray}, Filling -> {2 -> {3}},
   Exclusions -> False, PerformanceGoal -> "Speed",
   PlotLegends -> {"Prediction", "Confidence Interval"}],
  ListPlot[List @@@ data, PlotStyle -> Red, PlotLegends -> {"Data"}]]
```

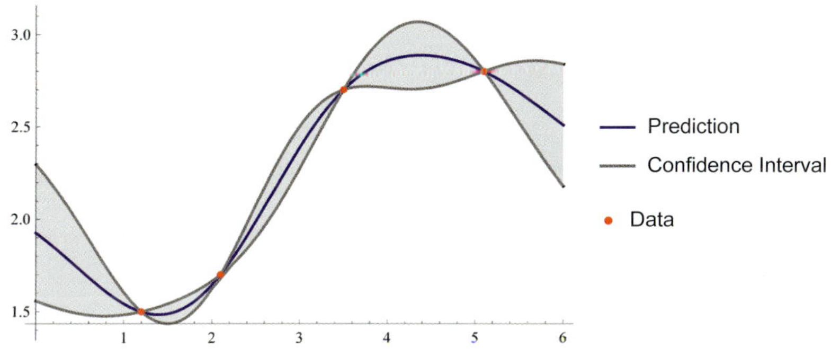

6.11. Application: Development of a Package for Quality Control

```
Clear["Global`*"]
```

Mathematica includes numerous statistical functions. If you need to perform a statistical computation, chances are there's already a built-in function for it.

However, in certain cases you may not be able to find the desired function. In those situations *Mathematica* enables us to easily define the desired function, a big advantage since that means we are not limited to the default statistical functionality of the software.

We are going to see a couple of examples of such custom functions and how we can put them together in a package.

6.11.1 Tolerance Interval Calculations in Normal Populations

In quality control sometimes it is necessary to define a tolerance interval (different from a confidence interval).

Let's consider a population from which we take a sample of size n to measure a certain characteristic y (diameter, density, weight, etc.). We calculate its average $\overline{y}$ and its standard deviation s. This is a typical situation in manufacturing were customers may require certain part features to be within a tolerance interval.

We define the p-th percentile as the value Y_p below which there's a percentage p of the population, with a probability or confidence level $1 - \alpha$, for example: to calculate the value below which we may find 95% of the population with a confidence level of 95% (that is with α = 5%).

We fix α and estimate Y_p as the unilateral upper limit using the following expression where t' is the noncentral Student's t distribution (`NoncentralStudentTDistribution`):

$$Y_p = \overline{y} + s\, k_{n,\alpha,p}\ ;\ \text{being}\ k_{n,\alpha,p} = \frac{t'\left(n-1,\ z(p)\sqrt{n}\ ;\ 1-\alpha\right)}{\sqrt{n}}$$

- The $k_{n,\alpha,p}$ factor can be obtained as shown. Notice the use of `Module` to define **z** as a local variable (if that concept is not clear, please refer to Chapter 3).

```
kUnilateral[n_, α_, p_] :=
 Module[{z}, z = Quantile[NormalDistribution[0, 1], p];
  (1/Sqrt[n]) *
   Quantile[NoncentralStudentTDistribution[n - 1, z*Sqrt[n]],
   1 - α]]
```

- Next we apply the previous function to compute the values of k below which 95% of the population falls with a probability of 95%, usually known as k(95/95), for sample sizes ranging from 5 to 10.

```
Table[{r, kUnilateral[r, 0.05, 0.95 ]} , {r, 5, 10}]
```

```
{{5, 4.20268}, {6, 3.70768}, {7, 3.39947},
 {8, 3.18729}, {9, 3.03124}, {10, 2.91096}}
```

Sometimes we may be interested in calculating the central limits (bilateral) within which the p-th percentile of the population resides with a probability of $1-\alpha$. These limits are computed as follows:

$$Y_p = \left\{\overline{y} + s\, k_{n,\alpha,p},\ \overline{y} - s\, k_{n,\alpha,p}\right\};\ \text{with}\ k_{n,\alpha,p} = \frac{t'\left(n-1,\ z(p)\sqrt{n}\ ;\ 1-\frac{\alpha}{2}\right)}{\sqrt{n}}$$

The previous method is computationally inefficient so it's not recommended for sample sizes $n > 10$. We can get an excellent approximation with:

$$k_{n,\alpha,p} \approx \left(\frac{1}{2n}+1\right) z\left(\frac{p+1}{2}\right)\sqrt{\frac{n-1}{\chi^2\,(n-1;\alpha)}}$$

```
kBilateral[n_, α_, p_] :=
 Module[{z, ji2}, z = Quantile[NormalDistribution[0, 1], (1 + p)/2];
  ji2 = Quantile[ChiSquareDistribution[n - 1], α];
  z Sqrt[(n - 1)/ji2] (1 + 1/(2 n))]
```

- We use the approximation to calculate the bilateral $k(95/95)$ for a sample of size $n = 1000$.

```
kBilateral[1000, 0.05, 0.95]
```

```
2.03608
```

Example: Let's take 30 experimental measurements representing population samples. We compute certain characteristic for which the tolerance limit is that 95% of the population falls below 20 with a 90% probability.

- We simulate the population samples assuming a normal distribution of mean 12 and standard deviation 3 (this information is usually unknown).

```
SeedRandom[314]
```

```
data = RandomVariate[NormalDistribution[12, 3], 30];
```

- We represent them graphically along with the density of a normal probability distribution.

```
Show[
 Histogram[data, 5, "ProbabilityDensity"],
 Plot[PDF[NormalDistribution[12, 3], x], {x, 0, 20}, PlotStyle → Thick]]
```

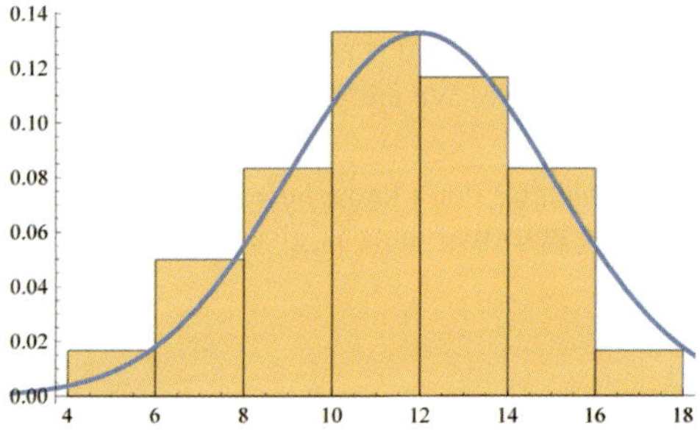

As mentioned previously, to be able to apply the method above, the sample must come from a normal distribution.

- First we test our normality assumption using the command DistributionFitTest.

```
ℋ = DistributionFitTest[data, Automatic, "HypothesisTestData"];
```

- The next table shows the results for different normality tests:

```
ℋ["TestDataTable", All]
```

	Statistic	P-Value
Anderson-Darling	0.137984	0.986886
Baringhaus-Henze	0.0636638	0.923334
Cramér-von Mises	0.0203191	0.970674
Jarque-Bera ALM	0.434132	0.784564
Kolmogorov-Smirnov	0.0741888	0.948625
Kuiper	0.098373	0.960911
Mardia Combined	0.434132	0.784564
Mardia Kurtosis	−0.598786	0.549315
Mardia Skewness	0.160603	0.688601
Pearson χ^2	3.06667	0.689708
Shapiro-Wilk	0.988989	0.985027
Watson U^2	0.0191112	0.985015

We see that many normality tests have been applied. P-Values above 0.05 (as in this case) indicate that it is reasonable to assume the data comes from a normal distribution.

- Next we check that the inequality $\overline{y} + k_{30,0.1,0.95}\, s < 20$ is satisfied:

```
{ȳ, s} = {Mean[data], StandardDeviation[data]}
```

```
{11.4702, 2.88186}
```

```
ȳ + kUnilateral[30, 0.10, 0.95] s < 20
```

```
True
```

Therefore we would accept the population.

6.11.2 Package Development for Sample Size Calculations

When creating customized functions that are going to be used frequently, it's convenient to put them in a package.

The most common problem in sampling for quality control is to determine the sample size and find out how many units in the sample are defective. We are going to explain how to do it in the case of simple random samples.

Let's assume a population of size N, from which we don't know how many units are defective. The probability P_a that there are c or fewer defective units is given by the hypergeometric distribution:

$$P_a = \sum_{x=0}^{c} \frac{\binom{D}{x}\binom{N-D}{n-x}}{\binom{N}{n}}$$

where:

D = Number of defective units in the batch (unknown).
f = Fraction of defective units: $f = D/N$, f is unknown, we assume the maximum acceptable fraction, p_A, of the population to be outside the tolerance limits.
α = significance level (Probability of committing a type I error that we are willing to consider).
n = sample size

In the particular case of the acceptance criteria being c = 0:

$$P_a = \frac{\binom{(1-f)N}{n}}{\binom{N}{n}}$$

- P_a can be calculated using the following function:

```
ProbAlpha[N_Integer, n_, f_] := FunctionExpand[Binomial[(1 - f) *N, n] /
    Binomial[N, n]];
```

- For a given population of size N, we want to calculate the sample size n that will allow us to state, within a certain confidence level, that the fraction of defective units in the population is lower or equal to f. Such a sample size can be computed in *Mathematica* as follows (we assume a minimum population size of N = 100):

```
SizeSample[N_Integer, alpha_, f_] :=
  Ceiling[
   n /. FindRoot[ProbAlpha[N, n, f] == alpha, {n, 20}]] /; N >= 100;
```

Let's put the two previous functions inside a package. The structure of a package was previously described in Chapter 4. For further information you can also access the Wolfram documentation: tutorial/SettingUpWolframLanguagePackages.

- Before we create the package, to avoid potential conflicts with the variables already defined, we proceed as usual:

```
Remove["Global`*"]
```

- We are going to type everything in a single cell. We name the package StatisticalPackage and include a brief description for the two functions defined:

```
BeginPackage["StatisticalPackage`"]

ProbAlpha::usage = "ProbAlpha[N, n, f] returns the probability
that a sample of size n, from a population of N elements of
which a fraction f is defective, will contain n defective elements.";

SizeSample::usage = "SizeSample[N, alfa, f] returns the sample
size n required to have a confidence level 1-alpha that the
population N contains a fraction of defective elements lower or
equal to f.";

Begin["`Private`"]

ProbAlpha[N_Integer, n_, f_] :=
  FunctionExpand[Binomial[(1 - f) *N, n] / Binomial[N, n]];

SizeSample[N_Integer, alpha_, f_] :=

  Ceiling[n /. FindRoot[ProbAlpha[N, n, f] == alpha, {n, 20}]] /; N >= 100;

SizeSample[N_Integer, alpha_, f_] := "N must be at least 100" /; N < 100;

End[]

EndPackage[]
```

We show a few examples of the package usage.

- Calculate the probability that in a sample of size 258 from a population of 1,000 units with 1% defective there will be one defective unit.

```
ProbAlpha[1000, 258, 0.01]

0.0497958
```

- Calculate the sample size required to be 95% confident that in a population of 1,000 units the fraction of defective elements is at most 1%.

```
SizeSample[1000, 0.05, 0.01]

258
```

  That is, we need to take a sample of 258 units and there should be no defective ones to be 95% confident that the maximum fraction of defective units in our population is 1% or lower.

- The package returns a warning message if the population size N is less than 100.

```
SizeSample[50, 0.05, 0.01]

N must be at least 100
```

Now we have to save the package using the name given in **BeginPackage**["StatisticalPackage`"].

To do that we copy the cell containing the package code and paste it in a new notebook. Then we change the cell format to code, initialize the cell and finally from the toolbar click on: **File ▸ Save As...** and in the **Save as type:** dropdown window choose *Mathematica* Package(*.m) with the name "StatisticalPackage".

A package is just a text file so we can modify it with any text editor.

- We copy the package in the subdirectory Addons\Applications located in:

```
$InstallationDirectory
```

```
C:\Program Files\Wolfram Research\Mathematica\11.0
```

- From then on, every time we want to use it we just need to load it first.

```
Needs["StatisticalPackage`"]
```

- To see the information about the functions we type:

```
?"StatisticalPackage`*"
```

▼ StatisticalPackage`

ProbAlpha	SizeSample

If we are going to create several packages it would be convenient to create a separate folder in Addons\Applications in which to place them.

6.12 Additional Resources

To access the following resources, if they refer to the help files, write their locations in a notebook (e.g., guide/ProbabilityAndStatistics), select them and press <F1>. In the case of external links, copy the web addresses in a browser:

Probability and Statistics: guide/ProbabilityAndStatistics
Basic Statistics: tutorial/BasicStatistics
Descriptive Statistics: tutorial/DescriptiveStatistics
Continuous Distributions: tutorial/ContinuousDistributions
Discrete Distributions: tutorial/DiscreteDistributions
Descriptive Statistics: tutorial/DescriptiveStatistics
Convolutions and Correlations: tutorial/ConvolutionsAndCorrelations
Optimization: tutorial/ConstrainedOptimizationOverview
Reliability Analysis: guide/Reliability
Survival Analysis: guide/SurvivalAnalysis
Time Series: guide/TimeSeries
Machine Learning: guide/MachineLearning
Random Processes: guide/RandomProcesses
Introduction to RLink: RLink/tutorial/Introduction
Probability demonstrations: http://demonstrations.wolfram.com/topic.html?topic=Probability
Statistics demonstrations: http://demonstrations.wolfram.com/topic.html?topic=Statistics

7

Calculating π and Other Mathematical Tales

It's been known for thousands of years that the ratio between the length of a circle's circumference and its diameter is a constant, whose value we call π. What looks like a simple geometric ratio has surprising properties and unexpected connections to seemingly unrelated fields. Many famous mathematicians throughout history have created their own formulas to calculate the largest number of decimals of π in the most efficient possible way, something that they still do presently. We will finish the chapter with a reference to one of the most important unsolved problems in mathematics: the Riemann hypothesis. There's a million dollar prize waiting for the person who can solve it and something even more important: the possibility of being remembered for posterity.

7.1 The Origins of π

Approximately 4,000 years ago, Egyptians and Babylonians already knew that the ratio between the circle's circumference and its diameter, what we call π (pi) nowadays, had a constant value slightly higher than 3.

The first estimates were most likely based on experimental measurements. After drawing a circle and measuring its perimeter and diameter, you will notice that the perimeter is a bit more than three times the diameter.

The Old Testament includes an estimation of π in the First Book of Kings 7:23, referring to the construction of Solomon's temple. We can read: "And he made a molten sea, ten cubits from the one brim to the other: it was round all about, and his height was five cubits: and a line of thirty cubits did compass it round about"; that is, its circumference was 30 cubits and its diameter was 10 cubits, so the relationship between the circumference and the diameter is 3.

In the Great Pyramid of Giza there are ratios clearly showing that the Egyptians had obtained a good approximation to the value of π. The Rhind Papyrus (circa 1650 BC), one of the best sources about mathematics in Ancient Egypt, contains the following ratio as the value of π: $\frac{256}{81}$ =3.16049.

The Babylonians noticed that the length of a circle was slightly bigger than the one of an hexagon inscribed inside it getting a π value of $3\frac{1}{8} = 3.125$, quite close to the correct value of 3.14159....

The following example describes a function to inscribe a n-sided polygon inside a circle.

- The display of a polygon inscribed in a circle can be done using the `Circle` and `Polygon` functions inside the `Graphics` command. `Circle[]` without arguments indicates a circle of radius $r = 1$ and center at $\{x, y\} = \{0, 0\}$. `Polygon[` $\{x_1, y_1\}, \ldots, \{x_n, y_n\}$`]` represents a polygon whose vertices are given by the coordinates $\{x_1, y_1\}, \ldots, \{x_n, y_n\}$. We use `Inset` to place text inside the graphical object. In the function definition we include ?`EvenQ` and ?`Positive` to execute the function only if n is an even positive integer.

```
polig1[(n_?EvenQ)?Positive] := Graphics[{Circle[],
   {LightGray, Polygon[Table[{Cos[2 π k /n], Sin[2 π k /n]}, {k, n}]]},
   Inset["inscribed polygon", {0, 0}]}, ImageSize → 130]
```

- We can add a warning message that will be displayed if the requirements for n are not met. This function must be defined immediately after the previous one. This way, **polig1** will check first whether n is a positive even number and if not, it will display the warning message. If we reverse the order in which we define the functions, it will not work.

```
polig1[n_] := "n must be an even positive number."
```

```
polig1[6]
```

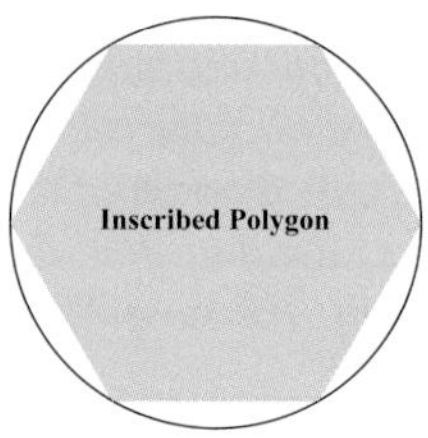


```
polig1[5]
```

```
n must be an even positive number.
```

```
polig1[-2]
```

```
n must be an even positive number.
```

An alternative way to display a 2D figure without using *Mathematica* functions is to use the Drawing Tools, available in the **Graphics** menu or by pressing CTRL+D in Windows or CMD+T in OS X.

7.2 Archimedes' Approximation

Archimedes (287–212 B.C.) is one of the most famous individuals in the history of science. Even nowadays some of his discoveries are still being studied in schools and his famous palimpsest, the oldest surviving copy of his works, has been the subject of extensive research since 1999 (http://www.archimedespalimpsest.org). He lived at the end of the Ancient Greece era, the same period as other famous mathematicians such as Euclid (ca. 325–ca. 270 B.C.) and Eratosthenes (ca. 284–ca. 192 B.C.). They were all located in a small geographical region consisting of several cities around the Aegean sea, a lucky coincidence that has not happened again in the history of humankind. Archimedes' life is the subject of many legends of which the best-known one is probably the story of how he was able to set on fire a large fleet of Roman galleys that were trying to attack Syracuse. His extraordinary intelligence was acknowledged even by his own enemies. Plutarch tells us that during the siege of Syracuse a Roman soldier found the old man so absorbed in his geometric drawings that he ignored his orders and was killed as a result. Marcellus, the person in charge of the Roman troops, had given express orders to spare his life and when he heard the news he punished the soldier, mourned Archimedes death and buried him with honors.

In Archimedes' tomb there is a cylinder enclosing a sphere, a reference to one of his greatest findings, that the sphere has 2/3 the volume of its circumscribing cylinder. The same 2/3 ratio applies to all surfaces.

Next we are going to show how to inscribe a sphere inside a cylinder.

- To show a cylinder enclosing a sphere, we use the syntax `Graphics3D`[{{*properties*, `Cylinder`}, {*properties*, `Sphere`}}, options]. We use `Opacity` to define the degree of transparency in the cylinder. We can also choose the color and other relevant details using options. Since we are not using any arguments for either `Sphere []` or `Sphere []`, the commands assume by default that $r = 1$, $\{x, y, z\} = \{0, 0, 0\}$ and, in the case of the cylinder, $h = 1$.

```
Graphics3D[{{Opacity[0.2], Blue, Cylinder[]}, {Green, Sphere[]}},
 Boxed → False, ImageSize → 200]
```

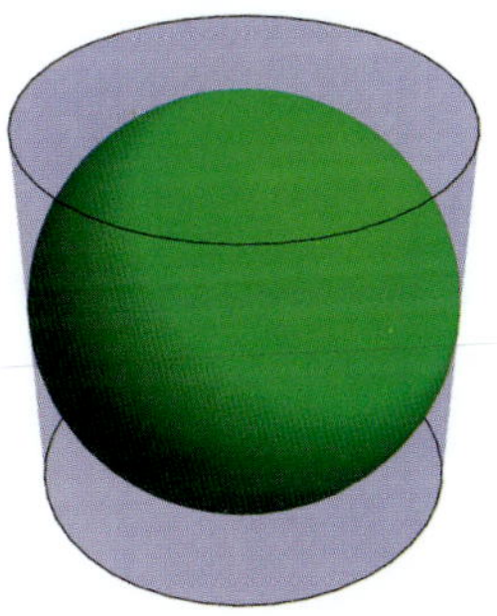

- The same graph can be done with `ContourPlot3D`, a *Mathematica* command used for any 3D function for which we know its analytic expression. `PlotLabel`, a general graphics option, adds a legend to the graph. In this example we use the equations of both geometric figures as legends and display them on separate lines with `TableForm`.

```
ContourPlot3D[{x^2 + y^2 + z^2 == 1, x^2 + y^2 == 1}, {x, -1, 1}, {y, -1, 1},
 {z, -1, 1}, ContourStyle → {Green, {Blue, Opacity[0.2]}}, Mesh → None,
 Axes → False, Boxed → False, ImageSize → 200, PlotLabel → TableForm[
   {"Sphere:" x^2 + y^2 + z^2 == 1, "Cylinder:" x^2 + y^2 == 1 ∧ 0 ≤ z ≤ 1}]]
```

Sphere: $x^2 + y^2 + z^2 = 1$
Cylinder: $x^2 + y^2 = 1 \wedge 0 \leq z \leq 1$

The method that follows enables us to verify that the ratio between the volume of a sphere and that of its circumscribing cylinder is 2/3 (without using the formulas for their respective volumes).

We first compute the volume of the sphere, V_{sphe}, and of the cylinder ,V_{cyl}, with $r = 1$ by integrating their analytical expressions and then calculate $V_{\text{sphe}}/V_{\text{cil}}$. Archimedes obviously did not use this method since he didn't know integral calculus.

- We use `Boole` to indicate that the integral of the points that do not satisfy the condition is 0.

```
volsphe = Integrate[ Boole[x^2 + y^2 + z^2 <= 1],
   {x, -1, 1}, {y, -1, 1}, {z, -1, 1}]
```

$\frac{4\pi}{3}$

```
volcyl =
 Integrate[ Boole[x^2 + y^2 ≤ 1], {x, -1, 1}, {y, -1, 1}, {z, -1, 1}]
```

2π

```
volsphe / volcyl
```

$\frac{2}{3}$

One of Archimedes's biggest discoveries was a new method for computing the value of π. He realized that by dividing a circle into identical sectors, the sum of their areas would resemble that of a sum of triangles whose total area would approximate the area of the circle. Geometrically, the previous method is equivalent to placing the sectors together inverting every other one. If each sector is replaced by its inscribed triangle, that Archimedes knew how to calculate, the resulting area would be very close to that of the circle. Furthermore, when $r = 1$, since the area of the circle is $A_c = \pi r^2$, then $A_c = \pi$, and from that formula we can deduce that the value of π is going to be between the value area of the inscribed polygon, A_{ins} and that of the circumscribed polygon A_{cir}, that is, $A_{\text{ins}} \leq A_c = \pi \leq A_{\text{cir}}$.

- The previous procedure can be easily modeled in *Mathematica*. We use `GraphicsRow` to display all the figures in the same row. The higher the number of triangles, the closer the resemblance of the resulting graphical object to a rectangle.

```
GraphicsRow[
 Table[Graphics[{EdgeForm[Opacity[1]], Table[{Hue[i/n, .5], Disk[
      {i Sin[π/n], ((-1)^(i + 1) + 1) Cos[π/n]/2}, 1,
      π {(-1)^i/2 - 1/n, (-1)^i/2 + 1/n}]},
    {i, n}]}], {n, 4, 12, 2}]]
```

Using modern notation to compute A_{ins}, we decompose the resulting polygon into n identical triangles of height $h = \text{Cos}[a/2]$ and base $b = 2\,\text{Sin}[a/2]$, where a is the $2\pi/n$ angle. A_{ins} will be the sum of the areas of those triangles.

That is: $A_{ins} = n \times 1/2 \times b \times h = n/2\,(2\,\text{Sin}[a/2]\,\text{Cos}[a/2]) = n/2\,\text{Sin}[a]$.

A_{cir} can also be obtained by decomposing the circumscribing polygon into n triangles. In this case, the height of each triangle will be 1, the same as the radius of the circumference, and its base $2\,\text{Tan}[a/2]$.

Then the area $A_{cir} = n \times 1/2 \times 2 \times \text{Tan}[a/2] = n\,\text{Tan}[a/2]$.

- The following function applies the previous method to compute the boundaries of π using the number of sides of the chosen polygon as the argument.

```
piPolig[n_] := Module[{a}, a = 2 N[π]/n;
  ToString[1/2 n Sin[a]] <> " ≤ π ≤ " <> ToString[n Tan[a/2]]]
```

- In the case of $n = 6$.

```
piPolig[6]
```

```
2.59808 ≤ π ≤ 3.4641
```

Archimedes' original method is very ingenious. He starts by inscribing and circumscribing a circle with hexagons. Then, since hexagons consist of 6 equilateral triangles, the perimeter of the inscribed hexagon is $6r$, and that of the circumscribing one, according to the Pythagorean Theorem, $4\sqrt{3}\,r$. Finally, dividing both results by $2r$, he gets the following approximation for the value of π: $3 \le \pi \le 3.4641$.

$$P_{2n} = (2\,p_n\,P_n)/(p_n + P_n)$$

$$p_{2n} = \sqrt{p_n\,P_{2n}}$$

where P_n is the perimeter of the circumscribing polygon and p_n is the perimeter of the n-side inscribed polygon. He can then calculate the perimeters of 12-, 24-, 48- and 96-sided polygons. For example, in the case of $n = 96$, he obtains the following approximation: $6336/2017 \le \pi \le 9376/9347$, that can be expressed as $3+10/71 \le \pi \le 3+1/7$.

- A method for displaying simultaneously both polygons and the circle is as follows:

```
polig2[n_] := Module[{a, V, v, i}, a = 2π/n;
  V = Table[Sec[a/2] {Cos[i a], Sin[i a]}, {i, 0, n - 1}];
  v = Table[{Cos[i a], Sin[i a]}, {i, 0, n - 1}];
  Graphics[{Opacity[0.4], Polygon[V],
    {Red, Disk[]}, {Green, Polygon[v]}}, ImageSize → 150]]
```

```
polig2[6]
```

- An alternative but slightly more difficult way of doing the same, is to use `GeometricTransformation` and `RotationMatrix`:

```
polig3[n_] := Graphics[{Circle[],
   Table[GeometricTransformation[Line[{{1, 0}, {Cos[2π/n], Sin[2π/n]}}],
     RotationMatrix[a]], {a, 2π/n*Range[n]}],
   Table[GeometricTransformation[Line[{{1, Tan[π/n]}, {1, Tan[-π/n]}}],
     RotationMatrix[a]], {a, 2π/n*Range[n]}]}, ImageSize → 150]
```

```
polig3[8]
```

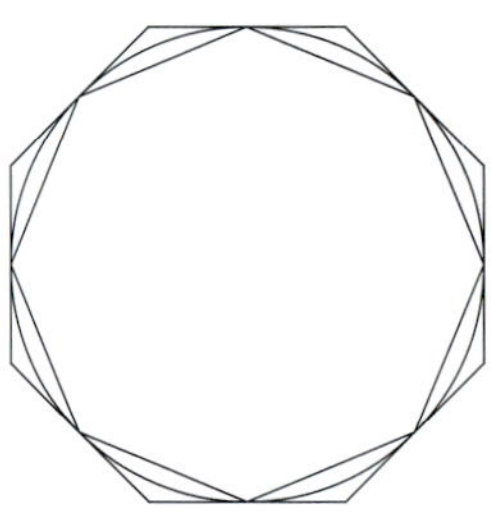

For further details we recommend the following demonstration:
http://demonstrations.wolfram.com/ArchimedesApproximationOfpi/

7.3 π with More Than One Billion Decimals

A 16-digit approximation of π, like the one available in any calculator, is more than enough for practical computations.

For example, the calculation error when measuring the earth circumference with such a number would be much smaller than 1 mm.

- The average distance from the earth to the sun is:

```
sunDist = AstronomicalData["Sun", "Distance"]
```

$1.49597870692 \times 10^{11}$ m

- The difference between using the exact value of π and its approximation with a 10^{-16} precision is insignificant.

```
2 π 10^-16 sunDist
```

0.0000939951143117 m

Despite this fact, the number of functions created to compute π with the largest number of decimals in the fastest possible way since Archimedes is remarkable. Some of the best mathematicians in history have developed their own approaches. In the following paragraphs we'll show some of them.

Contrary to expectations given the original definition of π as the ratio between the perimeter of a circumference and its diameter, many of its later definitions do not imply a geometric relationships. The original one was of a practical nature. Other non-geometrical definitions are also valid. As a matter of fact, we could take any of them as the definition of π. Next we present some examples.

- The formula below is probably the easiest way (but not the most effective one) to calculate π. Notice that *Mathematica* is able to obtain an exact result for infinite series.

$$4\sum_{j=0}^{\infty}\frac{(-1)^j}{2j+1}$$

π

- The next one shows π as the result of an integral computation. It's due to Leibniz, whose integration methods won over those from his contemporary fellow mathematician Newton.

$$\int_0^{\infty}\frac{1}{t^2+1}\,dt$$

$\frac{\pi}{2}$

- There are very curious methods such as the ones that compute π using a series of continuous fractions. Here's one example:

```
4/π == 1 + 1/(3 + 2^2/(5 + 3^2/(7 + 4^2/(9 + 5^2/(11 + 6^2/(13 + ...))))))
```

$$\frac{4}{\pi} == 1+\cfrac{1}{3+\cfrac{4}{5+\cfrac{9}{7+\cfrac{16}{9+\cfrac{25}{11+\cfrac{36}{13+\ldots}}}}}}$$

One of the most surprising formulas to calculate π comes from Ramanujan (1887–1920), the famous Indian mathematician.

Born to a poor family, he didn't have a formal education, learning mathematics completely on his own (The movie *The Man Who Knew Infinity* is based in his life). When he was in his teens, he read a book containing only theorems and a list of problems with their respective solutions. Since he could mentally arrived at the answer without the need to write down intermediate calculations, he thought that would be the ideal way to proceed: to figure out mentally the solution and how to get it and just write down the final result. Accordingly, he made a list of several identities that according to him where self-evident and sent them to several mathematicians. One of them, Hardy, quickly realized that the author of the list was a genius and he invited him to Oxford. From his time in England we have the following famous anecdote: One day Hardy was telling Ramanujan that he had taken a cab with a very boring license plate number 1729, to which Ramanujan replied that on the contrary that number was very interesting since it was the smallest one that could be decomposed into the sum of two different cubes.

- In fact:

```
9^3 + 10^3 == 1^3 + 12^3 == 1729.
```

```
True
```

- It's not by chance then that in 1914 Ramanujan found the following surprising formula:

$$\text{ramapi[n\_]} := \frac{2\sqrt{2}}{9801}\sum_{j=0}^{n}((4\,j)!\,(26390\,j + 1103))/((4^{4\,j}\,j!^4)\,99^{4\,j})$$

```
ramapi[∞]
```

$$\frac{1}{\pi}$$

- To check how fast the above formula converges we proceed as follows: Since π = 1/ramapi[∞], we increase the number of digits in the expression π - 1/ramapi[n], using `N[]` to force *Mathematica* to display the numeric approximation. We can see that with just the first 3 terms, the approximation is accurate to 30 decimal places:

$$\text{TableForm}\left[\text{Table}\left[\left\{n, N\left[\pi - \frac{1}{\text{ramapi[n]}}, 10\right]\right\}, \{n, 1, 5\}\right]\right]$$

1	$-6.395362624\times10^{-16}$
2	$-5.682423256\times10^{-24}$
3	$-5.238896280\times10^{-32}$
4	$-4.944187579\times10^{-40}$
5	$-4.741011769\times10^{-48}$

- If we add a 6th term we'll get an error message. The reason is that the series convergence is so fast that we need to substantially increase the computation precision. To do that we change the value of the variable `$MaxExtraPrecision` and limit the scope of the new assignment with `Block` to the previous expression only; otherwise, the new value would be applied to all calculations done after the latest definition.

$$\text{Block}\left[\{\$\text{MaxExtraPrecision} = 10000\},\ \text{TableForm}\left[\text{Table}\left[\left\{n, N\left[\pi - \frac{1}{\text{ramapi[n]}}, 10\right]\right\}, \{n, 1, 10\}\right]\right]\right]$$

1	$-6.395362624\times10^{-16}$
2	$-5.682423256\times10^{-24}$
3	$-5.238896280\times10^{-32}$
4	$-4.944187579\times10^{-40}$
5	$-4.741011769\times10^{-48}$
6	$-4.598865016\times10^{-56}$
7	$-4.499979208\times10^{-64}$
8	$-4.433276097\times10^{-72}$
9	$-4.391477319\times10^{-80}$
10	$-4.369600809\times10^{-88}$

Borwein, in 1989, developed an alternative iterative method that also converges very fast.

- Before showing it, let's recall how iteration works by again using the example described in Chapter 3 for calculating $n!$:

```
fact[1] = 1;
fact[n_] := n fact[n - 1];
```

- We use to it to compute 20!, and finally we delete the definition of **fact** to avoid any potential conflicts in future calculations:

```
fact[20]
```

```
2432902008176640000
```

```
Clear[fact]
```

Borwein's algorithm is an iteration based on the following limit: $\lim_{n\to\infty} a_n = 1/\pi$.

- The iteration is as follows (we use functional programming. If you try to program the same algorithm using a procedural approach, you'll soon realize that it would be significantly more difficult):

$$y_0 = \sqrt{2} - 1;$$

$$y_{n\_} := \frac{1 - \sqrt[4]{1 - y_{n-1}^4}}{1 + \sqrt[4]{1 - y_{n-1}^4}}$$

$$a_0 := 6 - 4\sqrt{2}$$

$$a_{n\_} := (1 + y_n)^4 a_{n-1} - 2^{2(n-1)+3} y_n (1 + y_n + y_n^2)$$

- Let's find the precision after only 3 iterations.

$$\text{Block}\left[\{\$\text{MaxExtraPrecision} = 10000\},\ \text{TableForm}\left[N\left[\text{Table}\left[\frac{1}{a_n} - \pi, \{n, 0, 3\}\right], 22\right]\right]\right]$$

```
-0.2273790912166981896610
-7.376250956313298951297×10^-9
-5.472109145689941832749×10^-41
-2.308580714934390266821×10^-171
```

Another recent example is Bailey–Borwein–Plouffe's formula (1996) that converges to π when $n \to \infty$.

$$\text{bayley1}[n\_] := \sum_{k=0}^{n} \frac{1}{16^k}\left(\frac{4}{8k+1} - \frac{2}{8k+4} - \frac{1}{8k+5} - \frac{1}{8k+6}\right)$$

In the journal article (On the rapid computation of various polylogarithmic constants, *Mathematics of computation* 66–218, 1997) the authors describe a method in which by using the previous formula we can calculate directly in base 16 the nth digit of π. We can use this fact to verify whether a record is valid.

- Here we see the extraordinary convergence of the formula:

```
Block[{$MaxExtraPrecision = 10000},
 TableForm[N[Table[ bayley1[n] - π, {n, 0, 100, 10}], 22]]]
```

```
-0.008259320256459905129310
-1.088484728192381580254×10^-16
-2.814026831526008285669×10^-29
-1.189722050936588364083×10^-41
-6.227358005614701952094×10^-54
-3.675420077848600975942×10^-66
-2.343075101379046695626×10^-78
-1.576132159096758651901×10^-90
-1.103039240116468483363×10^-102
-7.957699394337805229911×10^-115
-5.880802408811441129071×10^-127
```

By August 2010 the approximation had reached 5 billion digits (5 10^{12}). To put that figure in perspective, if we wanted to print that many digits we would need a library of 5 million volumes (150 km worth of shelves) using normal PCs. One month later, Yahoo employees using a 1000-computer grid calculated the digit 2 10^{15} (2-quadrillionth digit) in binary.

- The formulas that we have seen so far represent just a small fraction of the existing ones. We can take a look at some of those formulas by using *Mathematica*'s free-form input. Press = and type Pi. To see the results, expand the output by clicking on the + symbol in the top right corner.

To verify that a sequence is correct, we compare part of it with the ones obtained by different methods. For example:

- We can combine `RealDigits` and [[...]] to show the digits located in positions n to m+n after the decimal point. We extract the digit *n+1* since 3, the first digit, is not decimal. We then use $n+m+5$ to increase the precision with 5 more digits than the required ones to calculate $m+n$.

```
pinth[n_, m_] := RealDigits[N[π - 3, n + m + 5]][[1, n ;; n + m]]
```

- For example, we can check that after position 762, the digit 9 appears six consecutive times.

```
pinth[760, 10]
```

```
{3, 4, 9, 9, 9, 9, 9, 9, 8, 3, 7}
```

7.4 Buffon's Method

The Frenchman Georges-Louis Leclerc (1707–1788), Comte de Buffon, was a naturalist whose encyclopedia *Histoire naturelle* had a great influence on his 19th century colleagues. Charles Darwin even cites him in his book *On the Origin of Species*. In mathematics, he's famous for his experimental method to calculate π.

The method, as shown in Figure 7.1, consists of randomly throwing a needle of length l onto a plane containing parallel lines at a fixed distance $d > l$ from each other.

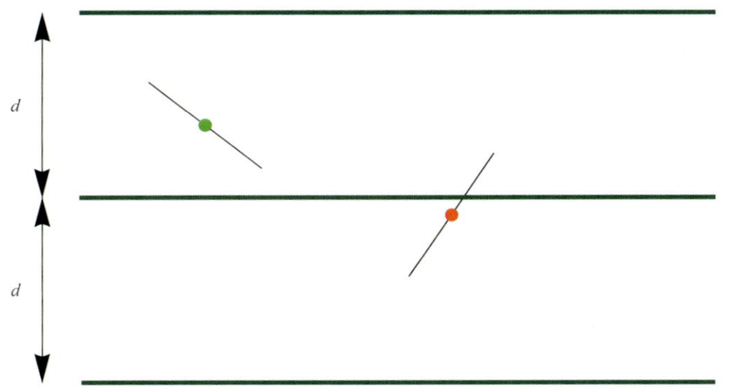

Figure 7.1 Visualizing Buffon's method for calculating π.

When throwing the needle, there are two possibilities: either the needle crosses one of the lines or it doesn't. The probabilities associated to each of those outcomes are respectively $p_0 = 1 - 2r\theta$, $p_1 = 2r\theta$, with $r = l/d$ and $\theta = 1/\pi$.

Let N_1 be the number of times that the needle crosses a line out of n throws. Then we can estimate p_1 using $\hat{p}_1 = N_1/n$ and consequently obtain θ. Therefore $\pi = 1/\theta$.

$$\hat{\theta} = \frac{\hat{p}_1}{2\,r} = \frac{N_1}{2\,r\,n}$$

The previous experiment has been performed on several occasions and it has always required thousands of throws to achieve a precision better than 2 digits. Nowadays we use simulations where the throws are replaced by computer-generated (pseudo)random numbers.

Simulation-based procedures, usually known as Monte Carlo methods, enable us to calculate π probabilistically in many different ways.

A very simple example to approximate the value of π is to first draw a circle of radius 1 circumscribed by a square with sides of length 2 and then to drop a ball m times onto the square counting how many times (n) the ball falls inside the circle. Assuming that the ball can land with equal probability on any point in the square, the probability that it will land inside the circle is given by the ratio between the area of the circle and that of the square, that is, $\frac{n}{m} \approx \pi r^2 / l^2$, since $l = d = 2$, then $4\,n/m \to \pi$.

- The above experiment can be simulated with *Mathematica.* We just need to generate m pairs of real numbers between -1 and 1 and count how many of them "land" inside a circle with radius $r = 1$ (we can use `Cases` to check that the pairs verify $x^2 + y^2 < 1$ and count with `Length` how many they are).

```
pib[m_] := Module[{data, n},
  data = RandomReal[{-1, 1}, {m, 2}];
  n = Length[Cases[data, {x_, y_} /; x^2 + y^2 < 1]];
  4. n / m]
```

- The method converges very slowly.

```
pib[100000]
```

```
3.13492
```

- The previous function can be extended to display the results graphically. Notice that text needs to go between quotation marks and that inside a string of text we can use \t or \n to insert a tab or a new line respectively.

```
Manipulate[Module[{data, n},
  data = RandomReal[{-1, 1}, {m, 2}];
  n = Length[Cases[data, {x_, y_} /; x^2 + y^2 < 1]];
  Text@Style[Column[{Graphics[
       {PointSize[0.006], Orange, Disk[{0, 0}, 1], Black, Point[data]}],
      Row[{"inside: ", n, "\toutside: ", m - n, "\ntotal:  ", m,
        "\tπ ≈ 4 × ", n, "/", m, " = ", 4. n / m}]}], "Label"]],
 {{m, 100, "Random pairs generated"}, 1, 10000,
  1, Appearance → "Labeled"}]
```

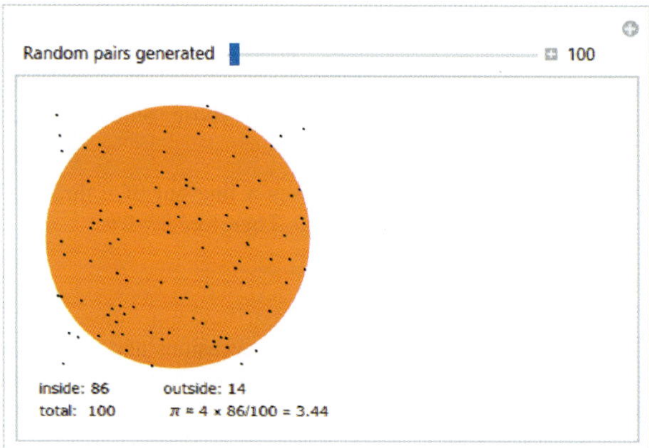

7.5 Application: Are the Decimal Digits of π Random?

It doesn't matter what method we use, if it's correct, we will always get the same sequence of π decimals even though such a sequence will never exhibit any periodicity since π is an irrational number (actually a transcendental one) and therefore it cannot be expressed as the ratio of two integers. However, if we explore a long enough sequence of its decimals, we will discover an interesting behavior: the frequency in which the digits appear is similar, that is, the numbers 0, 1, 2, ..., 9 will appear approximately with the same probability (this characteristic doesn't depend on expressing π in base 10. The same behavior happens regardless of the chosen base). Since this property is the same that random numbers originated from a discrete uniform distribution have, unless we knew in advance that certain digits were π decimals, we wouldn't be able to differentiate them from random numbers generated by a roulette with 10 positions, all equally likely. Numbers that have this property are called normal numbers. In the case of π, at least considering only the digits computed until now, it's been checked that is normal but it has not been proved that it would be normal for an infinite number of digits although everything suggests that it would.

- To convert the digits of π to list elements, we use the function `RealDigits` to decompose the decimal approximation of π into a list of numbers. For example: 3.14159... becomes:

```
RealDigits[N[π]]
```

```
{{3, 1, 4, 1, 5, 9, 2, 6, 5, 3, 5, 8, 9, 7, 9, 3}, 1}
```

- The 1 at the end indicates that the first element corresponds to the integer part. If we just want the decimal part we can subtract 3 from 3.14159... : 3.14159 ... − 3 =0.14159. For example, the first 10 decimal digits of π are:

```
RealDigits[N[π - 3, 10]][[1]]
```

```
{1, 4, 1, 5, 9, 2, 6, 5, 3, 5}
```

- The following expression enables us to find a specific π digit in a short time. In the example below we find the 10 000 000th digit.

```
Timing[First[RealDigits[Pi, 10, 1, -10^7]]]
```

```
{10.5938, {7}}
```

- This short expression tells us how often numbers 0–9 appear in the first 100 000 decimals of π.

```
Sort[Tally[RealDigits[N[π - 3, 100000]][[1]]]]
```

```
{{0, 9999}, {1, 10137}, {2, 9908}, {3, 10025}, {4, 9971},
 {5, 10026}, {6, 10029}, {7, 10025}, {8, 9978}, {9, 9902}}
```

The previous output seems to confirm that in base 10 all digits are present with approximately the same probability. This indicates that they are distributed following a random discrete uniform distribution.

- To test this hypothesis for the first 10000 π decimals we use a combination of the functions `DistributionFitTest` and `DiscreteUniformDistribution`.

```
vals = RealDigits[N[π - 3, 10000]][[1]];
```

```
h = DistributionFitTest[vals,
   DiscreteUniformDistribution[{μ, ω}], "HypothesisTestData"];
h[{"TestDataTable", All}]
```

	Statistic	P-Value
Pearson χ^2	9.318	0.408453

P-Values above 0.05 (as in this case) suggest that the data come from a discrete uniform distribution. If this were true then π would be normal.

- The command below returns the frequency of numbers 0–9 as a function of the total number of digits used in the calculation.

```
Manipulate[
 BarChart[BinCounts[RealDigits[N[π - 3, 10000]][[1, 1 ;; u]],
   {0, 10, 1}], ChartLabels -> Range[0, 9]],
 {{u, 100, "Total number of digits"}, 1, 10000,
  1, Appearance -> "Labeled"}]
```

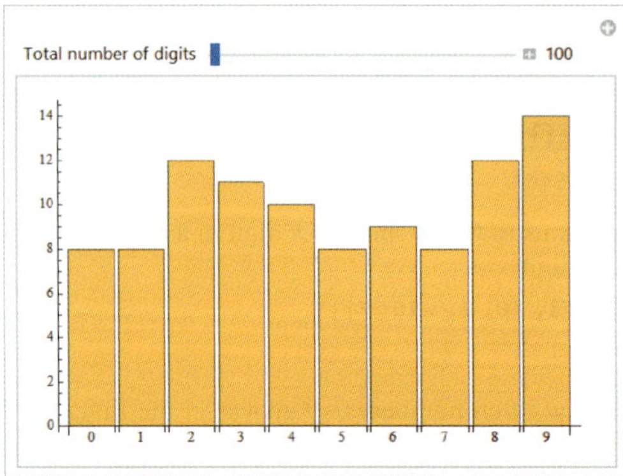

In a discrete uniform random distribution, all number combinations are possible. However, the probability of some of them happening is very small. For example: the probability that the same number, in this case any digit between 0 and 9, will repeat itself is very small, although sometimes it happens.

Let's create a command to identify when that happens with the digits of π.

- `Split` is very useful to divide lists into sublists consisting of identical elements.

```
Split[RealDigits[Pi, 10, 100][[1]]]
```

```
{{3}, {1}, {4}, {1}, {5}, {9}, {2}, {6}, {5}, {3}, {5}, {8}, {9}, {7}, {9},
 {3}, {2}, {3}, {8}, {4}, {6}, {2}, {6}, {4}, {3, 3}, {8}, {3}, {2}, {7},
 {9}, {5}, {0}, {2}, {8, 8}, {4}, {1}, {9}, {7}, {1}, {6}, {9}, {3},
 {9, 9}, {3}, {7}, {5}, {1}, {0}, {5}, {8}, {2}, {0}, {9}, {7}, {4},
 {9}, {4, 4}, {5}, {9}, {2}, {3}, {0}, {7}, {8}, {1}, {6}, {4}, {0},
 {6}, {2}, {8}, {6}, {2}, {0}, {8}, {9, 9}, {8}, {6}, {2}, {8}, {0},
 {3}, {4}, {8}, {2}, {5}, {3}, {4}, {2}, {1, 1}, {7}, {0}, {6}, {7}}
```

- Notice that some sublists contain more than one element. This happens when a digit repeats itself once or more consecutively. With the next function we can count the size of each sublist:

```
Length /@ Split[RealDigits[Pi, 10, 100][[1]]]
```

```
{1, 1, 1, 1, 1, 1, 1, 1, 1, 1, 1, 1, 1, 1, 1, 1, 1, 1, 1, 1, 1, 1, 1,
 1, 2, 1, 1, 1, 1, 1, 1, 1, 1, 2, 1, 1, 1, 1, 1, 1, 1, 1, 2, 1, 1, 1, 1,
 1, 1, 1, 1, 1, 1, 1, 1, 1, 2, 1, 1, 1, 1, 1, 1, 1, 1, 1, 1, 1, 1, 1,
 1, 1, 1, 1, 2, 1, 1, 1, 1, 1, 1, 1, 1, 1, 1, 1, 1, 1, 2, 1, 1, 1, 1}
```

We see that in 6 cases a digit repeated itself once and there were no instances of a number appearing consecutively 3 or more times.

- The next function counts the number of sublists with 1, 2, ..., n terms. Using `Drop` we eliminate the first element since it corresponds to the number of empty sublists. The function is then applied to the first 10^6 digits of π.

```
data1 =
  Drop[BinCounts[Length /@ Split[RealDigits[Pi, 10, 1000000][[1]]]], 1];
```

- We present the results in table format: 1 indicates that a digit does not appear in consecutive positions, 2 that the same digit appears in two consecutive positions and so on.

```
TableForm[{Table[i, {i, Length[data1]}], data1},
 TableHeadings → {{"repetitions", "frequency"}, None}]
```

repetitions	1	2	3	4	5	6	7
frequency	809776	81256	7976	830	77	12	1

- By adding `Position` we can identify the order in which a certain repetition happens. For example, to check the positions in which the sequence {9,9} appears within a 100 decimal digits approximation of π, we would type:

```
Position[Split[First@RealDigits[Pi - 3, 10, 100]], {9, 9}]
```

```
{{42}, {75}}
```

- To relate the sublist position to the number of digits, we proceed as follows:

```
Part[Accumulate[Length /@ Split[RealDigits[Pi - 3, 10, 100][[1]]]],
  {42, 75} - 1] + 1
```

```
{44, 79}
```

- This result indicates that the sequence 99 appears first in the 44th and 45th digits and the next one in the 79th and 80th ones. We can check it using the table below. In it, π decimals are grouped by rows containing 10 digits each.

```
TableForm[Partition[First@RealDigits[Pi - 3, 10, 100], 10],
 TableHeadings → {Range[10], Range[10]}]
```

	1	2	3	4	5	6	7	8	9	10
1	1	4	1	5	9	2	6	5	3	5
2	8	9	7	9	3	2	3	8	4	6
3	2	6	4	3	3	8	3	2	7	9
4	5	0	2	8	8	4	1	9	7	1
5	6	9	3	9	9	3	7	5	1	0
6	5	8	2	0	9	7	4	9	4	4
7	5	9	2	3	0	7	8	1	6	4
8	0	6	2	8	6	2	0	8	9	9
9	8	6	2	8	0	3	4	8	2	5
10	3	4	2	1	1	7	0	6	7	9

- Since *Mathematica* 10.1 the function `SequencePosition` can be used to find the position where the digit n will be repeated r consecutive times.
  In the example below, using the first 10 000 000 digits of π, we find the cases in which the number 9 appears consecutively exactly 7 times.

```
digitsPi = First[RealDigits[N[Pi, 10^7]]];
```

```
SequencePosition[digitsPi, PadRight[{9}, 7, 9]]
```

```
{{1722777, 1722783}, {3389381, 3389387},
 {4313728, 4313734}, {5466170, 5466176}}
```

According to the output in the first 10 000 000 digits of π in 4 occasions the number 9 appears consecutively 7 times. The first occurrence starts at the 1722777-th digit (or 1722776-th digit after the decimal point) and finishes at position 1722783-th.

We may be tempted to believe that any possible combination of numbers, no matter how unlikely, will eventually happen as long as we use enough digits. For example, the most recent π approximations have shown that there's a sequence in which 0 is repeated 12 consecutive times and the same happens with the numbers 1–9. There's even a position in which 8 appears repeated 13 consecutive times. The frequency of occurrence is consistent with a discrete uniform distribution: the probabilities, in base 10, that 0 will appear consecutively once, twice or 12 times are: 1/10, $1/10^2$ and $1/10^{12}$ respectively. This means that, using a π approximation of 10^{12}digits, we'd most likely see one sequence of 12 consecutive 0s. The same could be said for other sequences.

- In the example below we look for the pattern 3131313 in the first then million digits of Pi:

```
SequencePosition[digitsPi, {3, 1, 3, 1, 3, 1, 3}]
```

```
{{3662426, 3662432}}
```

The solutions found by the previous two examples can be compared at http://www.angio.net/pi/bigpi.cgi.

These types of patterns can lead us to a variant of what Borges proposed in his famous work *The Library of Babel*. The text describes a library containing all the books ever written and the ones that may be written in the future (actually, Borges mentions the number of pages and the number of lines per page but this is irrelevant for the purpose of our argument). Let's suppose that we could write a book where all its letters were *a*, followed by one containing only *bs* and so on until writing a final book with all *zs*. This way we would cover all the possible combinations of letters. The resulting number would be huge, immensely bigger than the number of atoms in the universe. Nevertheless it would be a finite number.

A variant of the previous story consists of replacing the order of the letters with probabilities. Let's supposed that we have a roulette with 35 positions representing the 26 letters of the English alphabet plus certain symbols such as punctuation marks, whitespace characters, etc. If we played such roulette enough times, we would end up not only with every possible word, sentence and paragraph but also with every possible book! The only problem would be the time and the space to store the results. We can speculate that the same applies to π. For example, if we express it in base 35 and assign the digit 1 to the letter *a*, the digit 2 to *b* and so on, with a number of digits big enough we will find the library of Babel (it doesn't matter how big the number needs to be since π has an infinite number of digits). This means that all books: past, present and future, have already been written and they are in the decimals of π.

Surprisingly, in contemporary physics when we speculate about the existence of parallel universes we arrive at what we might consider an extreme version of the previous speculation. If there is an infinite number of universes, all the possibilities that don't violate the laws of physics occur. Therefore, right now there's someone identical to you doing exactly the same as you're doing. According to this theory all the variants of your life will happen although in different universes.

7.6 The Strange Connection

π is a transcendental number. This means that a finite polynomial of the form $a_0 + a_1 x + a_2 x^2 + \ldots + a_n x^n = 0$, with the coefficients $a_0 \ldots a_n$ being rational numbers (or integers), in which π is one of its roots (solutions) doesn't exist. As a result, it can be proven that constructing a square with the same area of a given circle by a finite number of steps only using compass and straightedge, also known as squaring the circle, is not possible. There are other transcendental numbers with e being probably the second best-known one after π.

```
N[ⅇ, 100]
```

```
2.718281828459045235360287471352662497757247093699959574966967627724076630353547594571382178525166427
```

- As with π, the number e can be computed using several different methods (Click on + in the next cell).

Number E representations
Alternative representation
```
E == E^z /; z == 1
```

- When $e^{\pi\sqrt{163}}$ is expressed in decimal form, something funny happens (we use `NumberForm` combined with `ExponentFunction` to avoid the output being displayed in scientific notation).

```
NumberForm[N[e^(Pi Sqrt[163]), 30], ExponentFunction -> (Null &)]
```

```
262537412640768743.999999999999
```

We can see that the first 12 decimal digits are all 9s. This could give the false impression that all the decimal digits are 9 but that is not the case. It would be enough to increase the precision by one more digit to check that it is only a lucky coincidence.

In a separate category we include imaginary numbers, those whose square is a negative number as in the solution to the equation $x^2 + 1 = 0$.

- Notice that to express imaginary numbers, *Mathematica* uses the symbol i and not the Latin letter *i*.

```
Solve[x^2 + 1 == 0, x]
```

```
{{x → -i}, {x → i}}
```

Numbers consisting of a real part and an imaginary one are called complex numbers. As a matter of fact, all real numbers can be considered a special case of complex numbers where the imaginary element is 0. Sometimes strange connections exist among different numbers as in the case of the following equation discovered by Euler:

$$e^{i x} = \cos(x) + i \sin(x) \wedge x \in \mathbb{R}$$

This connection between different fundamental constants cannot be random. If we assume $x = \pi$, the previous expression becomes:

$$e^{i \pi} = -1$$

We can see that in just one expression there are two transcendental constants and the imaginary unit connected to each other and, the most surprising thing, the result is the integer -1. It's hard to believe that all these strange connections between numbers are random.

Pi is also connected to one of Riemann's Zeta functions: $\zeta(s) = \sum_{i=1}^{\infty} \frac{1}{k^s}$, that we will discuss in further detail in the next section. We can see that `Zeta` [2] converges to:

```
Zeta[2]
```

$$\frac{\pi^2}{6}$$

π also appears in fundamental notions in physics such as Heisenberg's uncertainty principle or Einstein's general relativity theory. It seems that π is much more than the ratio between the perimeter of a circumference and its diameter!

7.7 The Riemann Hypothesis

We've seen in the previous section that π, at least according to its latest approximations using billions of digits, is probably a normal number. However, there's no proof that this is the case if we consider an infinite number of digits. This is an open problem in mathematics although not the only or most important one. David Hilbert, one of the most influential mathematicians of the 19th and 20th centuries, at the Paris conference of the International Congress of Mathematicians in 1900 gave a task to mathematicians for the 20th century: the resolution of 23 problems (ten given in the conference and the rest published later). Although most of them have been already solved, the most famous unsolved one is the Riemann's hypothesis about which Bernhard Riemann himself stated that if he were to resuscitate after 500 years, the first question he would ask would be whether the "damn hypothesis" had been proven or not. This problem remains unsolved and is one of the seven "Millennium Problems", a list of problems named by the Clay Mathematics Institute that carry a prize of 1 million dollars for their complete mathematical solution.

The Riemann's Zeta function definition is:

Let $s = a + b\,i \in \mathbb{C}$, for $\text{Re}(s) > 1$, $\zeta(s) = \sum_{k=1}^{\infty} \frac{1}{k^s}$.

```
Clear["Global`*"]
```

- Given that s is a complex variable, $\zeta(s)$ is also complex. We can visualize its real component Re (s) and its imaginary one Im (s) with the following *Mathematica* command:

```
{Plot3D[Re[Zeta[a + b I ]], {a, -10, 10}, {b, -10, 10}, Mesh → None],
 Plot3D[Im[Zeta[a + b I ]], {a, -10, 10}, {b, -10, 10}, Mesh → None]}
```

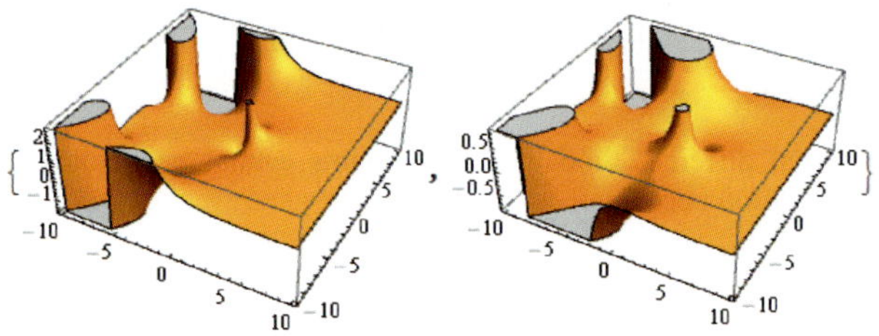

- It can be shown (using expansions of analytic functions, a complex analysis concept), that for negative even values, that is when $s = -2n$, $\zeta(s) = 0$. We can check it for some cases (in the example below when n ranges from 1 to 100 in steps of 10).

```
Table[Zeta[-2 n ], {n, 1, 100, 10}]
```

```
{0, 0, 0, 0, 0, 0, 0, 0, 0, 0}
```

- Graphically we can see that by plotting $s = a + b\,i$ with $b = 0$, $s = 0$ for $a = \{-2, \ldots, -2n\}$:

```
Plot[{Re[Zeta[a ]], Im[Zeta[a ]]}, {a, -30, -2}, PlotLegends → "Expressions"]
```

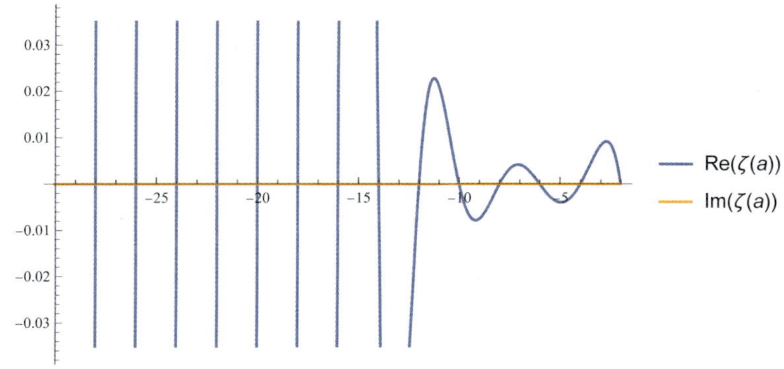

- Another way to visualize this is by plotting $1/\,|\zeta(s)|$, with $|\zeta(s)| = |x + y\,i|$. Remember that $|x+y\ i| = \sqrt{x^2 + y^2}$. Using the inverse expression, the zeros become asymptotes that look like columns. In the graph below, we can see these "columns" appearing when $s = -2\,n$. (This function and several others that follow are based on the ones described by Stan Wagon in his book *Mathematica in Action*).

```
Plot3D[1/Abs[Zeta[x + i y]], {x, -20, -2}, {y, -1, 1},
 MeshFunctions → (#3 &), MeshStyle → Blue, Mesh → 10,
 PlotPoints → 60, MaxRecursion → 3 ,
 Boxed → False, BoxRatios → {5, 2, 2}, AxesLabel → {"x", "y", None},
 Ticks → {Automatic, Range[-22, -2], Range[0, 4]}]
```

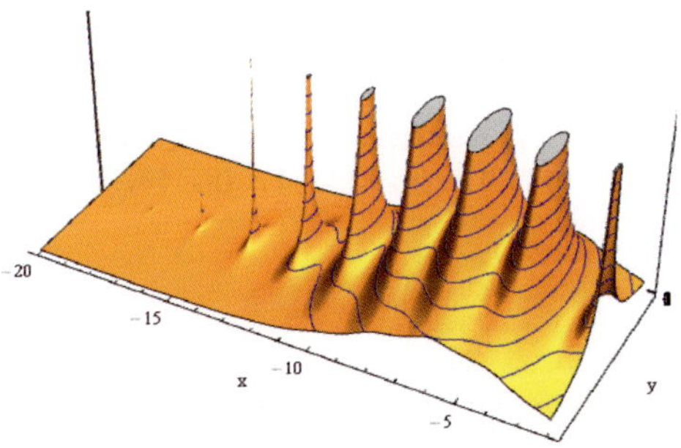

The $\zeta(-2\,n) = 0$ zeros, are named **trivial zeros**.

- Besides the trivial zeros, there are other complex values of s, with $0 < \mathrm{Re}(s) < 1$, for which the zeta function becomes also zero. These zeros are called **non-trivial zeros**. The Riemann's hypothesis states that all these non-trivial zeroes are located in the line $s = \frac{1}{2} + b\,i$ (known as the critical line) or put it in other words, that the real part of every non-trivial zero in the Riemann's Zeta function is 1/2. The graph below shows that this is true for the chosen interval.

```
Plot[{Re[Zeta[1/2 + b I]], Im[Zeta[1/2 + b I]]},
 {b, -30, 30}, PlotLegends → "Expressions"]
```

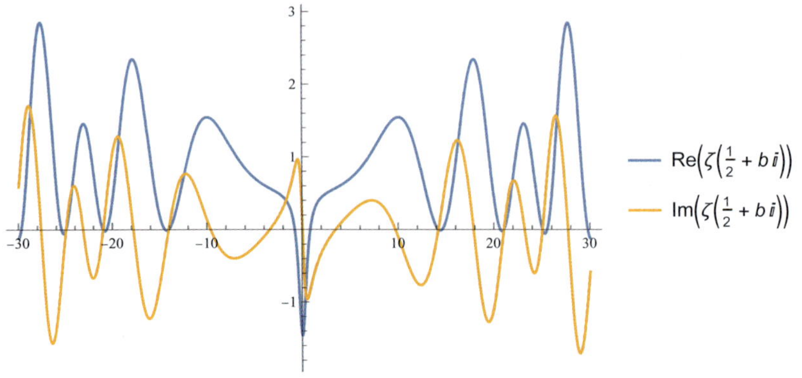

- In *Mathematica* the function **ZetaZero[k]** represents the *k-th zero* of $\zeta(s)$ in the critical line $s = \frac{1}{2} + i\,b$ with the smallest imaginary part. For example the first zero corresponds to:

```
s = N[ZetaZero[1], 15]
```

```
0.5000000000000 + 14.1347251417347 i
```

- We check that for that value $\zeta(s) = 0$.

```
Zeta[s] // Chop
```

```
0
```

- We've seen that the first zero is approximately at 14.1347251417347 i. If we would like to visualize all the existing zeros for a given interval, we can use the following function in terms of the imaginary component (remember that the real component is always 1/2).

```
Manipulate[
 Plot[RiemannSiegelZ[t], {t, 0, k}, Epilog -> {PointSize[.02], Red,
   Point[Table[{Im[ZetaZero[n]], 0}, {n, k}]]}], {k, 50, 100}]
```

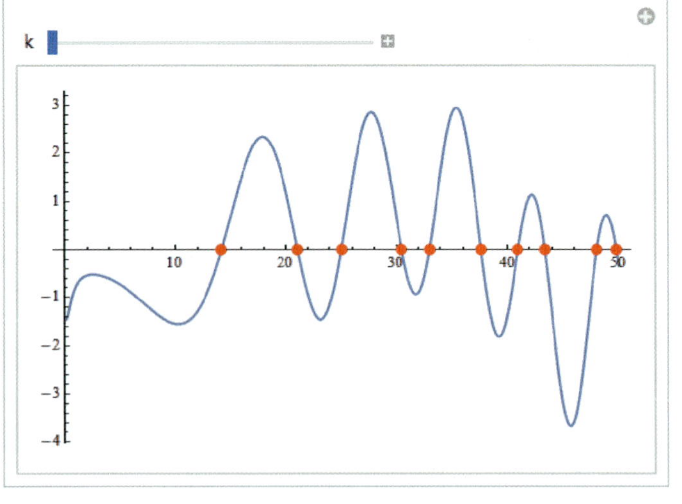

- Next, we plot 1/ |ζ(s)| below. We can see that all the zeros (columns) are clearly aligned.

```
Plot3D[1/Abs[Zeta[x + ⅈ y]], {x, -4, 5}, {y, -10, 40},
 MeshFunctions → (#3 &), MeshStyle → Blue, Mesh → 10,
 PlotPoints → 60, MaxRecursion → 3 , ViewPoint → {8, 1, 3} ,
 Boxed → False, BoxRatios → {5, 10, 2}, AxesLabel → {"x", "y", None},
 AxesEdge → {{1, -1}, Automatic, Automatic},
 Ticks → {Automatic, Range[0, 30, 10], Range[0, 4]},
 ClippingStyle → None, PlotRange → {0, 5}]
```

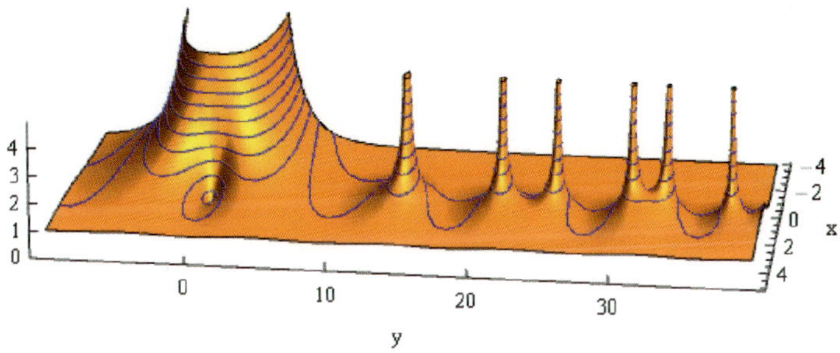

One of the main characteristics of the Zeta function is its connection with prime numbers. Remember that a prime number p is an integer greater than 1 that has no positive divisors other than 1 and itself. The set of prime numbers (that since Euclid we know is infinite) is called $\wp$.

It can be shown that $\zeta(s) = \sum_{k=1}^{\infty} \frac{1}{k^s}$, can also be computed using Euler's product.

- MathematicalFunctionData includes many mathematical functions including Euler's product:

```
myz = MathematicalFunctionData[
  Riemann zeta function ζ(s) (mathematical function) , "NamedIdentities"][[3]]
```

```
Euler's product →
 Function[{s, p}, Inactivate[ConditionalExpression[Zeta[s] == ∏_{k=1}^{∞} 1/(1 - p[k]^-s),
   Re[s] > 1 && p[k] == Prime[k]]]]
```

- These are:

$$\zeta(s) = \left(1 - \frac{1}{2^s}\right)\left(1 - \frac{1}{3^s}\right)\left(1 - \frac{1}{5^s}\right) + \ldots = \prod_{p \in \wp} \frac{1}{1 - p^{-s}}$$

- Prime numbers seem to be randomly distributed. For a given prime number, there's no general formula that will give us the next one.

```
Table[Prime[n], {n, 45, 60}]
```

```
{197, 199, 211, 223, 227, 229, 233,
 239, 241, 251, 257, 263, 269, 271, 277, 281}
```

- The graph below shows how many prime numbers are less than or equal to an arbitrary number n. We denote this function $\pi(n)$ (in the example $n = 200$).

```
Plot[PrimePi[n], {n, 1, 200}, AxesLabel → {"n", "π(n)"}]
```

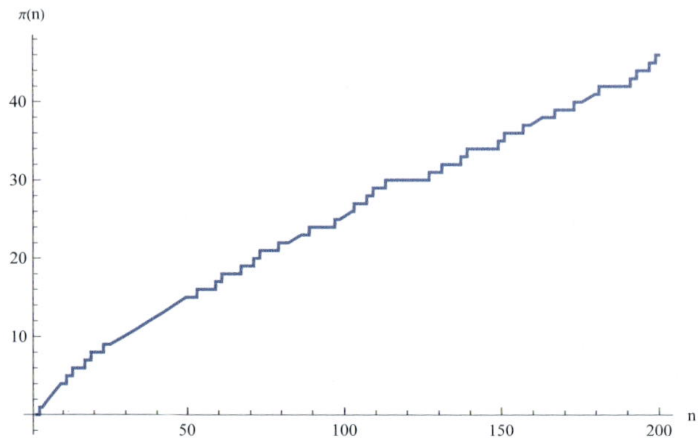

The trend is that the frequency of prime numbers decreases as n gets bigger.

There are several approximations to calculate $\pi(n)$ such as: $t/\text{Log}(t)$, or $\text{li}(n) = \int_0^z dt/\log t$ (in *Mathematica* **LogIntegral[x]**), but the best one is done by using Riemann's Zeta function. Specifically, the program uses the function $R(n) =$ **RiemannR[x]**.

For $x > 0$, $R(x) = \sum_n^\infty \mu(n)\,\text{li}(x^{1/n})/n$.

- In the following plot we compare the real value of $\pi(\text{n})$ with all these approximations.

```
Manipulate[Plot[{PrimePi[n], n/Log[n], LogIntegral[n], RiemannR[n]},
  {n, 1.5, p}, PlotLegends -> "Expressions"], {{p, 10000}, 100, 10^6}]
```

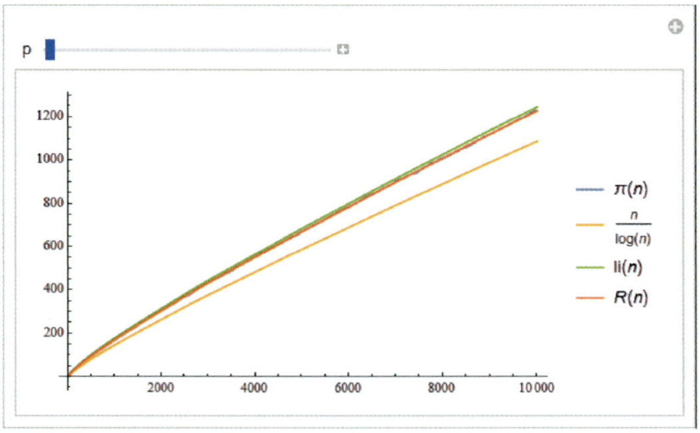

- We can visualize the comparison in an even better way by representing the difference between each of the alternative approximations and $\pi(n)$. In the case of R(n) the approximation is very accurate since, at least in the chosen interval, the error never gets too big (actually it never exceeds the real value by more than ± 3).

```
Plot[{n / Log[n] - PrimePi[n], LogIntegral[n] - PrimePi[n],
  RiemannR[n] - PrimePi[n]}, {n, 200, 1000}, PlotLegends -> "Expressions"]
```

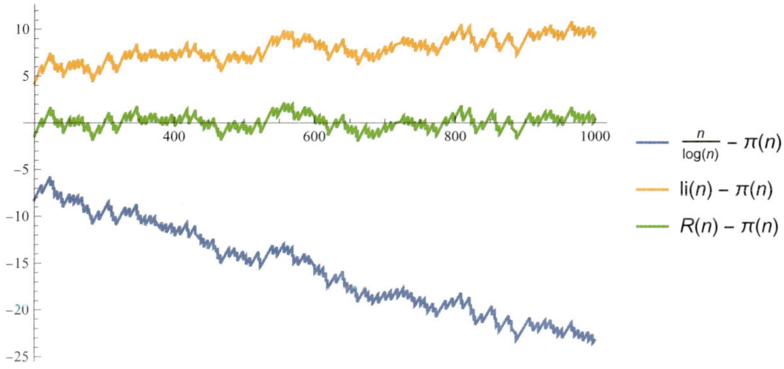

- The table below compares $\pi(n)$, with $n = 10^{13}$(10 trillion), with 4 of its approximations.

```
x = 10^13;
Grid[{{"π(n)", PrimePi[x]},
  {"Legendre:n/(Log[n]-1.08366) ", Round[x/ (Log[x] - 1.08366)]},
 {"Chebyshev:n/(Log[n]-1", Round[x / (Log[x] - 1)]},
 {"Gauss: li(n)", Round[LogIntegral[x]]},
 {"Riemann: R(n)", Round[RiemannR[x]]}
 }, Frame → All]
```

π(n)	346 065 536 839
Legendre:n/(Log[n]-1.08366)	346 621 096 885
Chebyshev:n/(Log[n]-1	345 618 860 221
Gauss: li(n)	346 065 645 810
Riemann: R(n)	346 065 531 066

7.8 Additional Resources

At mathworld.wolfram.com and functions.wolfram.com you can find a compilation of different methods used to compute π, their *Mathematica* implementations, references and hyperlinks to alternative sources:

http://mathworld.wolfram.com/Pi.html

http://functions.wolfram.com/Constants/Pi

To find out more about Buffon's method you can visit:

http://demonstrations.wolfram.com/BuffonsNeedleProblem

Throwing Buffon's Needle with Mathematica by Enis Siniksaran: http://www.mathematica-journal.com/issue/v11i1/BuffonsNeedle.html

The Wikipedia article on π is an excellent resource in terms of both its contents and its links:

https://en.wikipedia.org/wiki/Pi

Regarding the Riemann hypothesis, there are many references online, see for example:

http://mathworld.wolfram.com/RiemannZetaFunction.html

https://en.wikipedia.org/wiki/Riemann_zeta_function

8

Looking at the Sky

In this chapter we give a brief introduction to astronomy using some of the functions available in Mathematica. In the process we will also learn to model the orbits of eclipsing binary stars by creating and analyzing light curves, very useful when searching for extrasolar planets.

8.1 A Short Astronomical Walk

Contemplating the sky on a starry night, away from light pollution, is one of the most fascinating shows that we can enjoy. It's not strange then that for millennia humans have looked up at the sky and realized that the sun and the moon follow regular cycles. The majority of ancient cultures came up with mythological or religious explanations about celestial objects but they also used those objects for practical purposes such as the creation of calendars to schedule harvests. For example, the Egyptians associated the appearance of Sirius, the brightest star in the norther hemisphere (now we know it's a binary system), with the annual flooding of the Nile. Despite the reduced number of celestial bodies that can be seen with the naked eye, it's surprising how much knowledge of the sky that astronomers already had before the invention of the telescope.

The arrival of the telescope about 400 years ago (1609) heralded a revolution: the number of known stars went from thousands to millions, the moon, far from being a perfect sphere, was found to be covered with mountains and craters, the wandering stars (the planets) were more than mere bright dots, they actually had satellites, and Saturn had rings (it's not the only planet). Later on we learned to analyze the composition of the light coming from space. Many of the objects that looked like blurry spots under the telescope became galaxies with myriads of stars. It was discovered that those galaxies were moving apart from each other: the universe was expanding, an expansion that started around 14 billion years ago. It turned out that the visible stars and galaxies contain less than 5% of the mass of the Universe. The rest is dark energy and matter, we know it exists (at least the dark matter) but we don't really know what it is. The idea of the existence of dark energy ("a riddle wrapped inside a mystery") started with an amazing discovery (1998): The expansion of the universe is accelerating, a discovery that was worth a Nobel Prize in 2011. And the discovery race continues, promising new surprises: the important thing is still the journey.

For astronomical calculations, *Mathematica* has several functions: `CometData`, `ExoplanetData`, `GalaxyData`, `MinorPlanetData`, `MoonPosition` `Sunset`, `NebulaData`, `PlanetData`, `PulsarData`, `SiderealTime`, `StarClusterData`,

StarData, SunPosition, SupernovaData, Sunrise and many more. These functions replace and extend the AstronomicalData function available until *Mathematica* 9 although you can still use it with versions 10 and 11.

The syntax of all these functions is similar to that of the other computable functions. Example: **Function**["*name*", "*property*"] where "*name*" or entity (Entity) is an object and "*property*" is a characteristic or parameter of the object.

- For example: to refer to the solar system planets we can type:

```
PlanetData[]
```

```
{Mercury, Venus, Earth, Mars,
 Jupiter, Saturn, Uranus, Neptune}
```

- If we place the cursor on one of the list elements in the output, for example "Mercury", *Mathematica* displays Entity["Planet", "Mercury"]. This tells us that "Mercury" belongs to the "Planet" entity. We can also see it by showing the output in InputForm:

```
InputForm[PlanetData[]]
```

```
{Entity["Planet", "Mercury"], Entity["Planet", "Venus"],
 Entity["Planet", "Earth"], Entity["Planet", "Mars"],
 Entity["Planet", "Jupiter"], Entity["Planet", "Saturn"],
 Entity["Planet", "Uranus"], Entity["Planet", "Neptune"]}
```

- We can get the available properties for PlanetData by typing **EntityProperties**["Planet"]:

  In this example and the following one, we use Shallow or Short to avoid getting the complete output. If you want to see all the properties just remove Shallow when executing the command:

```
PlanetData["Properties"] // Shallow
```

```
{absolute magnitude H, age, albedo, alphanumeric name,
 altitude, next maximum altitude, angular diameter, angular radius,
 largest distance from the Sun, largest distance from orbit center, <<100>>}
```

- The next output informs us about the property "OrbitPeriod" for each one of the planets. Notice the use of "EntityAssociation" to link names with properties.

```
PlanetData[PlanetData[], "OrbitPeriod", "EntityAssociation"]
```

```
<|Mercury → 87.96926 days, Venus → 224.70080 days,
 Earth → 365.25636 days, Mars → 1.8808476 Julian years,
 Jupiter → 11.862615 Julian years, Saturn → 29.447498 Julian years,
 Uranus → 84.016846 Julian years, Neptune → 164.79132 Julian years|>
```

- We can use the free-form input: **Insert ▶ Inline Free-format input** to calculate for example the Mercury average orbit velocity:

```
{ Mercury average velocity ,
 EntityValue[Entity["Planet", "Mercury"], "AverageOrbitVelocity"] ,
 EntityValue[Entity["Planet", "Mercury"], "AverageOrbitVelocity"] }
```

```
{47.4 km/s , 47.4 km/s , 47.4 km/s }
```

- Sometimes, we may want to get the value without its corresponding unit. In those cases, we can use `QuantityMagnitude`.

```
QuantityMagnitude[PlanetData[ Mercury (planet) , average orbit velocity ]]
```

```
47.4
```

The result is given in light-years (ly), a common unit to express distances between stars.

- The parsec (pc), and its multiples (kiloparsec or kpc, and Megaparsec or Mpc), are used to measure very long distances such as distances between galaxies. Let's see the relationship between a parsec and a light year.

```
convert parsec to light year »
Result
  3.262 ly
```

- The table showing the equivalence to 1 light-year in different units:

```
Pane[TableForm[Map[{#, N@UnitConvert[Quantity["LightYears"], #]} &,
   {"Meters", "Parsecs", "SIBase", "SI", "Imperial", "Metric"}]] , 200]
```

Meters	9.46073×10^{15} m
Parsecs	0.306601 pc
SIBase	9.46073×10^{15} m
SI	9.46073×10^{12} km
Imperial	5.87863×10^{12} mi
Metric	63 241.1 au

- For some properties there may be no information available: (`Missing[NotAvailable]` or `Missing[NotApplicable]`). Remember that with `DeleteMissing` we can remove those properties whose headings contain the word "Missing".

```
DeleteMissing[
 MinorPlanetData[ Plutoids , "Density", "EntityAssociation"]]
```

```
<| 136199 Eris (2003 UB313) → 2.53 g/cm³ , Pluto → 2.095 g/cm³ |>
```

Some of the functions related to Astronomy need the precise location for which the information is desired. By default they will use our current time and position.

- Example: The following function returns the sunrise time on January 10, 2017, at a place located 40° N and 10° W, a location that is one hour ahead of GMT as we can see.

```
Sunrise[GeoPosition[{40, 10}], DateObject[{2017, 01, 10, 0, 0}]]
```

Tue 10 Jan 2017 08:43 GMT+2.

Dates and hours are of crucial importance in Astronomy. The existence throughout history, even nowadays, of different calendars makes it difficult to compare dates of astronomical and historical events. To avoid this problem we use Julian dates. A Julian date (not to be confused with the Julian calendar: Julian year) indicates the days that have gone by since January 1, 4713 B.C. at 12:00 PM UT (November 24, 4714 BC, in the proleptic Gregorian calendar). The name of the Julian date comes from Julius Scaliger (not Julius Caesar) who proposed it in 1583 after realizing that such a day was the least common multiple of 3 calendar cycles used by the Julian calendar. Using 12:00 PM as the starting time instead of 12:00 AM has the advantage that astronomical observations, usually done at night, are recorded on the same day. Starting with *Mathematica* 10.2 the function `JulianDate` is available. The function converts dates from our calendar, the Gregorian one, to Julian dates. It takes into account the modifications to the Julian calendar done by Gregorio XIII (1582) that improved the handling of leap years, but in which the year 0 doesn't exist.

- To find out the current date and time (if our computer is synchronized) type:

```
Date[]
```

```
{2016, 12, 4, 22, 7, 31.9273790}
```

- We can use the above mentioned formula to convert the actual date to a Julian date. We have added `Dynamic` to dynamically update the calculation with an interval defined with `UpdateInterval`, in this case 1 second.

```
Dynamic[AccountingForm[JulianDate[Date[]], 12], UpdateInterval → 1]
```

```
2457769.1177
```

We can stop the dynamic updating by unchecking it in the Evaluation menu: **Evaluation ▶ Dynamic Updating Enabled**.

Star positions in the sky don't follow an annual cycle exactly. They are subject to perturbations due to their own motions and other celestial movements such as the precession of equinoxes. Because of this, it's necessary to regularly set up dates to use when referring to celestial coordinates. These standard reference dates are known as epochs. The last one was set on January 1, 2000 at 11:58:55.816 AM. When referring to this epoch we use **J2000.0**. StarData and other astronomical functions refer to the position of stars using this latest epoch without taking into account either the precession of equinoxes or the stars' own motions. For practical purposes these adjustments are irrelevant but we need to consider them when making historical projections.

8.2 Stargazing

```
Clear["Global`*"]
```

An incredible experience is to look at the sky away from light-polluted cities, trying to imagine the hidden celestial mechanism generating the incredible view. We will soon realize the greatness of those night owl geniuses (Hipparchus, Ptolemy, Copernicus, Brahe, Kepler and others) that established the foundations of modern astronomy without any other means than their plain sight and basic instruments such as the astrolabe to measure the stars' positions. Everything changed in 1609 when Galileo used the telescope for astronomical purposes for the first time. Another equally important invention was the pendulum clock that enabled the measurement of time with great precision. Centuries later, telescopes mounted on satellites and atomic clocks heralded a new revolution.

8.2.1 Naked-Eye Astronomy

When looking at the sky, almost all of the objects that can be seen with the naked eye are stars. Throughout history, different cultures imagined that some stars resembled figures, named constellations. The same phenomenon occurred in civilizations that were not connected in any way. For example: The seven brightest stars from the Ursa Major were interpreted the same way in very different places. The inhabitants of the British islands said that those stars were the legendary King Arthur's cart. For the Germans, they represented a wagon pulled by three horses. The Greeks came up with a more imaginative story: In a Greek legend, the god Zeus and the mortal Callisto had a son named Arcas. Hera, Zeus' jealous wife, turned Callisto into a female bear. Arcas, unaware that the bear was actually her mother almost killed her. Afterward, Zeus also turned Arcas into another female bear. Callisto is Ursa Major and Arcas is Ursa Minor. Greek mythology is full of such stories to explain almost all the stars that appear near each other in the sky. Many of those myths have actually given names to the constellations. Over the centuries, many constellations were added until they eventually covered the entire visible sky from any place on Earth. There are 88 constellations in total. In 1930, the International Astronomical Union established the limits of each of them. Obviously, they are imaginary lines covering the entire firmament, including both hemispheres. As in maps, we don't need to keep all the constellations' details. They and the names of their most important stars are represented in a planisphere, a celestial map that shows us the sky each night for a given date and time.

- The function below returns the sky chart for our location at the time given in the input. We can get *Mathematica* to create it for us by typing "Sky chart at 22 GMT" in free-form input, right-clicking on the sky chart and selecting **Paste input for ► Subpod content**.

```
WolframAlpha["Sky chart at 22 GMT", {{"SkyMap:PlanetData", 1}, "Content"}]
```

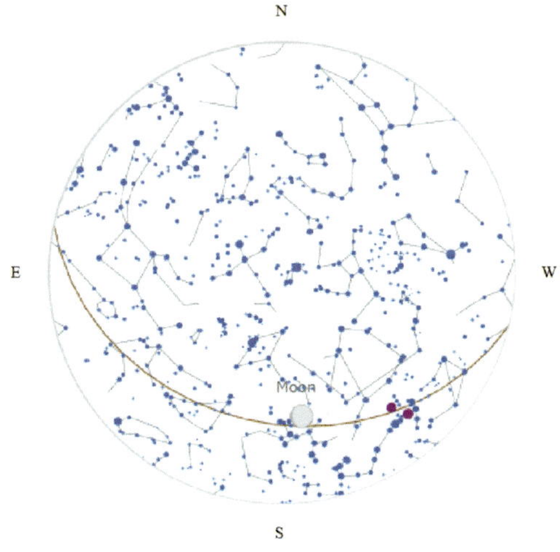

- With the instruments and methods available to astronomers nowadays, the information that we can obtain about stars is very extensive. In the next example we show the properties most commonly used when describing stars:

```
properties = {"AbsoluteMagnitude", "ApparentMagnitude",
   "RightAscension", "Declination", "Altitude", "Azimuth",
   "Constellation", "DistanceFromSun", "SpectralClass"};
```

- Let's find those properties for Polaris, the North Star:

```
StarData["Polaris", properties, "PropertyAssociation"]
```

```
<|AbsoluteMagnitude → -3.64,
 ApparentMagnitude → 1.97, RightAscension → 2h 31m 47.06s ,
 Declination → 89 degrees 15 arc minutes 51. arc seconds ,
 Altitude → 36 degrees 37 arc minutes 45. arc seconds ,
 Azimuth → 359 degrees 5.0 arc minutes , Constellation → Ursa Minor ,
 DistanceFromSun → 430.921 ly , SpectralClass → F7:Ib-IIvSB |>
```

The apparent magnitude gives how bright an astronomical object appears to an observer on regardless of its intrinsic brightness (http://scienceworld.wolfram.com/astronomy/ApparentMagnitude.html).

The smaller the magnitude, the bigger its luminosity. For the most luminous stars we use negative values. Most of the people looking at the sky with the naked eye and with little light pollution just see stars with apparent magnitude of less than 5.

- Here are the 10 closest stars to Earth, many of them not visible to the naked eye:

```
StarData[EntityClass["Star", "StarNearest10"]]
```

```
{Sirius , Lalande 21185 , Proxima Centauri ,
 Rigel Kentaurus A , Rigel Kentaurus B , Barnard's Star ,
 Luyten 726-8 A , Luyten 726-8 B , Wolf 359 , Sun }
```

Naturally, the closest one is the Sun. Usually for solar system bodies distances are shown by default in astronomical units (au), the average distance between the Earth and the Sun. The Earth rotates around the Sun following an ellipse, so its distance changes, although as its eccentricity is very small, in many calculations we can assume that it follows a circle. Its average value is 1 au.

- The command below returns the name of the star along with its distance to earth in (ly), except for the Sun, and its apparent magnitude.

```
{#, StarData[#, "DistanceFromEarth"],
   StarData[#, "ApparentMagnitude"]} & /@
 StarData[EntityClass["Star", "StarNearest10"]]
```

```
{{Sirius, 8.59093 ly, -1.44}, {Lalande 21185, 8.30216 ly, 7.49},
 {Proxima Centauri, 4.21809 ly, 11.01},
 {Rigel Kentaurus A, 4.38981 ly, -0.01},
 {Rigel Kentaurus B, 4.4001 ly, 1.35}, {Barnard's Star, 5.9339 ly, 9.54},
 {Luyten 726-8 A, 8.55731 ly, 12.57}, {Luyten 726-8 B, 8.55731 ly, 12.7},
 {Wolf 359, 7.78812 ly, 13.45}, {Sun, 1.01288 au, -26.72}}
```

We can immediately notice that most of them are not visible since their apparent magnitude is bigger than 5. With the exception of the Sun, the star with the biggest apparent magnitude is Sirius, a star that we can see during winter nights (in the Northern hemisphere) next to the Orion constellation. The closest one is Proxima Centauri that is only visible in the Southern hemisphere (Proxima Centauri rotates around the double system Centaurus A and B in a 500,000-year period. By chance, it's currently in its orbital position closest to the solar system).

- The number of stars visible without the use of technology is very small compared to the number of stars in our galaxy. The command below generates a plot representing the naked-eye stars and their distances to Earth in light years. (This type of query may take a long time to execute).

```
ListPlot[
 Sort[StarData[StarData[EntityClass["Star", "NakedEyeStar"], "Entities"],
   "DistanceFromSun"]], AxesLabel → {"stars", "ly"}]
```

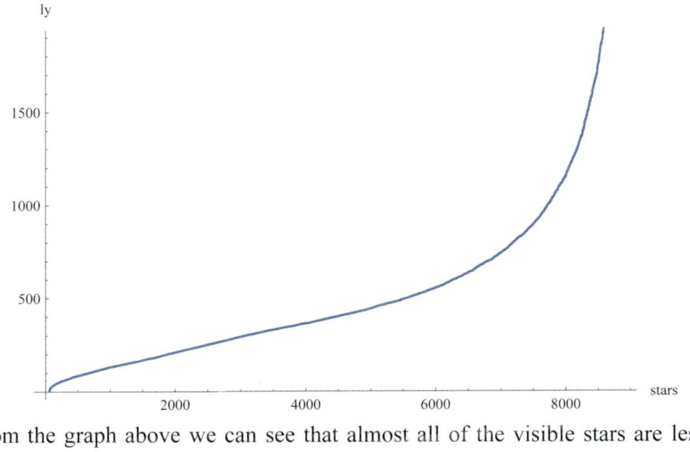

From the graph above we can see that almost all of the visible stars are less than 1,500 ly away. Our galaxy, the Milky Way, has a width of approximately 100,000 ly. Therefore, with our bare eyes we only see a very small fraction of the stars in our own galaxy, most of them the ones closest to Earth. The previous figure refers to the number of potentially visible stars without light pollution or moonlight. In practice, the actual number is 2 or 3 thousand, since we only see part of the sky depending on our position.

Besides stars with apparent magnitude 5 or less, the only other celestial bodies visible using only our eyesight are: the Moon, the five nearest planets and, sporadically, some comets. In the

Northern hemisphere, the only object visible that doesn't belong to our galaxy is the Andromeda galaxy. In short, we see an insignificant fraction of our own galaxy, which in turn is just one among billions of galaxies.

We've seen that among the properties of `StarData` and other astronomical functions are: "RightAscension", "Declination", "Altitude" and "Azimuth". These properties are commonly used to indicate the position of a celestial body. In equatorial coordinates, "RightAscension" and "Declination" are similar to latitude and longitude but they refer to the celestial sphere so they are independent of the observer (for further details see: *The Celestial Sphere*, http://demonstrations.wolfram.com/TheCelestialSphere, by Jeff Bryant). Another type of coordinates widely used are the alt-azimuthal (alt/az) that depend on the observer: The "Altitude"(alt) is the height of the star over the horizon. It goes from 0° to 90° and has a positive sign for stars located above the horizon and a negative sign for the ones located below it. The "Azimuth" (az) is the arc in the horizon measured counterclockwise from the South point until the object's vertical. Its value ranges from 0° to 360°.

- The next example shows Sirius' displacement on the first of each month at 20:00 h using alt/az coordinates.

```
sirius = Table[QuantityMagnitude[
     AstronomicalData["Sirius", {#, {2017, i, 1}, {20, 00}}]] & /@
   {"Azimuth", "Altitude"}, {i, 1, 12}]
```

```
{{138.2, 42.32}, {179.3, 53.3}, {217.9, 44.60}, {241.2, 22.46},
 {253.4, -3.222}, {262.1, -31.27}, {270.6, -59.0}, {329.6, -86.2},
 {87.8, -62.7}, {96.8, -35.01}, {105.3, -6.85}, {116.7, 19.11}}
```

```
Graphics[{Orange, Point[sirius]}, Frame → True,
 FrameLabel → {"azimuth", "altitude"}, AspectRatio → 1 / 2]
```

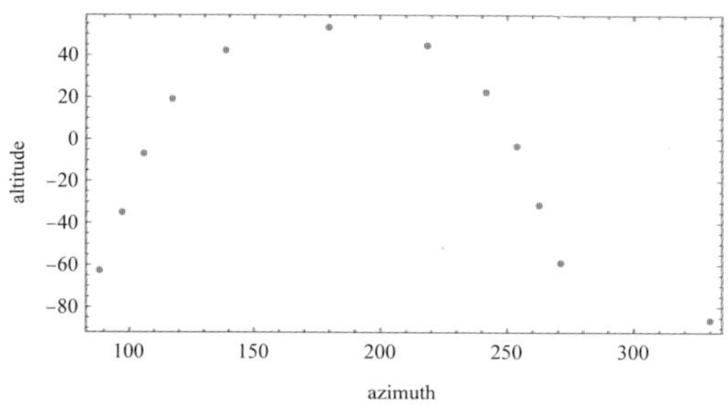

As we can see, there are negative values that correspond to the period of the year in which Sirius falls below the horizon and as a consequence is not visible. As a matter of fact, the Egyptians considered that the appearance of the star marked the beginning of a new year.

- In the previous example we didn't specify the observer's location. Remember that in cases like that the function uses the location given by `FindGeoLocation`.

```
FindGeoLocation[]
```

```
GeoPosition[{40.42, -3.68}]
```

- However, it's important to keep in mind that the function actually gives us the position of the server that we are using to connect to the Internet based on the IP address. Sometimes it may be far away from our real position. This would be the case if, for example, we were using a phone to surf the net. We can check it by showing the obtained position in a map:

```
GeoGraphics[
 GeoVisibleRegionBoundary[FindGeoLocation[]], GeoZoomLevel -> 12]
```

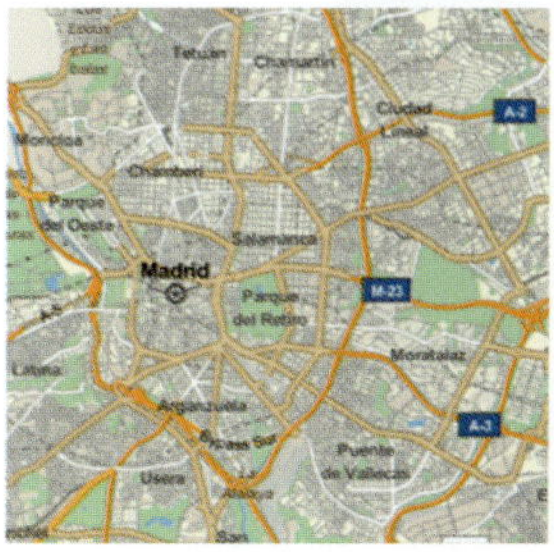

Naked-eye stars display an amazing regularity. It's possible to forecast the location of each star without prior knowledge of celestial mechanics. If we observe a star at a certain hour, for example at 24:00 h, ignoring the displacement due to the inclination of the Earth's axis with respect to the ecliptic, the next day it will be on the same spot four minutes earlier, that is at 23:56 h, completing a cycle in a year. The explanation that the Greek civilization and others provided for this fact was to assume that all the stars (which they called fixed, to distinguish them from the planets or wandering stars that did not exhibit such behavior) were glued to a dome that rotated with a daily cycle of 23 h and 56 min, the sidereal day.

```
sidereal days
```

```
1 sidereal day
```

- We can use `UnitConvert` to find out the actual duration. When executing the command above, the predictive interface will do the conversion automatically.

```
UnitConvert[1 sidereal day, MixedRadix["Hours", "Minutes", "Seconds"]]
```

```
23 h 56 min 4.09054 s
```

However, over very long periods we can notice discrepancies that at first sight may seem insignificant during a person's lifetime but become quite obvious after several generations. Hipparchus of Nice (c. 190 BC – c. 120 BC) compared the stellar maps available at the time (celestial cartography is over 2,500 years old!) and realized that there was a relative movement of the stars with respect to the ecliptic. This movement is nowadays known as the precession of the equinoxes. This is probably his most famous discovery. The displacement happens when the Earth's axis, moving along a circumference with respect to the ecliptic, rotates with a period of 25,771 years.

An excellent interactive demonstration (Figure 8.1) showing this effect can be downloaded from http://demonstrations.wolfram.com/PrecessionOfTheEquinoxes.

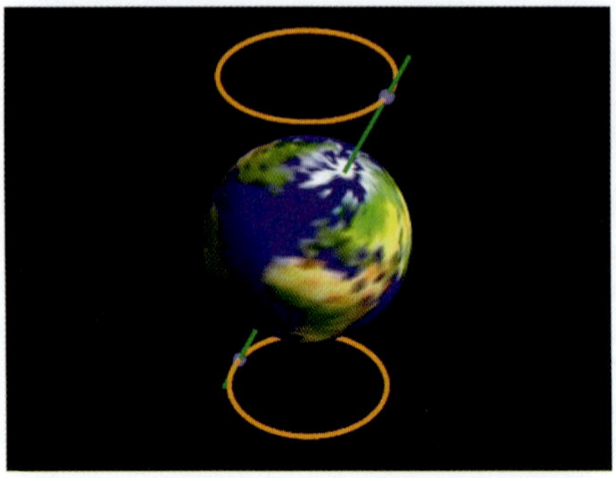

Figure 8.1 Precession of the Equinoxes demonstration.

As a result, the North celestial pole keeps on moving. Today it is close to the North Star but 4,800 years ago it was pointing toward Thuban (Alpha Draconis). William Shakespeare did not realize that, when in his play *Julius Caesar* stated: "But I am constant as the northern star, Of whose true-fix'd and resting quality There is no fellow in the firmament".

In reality, all the stars seen from the Earth have a slow displacement motion as a result of the precession and their own orbits in the galaxy. This last type of movement is known as proper motion.

- We can find the proper motion of the North Star as follows:

```
StarData["Polaris", "ProperMotion"]
```

```
11.75 mas/yr
```

"mas/yr" means milliarc seconds per year. It's a value that cannot be observed during a lifetime. We need to keep in mind that a telescope on the ground can rarely see details smaller than 1 arc second. However, during long periods even the shapes of the constellations change.

Astrology (a pseudoscience) attributed predictive powers to the constellations. For example, they were supposed to determine the future of people born under them. The firmament was divided into 12 signs corresponding approximately to the number of lunar cycles in a year. Each division was assigned a symbol called a zodiac sign. The names of the zodiac are associated with the first constellations named by the Greeks during the 5th century BC, even though the Babylonians were the first ones to mentioned them 4,000 years ago. The starting point was the constellation pointing toward Aries (the moment when the Sun moves from the south celestial hemisphere to the north one coinciding with the spring equinox) in the 5th century BC. Nowadays, we still use the same division although the constellations are now in a different location to where they were 2,500 years ago. The real astronomical zodiac dates correspond to the constellation located behind the solar disk during the spring equinox, in the direction opposite to the Earth's direction (the Sun actually passes through 13 constellations and not 12, the 13th one is Ophiuchus). The zodiac signs must be adjusted by a month to adapt to the current astronomical reality, so chances are you may have to revise your zodiac sign (although it wouldn't be very useful).

- The natural language input below returns the location of the Ursa Minor stars in the year 100,000 compared to their current position (dashed line). The rest of the stars experience similar changes.

ursa minor 100000 AD
Constellation region

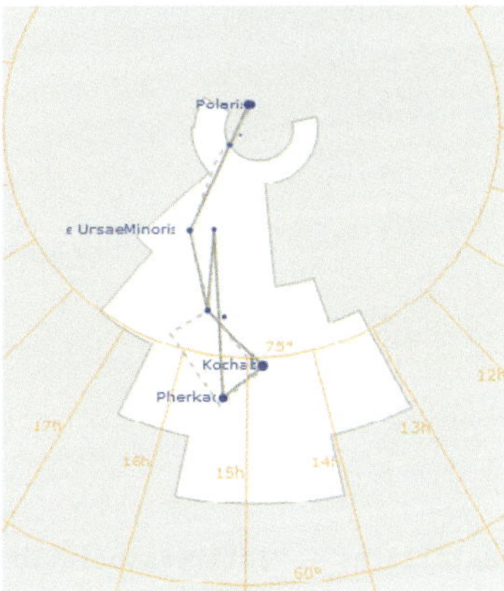

(current configuration shown dashed)

8.2.2 Wandering Planets

Since antiquity, people have realized that there are a small number of stars exhibiting a behavior different from the one of the fixed stars and their annual cycles. These stars were once called wandering stars. Nowadays we know that they correspond to the 5 visible planets: Mercury, Venus, Mars, Jupiter and Saturn (Uranus was not added to the list of planets until 1783 given how difficult it is to observe it with the naked eye).

To get information about planets we use the function `PlanetData` but before we learn more about this function, let's see an example first available in the Wolfram Cloud to showcase the use of `Dataset` and `CloudGet`. Remember that a dataset has a hierarchical structure and that this general structure is quite common in practice. In this case, the dataset (http://wolfr.am/7FxLgPm5) contains information about the mass and radius of each planet and its corresponding moons.

- Let's import the dataset:

```
planets = CloudGet["http://wolfr.am/7FxLgPm5"];
```

- Here we show the moons of Mars:

```
planets["Mars", "Moons"]
```

	Mass	Radius
Phobos	1.072×10^{16} kg	11.1 km
Deimos	1.5×10^{15} kg	6.2 km

- The next command displays the number of moons for each planet:

```
planets[All, "Moons", Length]
```

Mercury	0
Venus	0
Earth	1
Mars	2
Jupiter	63
Saturn	61
Uranus	27
Neptune	13

- Another way to obtain the Uranus moons is:

```
Length /@ PlanetData[PlanetData[], "Satellites", "EntityAssociation"]
```

```
<|Mercury → 1, Venus → 1, Earth → 1, Mars → 2,
  Jupiter → 63, Saturn → 61, Uranus → 27, Neptune → 14|>
```

- Note that there is a discrepancy in the number of Neptunian satellites. In fact, PlanetData uses curated information but in this case we are using http://wolfr.am/7FxLgPm5 as an example to show how to import a dataset. For further information we can check Wikipedia:

```
TextSentences[WikipediaData["Moons of Neptune"]][[;; 2]]
```

```
{Neptune has 14 known moons, which are
   named for minor water deities in Greek mythology.,
 By far the largest of them is Triton, discovered by William
   Lassell on October 10, 1846, just 17 days after the
   discovery of Neptune itself; over a century passed before
   the discovery of the second natural satellite, Nereid.}
```

As we mentioned previously, if we periodically observe a planet with the naked eye at the same time and from the same place and we record (sometimes we can take pictures, as we saw with the example of the Sun) its location with respect to the horizon, the resulting figure is called an analemma. In Chapter 5 we built the one corresponding to the Sun. Here we are going to do the same for Mars and Venus.

- We need first to define our position (latitude and longitude).

```
zone = "Location" -> GeoPosition[{40.96, -5.66}];
```

- Next, we display the Mars and Venus positions at 20:00 h each 5 days for the first 90 days of 2017.

```
analemmamars = Table[{
    QuantityMagnitude[PlanetData["Mars", EntityProperty["Planet", "Azimuth",
       {"Date" → DateObject[DateList[{2017, 1, i, 20}]], zone}]],
     "AngularDegrees"], QuantityMagnitude[
     PlanetData["Mars", EntityProperty["Planet", "Altitude",
       {"Date" → DateObject[DateList[{2017, 1, i, 20}]], zone}]],
     "AngularDegrees"]}, {i, 1, 90, 5}];
```

```
analemmavenus = Table[{
    QuantityMagnitude[
     PlanetData["Venus", EntityProperty["Planet", "Azimuth",
       {"Date" → DateObject[DateList[{2017, 1, i, 20}]], zone}]],
     "AngularDegrees"], QuantityMagnitude[
     PlanetData["Venus", EntityProperty["Planet", "Altitude",
       {"Date" → DateObject[DateList[{2017, 1, i, 20}]], zone}]],
     "AngularDegrees"]}, {i, 1, 90, 5}];
```

- With `GraphicsRow` we can display both graphs in the same row. The graph for Mars has been modified using `AspectRatio` to make its visualization easier.

```
GraphicsRow[{
  Graphics[{Orange, Point[analemmamars]},
   Frame → True, FrameLabel → {"azimuth", "altitude"},
   AspectRatio → 1/2, PlotLabel → Style["Mars Analemma", Bold]],
  Graphics[{Orange, Point[analemmavenus]}, Frame → True,
   FrameLabel → {"azimuth", "altitude"},
   PlotLabel → Style["Venus Analemma", Bold]]}, Frame → All]
```

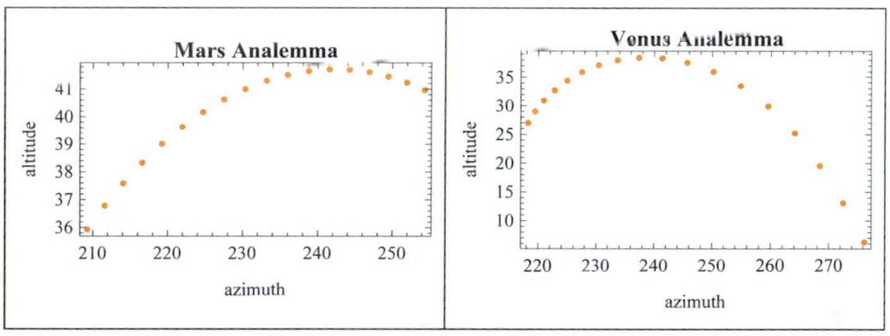

The points with negative altitude correspond to those times when the planet is below the horizon and therefore invisible. We cannot see the planet when it's behind the Sun either.

We can also see that in the case of Venus, the planet is only visible a few degrees over the horizon. This is because Venus is an interior planet and as such, when observed from the Earth, it will never be very high above the horizon. This means that we will never be able to see it next to the zenith since the sunlight would blind us. Obviously, this doesn't happen with exterior planets such as Mars.

- The graph below shows the orbits of the terrestrial planets or inner planets: Mercury, Venus, the Earth and Mars, with the Sun (not in scale) in the center.

Inner planets

terrestrial planets (planets)

terrestrial planets

```
InputForm[%]
```

```
EntityClass["Planet", "InnerPlanet"]
```

```
terrestialPlanets =
  PlanetData[EntityClass["Planet", "InnerPlanet"], "OrbitPath"];
```

```
Graphics3D[
 {{Yellow, Sphere[{0, 0, 0}, 0.1]}, terrestialPlanets}, Boxed → False]
```

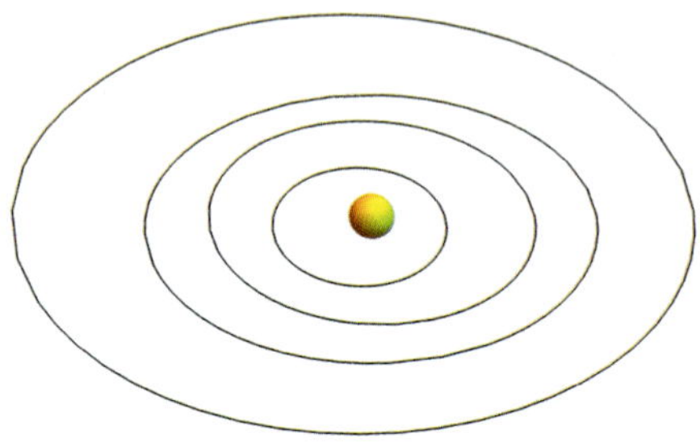

- The next four functions, based on examples included in the `PlanetData` help page, enable us to calculate and display graphically the distance from Earth and the apparent magnitude of the visible planets over a year:

```
visibleplanets[planet_, date_, property_] :=
 PlanetData[planet, EntityProperty["Planet", property, {"Date" → date}]]
```

```
ts[planet_, d1_, d2_, property_] := TimeSeries[
  {#, visibleplanets[planet, #, property]} & /@ DateRange[d1, d2, "Week"]]
```

```
DateListPlot[
 AssociationMap[ts[#, {2017, 1, 1}, {2017, 12, 31}, "DistanceFromEarth"] &,
  {"Mercury", "Venus", "Mars", "Jupiter", "Saturn"}],
 PlotLegends → "Expressions", FrameLabel → Automatic]
```

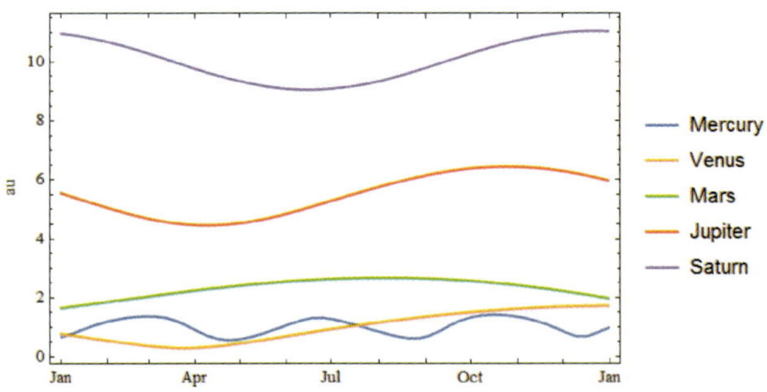


```
DateListPlot[
 AssociationMap[ts[#, {2017, 1, 1}, {2017, 12, 31}, "ApparentMagnitude"] &,
  {"Mercury", "Venus", "Mars", "Jupiter", "Saturn"}],
 PlotLegends → "Expressions", FrameLabel → Automatic]
```

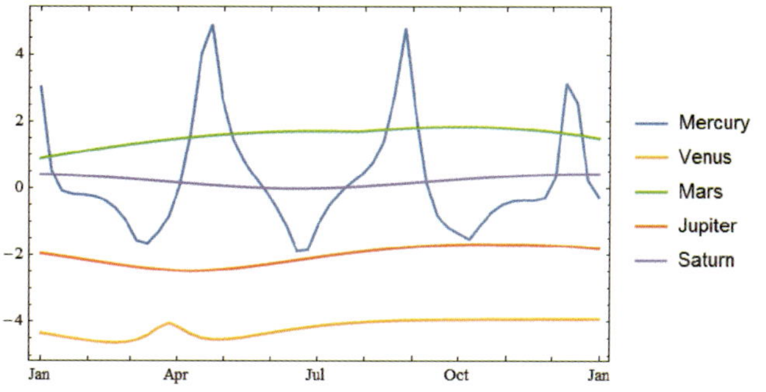

Even though naked-eye observation and the use of angle-measuring instruments (sextant, astrolabe, etc.) advanced the science of Astronomy, the arrival of the telescope and other instruments such as precision clocks was no less important. They opened new possibilities for exploring the firmament. Later on, in the 20th century, radio telescopes and spacial astronomy would arrive taking the science to a whole new level.

8.2.3. Dwarf Planets

We've seen that `PlanetData` doesn't include Pluto, demoted to the "Dwarf Planet" category in a controversial meeting of the International Astronomical Union on August 24, 2006. This action was motivated mainly by the discovery of planets beyond Pluto, in an area known as the Kuiper Belt that probably includes thousands of planetoids. Pluto is considered to be part of this belt. All the planets that didn't fit the new definition were categorized as dwarf planets. There are reasons to justify that Pluto doesn't belong to the same category as the classical planets but the dwarf label doesn't seem the most adequate one since there are dwarf planets that are most likely bigger than the classical planet Mercury.

- The celestial bodies that orbit around the Sun but are not planets have been included in `MinorPlanetData`.

```
MinorPlanetData["Classes"]
```

```
{minor planets, Amor asteroids, Apollo asteroids,
 asteroid belt, Aten asteroids, dwarf planets,
 Kuiper Belt objects, scattered disk objects, near-Earth asteroids,
 Plutoids, trans-Neptunian objects, Jupiter Trojan asteroids,
 centaur asteroids, inner main belt asteroids, main belt asteroids,
 Mars crossing asteroids, outer main belt asteroids}
```

- Next, we use a couple of functions to know how long it would take Pluto to complete one orbit starting in 1970-01-01 and how far away in astronomical units (au) it would be from the Earth each year.

```
MinorPlanetData["Pluto", "OrbitPeriod"]
```

```
247.92065 Julian years
```

```
DateListPlot[
 {{#, 01, 01 }, MinorPlanetData["Pluto", EntityProperty["MinorPlanet",
      "DistanceFromEarth", {"Date" → #}]] } & /@
  Range[1970, 1970 + 248, 10]]
```

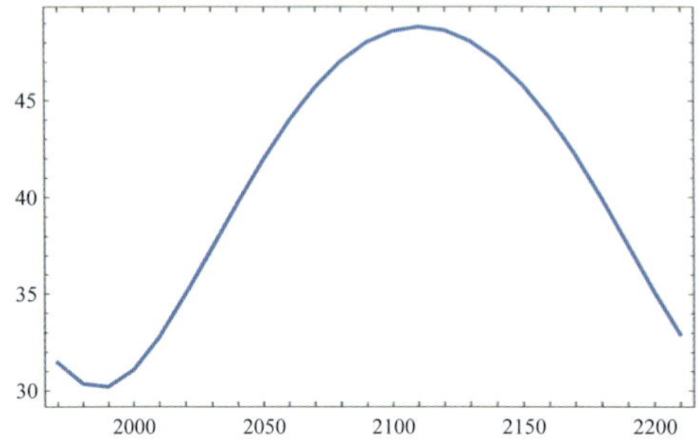

Notice that during our lifetimes Pluto will be further away each year. In July 2015 a probe, New Horizons, visited it for the first time.

- One of the most interesting characteristics of Pluto and other celestial bodies in the Kuiper Belt is that their orbits are normally slanted with respect to the ecliptic plane, next to the ones from the classical planets, as we can see with the following command:

```
Text[Grid[{MinorPlanetData[ Plutoids (minor planets) ],
   MinorPlanetData[ Plutoids (minor planets) , "Inclination"]},
  Frame → All, Background → LightYellow]]
```

136199 Eris (2003 UB313)	136108 Haumea (2003 EL61)	136472 Makemake (2005 FY9)	Pluto
44.143°	28.1950°	28.99899°	17.1°

In the coming years the number of newly discovered dwarf planets will most likely increase substantially.

- We dynamically simulate the movement of the dwarf planets starting on January 1, 2017 using `MinorPlanetData`. Notice the use of `Tooltip` to see the name when placing the mouse over a planet. The function is slow.

```
minorplanets = MinorPlanetData[EntityClass["MinorPlanet", "Plutoid"]];
Manipulate[
 Graphics3D[{(Tooltip[Sphere[#[[2]], 1], #[[1]]]) & /@
   Transpose[{minorplanets,
    QuantityMagnitude[
       MinorPlanetData[EntityClass["MinorPlanet", "Plutoid"],
     EntityProperty["MinorPlanet", "HelioCoordinates", {"Date" ->
      DateObject[{2017, 1, date}]}]]]}], ColorData[1, 1],
    MinorPlanetData[#, "OrbitPath"] & /@ minorplanets},
  PlotLabel -> DateString[DateList[{2017, 1, date}]]],
 {date, 1, 300 × 365.25}, SaveDefinitions → True]
```

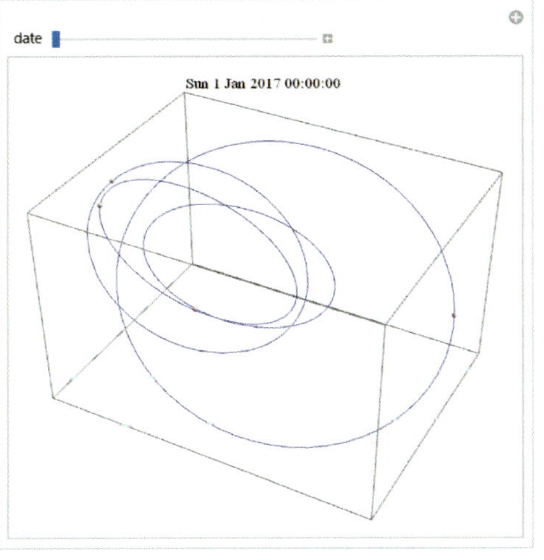

8.2.4 Exoplanets

To obtain information about planets outside of the solar system we can use the function `ExoplanetData`.

- According to a paper published in *Nature* on August 25, 2016, there's an Earth-sized planet orbiting our nearest neighboring star other than the sun, Proxima Centauri.

```
ExoplanetData[EntityClass["Exoplanet", {EntityProperty["Exoplanet",
"DistanceFromEarth"] -> TakeSmallest[1]}]]
```

```
{Alpha Centauri Bb}
```

- Next all the available information using `ExoplanetData` regarding this planet as of 2016-12-04 is shown:

```
DeleteMissing[ExoplanetData[ Alpha Centauri Bb (exoplanet) ,
  ExoplanetData["Properties"], "PropertyAssociation"]]
```

```
⟨| alphanumeric name → Alpha Centauri Bb, alternate names → {},
 altitude → -68 degrees -58 arc minutes -37. arc seconds ,
 apparent altitude → -68 degrees -59 arc minutes -0.×10^-1 arc seconds ,
 average orbit distance → 5.98×10^6 km ,
 azimuth → 195 degrees 22 arc minutes 37. arc seconds ,
 above the horizon → False, classes → {},
 component → b, constellation → Centaurus ,
 declination → -60 degrees -50 arc minutes -14. arc seconds ,
 discovery method → radial velocity, discovery year → 2012 ,
 distance from Earth → 4.4001 ly , distance from Sun → 4.40009 ly ,
 distance interval → Interval[{4.40311, 4.40311}] ly , orbital eccentricity → 0.,
 galactic latitude → -1 degrees -3 arc minutes -45.8321 arc seconds ,
 galactic longitude → 316 degrees 4 arc minutes 0.162094 arc seconds ,
 heliocentric XYZ coordinates → {-1.64501 ly , -2.79001 ly , -2.97835 ly },
 major axis → 1.20×10^7 km , mass → 6.75×10^24 kg ,
 mass interval → Interval[{6.72×10^24, ∞}] kg ,
 next maximum altitude time → Mon 5 Dec 2016 11:05 GMT+1. ,
 minor axis → 1.196×10^7 km , name → Alpha Centauri Bb,
 object type → Exoplanet, orbit center → Rigel Kentaurus B ,
 orbit circumference → 3.75734×10^7 km , orbital period → 77.71 h ,
 orbital period interval → Interval[{77.67, 77.75}] h ,
 longitude of periapsis ϖ → 55.0° ,
 primary star K semiamplitude → 0.51 m/s , right ascension → 14h 39m 34.9s ,
 semimajor axis → 5.98×10^6 km , semiminor axis → 5.98×10^6 km |⟩
```

Although the planet orbits (0.04 au) closer to its star than Mercury does to the Sun, the star itself is far fainter than the Sun. The surface temperature would allow the presence of liquid water. However, the conditions on the surface may be strongly affected by the ultraviolet and X-ray flares from the star far more intense than the Earth experiences from the Sun.

8.2.5 Galaxies and Nebulae

Nowadays, we know that the Sun is just one of the stars in a galaxy known as the Milky Way. However, the notion of galaxy was not introduced until the 1920s. Before then, the difference between galaxies and nebulae was not clear. Seen through the telescope both appeared like blurry objects with shapes resembling clouds. That's why the term nebulae was used to refer to both. Galaxies were considered a type of nebula and our galaxy was the entire Universe.

- Let's see the definition of galaxy:

```
galaxy definition
Result
```

1	noun	a splendid assemblage (especially of famous people)
2	noun	tufted evergreen perennial herb having spikes of tiny white flowers and glossy green round to heart-shaped leaves that become coppery to maroon or purplish in fall
3	noun	(astronomy) a collection of star systems; any of the billions of systems each having many stars and nebulae and dust

Based on the 3rd definition, nebulae are just part of galaxies. With telescopes we can only see the ones located in our own galaxy. The generic name nebulae actually includes two types of very different structures: the planetary nebulae ("PlanetaryNebula") generated by the accumulation of dust left after a supernova explosion, and nebulae ("Nebula") star nurseries, such as the Orion Nebula. As we can see the name is confusing as planetary nebulae have nothing to do with the formation of planets. Other very important galaxy elements are *clusters* or globular clusters (groups of stars). Information about galaxies, clusters and nebulae can be found using `GalaxyData`, `StarClusterData` and `NebulaData`.

- A very interesting nebula is Orion, visible specially during Winter in the northern hemisphere. It is an enormous gas cloud where new stars are constantly being born. The command below displays some of its properties ("//Short" should be removed to see the complete output).

```
DeleteMissing[
  Transpose[{NebulaData["Properties"], NebulaData["OrionNebula", #] & /@
     NebulaData["Properties"]}], 1, 1] // Short
```

```
{{alphanumeric name, M42}, <<28>>,
 {next transit time, Mon 15 Aug 2016 10:25 GMT+2.}}
```

- We can find out the diameter of our galaxy. To express it in terms of light-years we use `UnitConvert`.

```
{GalaxyData["MilkyWay", "Diameter"],
 UnitConvert[GalaxyData["MilkyWay", "Diameter"], "LightYears" ]}
```

$\{9.5\times10^{17}\ \mathrm{km},\ 1.0\times10^{5}\ \mathrm{ly}\}$

- We've seen that big distances, such as the ones between galaxies, are usually measured in parsec (pc), the equivalent of the distance corresponding to one arc second. The closest galaxies to our own are shown below, note the use of `Dataset`. They all form a group of galaxies known as the local group. Many of them are quite small.

```
lggaxies = Dataset[GalaxyData[ Local Group of galaxies (galaxies) ,
   "DistanceFromSun", "EntityAssociation"]];
```

- We select the galaxies closer than 100 kpc:

```
lggaxies[Select[# < 100 kpc &]]
```

Boötes Dwarf Galaxy	60.3591 kpc
Canis Major Dwarf Galaxy	7.65979 kpc
Draco Dwarf Galaxy	81.8065 kpc
Large Magellanic Cloud	49.9418 kpc
Milky Way	7.61113 kpc
Sagittarius Dwarf Elliptical Galaxy	19.9154 kpc
Sculptor Dwarf Galaxy	79.049 kpc
Sextans Dwarf Galaxy	85.7896 kpc
Small Magellanic Cloud	60.5736 kpc
Ursa Minor Dwarf Galaxy	65.8742 kpc
Willman I	36.767 kpc

- To check which galaxies are visible to the naked eye (Apparent Magnitude < 5) we sort the apparent magnitudes of the local group members (remember that the smaller the magnitude the higher the luminosity).

```
Dataset[DeleteMissing[GalaxyData[ Local Group of galaxies (galaxies) ,
   "ApparentMagnitude", "EntityAssociation"]]][Select[# < 5 &]]
```

Large Magellanic Cloud	0.9
M31	3.5
Sagittarius Dwarf Elliptical Galaxy	4.5
Small Magellanic Cloud	2.2

As mentioned earlier, with our own eyes it is very difficult to see magnitudes greater than 5. This means that in the northern hemisphere the only visible galaxy is M31 (Andromeda), the rest of the galaxies with high luminosity (and correspondingly low apparent magnitude) are only visible from the southern hemisphere.

8.3 Application: Determining the Color of the Stars

```
Clear["Global`*"]
```

The French philosopher Auguste Comte (1798–1857) believed that the composition of stars was going to be beyond human knowledge. In his *Cours de Philosophie* he used to say that stars would only be known as specks of light in the sky since they were very far away. However, the analysis of stellar light has enabled us to know reasonably well stars' composition, temperature and evolution.

A star is what in Physics is known as a black body and therefore has a light spectrum whose characteristics are related to its surface temperature following Planck's law.

- Using the free-from input we can access Planck's law formula. First we type "Planck law" using natural language, press + once Mathematica has understood our input, select the equation's pod, and after right-clicking on the subpod and choosing "Copy subpod content", paste the information on the cell below.

Planck law »

↳ Equation

$$L(\lambda) = \frac{2 h c^2}{\lambda^5 \left(e^{\frac{h c}{\lambda k T}} - 1\right)}$$

$L(\lambda)$	spectral radiance as function of wavelength
λ	wavelength
T	temperature
h	Planck constant ($\approx 6.62607 \times 10^{-34}$ J s)
c	speed of light in vacuum ($\approx 2.998 \times 10^{8}$ m/s)
k	Boltzmann constant ($\approx 1.38065 \times 10^{-23}$ J/K)

- The spectrum for a specific temperature can be represented using the BlackBodyRadiation package (guide/StandardExtraPackages).

```
<< BlackBodyRadiation`
```

```
BlackBodyProfile[2000 Kelvin, 4000 Kelvin, 5800,
 8000 Kelvin, PlotRange → {{0, 2.0×10^-6}, {0, 5×10^14}}]
```

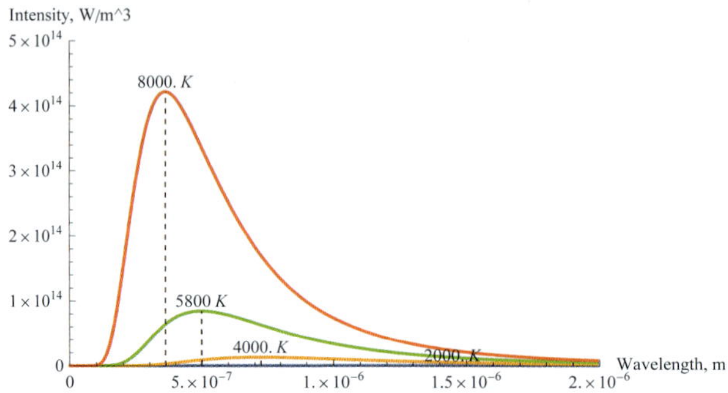

- An alternative way is to use `FormulaData`.

```
plancklaw1 = FormulaData[{"PlanckRadiationLaw", "Wavelength"}]
```

$$L[\lambda] == \frac{2 h c^2}{\left(-1 + e^{\frac{1 h c/k}{T \lambda}}\right) \lambda^5}$$

- Let's have a look at the inner formula composition:

```
InputForm[plancklaw1[[2]]]
```

```
Quantity[2, "PlanckConstant"*"SpeedOfLight"^2]/
 ((-1 + E^(Quantity[1, ("PlanckConstant"*"SpeedOfLight")/
      "BoltzmannConstant"]/(QuantityVariable["T", "Temperature"]*
     QuantityVariable["λ", "Wavelength"])))*
  QuantityVariable["λ", "Wavelength"]^5)
```

- To get the constant values in the International System (SI):

```
plancklaw = UnitConvert[plancklaw1[[2]], "SI"]
```

$$\frac{1.1910430 \times 10^{-16}\ \text{m}^2\text{W}}{\left(-1 + e^{\frac{0.0143878\ \text{m K}}{T \lambda}}\right) \lambda^5}$$

- Next, we build a function where the wavelength is computed as function of the temperature.

```
blackbody[T_] := plancklaw /.
  {QuantityVariable["λ", "Wavelength"] -> Quantity[k , "Meters"],
   QuantityVariable["T", "Temperature"] -> Quantity[T, "Kelvins"]}
```

```
Plot[{Callout[blackbody[8000], "8000 K", Above],
  Callout[blackbody[5800], "5800 K", Above],
  Callout[blackbody[4000], "4000 K", Above]},
 {k, 10^-7, 3 × 10^-6 }, AxesOrigin → {0, 0}, PlotRange → All]
```

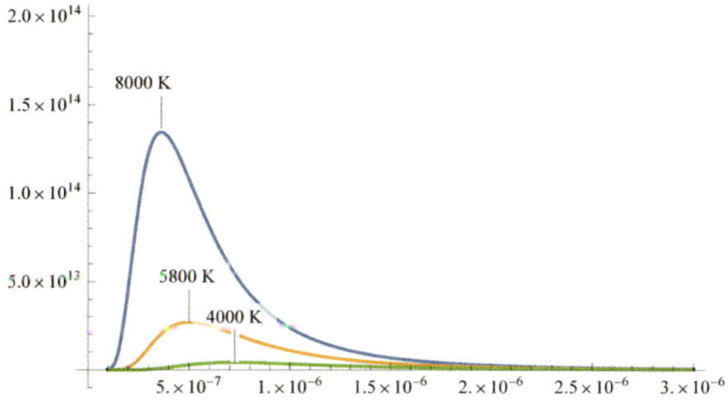

We are going to build a dynamic representation of Planck's law that shows the relationship between color and temperature.

- The next function associates a black body spectrum to its temperature (in degrees Kelvin).

```
BlackBody[T_] :=
 With[{h = 6.62607 × 10^-34 , c = 2.99792458 × 10^8, k = 1.38065 × 10^-23},
  Plot[(2 h c^2 / λ^5 ) / (Exp[h c / (λ k T)] - 1),
   {λ, 10^-9, 1000 × 10^-9}, MaxRecursion → 0, ColorFunction →
    (ColorData["VisibleSpectrum"][10^9 #] &), ColorFunctionScaling → False,
   Filling → Axis, Ticks → None, PlotLabel → Row[{T, " K"}]]]
```

- With `Manipulate` we create the dynamic representation:

```
Manipulate[
 BlackBody[T], {{T, 5800, "temperature"}, 3000, 10000},
 SaveDefinitions → True]
```

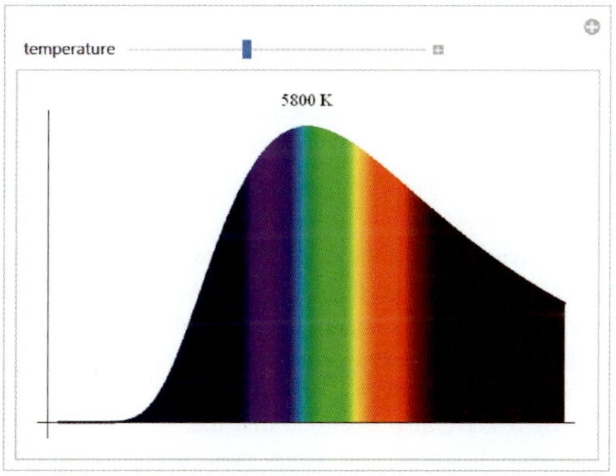

A more complete model by Jeff Bryant can be found in:

"Blackbody Spectrum" (http://demonstrations.wolfram.com/BlackbodySpectrum).

- The next function simulates the color of any given star. We use **StarData**[*"star name"*, *"EffectiveTemperature"*] to get the superficial temperature and then relate it to its color:

```
starColorPlot[star_] := Graphics3D[{ColorData["BlackBodySpectrum"][
    QuantityMagnitude[StarData[star, "EffectiveTemperature"]]], Sphere[]},
  Boxed → False, Lighting → {{"Ambient", Gray},
    {"Directional", White, ImageScaled[{0, 0, 1}]}}, PlotLabel → star]
```

- Let's apply it to several known stars:

```
starColorPlot /@ {"Rigel", "Sirius", "Sun", "Capella", "Betelgeuse"}
```

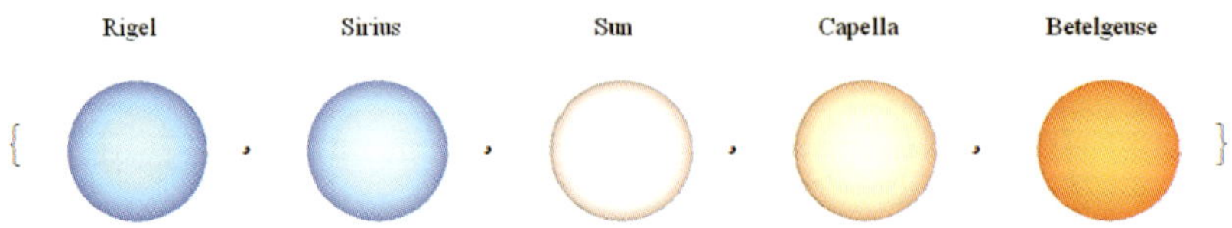

- Blue corresponds to hotter stars and orange to colder ones as shown below:

```
StarData[{"Rigel", "Sirius", "Sun", "Capella", "Betelgeuse"},
  "EffectiveTemperature"]
```

$\{1.5\times10^4\text{ K},\ 9.4\times10^3\text{ K},\ 5.78\times10^3\text{ K},\ 5.1\times10^3\text{ K},\ 3.6\times10^3\text{ K}\}$

8.4 The Measurement of Distances across the Universe

```
Clear["Global`*"]
```

Historically, the measurement of distances across the Universe was one of the fundamental problems in Astronomy. The first step toward solving it was to measure the Earth's size but before even that, someone had to realize that our planet is a sphere. It is believed that it was the Pythagoreans, mysterious people that didn't leave any written works behind, who for the first time, circa 430 B.C., came up with the idea that the Earth was round, based probably on two observations: a) Sailors traveling from Greece to Northern Africa had noticed that several constellations, such as Ursa Major, would appear higher in the horizon in Egypt than in Athens for a given date and time (although there are other geometric shapes that may explain this fact, the sphere is the one that provides the simplest explanation), b) The way the Moon hides during lunar eclipses can be easily explained if one assumes that it's a shadow projected by a sphere, the Earth, upon the Moon.

Once it was clear that the Earth was round, the next step was its measurement. It was Eratosthenes, who eventually became the head of the library of Alexandria, the first person to measure the Earth's size. The method he used is legendary and well-known among Astronomy enthusiasts, but it's worth describing it in the next few lines. He started measuring the distance between Alexandria and Syene (nowadays Aswan), that were located approximately on the same meridian (they really differ by 3°). He also assumed that the Sun was distant enough that its rays would hit the Earth in a parallel manner. He knew that during the summer solstice in Syene the light would illuminate the bottom of the wells in the city. At that moment the sunbeams would hit the ground perpendicularly while at the same time in Alexandria they would form an angle of approximately 1/50th of the length of a circumference (number deduced by the projection of the shadows). He then used a basic trigonometric identity to figure out that the distance between Alexandria and Syene was 1/50th of the Earth's circumference. Since that distance was 5,000 stadia, the Earth's circumference was 250,000 stadia. There's no agreement about the exact conversion rate to meters. Depending on the reference, if we take a value ranging from 185 m to 157.2 m, the error is between 17% and 1% of the actual size, quite an extraordinary approximation in any case. In 1492, Christopher Columbus assumed a considerably smaller figure while trying to convince Queen Isabella of Spain to finance his trip against the opinion of her advisors, who knew Eratosthenes' estimate. He persisted in his error and ended up discovering America.

Shortly before Eratosthenes' death, Hipparchus of Nice was born (c. 190 BC). He's not only famous due to his discovery of the precession of equinoxes, as we have already mentioned, but also because of his quite accurate measurement of the distance between the Earth and the Moon using the shadow projected by the former onto the later during Moon eclipses. He also initiated an empirical study of the apparent magnitude of the stars and their positions. For that purpose, he invented the ecliptic coordinates (different from the elliptic ones) and used the astrolabe (invention attributed to him). Many centuries later (from 1989 to 1993) a satellite, carrying its name, repeated his work with technology from the late 20th century. Its successor: Gaia (http://www.esa.int/Our_Activities/Space_Science/Gaia_overview), currently in mission, will measure the position of the objects in our galaxy with extreme precision.

We can compute the distance to the nearest stars using geometric methods (Figure 8.2). We measure the position of the star seen from a certain location and 6 months later we will observe its apparent displacement. This phenomenon is known as parallax, and its explanation is similar to what happens when we place our thumb between and object and our eyes. When closing either eye, we will notice the apparent movement of the object. In the case of a stellar parallax, each eye corresponds to the vertices of the base of an isosceles triangle created when joining the Earth position in two vertices A and B, separated by 6 months along the orbit of the star, with vertex C, the position of the star, considered fixed. The parallax is half the angle ACB.

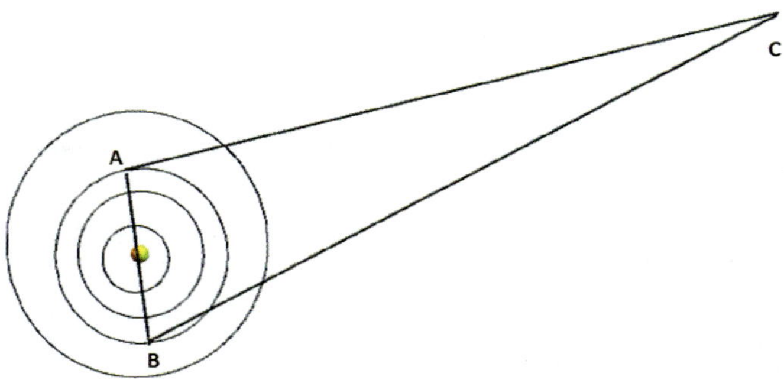

Figure 8.2 Calculating the distance to the nearest stars.

The distance, *d*, from the Sun to the star with parallax, *p*, in radians, can be computed as follows:

$d = \frac{1\text{ AU}}{\tan p} = \{\text{given that the angle is very small, } \tan p \simeq p\} \simeq \frac{1}{p}\text{ AU}.$

- Since the values of *p* are very small, it is convenient to express them in arcseconds, p_s.The parallax of Procyon is:

```
ps = StarData["Procyon", "Parallax"]
```

```
0.286"
```

- From that number we can calculate the distance in parsecs:

```
        1
─────────────────────
QuantityMagnitude[ps]
```

```
3.50
```

- We can get the same result directly:

```
UnitConvert[StarData["Procyon", "DistanceFromEarth"], "Parsecs"]
```

```
3.49329 pc
```

A big leap forward in the measurement of astronomical distances happened when Henrietta Swan Leavitt (1868–1921) noticed that certain stars, now known as Cepheid variables, exhibited regular changes in their luminosity from which their absolute magnitude could be calculated.

The relationship period/luminosity for the Cepheid variables has been revised several times since Henrietta Leavitt's original measurements. Currently, the following function is usually applied:

$$M = -2.78 \log (P) - 1.35$$

where M is the absolute magnitude of the star (the luminosity of the star if it was 1 pc away) and P is the average period in days.

- The previous expression can be written in *Mathematica* as:

```
abm[pd_] = -2.78 Log [10, pd] - 1.35;
```

Applying a similar criterion to the one used to determine the distance between a bulb with known brightness and the brightness experienced by an observer, we can deduct the following relationship between M, the apparent magnitude m (magnitude seen from the Earth, calculated as the average between the maximum and minimum observed magnitudes) and the distance D (in pc):

$$m - M = 5 \log (D/10) = 5 \log(D) - 5$$

- From the previous formulas we can establish the following equation to compute the distance using the period (pd) and the apparent magnitude (apm), both of which can be measured empirically:

```
distceph[pd_, apm_] := Solve[apm - abm[pd] - 5 Log[10, dist] + 5 == 0, dist]
```

This method can be applied to stars in our galaxy with the help of big ground telescopes or space ones. We can also use it to measure the distances to our nearest galaxies assuming we can find Cepheid variable stars in them.

Example: In Astroex (http://www.astroex.org), part of the European Southern Observatory (ESO), we can find a set of basic astronomy exercises using real data. In one of them we are given the results of the calculations done by Hubble in several Cepheid variables located in the M100 galaxy. According to the data, one star has a period of 53.5 days and an apparent magnitude of 24.5. Stars of such magnitude can only be observed with large telescopes.

- The absolute magnitude and distance, in pc, are:

```
{abm[53.5], dist /. distceph[53.5, 24.90]}
```

$\{-6.15482, \{1.62542\times10^{7}\}\}$

It's one of the most distant objects ever measured using this method.

For remote galaxies the method used is based on measuring the apparent luminosity of a type Ia supernovae and relating it to the real magnitude (that can be deduced using theory).

Another key parameter related to the previous one is the redshift experienced by light. This is an example of the Doppler effect caused by galaxy movements, mostly associated with the expansion of the universe.

- You can find information about redshift using `WolframAlpha` (just type in a *free-form input cell* "**`Cosmological redshift`**"). After expanding the cell and telling *Mathematica* to interpret it as a formula, you can even do some interactive calculations.

```
Cosmological redshift
↳ Basic dimensions
```

Astronomer Edwin Powell Hubble (1889–1953) (the famous telescope was named after him) realized that galaxies frequently presented redshift. The amount changed depending on the galaxy. Later on, this astronomical phenomenon became known as Hubble's law.

$$c\,z = H_0\,D$$

with:

z (redshift) $= (\lambda_{\text{ob}} - \lambda_e)/\lambda_e$, where λ_{ob} corresponds to the observed wavelength and λ_e to the emitted one; z is therefore a dimensionless number.
c is the speed of light.
D is the actual distance to the galaxy (in Mpc).
H_0 is the Hubble constant at the moment of observation (during our lifetime and for the duration of our civilization, this value will be practically constant).

Once z is known, we can calculate D using basic algebra:

$$D = \frac{c\,z}{H_0}$$

Hubble's law is considered a fundamental relation between recessional velocity and distance. However, the relation between recessional velocity and redshift depends on the cosmological model adopted, and is not established except for small redshifts.

- The Hubble constant, H_0, is regularly expressed in m/s/Mpc.

```
Hubble constant in km/s/Mpc
↳ Result

70 km/s/Mpc (kilometers per second per megaparsec)
```

A modified version of Hubble's law and other properties of the universe are now included in `UniverseModelData`.

- Determine the distance of a comoving object, in light years, for a redshift of 0.075:

```
f[t_?NumericQ] := QuantityMagnitude@
   UniverseModelData[Quantity[t, "LightYears"], "Redshift"];
FindRoot[0.075 == f[dist], {dist, 10^9}]
```

```
{dist → 1.00142×10^9}
```

- Determine the redshift of a comoving object $5\ 10^9$ light years away:

```
UniverseModelData[Quantity[5 × 10^9, "LightYears"], "Redshift"]
```

```
0.485194
```

- Let's apply this method to the galaxy IC1783.

```
zIC1783 = GalaxyData["IC1783", "Redshift"]
```

```
0.01117
```

```
FindRoot[zIC1783 == f[dist], {dist, 10^8}]
```

$\{\text{dist} \to 1.5599 \times 10^8\}$

- We compare this value with the best estimate available in `GalaxyData`.

```
UnitConvert[GalaxyData["IC1783", "DistanceFromEarth"], "ly"]
```

1.25501×10^8 ly

We can see that there are some discrepancies. This is because galaxies are also affected by the gravitational forces of their galactic neighbors, e.g., Andromeda influencing the Milky Way. In fact, Andromeda and our galaxy are approaching each other and will collide in approximately 4 billion years from now.

The possibilities of the astronomical functions in *Mathematica* don't end here. Feel free to continue exploring.

8.5 Application: Binary Systems and the Search for Extrasolar Planets

```
Quit[]
```

In this section we will show how we can use *Mathematica* for other Astronomy related matters. We will describe in particular several features developed by the author for the study of binary systems and, by extension, the search for extrasolar planets.

Since 1995, when the first extrasolar planet was found, more than 400 have been detected. NASA's Kepler spacecraft (http://www.nasa.gov/kepler) launched on March 6, 2009, was specifically designed to search for extrasolar planets. Probably, by the time you read these lines, their total number will have increased significantly.

To find exoplanets we mainly use two methods:

1) Radial velocity: A star with a planet will move in its own small orbit in response to the planet's gravity leading to variations in the radial velocity of the star with respect to Earth. These movements are only detectable for giant planets next to stars.

2) Transit photometry: When a planet crosses in front of its parent star's disk the observed brightness of the star drops by a small amount. This last method is becoming the most effective one. The principle is similar to the one used when studying binary star systems.

More than half of the stars that we see are actually systems of two or more stars. Normally it's not possible to distinguish between different stars in multiple systems (not to confuse them with naked-eye binaries that are usually stars that seem close to each other when seen from Earth but are actually located in different systems altogether). The detection is usually done with the transit photometry method. This approach is very effective when dealing with eclipsing binaries, systems of two stars with coplanar orbits when observed from Earth.

Next we are going to describe how to build this kind of system using *Mathematica* and study its brightness. In the example that follows, we use as reference *Eclipsing Binary Stars* by Dan Bruton: http://www.physics.sfasu.edu/astro/ebstar/ebstar.html.

Figure 8.3 represents a binary system (although we discuss the case of two stars, the method is similar if we consider a star and a planet). The coordinates f the centers are (x1, y1, z1) for star 1 and (x1, y1, z1) for star 2 (assuming homogeneous spheres, the centers of mass coincide with the geometric centers). We consider {0, 0, 0} as the coordinates for the center of mass of the system. These coordinates have been chosen so that an observer, located on Earth, is on the axis OZ. That is: the one that goes from the center of mass to Earth. For the OY axis, we take the perpendicular to OZ in the star 1 plane when it forms an angle $\theta = 0$. This happens when the star is closest to the observer.

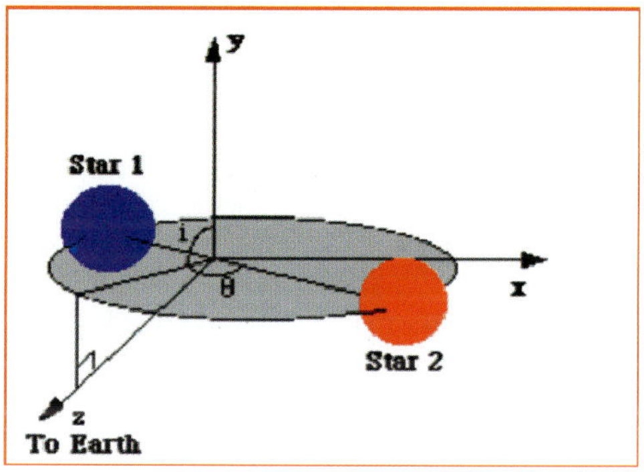

Figure 8.3 A binary system.

With:

i = orbital inclination: angle, in radians, between the line of vision and the orbital plane axis.
R = distance between the centers of both stars. It's expressed in arbitrary units, usually the average distance between the stars. In the case of a constant orbit R = 1.
m1, m2 = stars' masses (we can choose arbitrary units but it's important that they keep the equivalence q = m2/m1).
r1, r2 = radii. For convenient purposes we choose r1 > r2. In practice, we can choose stars 1 and 2 so that the condition is always met. It's convenient to express the radii in fractions of the major semi-axis. In the case of a circular orbit, obviously the major semi-axis is greater than or equal to the minor one and identical to the radius R.
θ = Phase, corresponding to the angle between the star and OZ, using $\theta = 0$, the angle where the center of star 2 is the closest to the observer. In a binary system with circular orbit, θ is a variable parameter (it's time dependent) while the previous parameters are constant (although that is not always the case, for example: There might be some intrinsic changes to the brightness or even the mass could change as in the case of a mass transfer between stars. However in the example that follows we'll use the simplest case, that happens to be the most common one). θ is known as the azimuthal angle and follows $\theta = \frac{2\pi(t-t0)}{P}$, with P being the orbital period.

In a binary system, the coordinates {x1, y1, z1} and {x2, y2, z2} for the location of the stars are given by:

$$x1 = \frac{-x}{1 + (1/q)},\ y1 = \frac{-y}{1 + (1/q)},\ z1 = \frac{-z}{1 + (1/q)};$$

$$x2 = \frac{x}{1 + q},\ y2 = \frac{y}{1 + q},\ z2 = \frac{z}{1 + q};$$

- We can convert them to spherical coordinates using the formulas:

```
x = R Sin[θ];
y = R Cos[i] Cos[θ];
z = R Sin[i] Cos[θ];
```

- Then, the position of each star as a function of R, i, θ, q with q = m2/m1 can be calculated with the following functions:

```
star1[R_, i_, θ_, q_ ] = {-x/(1 + (1/q)), -y/(1 + (1/q)), -z/(1 + (1/q))};
star2[R_, i_, θ_, q_ ] = {x/(1 + q), y/(1 + q), z/(1 + q)};
```

- We have used the previous equations to define a *Mathematica* function to visually display the movement of both stars as a function of R, i, θ, m1 and m2:

eclbin[R, i, θ, m2/m1, r1, r2] (The condition m2 < m1 is necessary).

```
eclbin[rr_, ii_, rho_, qq_, rr1_, rr2_] :=
 Module[{R, i, θ, q, r1, r2},
  Manipulate[Graphics3D[{ControlActive[Opacity[1], Opacity[.6]],
     Sphere[st1[R, i, θ, q], r1], Sphere[st2[R, i, θ, q] , r2],
     Opacity[1], RGBColor[.6, .74, .36], PointSize[.02],
     Point[{0, 0, 0}], Black}, PlotRange → {{-2, 2}, {-2, 2}, {-2, 2}},
    SphericalRegion → True, Boxed → False, Background → Black,
    ImageSize → {500, 340}, ViewPoint → {0, 0, 20}],
   {{R, rr, "R"}, 0.1, 5, Appearance → "Labeled"},
   {{i, ii, "i"}, 0, π, Appearance → "Labeled"},
   {{q, qq, "mass 2 / mass 1"}, 0.1, 1, Appearance → "Labeled"},
   {{r1, rr1, "radium 1"}, 0.1, .5, Appearance → "Labeled"},
   {{r2, rr2, "radium 2"}, 0.1, .5, Appearance → "Labeled"},
   {{θ, rho, "θ (rotation)"}, 0, 4 π, Appearance → "Labeled"},
   Initialization :>
    {st1[R_, i_, θ_, q_] := {-(R Sin[θ])/(1 + 1/q), -((R Cos[i] Cos[θ])/(1 + 1/q)),
       -((R Cos[θ] Sin[i])/(1 + 1/q))}, st2[R_, i_, θ_, q_] :=
      {(R Sin[θ])/(1 + q), (R Cos[i] Cos[θ]) / (1 + q), (R Cos[θ] Sin[i]) / (1 + q)}}]]
```

The initial moment $\theta = 0$ is the time when both stars are aligned and can be seen from the OZ axis (where the observer is located), that is when the interposition of the stars is at its maximum. In a real system, all the parameters would be fixed and only θ would change as a function of time.

- Let's consider the binary system defined by the following parameters R = 2, i = 2 Pi 85 /360, m1 = 0.6, m2 = 0.1, R1 = 0.4, R2 = 0.3.

```
eclbin[2, 2 Pi 85 / 360, 0, 0.1/0.6, 0.4, 0.3]
```

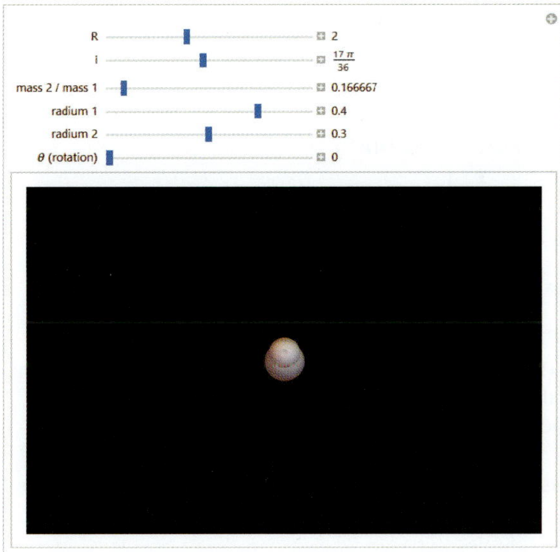

8.6 Light Curves

8.6.1 Light Curves in Binary Systems

```
Quit[]
```

The transit photometry method discussed in the previous section is based on the study of the changes in brightness when a star or a planet interposes itself between another star and the observer. As mentioned before, since stars are not visually distinguishable, even less a star and a planet, what we will see is a variation in the apparent brightness.

The method described here, with the arrival of astronomical CCD cameras, is within the reach of Astronomy fans. There have been instances where fans have even been able to detect exoplanets using this technique. This is a great example of how science can be done without the need for major resources.

The luminosity l of a star is defined as the amount of energy emitted by the star per unit of time. The flow F is the energy emitted per unit of surface and per unit of time, that is $\mathrm{F1} = \frac{l_1}{4\pi\,\mathrm{R1}^2}$ and $\mathrm{F2} = \frac{l_2}{4\pi\,\mathrm{R2}^2}$ where l_1, l_2 = luminosities (in arbitrary units, as long as they keep the ratio l_2/l_1).

Then an observer will see the system with a luminosity given by:

$$l = K(\mathrm{F1\,A1} + \mathrm{F2\,A2})$$

where A1 and A2 represent each of the areas of the stars' disks as seen by the observer and K that can be determined from the observer's detector area, on the OZ axis, and the distance between the Earth and the binary system.

To find these areas we need to know the apparent distance ρ as seen by the observer (located on the OZ axis). That distance can be calculated as follows:

$$\rho = \sqrt{(x_1 - x_2)^2 + (y_1 - y_2)^2}$$

- The next function computes ρ(R, i, θ, q):

```
rho[R_, i_, θ_, q_] = Module[{x, y, x1, x2, y1, y2}, x = R Sin[θ];
  y = R Cos[i] Cos[θ];
  {x1, y1} = {-x/(1 + (1/q)), -y/(1 + (1/q))};
  {x2, y2} = {x/(1 + q), y/(1 + q)};
  Sqrt[ (x2 - x1 )^2 + (y2 - y1 )^2]];
rho[2, 2 Pi 85 / 360, θ, 0.1/0.6 ]
```

$\sqrt{(0.0303845 \operatorname{Cos}[\theta]^2 + 4. \operatorname{Sin}[\theta]^2)}$

- With the example data (R = 2, i = 2 Pi 85 /360, $m_1 = 0.6$, $m_2 = 0.1$) the apparent distance ρ changes according to the graph below. Based on our problem formulation, the lowest apparent size corresponds to the maximum interposition that happens when θ = 0, Pi, 2 Pi, ..., *n* Pi.

```
Plot[rho[2, 2 Pi 85 / 360, θ, 0.1/0.6 ], {θ, 0, 4 Pi}, AxesLabel → {θ, ρ}]
```

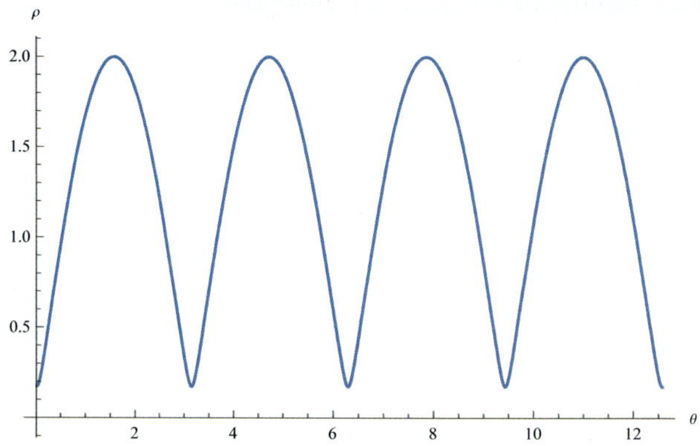

To calculate A1 and A2 it is necessary to take into account the current stage in the cycle.

- The two figures below show the star with the bigger radius (r_1) in red, and the one with the smaller radius (r_2) in blue.

```
Graphics[{Red, Disk[{0, 0}, 2], Blue, Disk[{4, 0}, 1]}]
```

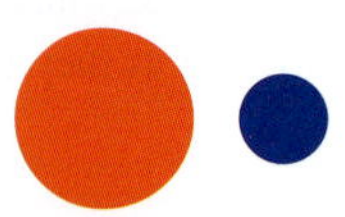

Without eclipse.- Corresponds to the period in which all the light emitted by both stars in the direction of the observer reaches him. Therefore, this is when the luminosity is at its maximum. In this stage, the following relationship holds: $\rho > r_1 + r_2$ with:

$A_1 = \pi\, r_1^2$ and $A_2 = \pi\, r_2^2$

Partial eclipse.- This happens when one star partially covers the other one. During this stage, $r_1 + r_2 > \rho > r_1 - r_2$.

- Figure 8.4 was created using the following function:

```
Graphics[{{Red, Circle[{0, 0}, 1]}, {Blue, Circle[{1.15, 0}, 0.5,
```

```
{-2 Pi/3, 2  Pi/3}]}, {Dashed, Blue, Circle[{1.15, 0}, 0.5]},
Line[{{0, 0}, {0.9, 0.425}, {0.9, -0.425}, {0, 0}}], Line[{{0.9,
0.425}, {1.15, 0} , {0.9, -0.425} }] }]
```

We added legends to the output of the function using *Mathematica*'s 2D Drawing Tools: CTRL+D in Windows or CTRL+T in OS X.

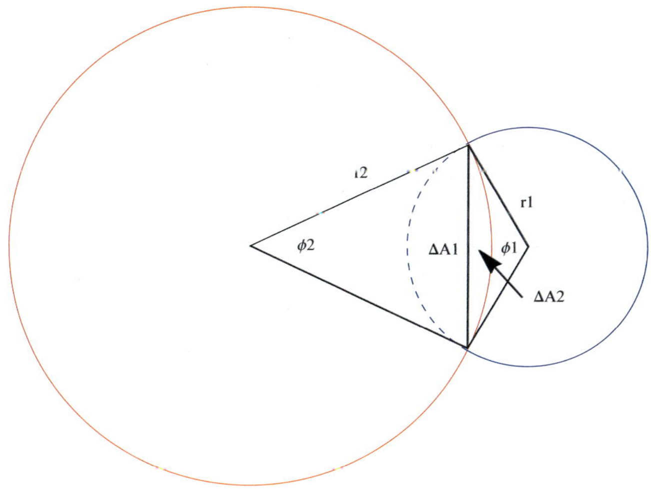

Figure 8.4 Partial eclipse diagram.

To calculate the amount of light that arrives from the eclipsed star, we have to subtract the eclipsed area from the total area. This means we have to compute ΔA1 and ΔA2 corresponding to the star 1 and star 2 segments respectively using the following formulas:

$$\Delta A1 = \frac{1}{2} R1^2 (\phi 1 - \mathrm{Sin}[\phi 1]);\ \Delta A2 = \frac{1}{2} R2^2 (\phi 2 - \mathrm{Sin}[\phi 2])$$

The following expressions enable us to compute $\phi 1$ and $\phi 2$:

$$R2^2 = R1^2 + \rho^2 - 2\, R1\, \rho\, \mathrm{Cos}\left[\frac{\phi 1}{2}\right],\ R1^2 = R2^2 + \rho^2 - 2\, R2\, \rho\, \mathrm{Cos}\left[\frac{\phi 2}{2}\right]$$

- The output below represents the different stages of a partial eclipse:

```
{Graphics[{Blue, Disk[{2, 0}, 1], Red, Disk[{0, 0}, 2]}],
 Graphics[{Blue, Disk[{-2, 0}, 1], Red, Disk[{0, 0}, 2]}],
 Graphics[{Red, Disk[{0, 0}, 2], Blue, Disk[{-2, 0}, 1]}],
 Graphics[{Red, Disk[{0, 0}, 2], Blue, Disk[{2.5, 0}, 1]}] }
```

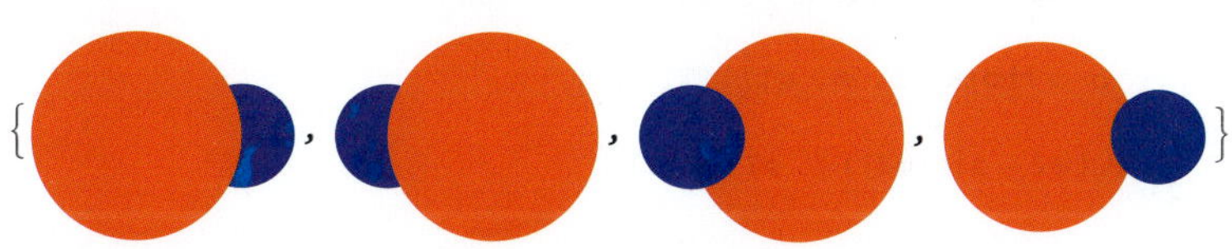

The two figures on the left correspond to the scenario where star 1 (red) is the closest one, that is z1 > z2. In this case the light that reaches the observer will be:

$A_1 = \pi r_1^2$ and $A_2 = \pi r_2^2 - \Delta A1 - \Delta A2$

$A_1 = \pi r_1^2$ and $A_2 = \pi r_2^2 - \Delta A1 - \Delta A2$

The rest of the figures corresponds to the case where star 1 (red) is the most distant one, that is z1 < z2 then:

$$A_1 = \pi r_1^2 - \Delta A1 - \Delta A2 \text{ and } A_2 = \pi r_2^2$$

$$A_1 = \pi r_1^2 - \Delta A1 - \Delta A2 \text{ and } A_2 = \pi r_2^2$$

Total or annular.- Corresponds to the period when a star is shaded by the other one completely. In this stage: $\rho \leq r_1 - r_2$ with: $A_1 = \pi r_1^2$ and $A_2 = 0$ for z1>z2 and $A_1 = \pi r_1^2 - \pi r_2^2$ and $A_2 = \pi r_2^2$;

All these conditions are summarized in the table below:

□	□	z1 > z2		z1 < z2	
Stage	Condition	A_1	A_1	A_1	A_1
Without eclipse	$\rho \geq r_1 + r_2$	πr_1^2	πr_2^2	πr_1^2	πr_2^2
Partial eclipse	$r_1 + r_2 > \rho > r_1 - r_2$	πr_1^2	$\pi r_2^2 - \Delta A1 - \Delta A2$	$\pi r_1^2 - \Delta A1 - \Delta A2$	πr_2^2
Total or annular	$\rho \leq r_1 - r_2$	πr_1^2	0	$\pi r_1^2 - \pi r_2^2$	πr_2^2

Now we are going to include all the criteria described earlier in the function **area[R, i, m2/m1, r1, r2, l1, l2, θ]**, where: m2 > m1 and r1 > r2. The function returns the area that the observer will measure.

Without eclipse. When $\rho \geq r_1 + r_2$:

```
area[R_, i_, q_, r1_, r2_, l1_, l2_, θ_] :=
 Module[{A1, A2}, {A1, A2} = {Pi r1^2 , Pi r2^2 };
   l1 A1 / r1^2 + l2 A2 / r2^2] /; rho[R, i, θ, q] ≥ r1 + r2
```

Partial eclipse. When $r_1 + r_2 > \rho > r_1 - r_2$:

- We first calculate ΔA1 and ΔA2

```
ΔA1[R_, i_, q_, r1_, r2_, θ_] := Module[{ϕ1, ρ, phi, phi1},
   phi1 = Last[Solve[r2^2 == r1^2 + ρ^2 - 2 r1 ρ Cos[phi/2], phi] /.
       ρ -> rho[R, i, θ, q]][[1, 2]];
   1/2 r1^2 (ϕ1 - Sin[ϕ1]) /. ϕ1 → phi1 ];
```

```
ΔA2[R_, i_, q_, r1_, r2_, θ_] := Module[{ϕ2, ρ, phi, phi1},
   phi1 = Last[Solve[r1^2 == r2^2 + ρ^2 - 2 r2 ρ Cos[phi/2], phi ] /.
       ρ -> rho[R, i, θ, q]][[1, 2]];
   1/2 r2^2 (ϕ2 - Sin[ϕ2]) /. ϕ2 → phi1];
```

- In this type of eclipse, for z1 > z2: $A1 = \pi r_1^2$ and $A2 = \pi r_2^2 - \Delta A1 - \Delta A2$ while for z1 < z2: $A1 = \pi r_1^2 - \Delta A1 - \Delta A2$ and $A2 = \pi r_2^2$.

```
area[R_, i_, q_, r1_, r2_, l1_, l2_, θ_] :=
 Module[{z1, z2, a1, a2, A1, A2 }, z1 = - R Cos[θ] Sin[i] / (1 + 1/q);
   z2 = R Cos[θ] Sin[i] / (1 + q);
   a1 = ΔA1[R, i, q, r1, r2, θ];
   a2 = ΔA2[R, i, q, r1, r2, θ];
   {A1, A2} =
    If[ z1 > z2, {Pi r1^2 , Pi r2^2 - a1 - a2}, {Pi r1^2 - a1 - a2, Pi r2^2}];
   l1 A1 / r1^2 + l2 A2 / r2^2 ] /; r1 + r2 > rho[R, i, θ, q] > r1 - r2
```

Total and / or annular eclipse. When $\rho \leq r_1 - r_2$:

- When r1 > r2 one of the stars will be directly in front of the other one, interposing it. Then a total eclipse will happen if z1 > z2 with $A1 = \pi r_1^2$ and A2 = 0, or an annular one will occur if z1 < z2 with A1' = A1 − A2 and A2' = A2.

```
area[R_, i_, q_, r1_, r2_, l1_, l2_, θ_] :=
 Module[{z1, z2, A1, A2}, z1 = - R Cos[θ] Sin[i] / (1 + 1/q);
   z2 = R Cos[θ] Sin[i] / (1 + q);
   {A1, A2} = If[ z1 > z2, {Pi r1^2 , 0}, {Pi r1^2 - Pi r2^2 , Pi r2^2}];
   l1 A1 / r1^2 + l2 A2 / r2^2] /; rho[R, i, θ, q] ≤ r1 - r2
```

Let's simulate the ideal case of an eclipsing binary. We can distinguish the repeating phases of an eclipse: one corresponds to the period without eclipse (t1 to t2), in t2 the eclipse starts and reaches its plenitude between t3 and t4, from t4 to t5 we are in the partial eclipse phase, from t6 to t9 is the transit phase, and when t9 = t1 the cycle starts again.

- We use **`Plot[area[2, 90 Degree, 0.1/0.6, 0.6, 0.3, 0.6, 0.2, 2 Pi t/8], { t, 0, 12}, AxesLabel -> {"t", "I"}, AxesOrigin -> {0, 0}]`** and include the legends using the Drawing Tools option in the Graphics menu: **Graphics ▶ Drawing Tools**.

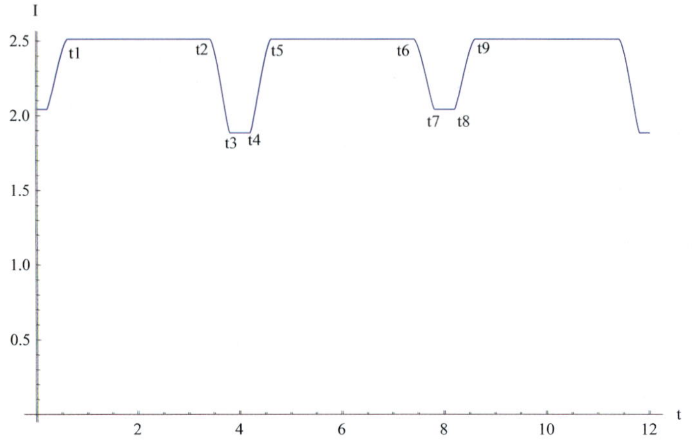

Figure 8.5 Light curve.

Example: With R = 2, i = 85º, m1 = 0.6, m2 = 0.1, R1 = 0.4, R2 = 0.3, l1 = 0.6, l2 = 0.2, we'd like to show I/Imax.

- We first calculate Imax.

```
max = NMaximize[area[2, 2 Pi 85 / 360, 0.1/0.6, 0.4, 0.3, 0.6, 0.2, θ],
    {θ, 0, 2 Pi}] [[1]] // Quiet
```

```
2.51327
```

- Then we display I/Imax.

```
Plot[area[2, 85 Degree, 0.1/0.6, 0.4, 0.3, 0.6, 0.2, 2 Pi θ] / max,
  { θ, 0, 1.5}, AxesLabel → {"Phase", "I/Imax"},
  AxesOrigin → {0, 0}] // Quiet
```

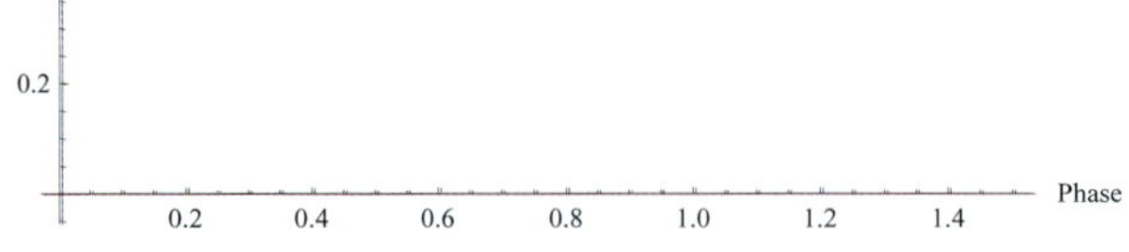

8.6.2 Application: Using Light Curves to Determine the Period-Luminosity Relation

Besides eclipsing variables or planets interposition, periodic changes in luminosity can happen for other reasons. Probably the best known example is the change in luminosity in variable stars, such as Cepheids. The usual method for building an experimental light curve consists of measuring the luminosity at different times t and deduce the period from them.

- The following expression simulates magnitude as a function of time (in practice, since we often don't know this function we try to deduce it from the experimental data). The chosen function is very simple and we only use it for didactic purposes).

```
mag[t_] = 5 + 0.14 Cos[0.3 t + 0.1];
```

- Let's assumed that we have already taken measurements at different times t_i.

```
data = Table[{t, Round[mag[t], 0.01]}, {t, 0, 200, 5}];
```

- We represent graphically the previous data. If we only had the previous data points it would be very difficult to figure out the underlying pattern (remove **Joined → True**), joining the dots makes it easier to observe the cyclical nature of the data. It's normal for many intermediate points to be missing making the analysis even more challenging.

```
ListPlot[data, Joined→ True]
```

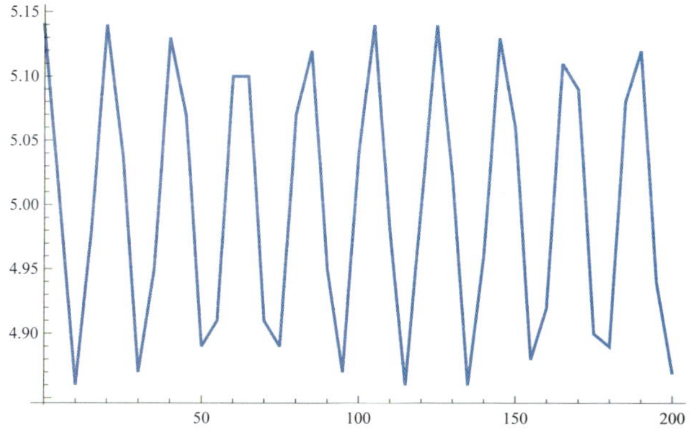

The analysis method frequently used is to represent the previous phenomenon in a phase diagram.

To calculate the phase we need to apply this equation:

$$\phi = \text{Decimal part of}\left[\frac{t - t_0}{P}\right]$$

where:

ϕ represents the phase in cycles.
t_0 (normally called epoch) represents the origin of the time variable expressed as a Julian Date (DJ).
t is the moment when the measurement is taken (normally we will measure the apparent magnitude, m).
P is the period, the time it takes for the cycle to repeat itself.

An equivalent expression to the previous one and the one that we will be using is:

$$\phi = \text{Mod}\left[\frac{t - t_0}{P}\right]$$

- With the function below *Mathematica* returns the phase given the initial data points and the period.

```
phi[list_, p_] := Module[{data}, data = list;
   Transpose[{Mod[data[[All, 1]] / p, 1], data[[All, 2]]}]];
```

In this example we are assuming that the period is known, but in practice that's what we are trying to find out. We can check that by using an approximate value of the period the representation is more diffuse.

- With real observations, the period has to be determined and that is not always easy. The next function will help us to compute it. We keep in mind that for period-magnitude curves, the order of representation in the axis OY is inverted. That is, the magnitudes are sorted in descending order not in the usual ascending one: $-3 > -2 > 0 > 1 > 2$.

```
curveL[data_, per_, kmax_] :=
 Module[{mag, a, b, k, i}, mag = Round[Transpose[data][[2]], 0.1];
  Manipulate[ListPlot[Join[phi[data, per + k] /. {a_, b_} -> {a, -b},
     phi[data, per + k] /. {a_, b_} -> {-(1 - a), -b}], AxesLabel ->
     {"Period", "Magnitude"}, AxesOrigin -> {-1, -Max[mag] - 0.1},
    Ticks -> {Automatic, Table[{i, -i}, {i, -Max[mag] - 0.1,
        -Min[mag] + 0.1, (Max[mag] - Min[mag]) / 10}]} ],
   {{k, 0, "Deviation from the period (in days)"}, 0, kmax}]]
```

- Let's assume a complete cycle (2π). We use the previous function to simulate the experimental measures and include a random component. We make the assumption that the period is 20 days (the duration of a complete orbit).

```
simulation = Table[
   {θ, area[2, 2 Pi 85 / 360, 0.1/0.6, 0.4, 0.3, 0.6, 0.2, 2 Pi θ / 20] +
     RandomReal[NormalDistribution[0, 0.005]]},
   {θ, 0, 60, 0.05}]; // Quiet
```

- We represent the data in a phase diagram:

```
simulation1 = Sort[phi[simulation, 20]];

ListPlot[simulation1, Joined → True]
```

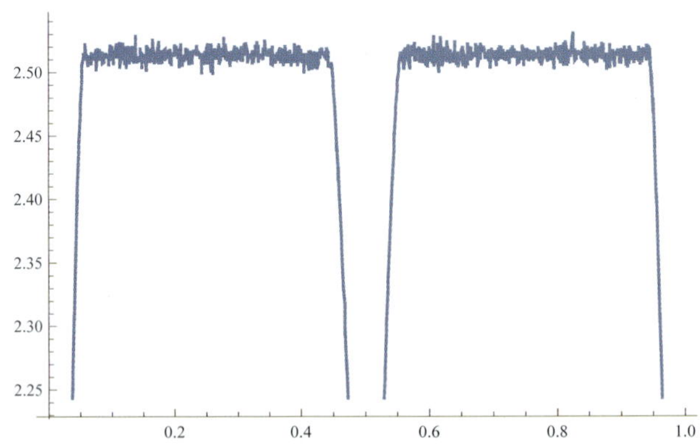

The previous luminosity data have to be multiplied by a parameter k to convert them into a magnitude measure. Remember that higher luminosity numerically implies lower magnitude. For that reason, to represent the data in a phase/luminosity diagram we assume $k = -1$.

- If you execute the following function and use the slider to choose a 5-day period, you'll see that the curve is very similar to the previous figure.

```
curveL[Map[Times[{1, -1}, #] &, simulation], 15, 20]
```

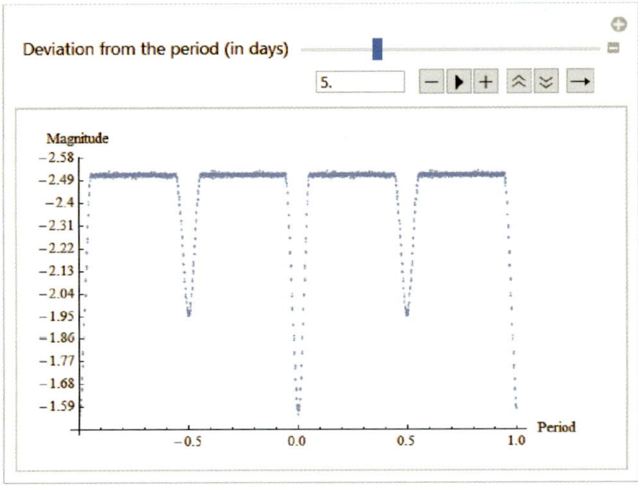

8.6.3 Application: Finding the Light Curve of a Known Eclipsing Binary

Let's apply the newly created function to the luminosity data from NSV 03199 obtained by Garcia-Melendo, E., Henden, A., 1998, IBVS, No. 4546. The data are in the file: NSV03199.xls in our data directory. It has been downloaded from: http://www.astrogea.org/VARIABLE/variables.htm.

- Let's import the luminosity measures:

```
vsdata =
  Drop[Import[FileNameJoin[{NotebookDirectory[], "Data", "NSV03199.xls"}],
    {"XLS", "Data", 1}], 1];
```

- The data have Julian dates so we use as reference JHD=2450510.46542.

```
vsdata1 = vsdata /. {a_, b_} -> {a - 2450510.46542, b};
```

- We cannot really see the cycle pattern:

```
ListPlot[vsdata1]
```

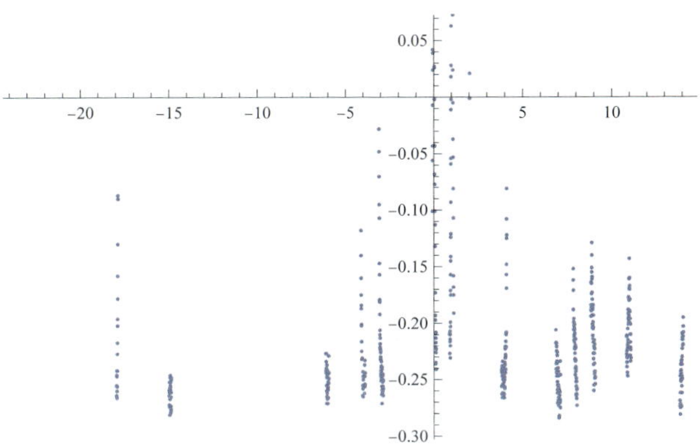

However, when using a period-magnitude curve, the periodic behavior corresponding to an eclipsing binary can be seen clearly.

- We find that the optimum period is 1.046400 days:

```
curveL[vsdata1, 1.046400, 0.5]
```

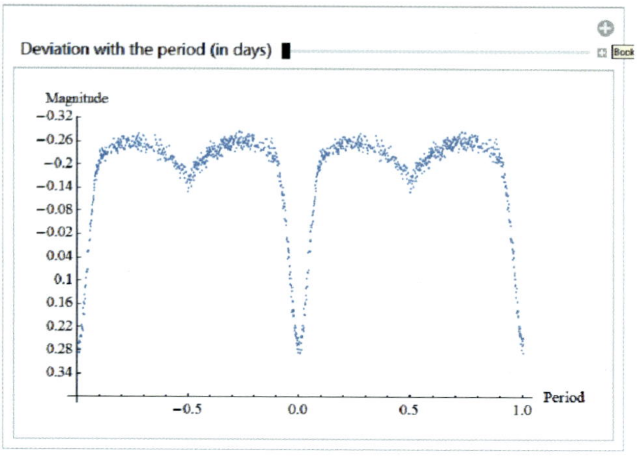

All the previous functions along with other examples in the book have been put together in a package that can be downloaded from diarium.usal.es/guillermo.

8.7. Additional Resources

Astronomy demonstrations:
http://demonstrations.wolfram.com/topic.html?topic=Astronomy

Eric Weisstein's World of Astronomy: http://scienceworld.wolfram.com/astronomy

To follow objects with variable luminosity (variable stars, supernovae and many more), the best place is the website of the American Association of Variable Star Observers (AAVSO): http://www.aavso.org

Author's website astronomy section (in Spanish):
http://diarium.usal.es/guillermo/astronomia

9

Nuclei and Radiations

Radioactivity is a natural phenomenon that allows us to peek into earth's past and the history of humankind through radioactive dating. Its effects when absorbed by human beings can also be measured. We will refer to this and other topics related to nuclear physics in this chapter. For that purpose, we'll use the Mathematica functions `IsotopeData` *and* `ParticleData`.

9.1 What Are Isotopes?

Matter consists of atoms that can be described in a simple way as a central part named the nucleus, containing neutrons and protons, surrounded by an electron cloud. Atomic elements have the same number of protons but the number of neutrons may change. Nuclei with the same number of protons, Z, and a different one for neutrons, N, are the isotopes of that element. Normally, an isotope is written as ^{A}X , where X is the symbol of the element and A is the mass number given by A = Z + N. For example: atmospheric carbon consists of 3 isotopes: Carbon–12 (or ^{12}C) with 6 protons and 6 neutrons, Carbon–13 (or ^{13}C) with 6 protons and 7 neutrons and Carbon–14 (or ^{14}C) that has 6 protons and 8 neutrons. All isotopes of the same element possess the same chemical properties but different physical ones.

In nature there are 92 elements (we can also find traces of elements with Z > 92, such as plutonium, but these are artificial elements), with an average of 2 or 3 isotopes each, although there are some elements with only one stable isotope and others with as many as 8. Isotopes can be stable or unstable. Stable nuclei, the majority on earth, last "forever", that is, they always keep the same number of neutrons and protons. Some theories predict the disintegration of protons, and therefore of any kind of atom, but they have not been proven yet. Furthermore, even if they could be validated, the disintegration speed would be so slow that for all practical purposes we could consider that stable nuclei have an almost eternal duration. Unstable nuclei (called radioisotopes) are those that over time (depending on the radioisotope the period can range from a fraction of a second to billions of years) are transmuted into other elements, called daughter isotopes. The new nucleus can in turn be stable or unstable. If it's unstable, the process continues until the daughter isotope is a stable one.

- With `IsotopeData`["*element*", "*property*"] we can obtain all the known isotopes for a given element. The element can be written as a complete word or preferably, as a symbol using standard notation. If we type "Symbol" in "*property*", we'll get the isotopes in ^{A}X notation. We use `Short` to display only one line, but the complete output includes more isotopes.

```
IsotopeData["C", "Symbol"] // Short
```

{^{8}C, ^{9}C, ^{10}C, <<9>>, ^{20}C, ^{21}C, ^{22}C}

The decay of a type of nucleus into another one usually happens due to the emission of particles α or β and radiation γ. In some radionuclides there are also emissions of neutrons and protons and even other types of reactions such as spontaneous fission (SF), common in some heavy nuclei. Fission is the division of a nucleus into two smaller ones, each one with approximately half the mass of the original.

Using "DecayModes" or "DecayModeSymbols" as properties we can get the type of decay for a certain isotope. One isotope can display more than one decay mode although normally one of them would be the most common one. Using "BranchingRatios" and "DecayEnergies" we can see, respectively, the probability associated to each decay mode and its energy measured in keV (kiloelectron-volt).

- The command below shows the different decay modes of Uranium–238, and the symbols, probabilities and energies associated to them.

```
Grid[IsotopeData["Uranium238", #] & /@
  {"DecayModes", "DecayModeSymbols", "BranchingRatios", "DecayEnergies"}]
```

AlphaEmission	SpontaneousFission	DoubleBetaDecay
α	SF	$2\beta^-$
1.00	5.45×10^{-7}	2.2×10^{-12}
4269.75 keV	Missing[NotAvailable]	1144.2 keV

Notice that the majority of the radioactive decays for U238 correspond to alpha emissions (the value 1.00 indicates that the probability is close to 100%). However, spontaneous fissions (SF), even though they have a small probability of happening, play a central role in chain reactions since they emit neutrons, essential to start a chain reaction.

In short, we can say that a radioactive isotope is like a father who has a daughter who in turn may have a granddaughter and so on until eventually there's a descendant without offspring, the stable nucleus. The disintegration happens at a constant pace and its characteristics depend on the particular isotope. In this process, particles (α , β, ...) with specific energies are emitted.

- You can use the free-form input to get the definition of chain reaction. Probably when you type "chain reaction definition" you will see several definitions. Just choose the most appropriate one depending on your needs.

chain reaction definition

Result

1	noun	a series of chemical reactions in which the product of one is a reactant in the next
2	noun	a self-sustaining nuclear reaction; a series of nuclear fissions in which neutrons released by splitting one atom leads to the splitting of others

- We can also get information by classes. We use `Shallow` to avoid displaying the complete output, which is quite large. If you want to see all the classes just remove the command from the next input.

```
IsotopeData["Classes"] // Shallow
```

{alpha emission, beta decay, beta delayed alpha emission, beta delayed deuteron emission, beta delayed fission, beta delayed four neutron emission, beta delayed neutron alpha emission, beta delayed neutron emission, beta delayed three neutron emission, beta delayed triton emission, <<37>>}

- Each class includes isotopes sharing a common characteristic. For example: if we want to know the isotopes with a beta emission associated to spontaneous fission (a very rare occurrence, only observed in 3 isotopes out of the more the 3,000 available), the command below gives us the answer:

```
IsotopeData[beta delayed fission (isotopes)]
```

{actinium-230, protactinium-236, protactinium-238}

9.2 Decay Constants, Decay Periods and Half-Lives

```
Clear["Global`*"]
```

The easiest case of radioactive decay is that of a radioactive isotope A that decays into a stable nucleus B, that is, A→B, where B is not radioactive. This process can be represented by a very basic compartmental model, such as the one shown in Figure 9.1 with $x(t)$ representing the quantity of a certain radioactive isotope A present at time t in a sample and λ being the decay constant, specific to each isotope. The quantity of atoms (measured in the most appropriate way: units, grams, becquerels, etc.) that decay per unit of time is proportional to the quantity in the sample at each moment, that is: $\lambda\, x(t)$.

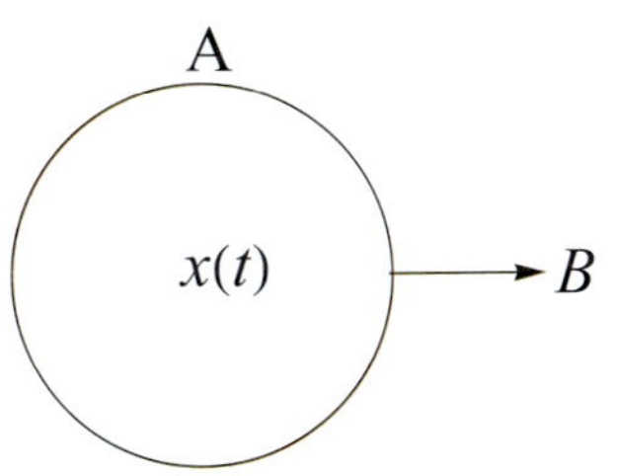

Figure 9.1 Radioactive decay of isotope A.

Let's assume that at $t = 0$ the sample contains x_0. This process can be mathematically modeled by a simple first-order differential equation:

$$\frac{dx}{dt} = -\lambda\, x(t), \text{ with initial condition : } x_0 \text{ at } t = 0 \tag{9.1}$$

- The solution to the previous differential equation can be obtained using `DSolve`. Notice that we use "=" and not ":=" Why? (normally is better to use ":=", but in this case we choose "=" since the next time we call the function it will already know the solution without having to calculate it again. In this case the time saving is negligible).

```
x1[t_, x0_, λ_] = x[t] /. DSolve[{x'[t] == -λ x[t], x[0] == x0}, x[t], t][[1]]
```

$e^{-t\lambda}$ x0

We type it again using the typical textbook format

$$x(t) = x_0 \, e^{-\lambda t} \tag{9.2}$$

- In the next example, we use λ and the total time t as our changing parameters. We also use the previously defined function within `Manipulate` using the command `Initialization`. Alternatively, we could have used `SaveDefinitions`.

```
Manipulate[Plot[y[t, λ, 1], {t, 0, t1}, Filling → Bottom],
 {{λ, 1}, 0, 5}, {{t1, 3}, 1, 10},
 Initialization :→ (y[t_, λ_, x0_] := x0 e^(λ (-t)))]
```

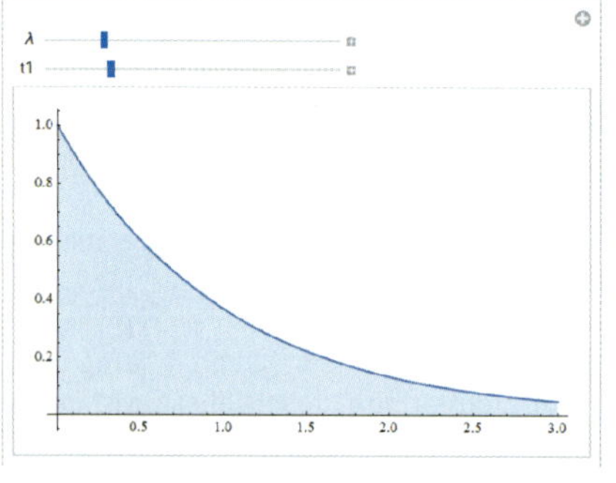

In many situations instead of the **decay constant**, λ, it's better to use the semi-decay period or **half-life**, $T_{1/2}$, defined as the time that it would take for the amount x_0 present in the sample at time $t = 0$ to become half, $x_0/2$.

- The relationship between λ and $T_{1/2}$ can be calculated as follows:

```
cteDes = λ /. Solve[x1[T, x0, λ] == x0/2, λ, Reals][[1]]
```

$\frac{\text{Log}[2]}{T}$

Another concept is the **mean lifetime**, the life expectation (duration) of the isotopes present in a sample. It's the same as calculating the arithmetic mean of the expected duration of the individual isotopes present in a sample.

- To calculate the mean of a continuous variable t we compute $\langle t\rangle = \int_0^\infty t\, f(t)\, dt \big/ \int_0^\infty f(t)\, dx$, and associate the result to the symbol τ. We specify that $\lambda \in \mathbb{R}{>}0$, otherwise we'll receive a message stating that the solution is only valid if $\lambda \in \mathbb{R}{>}0$.

```
τ = Integrate[t x₀ E^(-t λ), {t, 0, Infinity},
   Assumptions → λ ∈ Reals && λ > 0] / Integrate[x₀ E^(-t λ),
   {t, 0, Infinity}, Assumptions → λ ∈ Reals && λ > 0]
```

$\frac{1}{\lambda}$

- With IsotopeData we can find out the half-life and mean lifetime of Iodine–131. By default the command returns the time in seconds but we can convert it to other unit using UnitConvert.

```
IsotopeData["Iodine131", #] & /@ {"HalfLife", "Lifetime" }
```

```
{6.9338×10^5 s , 1.0003×10^6 s}
```

```
UnitConvert[
 IsotopeData["Iodine131", #] & /@ {"HalfLife", "Lifetime" }, "days"]
```

```
{8.0252 days , 11.578 days}
```

- In the example below we can choose the isotope among several options. We use QuantityMagnitude to get only the value (in this case, the mean lifetime), without the unit.

```
Manipulate[Plot[ y[t, 3600 λ , 1 ], {t, 0, t1}, Filling → Bottom],
 {{t1, 3, "Hours"}, 1, 100},
 {λ, {1 / QuantityMagnitude[IsotopeData["Iodine131", "Lifetime"]] →
    "Iodine-131", 1 / QuantityMagnitude[
      IsotopeData["Iodine125", "Lifetime"]] → "Iodine-125",
   1 / QuantityMagnitude[IsotopeData["Iodine128", "Lifetime"]] →
    "Iodine-128"}}, Initialization :→ (y[t_, λ_, x0_] := x0 e^(λ (-t)))]
```

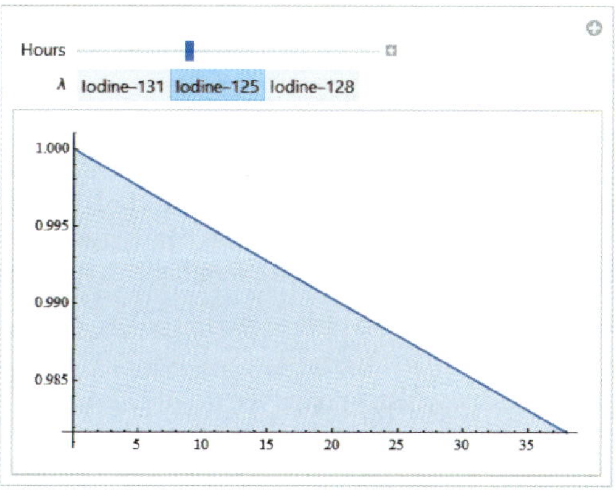

- The next example uses IsotopeData to show all the isotopes for iodine. We also include the option of selecting the unit for displaying the evolution over time for the chosen isotope. Since the decay rate for some of the isotopes is very fast we use a logarithmic scale.

```
Manipulate[LogPlot[ y[t, k * λ , 1], {t, 1, t1}, Filling → Bottom],
 {{t1, 3}, 1, 100 },
 {k, {1 -> "Seconds", 3600 -> "Hours", 3600 * 24 -> "Days"}},
 {{λ, "", "Isotope"},
  Thread[1 / QuantityMagnitude[IsotopeData["Iodine", "Lifetime" ]] ->
    IsotopeData["Iodine"]]}, Initialization :→ (y[t_, λ_, x0_] := x0 e^(λ (-t)))]
```

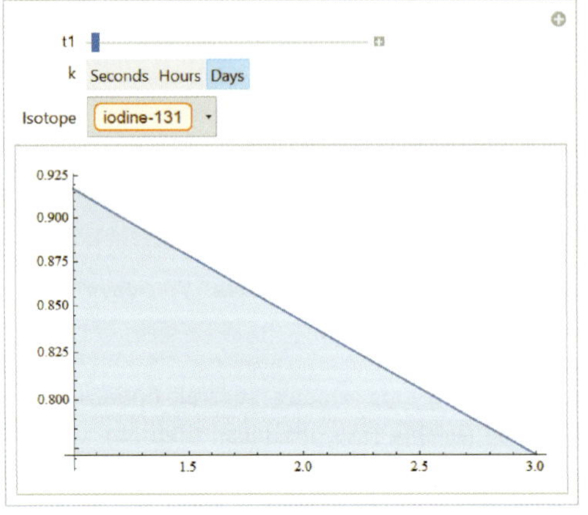

In summary: the radioactive decay constant λ, the half-life $T_{1/2}$, and the mean lifetime τ, are connected as follows: $T_{1/2}$ = Log[2]/$\lambda = \tau$ Log[2] , $\tau = 1/\lambda$. **Sometimes people use mean lifetime instead of half-life by mistake**.

For a given radioisotope, activity A, the decay rate in 1 second (1 nuclear transformation = 1 Becquerel o Bq), can be obtained from the decay constant, λ (in s^{-1}). The activity of n_{at} atoms will be $A = \lambda\ n_{at}$. Since what is known usually is the mass, m, we apply the conversion $n_{at} = m\ N_A / m_a$, where N_A is the Avogadro's number and m_a the atomic mass of the element expressed in the same units as m, normally in grams (g).

Iodine–131 example

Iodine–131 emissions from the Fukushima Daichi nuclear reactor due to the accident caused by the tsunami in March 2011, were estimated to be 1.5×10^{17}Bq as of the end of April 2011 (http://www.nisa.meti.go.jp/english/files/en20110412-4.pdf). Given this information, let's explore how much mass was emitted and how the emissions evolved over time.

Activity can be turned into mass using the formula $A = \lambda\ n_{at} = \lambda\ m\ N_A / m_a \rightarrow m = m_a A / (\lambda N_A)$.

- Iodine–131 decays into Xenon 131, a stable isotope that at room temperature is gas. We use `LayeredGraphPlot` to represent the decay chain.

```
LayeredGraphPlot[{IsotopeData["I131", "Symbol"] →
   IsotopeData[IsotopeData["I131", "DaughterNuclides"][[1]], "Symbol"]},
  Left, VertexLabeling → True]
```

- We need to know the decay constant, λ131, and the atomic mass. `IsotopeData` will give us the later (although we could have used 131 as an excellent approximation) and also return the mean lifetime τ (in seconds). Afterward we can calculate $\lambda = 1/\tau$.

```
{ mat131, λ131} = {IsotopeData["Iodine131", "AtomicMass"],
  1/IsotopeData["Iodine131", "Lifetime"] }
```

```
{130.906124609 u , 9.9967×10^-7 per second}
```

- Now we use the formula $m = m_a A / (\lambda N_A)$ where N_A (Avogadro's number) is given by *Mathematica* using its natural language capabilities ("Avogadro's Number") or with the function Quantity. UnitConvert shows the SI units, in this case just the number.

```
Quantity["Avogadro's Number"]

Avogadro numbers

NA = UnitConvert[1 N_A]

6.022141×10^23

1.5×10^17 QuantityMagnitude[mat131] / (QuantityMagnitude[λ131] * NA) "grams"

32.617 grams
```

It may seem strange at first sight that such a big radioactive activity would only result in a very small mass. What's actually happening is that the Bq is an extremely small unit. As a matter of fact, for many years instead of using the Becquerel to measure radioactivity, people used the Curie, Ci (1 Ci = 3.7 10^{10}Bq), unit that's still in use in some places.

- To find out in how many days the radioactivity of the Iodine–131 would become just 0.1% of its initial figure, we use the following command:

```
UnitConvert[t /. Solve[x1[t, 100, λ131] == 0.1, t], "days"]

{79.9774 days}
```

9.3 Decay Chains

```
Clear["Global`*"]
```

As we've seen, a radioactive isotope is like a father who keeps on having descendants until the offspring is a stable nucleus. Sometimes, the first descendant is stable right away, as with Iodine–131 → Xenon–131, but other isotopes, like Radium–226, shown at the end of this section, have more complicated decay chains. One of Radium–226's daughter isotopes, Radon–222 (Rn222), is responsible for a large share of the natural radiation that people are exposed to. Each time we breathe, a portion of the air includes radon gas. The content of Rn222 changes substantially from one area to another. For example: granitic houses with poor ventilation usually have a high content level of radon. The reason is that granite contains Radium–226 that becomes radon gas when it disintegrates.

- An isotope can decay following different branches. For example: Radium–226 decays with a half-life period of 5.0×10^{10} s into 3 radioisotopes: ^{222}Rn, ^{212}Pb and ^{226}Th. Almost all of it decays into ^{222}Rn and an insignificant part into ^{212}Pb. Occasionally, there can also be traces of Thorium–226.

```
ra226 = IsotopeData["Ra226", #] & /@
   {"DaughterNuclides", "BranchingRatios", "HalfLife"};

TableForm[ra226,
 TableHeadings -> {{"Isótopo", "fracción", "T (s)"}, None}]
```

Isótopo	radon-222	lead-212	thorium-226
fracción	1.00	2.6×10^{-11}	Missing[Unknown]
T (s)	5.0×10^{10} s		

- In different contexts it is likely that when using `IsotopeData` you may find that the output contains the word "Missing". If you want to eliminate it, you can use `DeleteMissing` or `DeleteCases`[*list*, *_h*] that eliminate all the elements in a list whose head starts with h, as shown in the example below:

```
is1 = IsotopeData["U238", "DaughterNuclides"]
```

```
{thorium-234, Missing[Variable], plutonium-238}
```

```
is2 = DeleteCases[is1, _Missing]
```

```
{thorium-234, plutonium-238}
```

The set of all branches produced by the decay of a parent isotope (like Uranium–238) is known as the radioactive decay chain. However, the majority of those disintegrations happen along only one branch. `IsotopeData`["isotope", "BranchingRatios"] shows the results after sorting the branches from bigger to smaller in terms of their likelihood of occurrence. When an isotope decays into one or several isotopes, we can choose to keep just one: the one with the highest probability. If we apply this criterium to each daughter isotope, we will end up with a single branch named the main branch. From a practical point of view, in many decay chains it is enough to study the main branch.

- The following function lets us choose the main branch, and within it, the number of daughter nuclides *n* that we are interested in. Notice the use of *iso_String*, and *n_Integer* to limit the input type to: "*isotope name*" (between double quotation marks), and *n* integer (number of daughter nuclides that we would like to display); in any other case the function will not be executed.

```
mainbranch[iso_String, n_Integer] :=
 NestList[First[IsotopeData[#, "DaughterNuclides"]] &, iso, n]
```

- We apply it to Uranium–238 and indicate 14 daughter nuclides, the entire decay chain (one of the longest ones).

```
mainbranch["U238", 14]
```

```
{U238, thorium-234, protactinium-234, uranium-234, thorium-230,
 radium-226, radon-222, polonium-218, lead-214, bismuth-214,
 polonium-214, lead-210, bismuth-210, polonium-210, lead-206}
```

- This problem was posted in "comp.soft-sys.math.mathematica". The function below shows the elegant solution proposed by one of the list members (Bob Hanlon) that returns a complete branch without the need to specify *n*. However, if we are interested in limiting the number of offsprings that we wish to see, it would be better to use **`mainbranch`**.

```
mainbranch1[x_String] :=
  NestWhileList[IsotopeData[#, "DaughterNuclides"][[1]] &,
   x, UnsameQ, 2, 100, -2] // Quiet;
```

- Let's apply it to Radon–226

```
ra226 = mainbranch1["Ra226"]
```

```
{Ra226, radon-222, polonium-218, lead-214, bismuth-214,
 polonium-214, lead-210, bismuth-210, polonium-210, lead-206}
```

- The function below generates a list of isotopes showing: their disintegration period, daughter nuclides and branching ratios.

```
isopro[isolist_] :=
 Module[{vals}, vals = Table[IsotopeData[#, prop], {prop,
       {"Symbol", "HalfLife", "DaughterNuclides", "BranchingRatios"}}] & /@
    isolist;
  Text[Grid[Prepend[vals, {"", "Disintegration\nperiod (sec)",
      "Daughter nuclides", "Branching ratios"}], Frame → All,
    Background → {None, {{{LightBlue, White}}, {1 → LightYellow}}}]] ]
```

- We use it to show the information related to the principal branch of Radon–226 obtained in the previous calculation.

```
isopro[ra226]
```

	Disintegration period (sec)	Daughter nuclides	Branching ratios
^{226}Ra	5.0×10^{10} s	{radon-222, lead-212, thorium-226}	{1.00, 2.6×10^{-11}, Missing[Unknown]}
^{222}Rn	330350. s	{polonium-218}	{1.00}
^{218}Po	185.9 s	{lead-214, astatine-218}	{1.00, 0.00020}
^{214}Pb	1610. s	{bismuth-214}	{1.00}
^{214}Bi	1.19×10^{3} s	{polonium-214, thallium-210, lead-210}	{1.0, 0.00021, 0.000030}
^{214}Po	0.0001643 s	{lead-210}	{1.00}
^{210}Pb	7.01×10^{8} s	{bismuth-210, mercury-206}	{1.00, 1.9×10^{-8}}
^{210}Bi	4.33×10^{5} s	{polonium-210, thallium-206}	{1.00, 1.32×10^{-6}}
^{210}Po	1.19557×10^{7} s	{lead-206}	{1.00}
^{206}Pb	∞	{}	{}

The functions below are part of the documentation related to `LayeredGraphPlot`. Sometimes we may get lucky and find examples in the documentation that perfectly match what we want to do. This is one of those occasions.

- The following functions let us input the parent isotope (the first in the chain) and get as a result all the daughter nuclides from all the branches except those that have an insignificant contribution ("Missing").

```
DaughterNuclides[s_List] :=
 DeleteCases[Union[Apply[Join, Map[IsotopeData[#, "DaughterNuclides"] &,
     DeleteCases[s, _Missing]]]], _Missing]
ReachableNuclides[s_List] :=
 FixedPoint[Union[Join[#, DaughterNuclides[#]]] &, s]
DecayNetwork[iso_List] :=
 Apply[Join,
  Map[Thread[# → DaughterNuclides[{#}]] &, ReachableNuclides[iso]]]
DecayNetworkPlot[s_] :=
 LayeredGraphPlot[Map[IsotopeData[#, "Symbol"] &, DecayNetwork[{s}], {2}],
  VertexLabeling → True, VertexRenderingFunction →
   (Text[Framed[Style[#2, 8], Background → LightYellow], #1] &)]
```

- We use it to show the decay chain of Rn222 (The Rn222 is a daughter isotope of Ra226, you can apply the same function to see the complete chain of the latter. You may have to adjust the output image size to see it properly. To do that just place your cursor in one of the corners of the image and resize it accordingly).

```
DecayNetworkPlot["Rn222"]
```

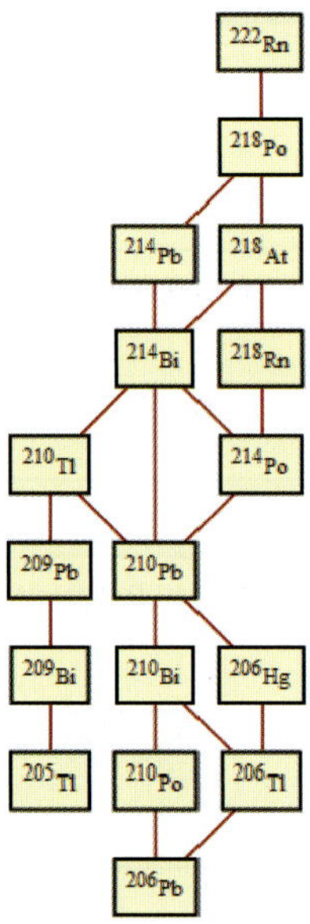

- The following expression returns all the isotopes that have the same number of neutrons:

```
isoneu[neut_] :=
 Select[IsotopeData[], IsotopeData[#, "NeutronNumber"] == neut &]
```

```
isoneu[14]
```

{ boron-19 , carbon-20 , nitrogen-21 , oxygen-22 , fluorine-23 , neon-24 , sodium-25 , magnesium-26 , aluminum-27 , silicon-28 , phosphorus-29 , sulfur-30 , chlorine-31 , argon-32 , potassium-33 , calcium-34 }

- Using the previous function we display the decay chains of all the isotopes with 4 neutrons.

```
LayeredGraphPlot[Table[(IsotopeData[#, "Symbol"] & /@ (i → #)) & /@
    IsotopeData[i, "DaughterNuclides"], {i, isoneu[4]}] // Flatten,
  DirectedEdges → True, VertexLabeling → True]
```

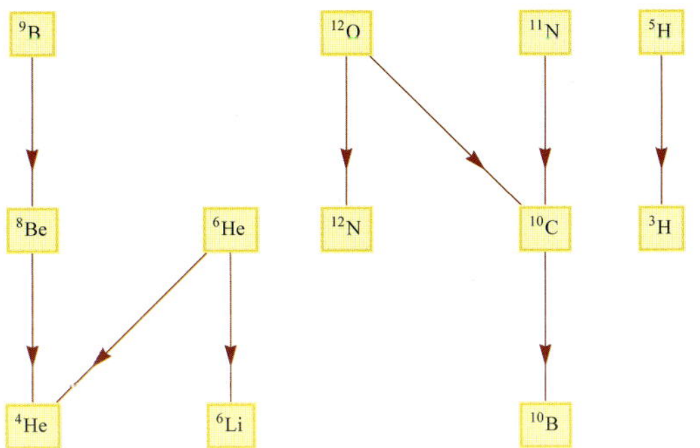

9.4 Application: Modeling the Evolution of a Chain of Isotopes over Time

A radioactive decay chain can be considered as a system of compartments in which the flow is one-directional, as shown in the figure below. This is an example of what is known in compartment modeling as a catenary model.

```
Graphics[{Circle[{0, 0}, 0.5] , Text["1", {0, 0}, {0, 0}],
  Text["λ1", {1.75, .4}, {0, 0}], Arrow[{{0.5, 0}, {3, 0}}],
           Circle[{3.5, 0}, 0.5] , Text["2", {3.5, 0}, {0, 0}],
  Circle[{7, 0}, 0.5] , Text["3", {7, 0}, {0, 0}],
  Arrow[{{4, 0}, {6.5, 0}}], Text["λ2", {5.25, .4}, {0, 0}],
  {Dashed, Arrow[{{7.5, 0}, {10.5, 0}}]}, Text["λi", {9, .4}, {0, 0}]},
 AspectRatio → Automatic, Axes → None]
```

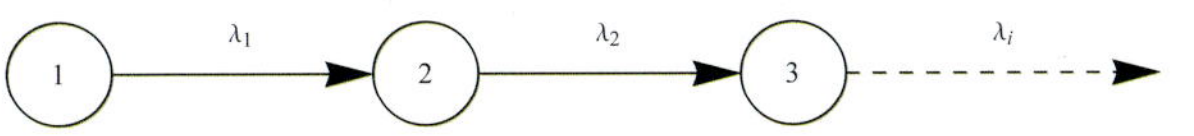

The first circle represents the first isotope in the chain, with a decay constant λ_1, the second circle refers to the next isotope, with a decay constant λ_2, and so on.

Let's consider the case of a chain of 3 isotopes, with $q_1(t)$, $q_2(t)$ and $q_3(t)$ the quantities present of each one at time t, satisfying the following set of differential equations:

$$q_1'(t) = -\lambda_1 q_1(t)$$
$$q_2'(t) = \lambda_1 q_1(t) - \lambda_2 q_2(t)$$
$$q_3'(t) = \lambda_2 q_2(t) - \lambda_3 q_3(t)$$

- As an initial condition let's assume that at $t = 0$ there's a quantity (atoms, gm, Bq) of isotope 1 in compartment 1 and nothing in the rest. This means: q1[0] = b1, q2[0] = q3[0] = 0 then:

```
eq = DSolve[{q1'[t] == -λ1 q1[t],
    q2'[t] == λ1 q1[t] - λ2 q2[t], q3'[t] == λ2 q2[t] - λ3 q3[t],
    q1[0] == b1, q2[0] == 0, q3[0] == 0}, {q1[t], q2[t], q3[t]}, t];
```

- The retention in compartments 1, 2 and 3 is given by:

```
{q1[t_], q2[t_], q3[t_]} = {q1[t], q2[t], q3[t]} /. eq[[1]] // Simplify
```

$$\left\{b1\, e^{-t\lambda 1},\ -\frac{b1\left(e^{-t\lambda 1}-e^{-t\lambda 2}\right)\lambda 1}{\lambda 1-\lambda 2},\ \left(b1\, e^{-t(\lambda 1+\lambda 2+\lambda 3)}\lambda 1\,\lambda 2\left(e^{t(\lambda 1+\lambda 2)}(\lambda 1-\lambda 2)+e^{t(\lambda 2+\lambda 3)}(\lambda 2-\lambda 3)+e^{t(\lambda 1+\lambda 3)}(-\lambda 1+\lambda 3)\right)\right)\Big/\left((\lambda 1-\lambda 2)(\lambda 1-\lambda 3)(\lambda 2-\lambda 3)\right)\right\}$$

- Grouping identical exponential terms we get:

$$q3\,(t) = \frac{b1\, e^{-t\lambda 1}\lambda 1\,\lambda 2}{(\lambda 1-\lambda 2)(\lambda 1-\lambda 3)} + \frac{b1\, e^{-t\lambda 3}\lambda 1\,\lambda 2}{(\lambda 1-\lambda 3)(\lambda 2-\lambda 3)} + \frac{b1\, e^{-t\lambda 2}\lambda 1\,\lambda 2}{(-\lambda 1+\lambda 2)(\lambda 2-\lambda 3)} =$$

$$b1\lambda 1\,\lambda 2\left(\frac{e^{-t\lambda 1}}{(\lambda 1-\lambda 2)(\lambda 1-\lambda 3)} + \frac{e^{-t\lambda 3}}{(\lambda 1-\lambda 3)(\lambda 2-\lambda 3)} + \frac{e^{-t\lambda 2}}{(-\lambda 1+\lambda 2)(\lambda 2-\lambda 3)}\right) =$$

$$= b_1\prod_{j=1}^{3-1}\lambda_j\sum_{j=1}^{3}\frac{e^{-\lambda_j t}}{\prod_{p=1,p\neq j}^{n}(\lambda_p-\lambda_j)}$$

Following the same approach, the previous equation can be extended for n daughter isotopes:

$$q_n(t) = b_1\prod_{j=1}^{n-1}\lambda_j\sum_{j=1}^{n}\frac{e^{-\lambda_j t}}{\prod_{p=1,p\neq j}^{n}(\lambda_p-\lambda_j)},\ \text{with initial conditions}: q_1(t) = b_1,\ q_2(0) =. .. = q_n(0) = 0 \quad (9.3)$$

This equation is known as Bateman's equation, in honor of the mathematician who first published it.

- We can type the equation as follows:

$$\texttt{q[n\_] := b}\left(\sum_{j=1}^{n} e^{-k_j t}\Big/\left(\left(\prod_{p=1}^{j-1}(k_p-k_j)\right)\prod_{p=j+1}^{n}(k_p-k_j)\right)\right)\prod_{j=1}^{n-1}k_j$$

- Here we apply it to the parent and the first 3 daughter isotopes (of the main branch) of Uranium–238.

```
u3 = mainbranch["U238", 3]
```

```
{U238, thorium-234, protactinium-234, uranium-234}
```

- We need the decay constants λ. Remember that $\tau = 1/\lambda$, with τ in seconds. However, since we prefer to measure t in days we multiply it by 24×3600.

```
λs = Thread[{k1, k2, k3, k4} → Evaluate[
    24 × 3600 / QuantityMagnitude[IsotopeData[#, "Lifetime"]] & /@ u3]]
```

$\{k_1\to 4.248\times 10^{-13},\ k_2\to 0.029,\ k_3\to 2.5,\ k_4\to 7.735\times 10^{-9}\}$

- Next we're going to calculate the evolution of the radioactivity, A, in the previous isotopes. We assume that at time $t = 0$ there's 1 gm of U238 in the sample, that expressed as the number of its atoms is N_A (Avogadro's number)/238, and 0 in the rest. Now we can compute A, using A = $\lambda\, N_A$.

```
{A1[t_], A2[t_], A3[t_], A4[t_]} =
  {k₁ q[1], k₂ q[2], k₃ q[3], k₄ q[4]} /. λs /.
   b → UnitConvert[1 N_A] / 238 // ExpandAll
```

$$\{1.075\times10^{9}\, e^{-4.248\times10^{-13}\,t},\ -1.1\times10^{9}\, e^{-0.029\,t}+1.1\times10^{9}\, e^{-4.248\times10^{-13}\,t},$$
$$1.3\times10^{7}\, e^{-2.5\,t}-1.1\times10^{9}\, e^{-0.029\,t}+1.1\times10^{9}\, e^{-4.248\times10^{-13}\,t},$$
$$-0.04\, e^{-2.5\,t}+3.\times10^{2}\, e^{-0.029\,t}-1.1\times10^{9}\, e^{-7.735\times10^{-9}\,t}+1.1\times10^{9}\, e^{-4.248\times10^{-13}\,t}\}$$

- We subtract 10^8 Bq from the activity of Pa234 with the only purpose of differentiating it from Th234. Otherwise both graphs would be superimposed since they are in equilibrium as we will see shortly.

```
Plot[{A1[t], A2[t] - 10^8, A3[t] - 10}, {t, 0, 180},
 PlotRange -> All, AxesLabel → {"t (days)", "A(t) in Bq"},
 PlotLegends -> Placed[{"U238", "Th234", "Pa234"}, Center]]
```

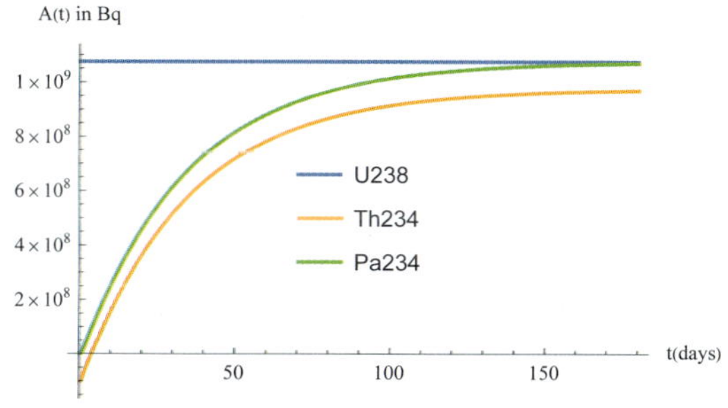

In the graph above we can see that after 180 days, A1≈A2≈A3. This is known as secular equilibrium. In this example it's enough to calculate the radioactivity of the isotope parent to know the daughter isotopes' radioactivity. This happens when $\lambda_1 < \lambda_2$.

- The next example shows that given enough time, in this case thousands of years, U238 and U234 will reach secular equilibrium. Normally we consider that the secular equilibrium has been reached after a period of time longer than 8 half-life periods of the daughter isotope.

```
Plot[{A1[t], A4[t]}, {t, 0, 10^9},
 PlotRange -> All, AxesLabel → {"t (days)", "A(t) in Bq"},
 PlotLegends → Placed[{"U238", "U234"}, Center]]
```

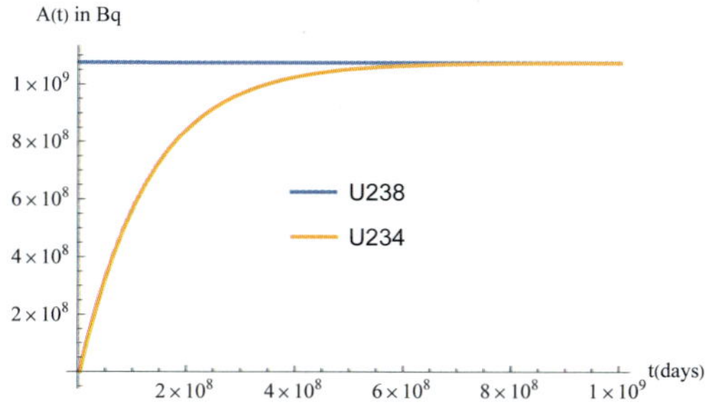

9.5 Application: Dating the History of Humankind

The fact that radioactive isotopes decay following a well-known physics law even under extreme conditions, makes them the ideal candidates for dating different type of samples. The actual isotopes used for that purpose will depend, among other factors, on the estimated age of the object that we want to date. We are going to discuss two methods: Carbon-14 dating is suitable for organic samples less than 50,000 years old and the lead-lead method is used when dating rocks thousands or even millions of years old.

Radiocarbon Dating

The Carbon-14 dating method (^{14}C) was discovered in 1947 by Willard Libby. It can be applied to samples that are no more than 50,000 years old. This method has become the most frequently used one for measuring the age of organic objects, especially archeological ones.

- The table below displays the isotopic abundance (proportion of isotopes) of the different carbon isotopes and their half-lives (If "HalfLife"< 1 s is not shown). An isotopic abundance of 0 indicates that the isotope doesn't exist in nature or is present only as traces. For example: the proportion of ^{14}C is very small although, as we will see, it plays a crucial role in dating.

```
vals = DeleteCases[Table[IsotopeData[#, prop],
      {prop, {"Symbol", "HalfLife", "IsotopeAbundance"}}] & /@
    IsotopeData["C"], {a_, b_ /; b < 1 s , c_}];
```

```
Text[
 Grid[Prepend[vals, {"", "Half-lives", "Isotope abundance"}], Frame → All,
  Background → {None, {{{LightBlue, White}}, {1 → LightYellow}}}]]
```

	Half-lives	Isotope abundance
^{10}C	19.29 s	0.
^{11}C	1220.0 s	0.
^{12}C	∞	0.9889
^{13}C	∞	0.0111
^{14}C	1.8×10^{11} s	0.
^{15}C	2.449 s	0.

This method is based on the following principle: 98.9% of the carbon in the Earth's atmosphere is Carbon–12, 1.1% Carbon–13, and the rest, 1.18 10^{-14} %, Carbon–14. This last isotope is the key to the method.

- The command below returns the decay scheme of ^{14}C. It decays into ^{14}N with a beta emission (1 electron) and has a half-life period of 1.8×10^{11} s (approximately 5,700 years)

```
c14 = IsotopeData["C14", #] & /@
  {"DaughterNuclides", "BranchingRatios", "DecayModes", "HalfLife"}
```

```
{{nitrogen-14}, {1.00}, {BetaDecay}, 1.8×10^11 s}
```

The nitrogen present in the atmosphere is subject to a continuous neutron bombardment from cosmic radiation. Some of these neutrons strike the nitrogen atoms with enough energy to transform them into ^{14}C that will eventually decay into ^{14}N.

If we assume that the cosmic radiation has not changed significantly during the last 50,000 years, the production rate of ^{14}C atoms will be constant. Since the disintegration period of

Carbon–14 is relatively short (approximately 5,700 years), the quantity of atoms that are produced will be similar to the ones that disintegrate, so the percentage of this isotope out of the total amount of carbon in the atmosphere will not have changed over the last millennia. Living organisms continuously absorb carbon when breathing or performing photosynthesis but when they die, this absorption stops. Two (Carbon–12 and Carbon–13) out of the three carbon isotopes are stable and will remain so; the other one, Carbon–14, will decay due to its radioactivity and its percentage of the total amount of carbon will get smaller. Measuring the content of ^{14}C in an organic sample and comparing it with the total amount of carbon in the entire sample will allow us to know its age. In reality, the method is not that simple since the proportion of ^{14}C in the atmosphere is not constant, but fluctuates with solar activity, introducing the need to make adjustments. Furthermore, we have to take into account ^{14}C originated from non-organic sources, such as nuclear explosions, mainly during the fifties and beginning of the sixties in the 20th century.

The evolution of Carbon–14 can be inferred by measuring its content in the trunks of very old trees. Tree trunks consist of layers or rings, with each one corresponding to one year in the life of the tree. By counting the number of rings starting from the bark we can know the age of a tree. This dating technique is known as dendrochronology. Using this method, scientists have been able to "calibrate" the ^{14}C method for samples up to 9,000 years old. This calibration has recently been extended to 40,000 years. This has been possible thanks to the comparison between this method and the measurement of U234 and Th230 in corals. When corals are born, their ^{234}U content is known. From that moment on ^{230}Th starts to accumulate. Comparing the content of C14 with Th230, we can determine the age of the coral. Since corals can be dated using the C14 method, it's possible to compare both techniques.

Example: There are 2 grams of carbon in a piece of wood found in an archeological site and we discover that it has an activity of 10 nuclear transformations per minute per gm. How old is the piece of wood? For simplification purposes we will assume that the C14 content in the wood at the time it was cut is the same as in wood recently cut: 15 nuclear transformations per minute per gm. However, in real cases, as explained before, we may have to make some adjustments regarding this assumption.

- This problem is a case of a decay chain with only one daughter isotope: $A = A_0 \, e^{-t k_1}$, with k_1 being the decay constant of C14. Therefore, the age of the sample would be:

```
Solve[{ A == A0 ⅇ^(-λ t)}, t, Reals] /.
 {λ -> 1/UnitConvert[IsotopeData["C14", "Lifetime"], "year"],
  A → 10, A0 → 15}
```

```
{{t → 3.3×10^3 yr}}
```

The Age of the Earth

To estimate the age of the Earth, we need to know its origins. The mainstream theory of the formation of the solar system and probably, of other planetary systems similar to ours, is as follows: there are certain areas in galaxies where clouds of dust and gas accumulate and once these clouds reach certain mass and density they may condense as a result of gravitational attraction forces. This phenomenon can benefit from, or depends on?, the explosion of a star (supernova) in its proximity. During the condensation, the clouds may break into smaller units that become protostars. These protostars continue contracting rapidly until their centers reach very high temperatures and densities causing the fusion (union) of the lightest atomic nuclei. This fusion emits enough energy to stop the gravitational collapse. At this moment, stars are born. Surrounding them there are still many fragments. Some of these fragments start clustering during successive collisions until they reach a certain size and become planets. This is probably the origin of the planets closest to the Sun, such as the Earth. The fragments that didn't become part of a bigger mass were left wandering around the solar system, originating some of the currently existing meteorites.

Over long periods of time, orogenic and sedimentary processes have destroyed the remains of the rocks that originally formed the Earth (primordial), making it impossible for us to date the Earth directly from them. Fortunately, we've been able to calculate its age using visitors from outer space: meteorites. Their analysis supports the idea that after the formation of the solar system, with some exceptions, the distribution of isotopes was homogeneous.

Among meteorites, the carbonaceous chondrite type is particularly interesting. Chondrites are made of chondrules, molten droplets created during the collisions that gave birth to planets, that have the isotopic composition of the planet-forming period. Furthermore, they didn't experience any further heating intense enough to melt them again. It's been possible to determine their age using dating techniques based on radioactive isotopes with long half-lives. The most commonly used methods are: the lead-lead method (Pb-Pb) and the rubidium-strontium one (Rb-Sr). Next, we're going to discuss the Pb-Pb method.

- The decay chains of Uranium–238 and Uranium–235 are of particular interest among natural isotopes. Their final nuclides are respectively: ^{206}Pb and ^{207}Pb.

```
Row[IsotopeData[#, "Symbol"] & /@ mainbranch1["U238"], "->"]
```

^{238}U -> ^{234}Th -> ^{234}Pa -> ^{234}U -> ^{230}Th -> ^{226}Ra -> ^{222}Rn -> ^{218}Po -> ^{214}Pb -> ^{214}Bi -> ^{214}Po -> ^{210}Pb -> ^{210}Bi -> ^{210}Po -> ^{206}Pb

```
Row[IsotopeData[#, "Symbol"] & /@ mainbranch1["U235"], "->"]
```

^{235}U -> ^{231}Th -> ^{231}Pa -> ^{227}Ac -> ^{227}Th -> ^{223}Ra -> ^{219}Rn -> ^{215}Po -> ^{211}Pb -> ^{211}Bi -> ^{207}Tl -> ^{207}Pb

- In nature, apart from ^{206}Pb and ^{207}Pb, we also have other stable lead isotopes in the following proportions:

```
pbestable = DeleteCases[ Transpose[{IsotopeData["Pb"],
     IsotopeData[#, "Stable"] & /@ IsotopeData["Pb"] }], {_, False}];
```

```
vals =
  Table[IsotopeData[#, prop], {prop, {"Symbol", "IsotopeAbundance"}}] & /@
   First[Transpose[pbestable]];
```

```
Text[Grid[Prepend[vals, {"Isotope", "Isotope Abundance"}], Frame → All,
  Background → {None, {{{LightBlue, White}}, {1 → LightYellow}}}]]
```

Isotope	Isotope Abundance
^{204}Pb	0.014
^{206}Pb	0.241
^{207}Pb	0.221
^{208}Pb	0.524

These proportions may vary widely depending on the origin of the sample.

- ^{208}Pb is a stable daughter isotope of the decay chain of Th232, and doesn't play any role in the dating method we are about to describe. ^{204}Pb, however, is a primordial isotope. This means that the amount of this nuclide present on Earth has not changed since the formation of our planet. Therefore it's useful for estimating the fraction of the other lead isotopes in a given sample that are also primordial since their relative fractions are constant everywhere.

```
Row[IsotopeData[#, "Symbol"] & /@ mainbranch1["Th232"], "->"]
```

^{232}Th ->
^{228}Ra -> ^{228}Ac -> ^{228}Th -> ^{224}Ra -> ^{220}Rn -> ^{216}Po -> ^{212}Pb -> ^{212}Bi -> ^{212}Po -> ^{208}Pb

At the moment of the formation of the solar system, U238 and U235, after going through an homogenization process they started to decay respectively into Pb-206 and Pb-207 adding to the existing amount in the rocks.

The quantity from the parent isotope that decays into a daughter isotope is: $N_h(t) = 1 - N_p(t)$ that is:

$$N_h = N_p\, e^{\lambda t}$$

As components of rocks or meteorites, the total number of atoms in the systems U238 + Pb206 and U235 + Pb207 remains constant.

$$\left({}^{207}\mathrm{Pb}\right)_P = \left({}^{207}\mathrm{Pb}\right)_I + \left({}^{235}\mathrm{U}\right)\left(e^{\lambda 235\, t} - 1\right)$$

$$\left({}^{206}\mathrm{Pb}\right)_P = \left({}^{206}\mathrm{Pb}\right)_I + \left({}^{238}\mathrm{U}\right)\left(e^{\lambda 238\, t} - 1\right)$$

with the P and I subscripts indicating respectively, the present and initial quantities, $\lambda 235$ and $\lambda 238$ the decay constants of U235 and U238, and t the elapsed time.

There will also be a certain amount of Pb204 that will not change. Under these hypotheses we obtain the following relationships:

$$\left(\frac{{}^{207}\mathrm{Pb}}{{}^{204}\mathrm{Pb}}\right)_P = \left(\frac{{}^{207}\mathrm{Pb}}{{}^{204}\mathrm{Pb}}\right)_I + \left(\frac{{}^{235}\mathrm{U}}{{}^{204}\mathrm{Pb}}\right)\left(e^{\lambda 235\, t} - 1\right)$$

$$\left(\frac{{}^{206}\mathrm{Pb}}{{}^{204}\mathrm{Pb}}\right)_P = \left(\frac{{}^{206}\mathrm{Pb}}{{}^{204}\mathrm{Pb}}\right)_I + \left(\frac{{}^{238}\mathrm{U}}{{}^{204}\mathrm{Pb}}\right)\left(e^{\lambda 238\, t} - 1\right)$$

Rearranging the previous equations we get:

$$\left(\frac{{}^{207}\mathrm{Pb}}{{}^{204}\mathrm{Pb}}\right)_P - \left(\frac{{}^{207}\mathrm{Pb}}{{}^{204}\mathrm{Pb}}\right)_I = \left(\frac{{}^{235}\mathrm{U}}{{}^{204}\mathrm{Pb}}\right)\left(e^{\lambda 235\, t} - 1\right)$$

$$\left(\frac{{}^{206}\mathrm{Pb}}{{}^{204}\mathrm{Pb}}\right)_P - \left(\frac{{}^{206}\mathrm{Pb}}{{}^{204}\mathrm{Pb}}\right)_I = \left(\frac{{}^{238}\mathrm{U}}{{}^{204}\mathrm{Pb}}\right)\left(e^{\lambda 238\, t} - 1\right)$$

After dividing the first one by the second one:

$$\left[\left(\frac{{}^{207}\mathrm{Pb}}{{}^{204}\mathrm{Pb}}\right)_P - \left(\frac{{}^{207}\mathrm{Pb}}{{}^{204}\mathrm{Pb}}\right)_I\right] \Big/ \left[\left(\frac{{}^{206}\mathrm{Pb}}{{}^{204}\mathrm{Pb}}\right)_P - \left(\frac{{}^{206}\mathrm{Pb}}{{}^{204}\mathrm{Pb}}\right)_I\right] = \frac{1}{{}^{238}\mathrm{U}/{}^{235}\mathrm{U}}\,\frac{\left(e^{\lambda 235\, t} - 1\right)}{\left(e^{\lambda 238\, t} - 1\right)} = K$$

- The ratio ^{238}U/ ^{235}U is:

```
IsotopeData["U238", "IsotopeAbundance"] /
 IsotopeData["U235", "IsotopeAbundance"]
```

```
137.80
```

However, in the scientific literature: ^{238}U/ ^{235}U =137.88 and this is the value we are going to use.

Since $\frac{1}{137.88}\frac{\left(e^{\lambda 235\, t}-1\right)}{\left(e^{\lambda 238\, t}-1\right)}$ is constant for samples from the same period, the relation $\frac{{}^{207}\mathrm{Pb}}{{}^{204}\mathrm{Pb}}/\frac{{}^{206}\mathrm{Pb}}{{}^{204}\mathrm{Pb}}$ is a straight line, as Figure 9.2 shows:

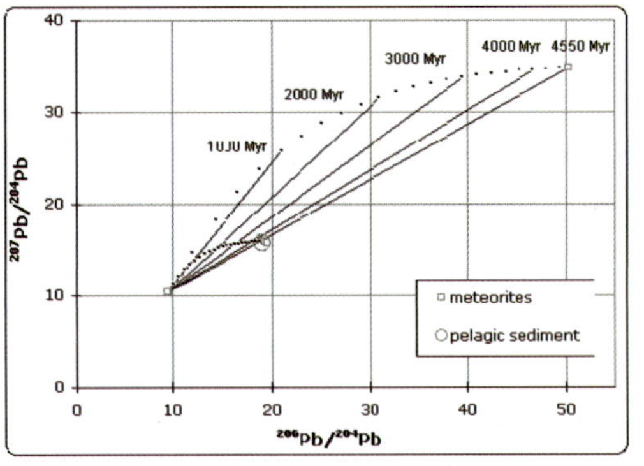

Figure 9.2 Paterson's Lead Isotope Isochron.

Source: http://en.wikipedia.org/wiki/File:Paterson_isochron _animation.gif

- The different lines obtained depend on the original U/Pb ratio. For samples coming from the oldest meteorites (see the graph), K = 0.61, meaning that *t*, in eons (billions of years) is:

```
K = 1/137.88 (e^(λ235 t) - 1)/(e^(λ238 t) - 1) /.
   {λ235 -> 1/QuantityMagnitude[UnitConvert[IsotopeData["U235",
        "Lifetime"], "eons"]], λ238 -> 1/QuantityMagnitude[
      UnitConvert[IsotopeData["U238", "Lifetime"], "eons"]]};
```

```
FindRoot[K == 0.61 , {t, 0.5}]
```

```
{t → 4.54709}
```

- We can compare this result with the one given by *Mathematica* using the free-form input and see that is quite similar.

```
Age of Earth »
PlanetData[Entity["Planet", "Earth"], "Age"]
```

4.54×10^9 yr

Orogenic phenomena have left us without remains of rocks from the time of the formation of our planet. The oldest ones have been found in zircon crystals from Mount Narryer, in Western Australia. They are estimated to be more than 4,000 years old, proving that at that time the earth already had a continental crust.

9.6 Application: Calculating Binding Energies

The mass of an atom is always smaller than the sum of its component particles. This difference in mass is known as binding energy and is responsible for keeping the particles together to form the element. To obtain some property for a subatomic particle we use `ParticleData`.

- For example, the iron isotope Fe56 consists of N neutrons, and Z protons, whose actual numbers can be calculated as follows:

```
{Nfe56, Zfe56} =
 IsotopeData["Fe56", #] & /@ {"NeutronNumber", "AtomicNumber"}
```

```
{30, 26}
```

- The sum of the particle masses that form the iron isotope (expressed in MeV/c^2): $m_A = \text{N}\, m_n + \text{Z}\, (m_p + m_e)$:

```
fe56particlesmass = (Nfe56 ParticleData["Neutron", "Mass"] +
    Zfe56 ParticleData["Proton", "Mass"] +
    Zfe56 ParticleData["Electron", "Mass"])
```

```
52595.320 MeV/c^2
```

- We calculate below the conversion factor of 1 amu (atomic mass unit) to 1 MeV:

AtomicMassUnit in MeV

Result

931.4941 MeV/c²

- We find the mass of the iron isotope Fe56 and express it in MeV:

```
fe56mass =
 UnitConvert[IsotopeData["Fe56", "AtomicMass"], 931.494 MeV/c^2 ]
```

```
52103.06 MeV/c^2
```

- Next we calculate the binding energy per nucleon B, in MeV as well:

```
(fe56particlesmass - fe56mass) / (Nfe56 + Zfe56)
```

```
8.7903 MeV/c^2
```

- We can get approximately the same value directly with (sometimes in MeV/c^2 we omit c^2 or in natural units we consider $c = 1$):

```
IsotopeData["Fe56", "BindingEnergy"]
```

```
8.790323 MeV
```

- Another commonly used concept is "excess of mass" calculated as:

```
(QuantityMagnitude[IsotopeData["Fe56", "AtomicMass"]] -
   IsotopeData["Fe56", "MassNumber"]) 931.494
```

```
-60.6054
```

- We can also calculate it directly:

```
IsotopeData["Fe56", "MassExcess"]
```

-60.605352 MeV

- The function below displays the binding energies as a function of the mass number. With `Tooltip` we can see the information related to a point when placing the cursor on it:

```
bindingenergy = Tooltip[{IsotopeData[#, "MassNumber"],
      IsotopeData[#, "BindingEnergy"]}] & /@ IsotopeData[];
```

```
ListPlot[bindingenergy, Frame → True,
 FrameLabel → {"Mass number A", "Binding energy per nucleon (B), in MeV"},
 Axes → False, ImageSize → {300, 250}]
```

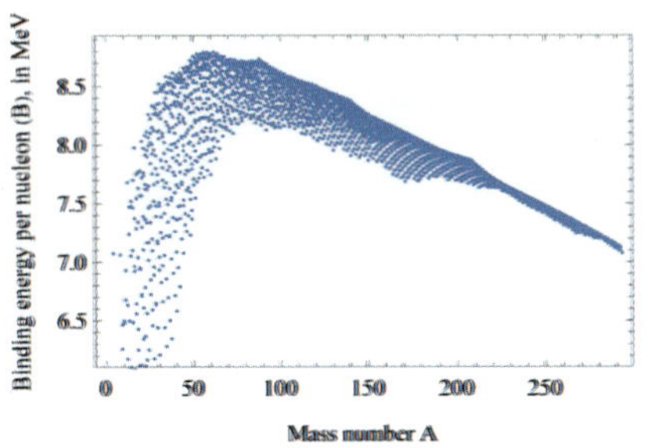

- The following command shows the binding energy per nucleon, as a function of the atomic number and the number of neutrons until Z = 92 (uranium):

```
bindingenergyZNB = Flatten[
   Table[{ z, a - z, IsotopeData[{z, a}, "BindingEnergy"]}, {z, 1, 92},
    {a, IsotopeData[#, "MassNumber"] & /@ IsotopeData[z]}], 1];
```

- Here we visualize it in 3D:

```
ListPlot3D[bindingenergyZNB, AxesLabel → {"A", "N", "B"}]
```

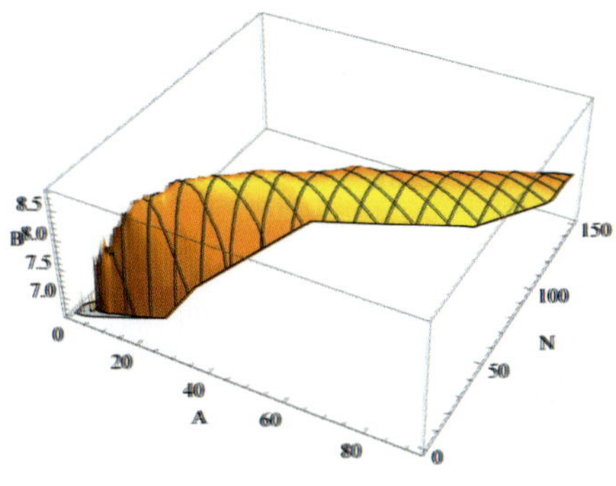

- Next we show an example with 224 nuclides with less than 15 protons each (Z < 15) and plot it using the function `ColorData`["Atoms"] that associates each element to one color. We can find the value of Z and N by placing the cursor inside the graph, right clicking with the mouse button and selecting “Get Coordinates”.

```
lessthan15 = Take[bindingenergyZNB, 224];
```

```
ListDensityPlot[lessthan15, FrameLabel → {"N", "Z"},
 ColorFunction → Map[ColorData["Atoms", ElementData[#, "Abbreviation"]] &,
   Transpose[lessthan15][[1]]]]
```

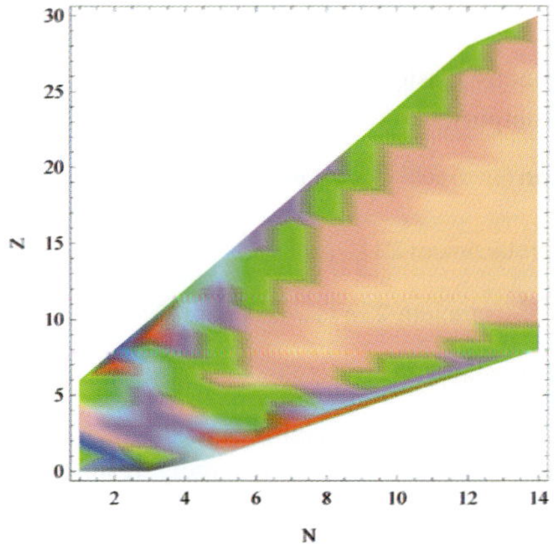

- In this case we present the same information but using a color code that associates redder colors to higher binding energies:

```
ListDensityPlot[lessthan15, FrameLabel → {"N", "Z"},
 InterpolationOrder → 1, ColorFunction → "Rainbow"]
```

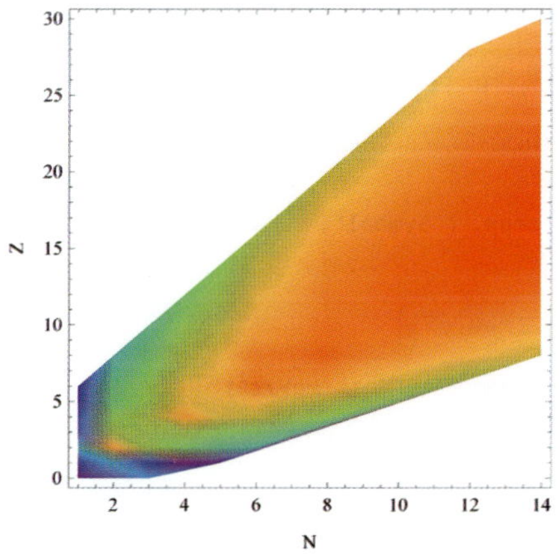

Decays and nuclear transformations are related to binding energies. There are two types of nuclear transformations that are particularly useful: fusion and fission.

Fusion is the process of joining nuclei. This process is the most predominant one in elements with low atomic numbers, specifically, until iron. The lower the atomic number the lower the binding energy associated to it. Fusion is the normal method by which stars generate energy.

Fission is the opposite process of fusion. It consists of splitting apart nuclei to turn them into lighter ones.

Some isotopes of uranium and plutonium represent a special case where the process accelerates when interacting with neutrons.

- Some very heavy nuclei may experience spontaneous fission. The next function displays the first five such elements:

```
Take[Transpose[Cases[Table[IsotopeData[#, prop],
      {prop, {"Symbol", "SpontaneousFission"}}] & /@
    IsotopeData[], Except[{_, False}]]][[1]], 5]
```

{^{230}Th, ^{232}Th, ^{231}Pa, ^{234}Pa, ^{230}U}

- Since `IsotopeData` includes "Spontaneous Fission" as a property, we can also get them as follows (we don't show the complete output):

```
IsotopeData[ spontaneous fission (isotopes) ] // Shallow
```

{thorium-230, thorium-232, protactinium-231, protactinium-234, uranium-230, uranium-232, uranium-233, uranium-234, uranium-235, uranium-236, <<193>>}

- Many of them are transuranic elements obtained by artificial means although there are also some isotopes in nature that exhibit this property. They are largely responsible for the presence of free neutrons that have lifetimes of just a few minutes:

```
ParticleData["Neutron", "Lifetime"]
```

885.6 s

- This can be explained by the fact that both, neutrons and protons, are not really elementary particles but are made of two types of quarks:

```
ParticleData[#, "QuarkContent"] & /@ {"Proton", "Neutron"}
```

{{{d, u, u}}, {{d, d, u}}}

- Let's see some these quarks' characteristics:

```
values =
 Table[ParticleData[#, prop], {prop, {"Symbol", "Charge", "Mass"}}] & /@
  {"DownQuark", "UpQuark"}
```

{{d, $-\frac{1}{3}$, 5.0 MeV/c^2}, {u, $\frac{2}{3}$, 2.2 MeV/c^2}}

```
Text[Grid[Prepend[values, {"", "Charge", "Mass(MeV/c^2)"}], Frame → All,
  Background → {None, {{{LightBlue, White}}, {1 → LightYellow}}}]]
```

	Charge	Mass(MeV/c^2)
d	$-\frac{1}{3}$	5.0 MeV/c^2
u	$\frac{2}{3}$	2.2 MeV/c^2

- Notice how small the mass of the quarks is compared to that of a neutron (two quarks down and one quark up) or a proton (two quarks up and one quark down):

```
{qd, qu} = ParticleData[#, "Mass"] & /@ {"DownQuark", "UpQuark"}
```

{5.0 MeV/c^2, 2.2 MeV/c^2}

```
{quarksproton, quarkneutron} =
 100 {(2 qu + 2 × 1 qd) / ParticleData["Proton", "Mass"],
    (2 qu + 1 qd) / ParticleData["Neutron", "Mass"]} "%"
```

```
{1.5 %, 1.0 %}
```

This means that the quarks only represent between 1% and 1.5% of the atomic nuclei mass with the rest coming from their interactions. This is the reason why quarks are not present as free particles in nature.

The mass of quarks and other particles such as electrons is attributed to the Higgs field (more accurately to the Brout–Engler–Higgs mechanism) whose existence was confirmed by the detection of the Higgs boson, officially announce by CERN on July 4, 2012. In that case, what about our mass? Does it come as well from the Higgs field? Only a small portion of it.

As we've seen, quarks only represent between 1% and 1.5% of the nuclei mass and that mass is explained by the Higgs field. Therefore, for a person weighing 80 kg, the Higgs field explains approximately 1 kg. What about the rest? It comes from gluons, g, that keep the photons and quarks joined. In this case we should refer to mass equivalence, $m = E/c^2$. These particles don't interact with the Higgs field or BEH but they interact with the gravitational field. However, without the Higgs field, electrons would move at the speed of light, atoms would not have formed and therefore we wouldn't exist.

The *Standard Model of Particle Physics* (https://en.wikipedia.org/wiki/Standard_Model), apart from quark d and quark q, also includes ten additional types of quarks and antiquarks, the leptons, the gauge bosons, and the Higgs boson. We will use the `EntityClass` function, which we have seen in Chapter 5, to represent some properties of these particles.

```
TextSentences[WikipediaData["Standard_Model"]][[;; 2]]
```

```
{The Standard Model of particle physics is a theory concerning
   the electromagnetic, weak, and strong nuclear interactions,
   as well as classifying all the subatomic particles known.,
 It was developed throughout the latter half of the 20th century,
   as a collaborative effort of scientists around the world.}
```

■ Some properties of the gauge bosons, quarks, and leptons are summed up in the next tables.

```
Dataset[EntityValue[EntityClass["Particle", "GaugeBoson"],
 {EntityProperty["Particle", "Lifetime"], EntityProperty["Particle",
  "Width"]}, "EntityPropertyAssociation"]]
```

g	mean lifetime	—
	width	—
photon	mean lifetime	∞ s
	width	0. MeV
Z	mean lifetime	2.6379×10^{-25} s
	width	2.4952×10^{9} eV
W−	mean lifetime	3.076×10^{-25} s
	width	2.140×10^{9} eV
W+	mean lifetime	3.076×10^{-25} s
	width	2.140×10^{9} eV

```
Dataset[EntityValue[EntityClass["Particle", "Lepton"],
  {"Symbol", "Charge", "Mass"}, "EntityPropertyAssociation"]]
```

	Symbol	Charge	Mass
e−	e	−1 e	510.9989 keV/c^2
ν_e	ν_{e}	0 e	Interval[{0.0009, 3.}] eV/c^2
ν_e-bar	$\bar{\nu}_{e}$	0 e	Interval[{0.0009, 3.}] eV/c^2
μ−	μ	−1 e	105.658369 MeV/c^2
μ+	$\bar{\mu}$	1 e	105.658369 MeV/c^2
ν_μ	ν_{μ}	0 e	Interval[{$9. \times 10^{-7}$, 2.0×10^{2}}] keV/c^2
ν_μ-bar	$\bar{\nu}_{\mu}$	0 e	Interval[{$9. \times 10^{-7}$, 2.0×10^{2}}] keV/c^2
e+	$\bar{e}$	1 e	510.9989 keV/c^2
τ−	τ	−1 e	1.77699 GeV/c^2
τ+	$\bar{\tau}$	1 e	1.77699 GeV/c^2
ν_τ	ν_{τ}	0 e	Interval[{$9. \times 10^{-10}$, 18.3}] MeV/c^2
ν_τ-bar	$\bar{\nu}_{\tau}$	0 e	Interval[{$9. \times 10^{-10}$, 18.3}] MeV/c^2

```
Dataset[EntityValue[EntityClass["Particle", "Quark"],
  {"Symbol", "Charge", "Mass"}, "EntityPropertyAssociation"]]
```

	Symbol	Charge	Mass
b	b	−13 e	4.200 GeV/c^2
b-bar	$\bar{b}$	13 e	4.200 GeV/c^2
c	c	23 e	1.250 GeV/c^2
c-bar	$\bar{c}$	−23 e	1.250 GeV/c^2
d	d	−13 e	5.0 MeV/c^2
d-bar	$\bar{d}$	13 e	5.0 MeV/c^2
s	s	−13 e	95. MeV/c^2
s-bar	$\bar{s}$	13 e	95. MeV/c^2
t	t	23 e	174.200 GeV/c^2
t-bar	$\bar{t}$	−23 e	174.200 GeV/c^2
u	u	23 e	2.2 MeV/c^2
u-bar	$\bar{u}$	−23 e	2.2 MeV/c^2

9.7 Additional Resources

In the Wolfram Demonstrations website you can find quite sophisticated examples about nuclear physics and isotopes:

Nuclear physics demonstrations:
http://demonstrations.wolfram.com/search.html?query=Nuclear
Isotopes demonstrations: http://demonstrations.wolfram.com/search.html?query=isotopes

Individual demonstrations worth mentioning:

Table of Nuclides: http://demonstrations.wolfram.com/TableOfNuclides by Enrique Zeleny
Isotope Browser: http://demonstrations.wolfram.com/IsotopeBrowser by Theodore Gray and Yifan Hu
Binding Energies of Isotopes:
http://demonstrations.wolfram.com/BindingEnergiesOfIsotopes by Stephen Wolfram and Jamie Williams

Articles in English and Spanish in the author's website:
http://diarium.usal.es/guillermo/publicaciones/

10

Modeling: Applications in Biokinetics

In this chapter we show how to use Mathematica for biokinetic and pharmacokinetic modeling, mainly in the context of building multi-compartment models and nonlinear models using differential equations. These models also have many applications in a wide variety of fields such as: chemical kinetics, clinical and nuclear medicine, ecology and internal dosimetry, among others. We will also cover the statistical analysis of experimental data with a special emphasis on how to estimate parameters and perform optimal experimental designs. The reader may find the contents of this chapter useful for learning about modeling in general, and in particular the fitting of non-linear data in systems of differential equations. Some of the functions described in the following sections are part of the Biokomod toolbox, created by the author to make this type of modeling in Mathematica easier. Software developers may find this chapter useful as well since it shows how someone with a good knowledge of the Wolfram Programming Language can create a collection of functions and put them together in a package to help people without much knowledge of either Mathematica or the mathematical methods behind the package functions solve relatively complex problems.

10.1 Compartmental and Physiological Modeling

10.1.1 An Introduction to Compartmental Analysis

Compartmental analysis has applications in chemical kinetics, clinical medicine, ecology, internal dosimetry, nuclear medicine and pharmacokinetics. It can be described as the analysis of a system when it is separated into a finite number of component parts which are called compartments. Compartments interact through the exchange of materials or energies. These materials or energies can be chemical substances, hormones, individuals in a population and so on. A compartmental system is usually represented by a flow or a block diagram.

- The code below generates a diagram representing a basic compartmental system:

```
Show[Graphics[{Arrow[{{
                    0, 1.5}, {0, 0.5}}],
   Text["Input", {0.4, 1}, {0, -1}], Circle[{0, 0}, 0.5] ,
   Text["Compartment 1", {0, 0}, {0, 0}], Arrow[{{0.5, 0.}, {2.5, 0.}}],
   Text["Flow from 1 to 2", {1.5, .4}, {0, 0}],
   Circle[{3, 0}, 0.5] , Text["Compartment 2", {3, 0}, {0, 0}],
   Arrow[{{3, -0.5}, {3, -1.5}}], Text["Output", {2.4, -1}, {0, -1}]},
  AspectRatio → Automatic, Axes → None]]
```

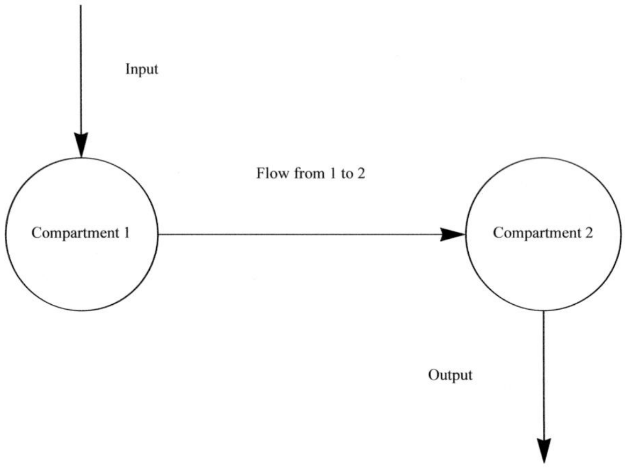

We adopt the convention of representing compartments with circles or rectangles. The flow in or out of the compartments is represented by arrows. The ith compartment of a system of n compartments is labeled i and the size (amount or content) of the component in compartment i as $x_i(t)$. The exchange between compartments, or between a compartment and the environment is labeled $k_{i\,j}$, where $i\,j$ represents the flow from i to j. If there is no ambiguity, k_i will be used in place of k_{ij}. Additionally, if $i < 10$ and $j < 10$, k_{ij} will be written in place of $k_{i\,j}$. The environment refers to the processes that are outside the system and is usually represented by zero, so $k_{i\,0}$ is the fractional excretion coefficient from the ith compartment to the environment. If we assume that the substance introduced into the system is a radioactive isotope, we must consider the radioactive decay, which is given by a constant rate represented by λ (this constant is specific for each isotope). The decay constant can be interpreted as an equal flow going out of the system in each compartment. The input from the environment into the jth compartment is called $b_j(t)$. With regards to the environment, we only need to know the flow, $b_j(t)$, into the system from the outside. The $k_{i\,j}$s are called fractional transfer rate coefficients. They are usually assumed to be constants but in some cases they can be functions of time (that is $k_{i\,j}(t)$).

Based on the definitions in the previous paragraph, we can describe the two-compartment model in Figure 10.1 as follows: $b_1(t)$ is the input from the environment to compartment 1, k_{12} the transfer rate coefficient from compartment 1 to compartment 2 and k_{21} from compartment 2 to 1, k_{20} the transfer rate coefficient from compartment 1 to the environment (output), and $x_1(t)$ and $x_2(t)$ represent the quantities in compartments 1 and 2 at time t:

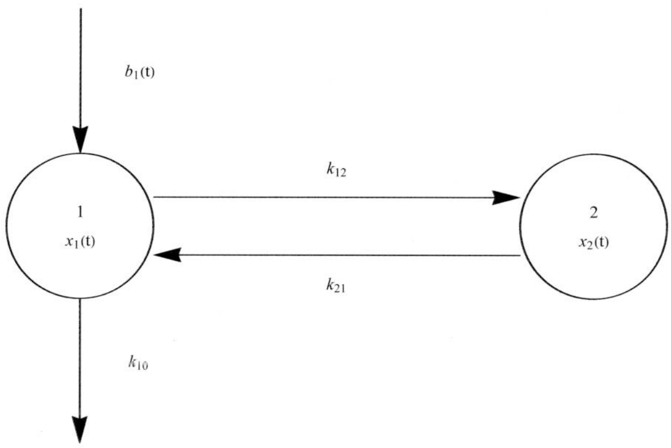

Figure 10.1 A two-compartment model.

A Two-compartment Model with Input and Output from Compartment 1

The variables $x_1(t)$ and $x_2(t)$ are called the state variables of the system and their evolution over time is described by a system of ordinary differential equations (SODE). In this case:

$$\begin{aligned} \dot{x}_1(t) &= -k_{10}\,x_1(t) - k_{12}\,x_1(t) + k_{21}\,x_2(t) + b_1\,(t) \\ \dot{x}_2(t) &= \quad k_{12}\,x_1(t) - k_{21}\,x_2(t) \end{aligned} \tag{10.1}$$

or in matrix notation:

$$\begin{pmatrix} \dot{x}_1\,(t) \\ \dot{x}_2\,(t) \end{pmatrix} = \begin{pmatrix} -k_{10} - k_{12} & k_{21} \\ k_{12} & -k_{21} \end{pmatrix} + \begin{pmatrix} b_1(t) \\ 0 \end{pmatrix}$$

We also need to know the initial conditions. In this case, the contents of $x_1(t)$ and $x_2(t)$ at $t = 0$.

$$x_1(0) = x_{10};\ x_2(0) = x_{20}$$

This SODE can be solved using `DSolve` or `DSolveValue`.

- For instance, in day^{-1}, for $k_{10} = 0.1$, $k_{12} = 0.2$, $k_{21} = 0.3$, $b_1(t) = 0.2\,t\,\text{Exp}[-2.1\,t]$, with $x_1(0) = x_2(0) = 0$. The solution is:

```
{x1[t_], x2[t_]} =
DSolveValue[{x1' [t] == -0.1 x1[t] - 0.2 x1[t] + 0.3 x2[t] + 0.2 t Exp[-2.1 t] ,
    x2' [t] == 0.2 x1[t] - 0.3 x2[t], x1[0] == 0, x2[0] == 0},
   { x1[t], x2[t]}, t] // ExpandAll // Chop
```

$\{-0.0652664\,e^{-2.1\,t} + 0.0413534\,e^{-0.544949\,t} + 0.0239131\,e^{-0.055051\,t} - 0.113208\,e^{-2.1\,t}\,t,$
$0.0142399\,e^{-2.1\,t} - 0.0337649\,e^{-0.544949\,t} + 0.0195249\,e^{-0.055051\,t} + 0.0125786\,e^{-2.1\,t}\,t\}$

We have used "**ExpandAll // Chop**" to eliminate insignificant terms.

- The evolution of $x_1(t)$ and $x_2(t)$ over time can be visualized as follows (we show two graphs using different ways to include the legends. In the graphic on the right we use `Callout`, a new function in *Mathematica* 11):

```
GraphicsRow[
 {Plot[{x1[t], x2[t]}, {t, 0, 30}, AxesOrigin→ {0, 0}, AxesLabel→
    {"Days", "x"}, PlotLegends→Placed[{"x₁(t)", "x₂(t)"}, Center],
   PlotLabel→Style["Using PlotLegends", Bold]],
  Plot[{Callout[x1[t], "x₁(t)", Above], Callout[x2[t], "x₂(t)", 4]},
   {t, 0, 30}, AxesOrigin→ {0, 0}, AxesLabel→ {"Days", "x"},
   PlotLabel→Style["Using Callout", Bold]]}, ImageSize→ 400 ]
```

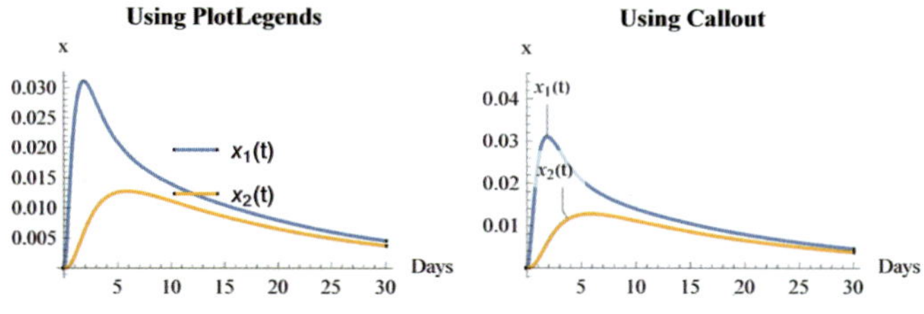


```
Clear[x1, x2]
```

10.1.2 Developing Functions for Solving Compartmental Systems

Instead of using `DSolveValue` or `DSolve` directly, we are going to develop some functions for solving multi-compartment models using a friendly notation.

Figure 10.2 shows a more complex compartmental model: The general multi-input multi-output (MIMO) three-compartment model. We have assumed a radioactive decay with a disintegration constant λ (it can be considered as an output from each compartment):

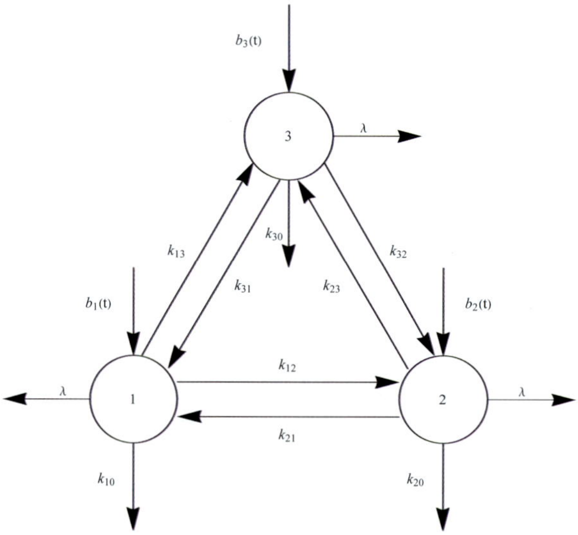

Figure 10.2 The generic three-compartment model.

$$\begin{aligned}\dot{x}_1(t) &= -(\lambda + k_{10} + k_{12} + k_{13})\, x_1(t) + k_{21}\, x_2(t) + k_{31}\, x_3(t) + b_1(t)\\ \dot{x}_2(t) &= k_{12}\, x_1(t) - (\lambda + k_{20} + k_{21} + k_{23})\, x_2(t) + k_{32}\, x_3(t) + b_2(t)\\ \dot{x}_3(t) &= k_{13}\, x_1(t) + k_{23}\, x_2(t) - (\lambda + k_{30} + k_{31} + k_{32})\, x_3(t) + b_3(t)\end{aligned} \quad (10.2)$$

The model can be expressed in matrix notation as follows:

$$\dot{\mathbf{x}}(t) = \mathbf{A}\,\mathbf{x} + \mathbf{b}(t), \; t \geq 0$$

where $\mathbf{x}(t)$ is a column vector representing the contents of the compartments:

$$\mathbf{x}(t) = \{x_1(t),\; x_2(t),\; x_3(t)\}^T$$

$\boldsymbol{b}(t)$ is a column vector representing the inputs to compartments 1, 2, and 3:

$$\mathbf{b}(t) = \{b_1(t),\; b_2(t),\; b_3(t)\}^T$$

where $\mathbf{A}$ is given by:

$$A = \begin{pmatrix} a_{11} & a_{12} & a_{13} \\ a_{21} & a_{22} & a_{23} \\ a_{31} & a_{32} & a_{33} \end{pmatrix}$$

with:

$a_{11} = -(k_{10} + k_{12} + k_{13})$; $a_{12} = k_{21}$; $a_{13} = k_{31}$;
$a_{21} = k_{12}$; $a_{22} = -(k_{20} + k_{21} + k_{23})$; $a_{23} = k_{32}$;
$a_{31} = k_{13}$; $a_{32} = k_{23}$; $a_{33} = -(k_{30} + k_{31} + k_{32})$;

This pattern can be expanded to systems of n compartments or n state variables (in the case of physiological models). The equation for any compartment i is given by:

$$\begin{aligned}\dot{\mathbf{x}}(t) &= \mathbf{A}\,\mathbf{x} + \mathbf{b}(t), \; t \geq 0\\ \mathbf{x}(0) &= \mathbf{x}_0\end{aligned} \quad (10.3)$$

where :

$\mathrm{x}(t) = \{x_1(t),\; x_2(t),\; \ldots,\; x_n(t)\}^T$ is a column vector and x_i(t) denotes the amount or content of materials or energies in compartment i at time t.

$\mathbf{A}$ is a $n \times n$ is usually known as the compartmental matrix or system matrix.

$\mathbf{b}(t) = \{b_1(t),\; b_2(t),\; \ldots,\; b_n(t)\}^T$ is a column vector where $\{b_i(t)\}$ is the input rate into compartment i from an outside system.

$\mathbf{x}(0) = \{x_1(0),\; x_2(0),\; \ldots,\; x_n(0)\}^T$ are the initial conditions, so $x_i(0)$ represents the amount or content of materials or energies in compartment i at time $t = 0$.

These models are known as Systems of Ordinary Differential Equations with Constant Coefficients (SODECC).

In the rest of this section, we're going to develop several functions that will enable users, even those with minimal knowledge of differential equations, to solve SODEs related to compartmental and physiological systems. If you're interested in programming, you may want to pay close attention to how the functions have been created. Otherwise, you just need to follow the examples to see how to use the commands properly.

The flow diagram in Figure 10.3 represents the iodine model (obtained from ICRP 78) where compartment 1 is the blood, compartment 2 is the thyroid, compartment 3 is the rest of the body and compartment 4 is the bladder. Additionally, $3 \rightarrow 0$, i.e. a transfer from compartment 3 to the environment, represents the output to the gastrointestinal tract (GIT) and $4 \rightarrow 0$ represents the output, via urine excretion, to the environment.

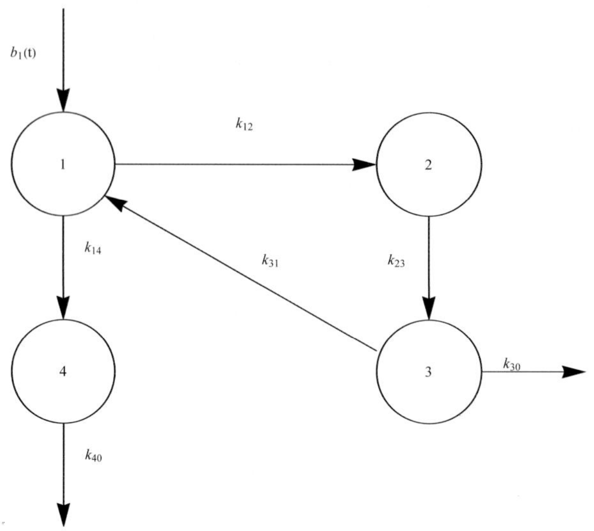

Figure 10.3 The Iodine ICRP model.

The Iodine ICRP Model

We create the function **CompartMatrix**[*n*, *matrixA*, λ] that returns the matrix of coefficients of a compartmental system where *n* is the number of compartments of the system, *matrixA* represents the transfer rates and λ is the radioactive decay constant (by default λ = 0, which means that it is not a radioactive isotope.).The transfer rates $\{k_{ij}\}$ from compartment *i* to compartment *j* are written: $\{\{1, 2, k_{12}\}, ..., \{i, j, k_{ij}\}, ...\}$, by default $k_{ij} = 0$.

```
CompartMatrix[(n_)?IntegerQ, matrixA_?MatrixQ, lambda_: 0] :=
 Module[{k, A},
  Apply[Set, ({k @@ Take[#1, 2], Last[#1]} &) /@ matrixA, {1}];
  k[i_, j_] := 0;
  A = DiagonalMatrix[Table[-Sum[k[i, j], {j, 0, n}], {i, 1, n}]] +
    Table[k[j, i], {i, 1, n}, {j, 1, n}] - lambda*IdentityMatrix[n]];
```

Notice the use of: `?IntegerQ`, matrixA_`?MatrixQ` to check that the input contains an integer and a matrix as the first two arguments. Otherwise, the function will not be executed. By using the structure: lambda_:0, we give lambda (the disintegration constant) a default value of 0 in case the user doesn't enter any value for it.

- Using this function we can construct the Iodine model.

```
(iodine131matrix = CompartMatrix[4, {{1, 2, k12}, {1, 4, k14}, {4, 0, k40},
     {2, 3, k23} , {3, 0, k30} , {3, 1, k31}}, λ]) // MatrixForm
```

$$\begin{pmatrix} -\lambda - k_{12} - k_{14} & 0 & k_{31} & 0 \\ k_{12} & -\lambda - k_{23} & 0 & 0 \\ 0 & k_{23} & -\lambda - k_{30} - k_{31} & 0 \\ k_{14} & 0 & 0 & -\lambda - k_{40} \end{pmatrix}$$

When the coefficients a_{ij} of a SODE are associated with physiologically meaningful values corresponding to measured physiological parameters or a function of them, we should use a physiological model instead of a compartmental one. In those cases, we will directly enter the values for a_{ij} to the **A** matrix, instead of the k_{ij} values.

- We build the function **CoeffMatrix**[*n*, *matrixA*, λ] to return the matrix of coefficients for *n* retention variables where the individual elements are the coefficients of the matrix: {{1, 2, a_{12}}, ..., {*i*, *j*, a_{ij}}, ...} for the ith row and the jth column. By default, $a_{ij} = 0$.

```
CoeffMatrix[(n_)?IntegerQ, (matrixA_)?MatrixQ] :=
  Module[{i, j, k},

   Normal[SparseArray[matrixA /. {i_, j_, k_} -> {i, j} -> k, {n, n}]]];
```

- Example:

```
CoeffMatrix[2, {{1, 2, a12}, {2, 1, a21}}] // MatrixForm
```

$$\begin{pmatrix} 0 & a_{12} \\ a_{21} & 0 \end{pmatrix}$$

- We also create the function **ShowODE** to represent the individual differential equations in a SODE for a compartmental or physiological model in classical mathematical notation; *matrixA* is the coefficients matrix; *initcond* are the initial conditions: {c_1, …, c_n}; input are the inputs {b_1 (*t*), …, $b_n(t)$}; *x* is the symbol that represents the retention variables.

```
ShowODE[matrixA_?MatrixQ, initcond_List, input_List, t_, x_] :=
  Module[{n, n1, A, B, c}, n = Dimensions[matrixA][[1]]; n1 = Range[n];
   A = (Derivative[1][Subscript[x, #1]][t] &) /@ n1;
   B = (Subscript[x, #1][t] &) /@ n1.Transpose[matrixA] + input;
   c = (Subscript[x, #1][0] &) /@ n1;
   Join[Thread[A == B], Thread[c == initcond]]];
```

- An iodine model with the initial conditions: {$x_1(0) = \text{x}1$, $x_2(0) = 0$, $x_3(0) = 0$, $x_4(0) = 0$} and a continuous input to compartment 1 given by b_1(t), can be modeled using the following SODE:

```
ShowODE[iodine131matrix ,
   {x1, 0, 0, 0}, {b1[t], 0, 0, 0}, t, x] // TableForm
```

```
x1'[t] == b1[t] + (-λ - k12 - k14) x1[t] + k31 x3[t]
x2'[t] == k12 x1[t] + (-λ - k23) x2[t]
x3'[t] == k23 x2[t] + (-λ - k30 - k31) x3[t]
x4'[t] == k14 x1[t] + (-λ - k40) x4[t]
x1[0] == x1
x2[0] == 0
x3[0] == 0
x4[0] == 0
```

```
iodine131matrix = CompartMatrix[4, {{1, 2, k12},
    {1, 4, k14}, {4, 0, k40}, {2, 3, k23} , {3, 0, k30} , {3, 1, k31}}, λ]
```

```
{{-λ - k12 - k14, 0, k31, 0}, {k12, -λ - k23, 0, 0},
  {0, k23, -λ - k30 - k31, 0}, {k14, 0, 0, -λ - k40}}
```

- The next function computes the SODE returning the analytical solution $x_i(t)$. We can use t1 = *t* to get the solution as a function of *t* instead of as a numerical value.

```
SystemDSolve[matrixA_?MatrixQ, incond_List, input_List, t_, t1_, x_,
   opts___?OptionQ] :=
  Module[{B, A1, R, X, n, i}, A1 = Rationalize[matrixA];
   n = Dimensions[A1][[1]];
   R =
    Join[Thread[
      Table[Derivative[1][Subscript[x, i]][t], {i, 1, n}] ==
       Table[Subscript[x, i][t], {i, 1, n}].Transpose[A1] +
        Rationalize[input]],
     Thread[Table[Subscript[x, i][0], {i, 1, n}] ==
       Rationalize[incond]]];
   B = Table[Subscript[x, i][t], {i, 1, n}];
   X = Flatten[DSolve[R, B, t]]; Chop[ExpandAll[X]] /. t -> t1];
```

In the function above, notice the use of `Rationalize`. With this function, all the decimal terms that meet certain requirements (see the documentation for further details), are rationalized to avoid problems with numerical approximations as shown in the examples in Section 3.11. By using `Rationalize`[..., 0], the decimal terms will always be rationalized. The downside of this approach is that the calculation time will increase. Also, in the last step, we have used `Chop` and `ExpandAll` to eliminate the terms with values very close to 0.

- To apply the previous function, we need to know the transfer constants, the initial condition values and the input function. According to the ICRP 78, the transfer rates for iodine, in days^{-1}, are $k_{12} = 0.3\,\mathrm{Log}(2)/0.25$, $k_{14} = 0.7\,\mathrm{Log}(2)/0.25$, $k_{23} = \mathrm{Log}(2)/80$, $k_{30} = 0.2\,\mathrm{Log}(2)/12$, $k_{31} = 0.8\,\mathrm{Log}(2)/12$ and $k_{40} = 12$; we also assume as the initial condition: {1, 0, 0, 0} and as the input function: {1 + 0.5 Cos(0.3 t), 0, 0, 0}.

```
iodine131matrix1 = iodine131matrix /. {k12 -> 0.3 Log[2]/0.25, k14 -> 0.7 Log[2]/0.25,
    k23 -> Log[2]/80, k30 -> 1/12 × 0.2 Log[2], k31 -> 1/12 × 0.8 Log[2], k40 -> 12};
```

- In the example we refer to Iodine–131, whose radioactive half-life can be obtained using `IsotopeData` (by default the output is given in seconds but in our example the k_{ij} are in days^{-1}).

```
λ = 1 / QuantityMagnitude[
   UnitConvert[IsotopeData["Iodine131", "Lifetime"], "day"]]
```

```
0.086371
```

- The retention function in each compartment can be computed using **`SystemDSolve`**[...] (Delete ";" to see the output):

```
{x1[t1_], x2[t1_], x3[t1_], x4[t1_]} =
  {x1[t1], x2[t1], x3[t1], x4[t1]} /. SystemDSolve[ iodine131matrix1,
    {1, 0, 0, 0}, {1 + 0.5 Cos[0.3 t], 0, 0, 0}, t, t1, x];
```

- The evolution of the iodine retention in the blood (compartment 1) and in the thyroid (compartment 2) is plotted below:

```
Plot[{x1[t], x2[t]}, {t, 1, 200}, PlotStyle →
  {{RGBColor[1, 0, 0]}, {AbsoluteDashing[{4, 4}], RGBColor[0, 0, 1]}},
 AxesLabel → {"Days", "r"}, PlotLegends →
  Placed[{"Blood", "Thyroid"}, Center, (Framed[#, RoundingRadius → 5] &)],
 PlotLabel → Style[Framed["b1(t)= 1 + 0.5 Cos[0.3 t]"],
   12, Blue, Background → Lighter[Yellow]]]
```

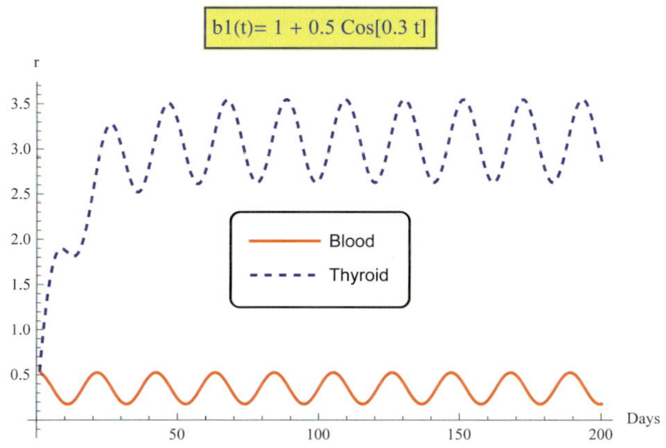


```
Clear[x1, x2, iodine131matrix1]
```

- The following function computes numerically the SODE using `NDSolve` and returns the solution in the interval $\{t, t_{\min}, t_{\max}\}$:

```
SystemNDSolve[matrixA_, initcond_, input_, {t_, tmin_, tmax_}, t1_, x_,
  opts___] :=
 Module[{B, A1, R, P, n, i, opt1},
  A1 = matrixA; n = Dimensions[A1][[1]];
  R =
   Join[Thread[
     Table[Derivative[1][Subscript[x, i]][t], {i, 1, n}] ==
      Table[Subscript[x, i][t], {i, 1, n}].Transpose[matrixA] +
       input],
    Thread[Table[Subscript[x, i][0], {i, 1, n}] == initcond]];
  B = Table[Subscript[x, i][t], {i, 1, n}];
  Flatten[NDSolve[R, B, {t, tmin, tmax}, opts]] /. t -> t1];
```

- This function can be used when **SystemDSolve** does not find a solution, for instance, if some transfer coefficients are variables:

```
iodine131matrix2 = iodine131matrix /. {k12 -> (0.3 Log[2])/0.25 (1 + Cos[0.2 t]),
    k14 -> (0.7 Log[2])/0.25, k23 -> Log[2]/80 (1 + Cos[0.3 t]),
    k30 -> 1/12 × 0.2 Log[2], k31 -> 1/12 × 0.8 Log[2], k40 -> 12};
{q1[t1_], q2[t1_], q3[t1_], q4[t1_]} =
  {q1[t1], q2[t1], q3[t1], q4[t1]} /. SystemNDSolve[ iodine131matrix2,
    {1, 0, 0, 0}, {1 + 0.5 Cos[0.3 t], 0, 0, 0}, {t, 0, 100} , t1, q];
```

```
Plot[{Callout[q1[t], "Blood", Above], Callout[q2[t], "Thyroid", Above]},
 {t, 1, 100}, AxesOrigin → {0, 0}, AxesLabel → {"Days", "r"}]
```

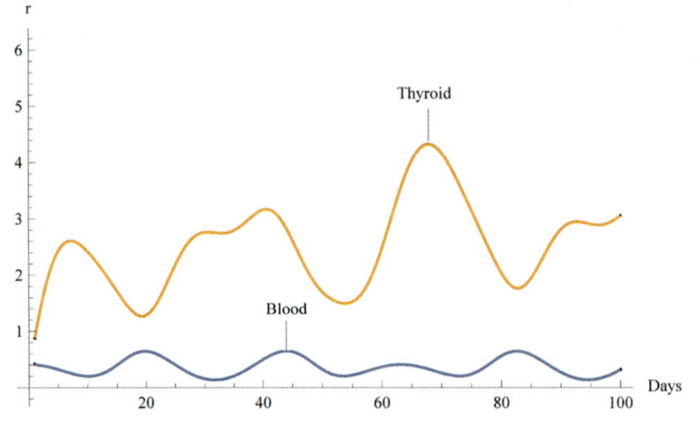


```
Clear[q1, q2, q3, q4, iodine131matrix2]
```

10.1.3 Laplace Transforms and Identifiability Analysis

The transfer rates $k_{i\,j}$ are usually estimated using experimental data as shown in the next section. The problem that often arises is that there is no unique value of $k_{i\,j}$ that satisfies the model but a finite number of values. This issue is addressed by a group of mathematical methods named identifiability analysis. The Laplace transforms are very useful in this kind of analysis.

The solution of a SODE with constant coefficients (eqn. 10(3)) can be found as follows:

$$x(t) = x_0\, e^{A\,t} + \int_0^t e^{A\,(t-\tau)}\, b(\tau)\, d\,\tau$$

Applying Laplace transforms we get:

$$X(s) = (s\,I - A)^{-1}\, x_0 + (s\,I - A)^{-1}\, B(s)$$

where $X(s)$ and $B(s)$ are the Laplace transforms of $x(t)$ and $b(t)$ respectively.

- This equation can be written in *Mathematica* as:

```
SystemLTSolve[matrixA_List, initcond_List, input_List, t_, s_, X_] :=
 Module[{B, A1, R, R1, P, r, n}, A1 = matrixA; n = Dimensions[A1][[1]];
  B = s*IdentityMatrix[n] - A1; R = Inverse[B]; P = Apart[R];
  r = LaplaceTransform[input, t, s];
  R1 = Dot[P, initcond] + Dot[P, r];
  Thread[Subscript[X, #1][s] & /@ Range[n] -> R1]];
```

Example: In Figure 10.4 (Godfrey, 1983, Example 6.5) the response of the central compartment 1 to a single input of 1 unit at $t = 0$ has been fitted (no noise is assumed) to the function $x_{1\,\exp}(t) = 0.7\exp(-5\ t) + 0.2\exp(-t) + 0.1\exp(-0.1\ t)$. The problem consists of determining the unknown transfer rates, $\{k_{10}, k_{12}, k_{13}, k_{21}, k_{31}\}$, assumed to be constants.

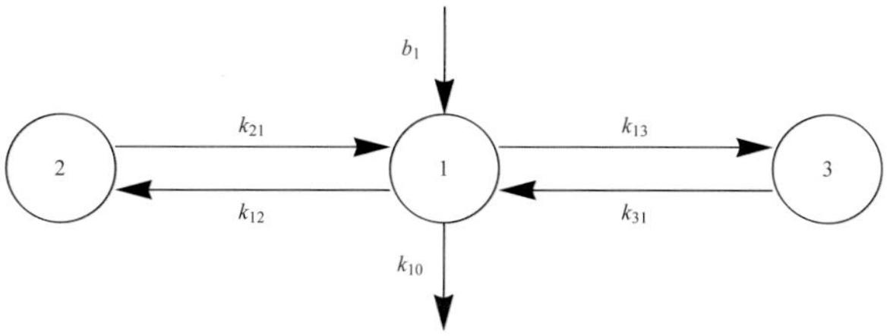

Figure 10.4 Catenary three-compartment model.

- The Laplace transform $X(s)$ of an impulsive input of 1 (We can use any mass unit, so if 1 is in mg the solution will be in mg) in compartment 1 at $t = 0$ can be modeled using **SystemLTSolve** with the initial conditions {1, 0, 0} and $b(t) = \{0, 0, 0\}$. To solve the problem, we will also need to define the compartmental matrix of the system.

```
tricompartment = CompartMatrix[3,
   {{1, 2, k12}, {2, 1, k21} , {1, 3, k13}, {3, 1, k31}, {1, 0, k10}}];

{X1[s_], X2[s_], X3[s_]} = {X1[s], X2[s], X3[s]} /.
   SystemLTSolve[tricompartment , {1, 0, 0}, {0, 0, 0}, t, s, X] // Simplify

{((s + k21) (s + k31)) /
   (k10 (s + k21) (s + k31) + s (k12 (s + k31) + (s + k21) (s + k13 + k31))),
  (k12 (s + k31)) / (k10 (s + k21) (s + k31) +
     s (k12 (s + k31) + (s + k21) (s + k13 + k31))), (k13 (s + k21)) /
   (k10 (s + k21) (s + k31) + s (k12 (s + k31) + (s + k21) (s + k13 + k31)))}
```

- The Laplace transform $X_{1\,\exp}(s)$ of $x_{1\,\exp}(t) =$ 0.7exp(−5 t) + 0.2exp(−t) + 0.1exp(−0.1 t) is:

```
X1exp = LaplaceTransform[0.7 Exp[-5 t] + 0.2 Exp[-t] + 0.1 Exp[-0.1 t], t, s]

  0.1       0.2     0.7
------- + ----- + -----
0.1 + s   1 + s   5 + s
```

- Now, the transfer rate $\{k_{10}, k_{12}, k_{13}, k_{21}, k_{31}\}$ can be obtained solving $X_{1\,\exp}(s) = X_1(s)$. The following procedure is applied:

```
SolveAlways[X1exp == X1[s], s]

{{k10 → 0.746269, k12 → 1.25488, k13 → 1.70885, k21 → 0.324354, k31 → 2.06565},
 {k10 → 0.746269, k12 → 1.70885, k13 → 1.25488, k21 → 2.06565, k31 → 0.324354}}
```

  In this case, there are two sets of possible solutions. Notice that neither set contains any negative rate constants, which would permit the rejection of such set.

All the functions described so far are part of *Sysmodel,* a package included in Biokmod that we will cover in Section 10.4.

```
Clear[X1, X2, X3, X1exp]
```

10.1.4 Michaelis–Menten (M–M) Models

Most drugs are metabolized by enzymes. The enzyme levels will change under different physiological cases (e.g., when many drugs are administered at the same time, they might have affinity for the same enzymes and thus compete for the same enzyme sites) or pathological states (e.g., in case of disease). This change in enzyme levels produces fluctuations in the concentration levels of drugs in the blood. For example, when given low-concentration/tolerable doses of Phenytoin, an anti-epileptic drug, its degradation by the corresponding metabolizing enzymes will follow first order or linear kinetics, that is, the rate of degradation will be directly proportional to the drug or substrate concentration. But when given a higher concentration of Phenytoin, as in the case of drug overdose or accidental poisoning, the metabolism will initially follow first order kinetics for some time until all enzyme sites get occupied/saturated, and then it will change to zero order kinetics, that is, the rate of change will become constant or independent of the drug concentration.

This mixed-order reaction (combining linear with nonlinear kinetics when there are no more sites available for drug binding) can be depicted by the Michaelis–Menten (M–M) equation. A example of the M–M model is shown in Figure 10.5. It describes the disposition in the animal or human body of a drug distributed according to a mammillary model, consisting of a central component with peripheral components connecting to it. This is a convenient model for many drugs for which the equilibrium between drug concentrations in different tissues is not achieved rapidly. It tries to describe mathematically the concentration data after the administration of a drug by depicting the body as a two-compartment open model where the drug is both introduced into, and exits from, the central compartment, named compartment 1, which can be sampled through the blood (plasma or serum). It may consist of organs or tissues which, being highly perfused with blood, reach a rapid equilibrium distribution with the blood. The peripheral compartment, compartment 2, usually cannot be sampled. It may consist of organs or tissues which, being poorly perfused with blood, reach equilibrium distribution with the blood only slowly. The concentration in compartment i is denoted x_i.

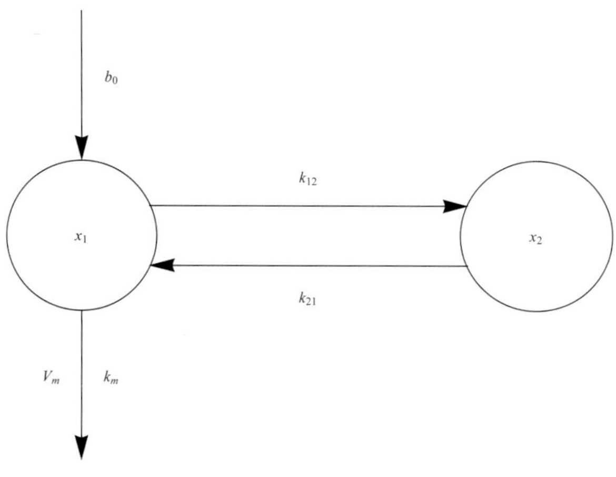

Figure 10.5 A Michaelis–Menten (M–M) model.

In this case, the drug movement between compartments will be considered as a linear kinetic process described by the transfer coefficients k_{12} and k_{21}. However, the elimination process will be nonlinear, as it happens for instance in the hepatic metabolism, and the elimination rate of the drug can be mathematically expressed by the M–M equation with parameters V_m = maximum transformation speed and k_m = M–M constant. The drug administration will be assumed to be an impulsive input (bolus) b_0. Thus:

$$\begin{aligned} \dot{x}_1(t) &= -k_{21}\,x_1(t) + k_{12}\,x_2(t) - \frac{V_m\,x_1(t)}{k_m + x_1(t)} \\ \dot{x}_2(t) &= k_{21}\,x_1(t) - k_{12}\,x_2(t) \\ x_1(0) &= b_0,\; x_2(0) = 0 \end{aligned} \tag{10.4}$$

In our example: $b_0 = 1$, Vm = 0.1, km = 0.3, and k_{12} and k_{21} are the unknown parameters. They can be estimated using experimental data.

- We define a model $x_1(t,\ k_{12}, k_{21})$ and then use the `ParametricNDSolve` function to obtain a numerical solution as a function of k12 and k21:

```
Vm = 0.1; km = 0.3;
eq1 = x1'[t] == -k12 x1[t] + k21 x2[t] - (Vm x1[t])/(km + x1[t]);
eq2 = x2'[t] == k12 x1[t] - k21 x2[t];
sol = ParametricNDSolve[{eq1, eq2, x1[0] == 1, x2[0] == 0},
  {x1, x2}, {t, 0, 100}, {k12, k21}]
```

```
{x1 → ParametricFunction[ Expression: x1  Parameters: {k12, k21} ],

 x2 → ParametricFunction[ Expression: x2  Parameters: {k12, k21} ]}
```

- $x_1(t,\ k_{12}, k_{21})$ and $x_2(t,\ k_{12}, k_{21})$ can be represented as functions of k_{12} and k_{21}.

```
Manipulate[Plot[Evaluate[{x1[k12, k21][t], x2[k12, k21][t]} /. sol],
  {t, 0, 30}, PlotRange → All, PlotLegends →
   Placed[{"x1", "x2"}, Center, (Framed[#, RoundingRadius → 5] &)] ],
 {{k12, 0.03, "k12"}, 0.01, 0.1}, {{k21, 0.02, "k21"}, 0.01, 0.1}]
```

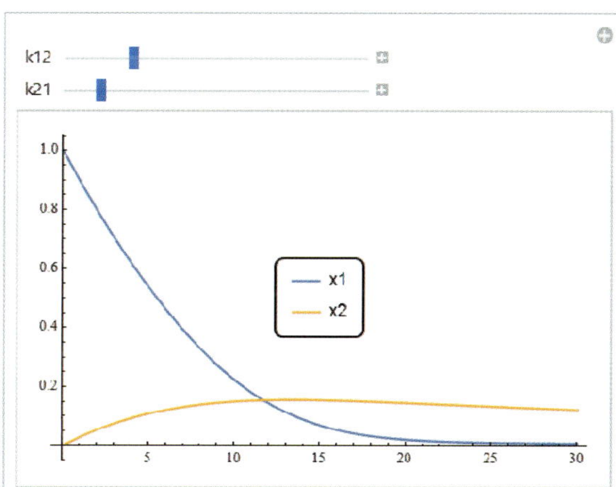

10.1.5 A Nonlinear Physiological Model

In this section we present an example of a physiological nonlinear model (Robert & Rowland, 1986):

An hepatic drug dispersion model in the liver can be represented as a cylindrical vessel (Figure 3.6). Let's consider an element of differential thickness dz at distance z along its length L. Inside this element there is blood, with a volume $V_{B,z}$ flowing at rate Q through a cross-sectional area A and a velocity v = Q/A, containing a drug concentration $C_B(z, t)$ part of which is unbounded at a concentration Cu in the plasma.

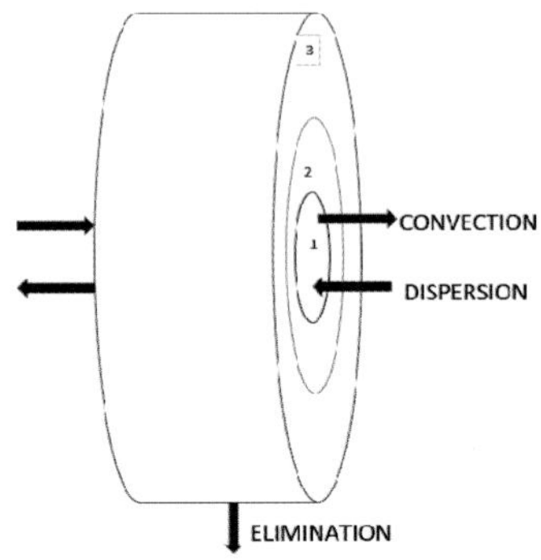

Figure 10.6 Schematic cross section of the liver:(1) Blood, (2) Space of Disse, (3) The hepatocyte.

The model is represented by a partial differential equation where the axial dispersion flux is given by term (1) and the convective flux by term (2). $\text{fu}_B = C_u / C_B$, CL_{int} is a clearance constant, ρ is a permeability constant and V_H is the apparent volume of the distribution of the drug.

$$V_B D \overbrace{\frac{\partial C_B(z,t)}{\partial z^2}}^{(1)} - V_B v \overbrace{\frac{\partial C_B(z,t)}{\partial z}}^{(2)} - \text{fu}_b\, \text{CL}_{\text{int}}\, \rho\, C_B(z,t) = V_H \frac{\partial C_B(z,t)}{\partial t}$$

To scale the equations and make the variables dimensionless we make the following replacement:

$$Z = \frac{z}{L};\ d_n = \frac{D}{v L} = \frac{D A}{Q L};\ r_n = \frac{\text{fu}_b\, \text{CL}_{\text{int}}\, \rho}{Q};\ c(z,t) = C_B(z,t)$$

After which we obtain:

$$\frac{\partial^2 c}{\partial z^2} - \frac{1}{d_n}\frac{\partial c}{\partial z} - \frac{r_n}{d_n} c = \frac{1}{d_n}\frac{\partial c}{\partial t} \qquad (10.5)$$

with the following the initial conditions (IC):

$t = 0,\ c(z, 0) = 0,\ 0 \le z \le 1$

- In *Mathematica* notation, the above result can be written as:

```
roweq = D[c[z, t], {z, 2}] - 1/d D[c[z, t], z] - r/d c[z, t] == 1/d D[c[z, t], t]
```

$$-\frac{r\,c[z,t]}{d} - \frac{c^{(1,0)}[z,t]}{d} + c^{(2,0)}[z,t] == \frac{c^{(0,1)}[z,t]}{d}$$

- We can solve it using the Laplace transform method turning the partial differential equation into an ordinary differential equation:

```
rowlandeqLT = LaplaceTransform[roweq, t, s]
```

$$-\frac{1}{d}r\,\text{LaplaceTransform}[c[z,t],t,s]-\frac{1}{d}\,\text{LaplaceTransform}\left[c^{(1,0)}[z,t],t,s\right]+\text{LaplaceTransform}\left[c^{(2,0)}[z,t],t,s\right]==\frac{1}{d}(-c[z,0]+s\,\text{LaplaceTransform}[c[z,t],t,s])$$

- In the above output, the replacement **LaplaceTransform**[f(z,t), t,s] → LT[z], takes also into account the initial condition (c[z,0]=0):

```
rowlandeqLT1 = rowlandeqLT /. {LaplaceTransform[c^(1,0)[z, t], t, s] → cT'[z],
    LaplaceTransform[c^(2,0)[z, t], t, s] → cT''[z],
    LaplaceTransform[c[z, t], t, s] → cT[z], c[z, 0] → 0}
```

$$-\frac{r\,cT[z]}{d}-\frac{cT'[z]}{d}+cT''[z]==\frac{s\,cT[z]}{d}$$

- The above equation is an ordinary differential equation that can be solved using DSolve:

```
sol1 = cT[z] /. DSolve[ rowlandeqLT1 , cT[z], z]
```

$$\left\{e^{\frac{1}{2}\left(\frac{1}{d}-\frac{\sqrt{1+4\,d\,r+4\,d\,s}}{d}\right)z}C[1]+e^{\frac{1}{2}\left(\frac{1}{d}+\frac{\sqrt{1+4\,d\,r+4\,d\,s}}{d}\right)z}C[2]\right\}$$

The coefficients C_1 and C_2 can be determined using the boundary conditions, which in this case are: at $z=0$, $c=\delta(z)$, and at $z=\infty$, $c=0$, that is, $c(0,t)=\delta(z)$ and $c(\infty,t)=0$.

- For convenience we replace C_1 by c1 and C_2 by c2.

```
sol2 = sol1[[1]] /. {C[1] → c1, C[2] → c2}
```

$$c1\,e^{\frac{1}{2}\left(\frac{1}{d}-\frac{\sqrt{1+4\,d\,r+4\,d\,s}}{d}\right)z}+c2\,e^{\frac{1}{2}\left(\frac{1}{d}+\frac{\sqrt{1+4\,d\,r+4\,d\,s}}{d}\right)z}$$

- c1 can be computed using the boundary condition $c(0,t)=\delta(z)$:

```
solc1 = Solve[Evaluate[sol2 /. z → 0 ] ==
    LaplaceTransform[DiracDelta[z], z, s] , c1][[1]]
```

{c1 → 1 - c2}

- c2 can be computed using the boundary condition $c(\infty,t)=0$:

```
solc2 = Solve[Evaluate[sol2 /. solc1] == 0, c2][[1]] // Simplify
```

$$\left\{c2\to-\frac{1}{-1+e^{\frac{\sqrt{1+4\,d\,(r+s)}\,z}{d}}}\right\}$$

```
Limit[solc2[[1, 2]], z→ Infinity, Assumptions → s > 0 && r > 0 && d > 0]
```

0

- Hence:

```
sol3 = Evaluate[sol2 /. c1 → 1 - c2] /. c2 → 0
```

$$e^{\frac{1}{2}\left(\frac{1}{d}-\frac{\sqrt{1+4\,d\,r+4\,d\,s}}{d}\right)z}$$

```
InverseLaplaceTransform[sol3, s, t] // Simplify
```

$$\text{ConditionalExpression}\left[\frac{e^{-\frac{t^2+4\,d\,r\,t^2-2\,t\,z+z^2}{4\,d\,t}}\,z}{2\,d^2\sqrt{\pi}\sqrt{\frac{t^3}{d^3}}},\ d\,z>0\right]$$

```
csol[z_, t_] = z e^(-(4 d r t^2 + t^2 - 2 t z + z^2)/(4 d t)) / (2 Sqrt[π] d^2 Sqrt[t^3/d^3]);
```

$$\text{csol}[z\_,\ t\_] = \frac{z\, e^{-\frac{4 d r t^2 + t^2 - 2 t z + z^2}{4 d t}}}{2\sqrt{\pi}\, d^2 \sqrt{\frac{t^3}{d^3}}};$$

- The plot below displays $c(z, t)$ for different values of d and $r = 0$:

```
Plot[Evaluate[csol[1, t] /. {d → {0.01, 0.1, 0.2, 1, 2}, r → 0}],
 {t, 0.01, 4}, PlotRange → {0, 4}, PlotLegends → {0.01, 0.1, 0.2, 1, 2}]
```

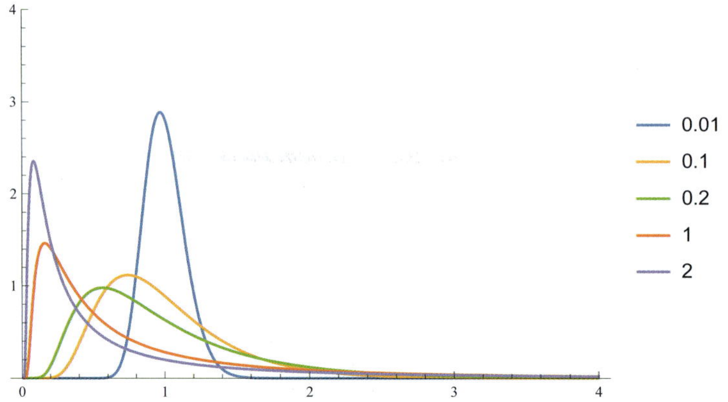


```
Manipulate[ContourPlot[Evaluate[csol[z, t] /. {d → d1, r -> r1}],
  {z, 0, 30}, {t, 0, 30}, FrameLabel → {"z", "t"},
  PlotLabel → Style[c(z,t), Blue, 20], PlotLegends → Automatic],
 {{d1, 0.2, "d"}, 0.01, 1}, {{r1, 0.2, "r"}, 0, 0.5}]
```

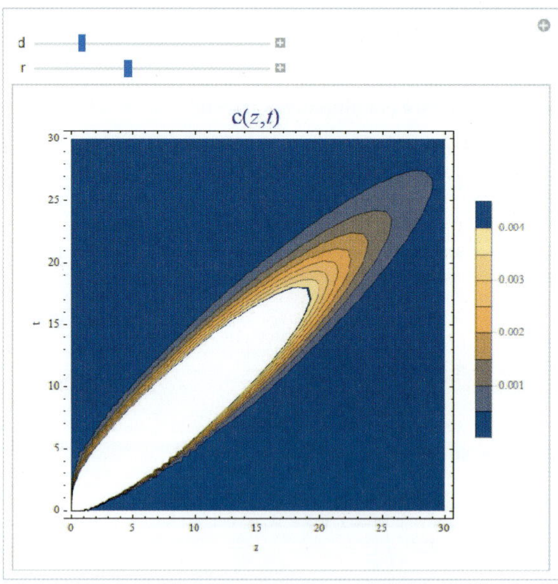

10.2 Application: Fitting a Model

In this section, with the help of examples, we describe how to fit biokinetic parameters using experimental data. For more details about the model see A. Sánchez-Navarro, 1999. We will use the functions developed in Section 10.1.

Let's suppose that we have a physiological model represented by the following SODE:

$$\begin{aligned}
\dot{x}_1(t) &= -\frac{Q+PS}{V_p}\,x_1(t) + \frac{PS}{V_p}\,x_2(t) + \frac{Q}{V_p}\,b_1(t)\\
\dot{x}_2(t) &= \frac{PS}{V_{\rm Tu}}\,x_1(t) - \left(\frac{PS}{V_{\rm Tu}} + k_{\rm on}\right)x_2(t) + k_{\rm off}\,\frac{V_{\rm Tb}}{V_{\rm Tu}}\,x_3(t)\\
\dot{x}_3(t) &= k_{\rm on}\,\frac{V_{\rm Tu}}{V_{\rm Tb}}\,x_2(t) - k_{\rm off}\,x_3(t)
\end{aligned} \tag{10.6}$$

where the drug is transported from V_p (vascular volume), by perfusate flow Q, to V_{Tu} (tissue water space) across a permeability barrier (permeability-surface product PS) and its binding is described by binding/unbinding constants k_{on}/k_{off}.

- The matrix A for this model, which will be used later, is:

```
physiomodel = CoeffMatrix[3, {{1, 1, -(Q/Vp + PS/Vp)}, {1, 2, PS/Vp}, {2, 1, PS/VTu},
   {2, 2, -(PS/VTu + kon)}, {2, 3, koff VTb/VTu}, {3, 2, kon VTu/VTb}, {3, 3, -koff}}]
```

$$\left\{\left\{-\frac{PS}{V_p}-\frac{Q}{V_p},\ \frac{PS}{V_p},\ 0\right\},\ \left\{\frac{PS}{V_{Tu}},\ -k_{on}-\frac{PS}{V_{Tu}},\ \frac{k_{off}\,V_{Tb}}{V_{Tu}}\right\},\ \left\{0,\ \frac{k_{on}\,V_{Tu}}{V_{Tb}},\ -k_{off}\right\}\right\}$$

- After taking into account that the input function is $\left\{\frac{Q}{V_p}\,b_1(t),\,0,\,0\right\}$ and the vector of the initial conditions is {0, 0, 0}, the SODE can be written as follows:

```
ShowODE[physiomodel, {0, 0, 0}, {Q/Vp b1[t], 0, 0}, t, x] // TableForm
```

$$\begin{aligned}
&x_1'[t] == \frac{Q\,b_1[t]}{V_p} + \left(-\frac{PS}{V_p}-\frac{Q}{V_p}\right)x_1[t] + \frac{PS\,x_2[t]}{V_p}\\
&x_2'[t] == \frac{PS\,x_1[t]}{V_{Tu}} + \left(-k_{on}-\frac{PS}{V_{Tu}}\right)x_2[t] + \frac{k_{off}\,V_{Tb}\,x_3[t]}{V_{Tu}}\\
&x_3'[t] == \frac{k_{on}\,V_{Tu}\,x_2[t]}{V_{Tb}} - k_{off}\,x_3[t]\\
&x_1[0] == 0\\
&x_2[0] == 0\\
&x_3[0] == 0
\end{aligned}$$

We want to estimate the values, $k_{\rm on}$ and $k_{\rm off}$ given $V_{\rm Tu} = 6.411$, V_p, 0.973, $V_{\rm Tb} = 1$, $PS = 2.714$ and $Q = 3$. We build the coefficients matrix (note that we use **CoeffMatrix** and not **CompartMatrix**) as a function of $k_{\rm off}$ and $k_{\rm on}$ replacing $V_{\rm Tu}$, V_p, $V_{\rm Tb}$, PS, Q with their actual values:

```
physiomodel1[kon_, koff_] :=
  CoeffMatrix[3, {{1, 1, -(Q/Vp + PS/Vp)}, {1, 2, PS/Vp}, {2, 1, PS/VTu}, {2, 2,
    -(PS/VTu + kon)}, {2, 3, koff VTb/VTu}, {3, 2, kon VTu/VTb}, {3, 3, -koff}}] /.
```

```
{VTu → 6.411, Vp → 0.973, PS → 2.714, Q → 3, VTb → 1, kon → kon, koff → koff};

ShowODE[physiomodel1[kon, koff], {0, 0, 0},
  {Q/Vp b1[t] /. {Vp → 0.973, Q → 3}, 0, 0}, t, x] // TableForm
```

$x_1'[t] == 3.08325\, b_1[t] - 5.87256\, x_1[t] + 2.78931\, x_2[t]$
$x_2'[t] == 0.423335\, x_1[t] + (-0.423335 - k_{on})\, x_2[t] + 0.155982\, k_{off}\, x_3[t]$
$x_3'[t] == 6.411\, k_{on}\, x_2[t] - k_{off}\, x_3[t]$
$x_1[0] == 0$
$x_2[0] == 0$
$x_3[0] == 0$

- The mathematical expression for $b_1(t)$ was unknown initially but an experiment was made where $b_1(t)$ was given by the best fit of the input function to experimental data $\{\{t_1, b_1\}, \ldots, \{t_n, b_n\}\}$ obtained via sampling from an arterial catheter. Here are the data:

```
dataCateter =
  {{0., 0.}, {0.05, 402.7}, {0.1, 430.3}, {0.15, 375.4}, {0.2, 292.4},
   {0.25, 202.2}, {0.3, 148.4}, {0.35, 96.4}, {0.4, 64.9}, {0.45, 41.7},
   {0.5, 25.3}, {0.55, 17.8}, {0.6, 8.8}, {0.65, 6.6}, {0.7, 3.2},
   {0.75, 2.5}, {0.8, 1.4}, {0.85, 0.9}, {0.9, 0.5}, {1., 0.2}, {1.1, 0.07},
   {1.2, 0.03}, {1.3, 0.01}, {1.4, 0.003}, {1.45, 0.001}, {1.5, 0.001}};

ListPlot[dataCateter, Joined → True, PlotRange → All, ImageSize → Small]
```

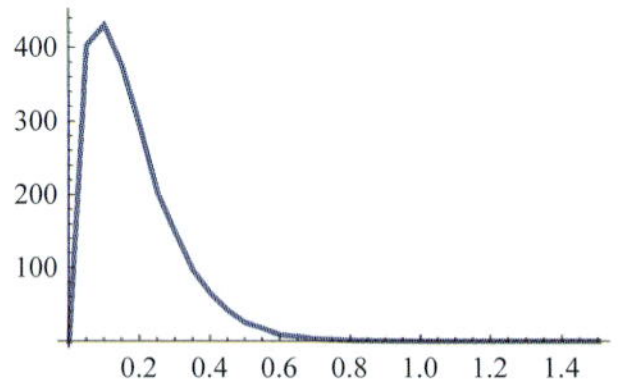

- The shape of the graph suggests that the function below could provide a good fit:

```
b[t_] = a t Exp[c t] /. FindFit[dataCateter, a t Exp[c t] , {a, c}, t]
```

$13\,610.1\, e^{-11.216\, t}\, t$

- Here are the experimental data and the fitted function:

```
Plot[b[t], {t, 0, 1}, Epilog →
  {Hue[0], PointSize[0.02`], Point /@ dataCateter}, ImageSize → Small]
```

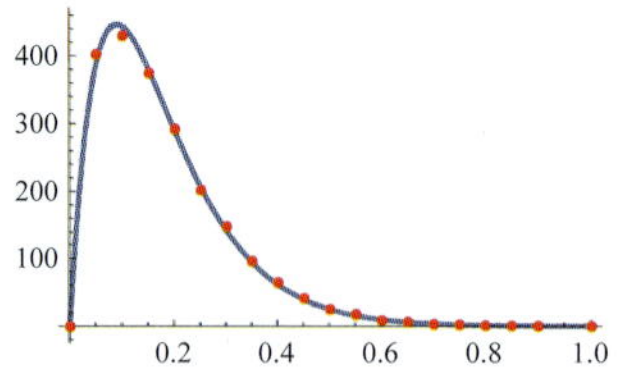

- Now, we have the input function in the form required by **SystemLTSolve**:

```
inputb[t_] = {Q/Vp b[t] , 0, 0} /. {Vp → 0.973, Q → 3}
```

$\{41\,963.3\, e^{-11.216\, t}\, t, 0, 0\}$

- We also measured $x_1(t)$ by sampling from the following experimental data $\{\{t_1, x_1(t_1)\}, \ldots, \{t_n, x_1(t_n)\}\}$.

```
dataPhysio = {{0.03, 14.50}, {0.08, 61.06}, {0.28, 120.93}, {0.33, 109.01},
   {0.38, 98.08}, {0.48, 69.66}, {0.55, 51.51}, {0.65, 34.77},
   {0.75, 23.21}, {0.85, 15.59}, {0.95, 12.16}, {1.05, 9.598},
   {1.15, 8.278}, {1.25, 6.842}, {1.35, 5.871}, {1.45, 5.297},
   {1.6, 4.886}, {1.8, 3.846}, {2., 3.317}, {2.2, 2.899}, {2.4, 2.627},
   {2.6, 2.289}, {2.8, 1.998}, {3., 1.930}, {3.4, 1.589}, {3.75, 1.308},
   {4.25, 1.112}, {4.75, 1.064}, {5.25, 0.938}, {6.75, 0.842},
   {7.25, 0.831}, {7.75, 0.778}, {8.25, 0.818}, {11., 0.739}};
```

- Since we wish to obtain the numeric values of k_{on} and k_{off} fitting $x_1(t)$ to the data, we apply the Laplace transforms obtaining $X_1(s, k_{on}, k_{off})$:

```
X1[kon_, koff_] = X₁[s] /.
   SystemLTSolve[physiomodel1[kon, koff] , {0, 0, 0}, inputb[t], t, s, X];
```

- $x_1(t)$ can be obtained applying the `InverseLaplaceTransform` as a function of k_{on} and k_{off}. Notice that x_1 (t) has only an analytical solution when both k_{on} and k_{off} are numbers.

```
x1fit[kon_?NumericQ, koff_?NumericQ] :=
  Block[{x1}, x1[t_] = InverseLaplaceTransform[X1[kon, koff], s, t];
   Plus @@ Apply[(x1[#1] - #2)^2 &, dataPhysio, {1}]];
```

- We can now use `FindMinimum` to find k_{on} and k_{off}:

```
FindMinimum[x1fit[kon, koff], {kon, 0.1, 1}, {koff, 0.1, 1}]

{72.3264, {kon → 0.713783, koff → 0.11072}}
```

- The solution, including extensive statistical information, can be obtained by using **SystemNDSolve**[...] and `NonlinearModelFit`. The first step consists of defining a *Mathematica* function that gives the solution of the model as a function of the parameters to be fitted. We need the solution of the model for $x_1(t)$, the variable for which we have experimental data. Once again, notice that the model can only be solved if the parameters take numeric values.

```
model[t1_?NumericQ, kon_?NumericQ, koff_?NumericQ] :=
 x₁[t1] /. SystemNDSolve[physiomodel1[kon, koff],
   {0, 0, 0}, inputb[t] , {t, 0, 12}, t1, x]
```

- We apply `NonlinearModelFit` and generate the desired statistical reports (the list of all available reports can be obtained with the command **nlm["Properties"]**). It will usually take a long time.

```
nlm = NonlinearModelFit[dataPhysio , model[t, kon, koff],
  {{kon, 0.71}, {koff, 0.11}}, {t}, Method → Gradient]

FittedModel[ model[t, 0.71354, 0.110548] ]
```

- The next command displays the experimental data and the fitted function. Using the function `Inset` we have added the "ParameterTable" report inside the graph.

```
Show[Plot[nlm[t], {t, 0, 12}, PlotRange -> All],
 ListPlot[dataPhysio], Epilog → Inset[nlm["ParameterTable"], {10, 50}]]
```

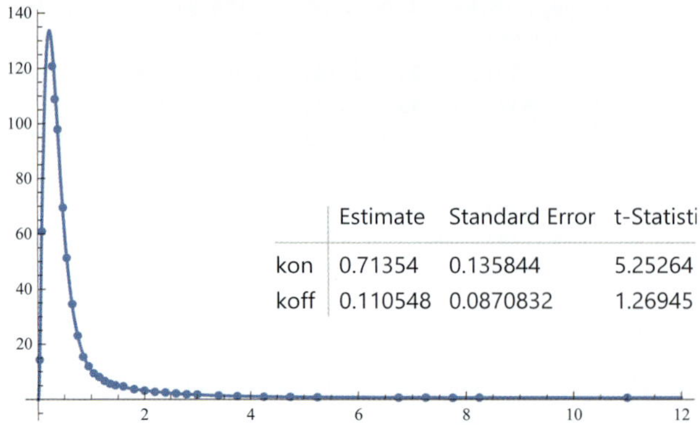

- Here's the visualization of the predicted values against the observed ones. The plot also incorporates the difference between actual and predicted responses using the "FitResiduals" property from the `NonlinearModelFit` function.

```
ListPlot[Transpose[nlm[{"Response", "PredictedResponse"}]],
 FrameLabel → {"Observed", "Predicted"}, Frame → True, Axes → False,
 Epilog → Inset[ListPlot[nlm["FitResiduals"], Filling → Axis], {25, 50}]]
```

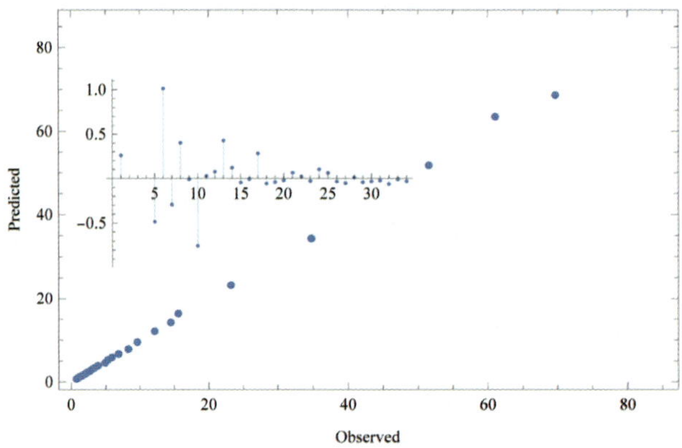

The previous functions can be easily modified to fit other models with two or more parameters.

- To avoid any problems in the next section with previously defined variables we quit the session.

```
Quit[]
```

10.3 Optimal Experimental Designs (OED)

10.3.1 OED Introduction

Many scientific processes, specially in pharmacokinetics (PK) and pharmacodynamics (PD) studies, as we have seen before, are defined by systems of ordinary differential equations (ODEs). If there are unknown parameters that need to be estimated, the optimal experimental design (OED) approach offers a way to obtain optimum bias-free and minimum-variance estimators for those parameters based on specific criteria.

The unknown parameters are usually estimated by fitting experimental data. A typical model has a solution given by a function $r(t, \beta)$, where $\beta = \{\beta_1, ..., \beta_p\}$ are the unknown parameters that need to be fitted. The objective is to choose the best moments $\{t_0, ..., t_i, ..., t_n\}$ based on the data, and this can be done using optimal experimental designs (OED). Here, "best" depends on the objectives of the practitioner, which are usually linked to a particular optimality criterion such as obtaining accurate estimators of the parameters, minimizing the variance of the predicted response or optimizing a different characteristic of the model. Many optimality criteria are based on the Fisher information matrix (FIM). The FIM has a simple expression for linear models and observations that come from exponential-family distributions. For more complex models, some formulas can be obtained from the general information matrix. The usual procedure for a nonlinear model is to linearize it and apply the well-known toolbox for linear models. The most widely-used criterion is D-optimality, which focuses on the determinant of the FIM.

In the following pages, we're going to describe one example using the D-optimality criterion to obtain the $\{t_0, ..., t_i, ..., t_n\}$ values. For further information we refer the interested reader to: J. Lopez-Fidalgo, J. Rodriguez-Diaz, JM. Sanchez and G. Santos-Martin, Optimal designs for compartmental models with correlated observations, *Journal of Applied Statistics*: 32(10), pp.1075-1088, 2005. A *D*-optimal design will be a design that maximizes the determinant of the FIM.

- We first define $r(t, \beta)$. In our example: $r(t, \beta_1, \beta_2) = \beta_1 \left(e^{-2.0t-0.09\beta_1} + e^{-0.001t-0.2\beta_2}\right)$ being $\beta = \{\beta_1, \beta_2\}$ the unknown parameters.

```
r[β1_, β2_] = β1 (e^(-2.0t-0.09β2) + e^(-0.001t-0.2β2)); β = {β1, β2};
```

- Then, we compute $\nabla(r(t, \beta_1, \beta_2) = \{\frac{\partial r(t)}{\partial \beta_1}, \frac{\partial r(t)}{\partial \beta_2}\}$:

```
g[t_] = ∇β r[β1, β2]
```

$\{e^{-0.001t-0.2\beta 2} + e^{-2.t-0.09\beta 2}, \left(-0.2\, e^{-0.001t-0.2\beta 2} - 0.09\, e^{-2.t-0.09\beta 2}\right)\beta 1\}$

- Next, we specify the number of points to be used in the optimal design. We assume a 3-point design with the first one, t_0, defined by the user.

```
n = 3;
```

- For computational purposes, we prefer to use the distance $d_i = t_i - t_{i-1}$, instead of t_i, then $t_i = \sum_i d_i$ with $d_0 = t_0$ defined by the user. That is, the points to be estimated are:

```
dd = Table[d_i, {i, n}]
```

$\{d_1, d_2, d_3\}$

```
tm = Accumulate[Flatten[{d_0, dd}]]
```

$\{d_0, d_0 + d_1, d_0 + d_1 + d_2, d_0 + d_1 + d_2 + d_3\}$

- We evaluate $\nabla(\mathrm{r}(t), \beta)$ at points t:$\{t_0, \ldots, t_n\}$, obtaining $X = \{X_1, \ldots, X_p\}$ with $X_1 = \{\frac{dr(t_0)}{d\beta_1}, \ldots, \frac{dr(t_n)}{d\beta_1}\}, \ldots, X_p = \{\frac{dr(t_0)}{d\beta_p}, \ldots, \frac{dr(t_n)}{d\beta_p}\}$:

```
X = g[tm];
```

- A typical choice for the covariance matrix is $\Gamma = \{l_{ij}\}$ with $l_{ij} = \exp\{\rho|t_j - t_j|\}$ (meaning the relationship between samples decays exponentially as the time-distance between them increases), that is:

```
ff[i_, j_] := Which[i == j, 1, i < j, e^(-ρ Σ_{k=i}^{j-1} d_k), i > j, e^(-ρ Σ_{k=j}^{i-1} d_k)];
```

```
Γ = Array[ff, {n + 1, n + 1}]
```

$\{\{1, e^{-\rho d_1}, e^{-\rho (d_1+d_2)}, e^{-\rho (d_1+d_2+d_3)}\}, \{e^{-\rho d_1}, 1, e^{-\rho d_2}, e^{-\rho (d_2+d_3)}\}, \{e^{-\rho (d_1+d_2)}, e^{-\rho d_2}, 1, e^{-\rho d_3}\}, \{e^{-\rho (d_1+d_2+d_3)}, e^{-\rho (d_2+d_3)}, e^{-\rho d_3}, 1\}\}$

- We assume that all the measures have approximately the same uncertainty level, that is $\sigma^2 \simeq \sigma_i^2$, then covariance matrix is $\Sigma = \sigma^2\ \Gamma$. In our example we assume $\rho = 1$ and $\sigma = 2$.

```
σ = 2; ρ = 1; Σ = σ^2 * Γ;
```

- We also need to give the initial values to β_1 and β_2. The ultimate purpose is to find their real values empirically. However, the method requires an initial guess, (we can use previous experiments, information derived from similar experiments, etc.) and to find out how sensible our choice of values has been we can perform a sensitivity analysis. We assume the following values:

```
β1 = 100; β2 = 5;
```

- The moment $d_0 = t_0$ for taking the first measurement is defined by the user. In this case we take the measurement at t_0= 0.5.

```
d0 = 0.5;
```

- Then we can obtain the information matrix (FIM): M = $X^T\ \Sigma^{-1}\ X$.

```
m = X . Inverse[Σ]. Transpose[X];
```

- A D-optimal design will be a design that maximizes det|M|, the determinant of the information matrix. This can be done using `Maximize` as follows (in the next example we show that usually it is better to use `FindMaximum` or `NMaximize`):

```
deter1[x_] := Det[m /. Thread[dd -> x]];
```

```
sol = Maximize[{deter1[dd], Thread[dd > 1]}, dd]
```

- This indicates that the samples should be taken in $\{t_i\}$. The same solution can be directly obtained using Biokmod (Section 10.4.4):

```
optimumt = tm /. sol[[2]]
```

```
{0.5, 1.98704, 8.16931, 15.083}
```

- In practice, since it's very difficult to take samples at exactly the precise time, we need to measure the robustness of our calculations: how sensitive they are to small changes. For that purpose, we calculate their efficiency using the following formula:

$\sqrt[N]{\frac{\mathrm{Det}[m[d]]}{\mathrm{Det}[m[max]]}}$ (N is the total number of estimated points).

For example, to measure the efficiency of d_3:

```
d3 = Table[i, {i, -3 + d3 /. sol[[2, 3]], 3 + d3 /. sol[[2, 3]], 0.1}];
```

```
ListPlot[(Det[m /. Thread[{sol[[2, 1]], sol[[2, 2]] , d3 -> #}]] /
      Det[m /. sol[[2]]]) ^ (1/(n - 1)) & /@ d3,
  Filling -> Axis, ColorFunctionScaling → True]
```

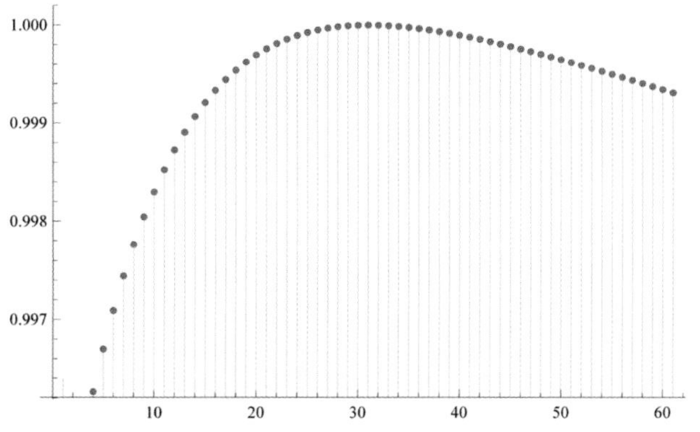

- The optimum point is the one with efficiency 1. In our case, this corresponds to **Det**[m[max]], obtained when t:

```
optimumt[[4]]
```

```
15.083
```

- As usual, to avoid any problems in the next section with previously defined variables we quit the session.

```
Quit[]
```

10.3.2 Michaelis–Menten Models (M–M) and OED

We will now apply OED to the model described in Section 10.1.4 to choose the best moments $\{t_1, ..., t_i\}\}$ to take the samples. In this case, the problem is that we do not have an exact solution of the model so we cannot obtain an analytical expression for $\{x_1(t, k_{12}, k_{21}), x_2(t, k_{12}, k_{21})\}$. Because of that, we apply the *D*-optimal method described in the previous section. We want to design an experiment to take the samples in compartment 1.

- We repeat the process discussed in 10.1.4 to obtain $x_1(t, k_{12}, k_{21})$:

```
Vm = 0.1; km = 0.3;
eq1 = x1'[t] == -k12 x1[t] + k21 x2[t] - (Vm x1[t])/(km + x1[t]);
eq2 = x2'[t] == k12 x1[t] - k21 x2[t];
sol = ParametricNDSolve[{eq1, eq2, x1[0] == 1, x2[0] == 0},
    {x1, x2}, {t, 0, 100}, {k12, k21}] ;
```

- We then proceed to compute numerically $\nabla x_1(t, k_{12}, k_{21}) = \{\frac{\partial x_1(t)}{\partial k_{12}}, \frac{\partial x_1(t)}{\partial k_{21}}\}$. For convenience, we call $k_{12} = a$ and $k_{21} = b$:

```
fa[a1_?NumberQ, b_?NumberQ, t_?NumberQ] :=
  D[x1[a, b], a][t] /. a → a1 /. sol
fb[a_?NumberQ, b1_?NumberQ, t_?NumberQ] :=
  D[x1[a, b], b][t] /. b → b1 /. sol
```

```
X1[a_, b_, ti_] := {fa[a, b, ti], fb[a, b, ti]}
```

- Next, we define the number of points to be used in the optimal design. We assume a 5-point design:

```
n = 5 - 1;
```

- After that, we compute $\Gamma = \{l_{ij}\}$ with $l_{ij} = \exp\{\rho|t_j - t_j|\}$ (remember that $d_i = t_i - t_{i-1}$ is used instead of t_i:

```
dd = Table[d_i, {i, n}]; tt = FoldList[Plus, Subscript[d, 0], dd]
```

```
{d_0, d_0 + d_1, d_0 + d_1 + d_2, d_0 + d_1 + d_2 + d_3, d_0 + d_1 + d_2 + d_3 + d_4}
```

```
ff[i_, j_] := Which[i == j, 1, i < j, e^(-ρ Sum_{k=i}^{j-1} d_k), i > j, e^(-ρ Sum_{k=j}^{i-1} d_k)];
```

```
Γ = Array[ff, {n + 1, n + 1}];
```

- Once we know Γ, we compute the covariance matrix $\Sigma = \sigma^2\ \Gamma$. We assume $\rho = 1, \sigma = 1$:

```
ρ = 1; σ = 1;  Σ = σ^2 * Γ;
```

- The next to last step is the calculation of the information matrix $M = X^T\,\Sigma^{-1}\,X$. Initially, we assign arbitrary values to k_{12} and k_{21}. As mentioned in the previous section, their real values are unknown but we have to make an initial estimation based on the information that we have. In this case we assume $k_{12} = 0.03$ and $k_{21} = 0.02$:

```
m1[ti_] := Transpose[Map[X1[0.03, 0.02, #] &, ti] ].
  Inverse[Σ].Map[X1[0.03, 0.02, #] &, ti]
```

- `Maximize` tries to find a global maximum using analytical methods that may take too long to compute. For this reason, it may be more convenient to use `NMaximize` or `FindMaximum`, *Mathematica* functions that take a vector of numbers as their inputs. In our example, with $n = 4$.

```
obj[d0_?NumericQ, d1_?NumericQ,
  d2_?NumericQ, d3_?NumericQ, d4_?NumericQ] :=
 Det[m1[{d0, d1 + d0, d0 + d1 + d2, d0 + d1 + d2 + d3, d0 + d1 + d2 + d3 + d4}]]
```

- Finally, we find the solution. Remember that a D-optimal design is a design that maximizes det|M|, the determinant of the information matrix. We can show the intermediate calculations with `StepMonitor`:

```
sol1 = FindMaximum[{obj[d_0, d_1, d_2, d_3, d_4],
   0 < d_0 < 10, 0 < d_1 < 10, 0 < d_2 < 10,  0 < d_3 < 10,  0 < d_4 < 10},
  {{d_0, 3}, {d_1, 3}, {d_2, 3}, {d_3, 3}, {d_4, 3}} , StepMonitor :>
   Print[{"d0:", d_0, "d1:", d_1, "d2:", d_2, "d3:", d_3, "d4:", d_4}]]
```

```
{d0:, 3., d1:, 3., d2:, 3., d3:, 3., d4:, 3.}

{d0:, 3.96374, d1:, 3.00914, d2:, 3.56722, d3:, 3.59301, d4:, 3.19179}

{d0:, 4.30937, d1:, 2.84337, d2:, 3.98828, d3:, 3.71813, d4:, 3.38737}

{d0:, 4.34705, d1:, 2.80134, d2:, 4.16208, d3:, 3.66308, d4:, 3.45106}

{d0:, 4.34836, d1:, 2.79967, d2:, 4.1764, d3:, 3.65599, d4:, 3.45459}

{d0:, 4.34836, d1:, 2.79963, d2:, 4.17644, d3:, 3.65596, d4:, 3.45455}

{d0:, 4.34836, d1:, 2.79963, d2:, 4.17644, d3:, 3.65596, d4:, 3.45455}

{12.8081, {d_0 → 4.34836, d_1 → 2.79963, d_2 → 4.17644, d_3 → 3.65596, d_4 → 3.45455}}
```

- This indicates that the samples should be taken at $\{t_i\}$:

```
bestsampling = tt /. sol1[[2]]
```

```
{4.34836, 7.14799, 11.3244, 14.9804, 18.4349}
```

Sensitivity Analysis of k_{12}, k_{21}

As we have explained before, to solve D-optimal designs we need to give the initial values to the starting points k_{12} and k_{21}. To learn more about how different values for those points would impact the design, we need to perform a sensitivity analysis. One way to proceed would be to calculate the efficiency, as we did in the example of Section 10.3.1, but in this example we will do it using derivatives.

- To analyze the sensibility of x_1 with respect to k_{12}, we keep k_{21} constant at 0.02 and then calculate $\frac{\partial x1(k_{12}, 0.02)}{\partial k_{12}}$, that we call $s(t,\ k_{12})$, to see how it changes over time. The closer the derivative is to zero, the less sensitive x_1 will be to changes in initial values of k_{12}. We can do the same for k_{21}, calculating in this case $s(t,\ k_{21}) = \frac{\partial x1(0.03, k_{21})}{\partial k_{21}}$ (Notice that *Mathematica* actually calculates the numerical derivatives of x1[k12, 0.02][t] and x1[0.03, k21][t], since both are numerical expressions).

```
Plot[{D[x1[k12, 0.02][t], k12] /. k12 → 0.03 /. sol,
  D[x1[0.03, k21][t], k21] /. k21 → 0.02 /. sol},
  {t, 0, 20}, AxesLabel → {"t", "s(t)"},
  PlotLegends → {"∂x1(k12, 0.02)/∂k12", "∂x1(0.03, k21)/∂k21"}]
```

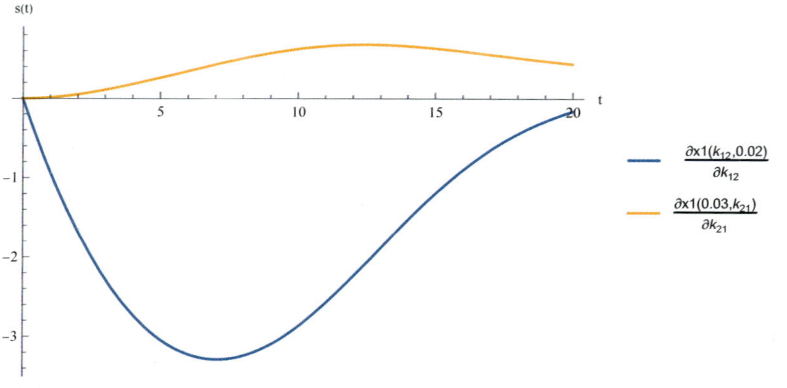

- The derivative gives an idea of the absolute sensibility but we are more interested in the relative sensitivity (rs). This sensibility can be analyzed using the expressions $\mathrm{rs}(t,\ k_{12}) = x_1(0.03,\ 0.02)(t) \pm \Delta\, s(t,\ k_{12})$ and $\mathrm{rs}(t,\ k_{12}) = x_1(0.03,\ 0.02)(t) \pm \Delta\, s(t,\ k_{21})$. A good choice for Δ is 0.1.

```
GraphicsRow[{ Plot[
   Evaluate[(x1[0.03, 0.02][t] + {0, .1, -.1} D[x1[k12, 0.02], k12][t] /.
      k12 → 0.03) /. sol], {t, 0, 20},
   Filling → {2 → {3}}, AxesLabel → {"t", "rs(t, k12)"}], Plot[
   Evaluate[(x1[0.03, 0.02][t] + {0, .1, -.1} D[x1[k21, 0.02], k21][t] /.
      k21 → 0.02) /. sol], {t, 0, 20}, Filling → {2 → {3}},
   AxesLabel → {"t", "rs(t, k21)"}]}, ImageSize → {500, 200}]
```

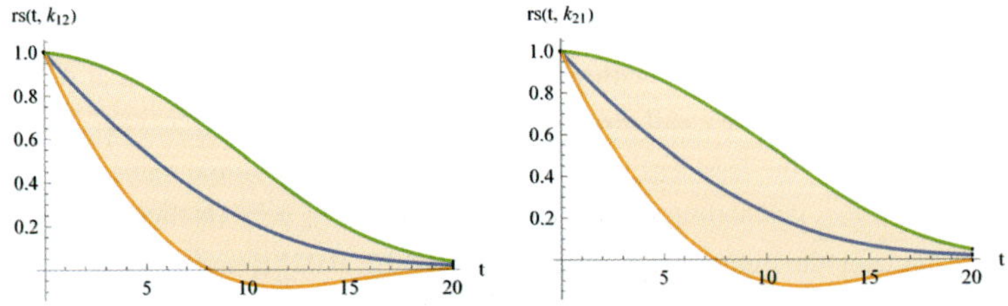

The left graph indicates that the sensitivity of the solution with respect to k_{12} increases with t until t = 10. We can see the pattern by looking at the difference in the measurement of rs(t, k_{12}) between the 3 potential values of k_{12} around t = 10. After that, it decreases. Basically, this means that measurements taken at the beginning or at the end of the interval {0, 20} would not change significantly for different values of k_{12}. However, measurements taken around t = 10 may show large discrepancies if our initial estimate for the constant was not the correct one. For k_{21}, we obtain a similar conclusion (right graph).

- Once again, we quit the session before proceeding to the next section.

```
Quit[]
```

10.4 Biokmod: Applications to ICRP Models

In this section we're going to learn more about Biokmod, a *Mathematica* toolbox developed by the author for modeling biokinetic systems that consists of the following:

(i) A package (Sysmodel) to solve systems of ordinary linear differential equations (SOLDE) with special applications to compartmental and physiological models; (ii) A package for fitting data that can be used to fit transfer rates using experimental data; (iii) A package for optimal design; (iv) Some packages for solving the current ICRP models specially the calculations related to internal doses and bioassay data assessments.
The tutorials included cover: (i) Compartmental and physiological modeling (linear and nonlinear), (ii) Random intakes in occupational exposures and their implications in bioassays, (iii) Analytical methods to evaluate the uncertainties associated with biokinetic model parameters, (iv) Nonlinear regression techniques for bioassay data fitting, (v) Optimal design for bioassay programs.

The toolbox can be downloaded from: http://diarium.usal.es/guillermo/biokmod/. To install: Extract the file "biokmodXX.zip" with a zip utility (e.g., Winzip). It will create a folder called Biokmod. Copy it to the AddOns\Applications directory in the *Mathematica* installation folder ($InstallationDirectory).

10.4.1 Solving ICRP Models

Biokmod includes the ICRP models. The ICRP or International Commission on Radiological Protection (http://www.icrp.org) is an independent international organization devoted to preventing the effects associated with exposure to ionizing radiation, and to protect the environment. The ICRP, among other things, publishes models for analyzing the evolution of radioactive particles in the human body and the environment.

All ICRP 66 and 78 models can be represented by compartmental systems with constant coefficients. The generic model used by ICRP 66 and 78 is shown in Figure 10.7: The particles are deposited in some compartments of the Respiratory Tract (RT). From the RT, the flow goes to the ST (Stomach) or to B (Blood). "Rest of Body" represents the systemic compartments. The detailed flow diagram is specific to each kind of element. The dashed arrows indicate the direction of the flow, depending on the characteristics of the element. The particles are eliminated through fecal or urine excretion. The disintegration can be considered the output from each compartment to the environment; it is given by the disintegration constant of the isotope. This model has been modified by ICRP 130 , but it has not been yet, as of December 2016, included in BIOKMOD. However, the package Sysmodel can be used to solve the models described in the ICRP 130.

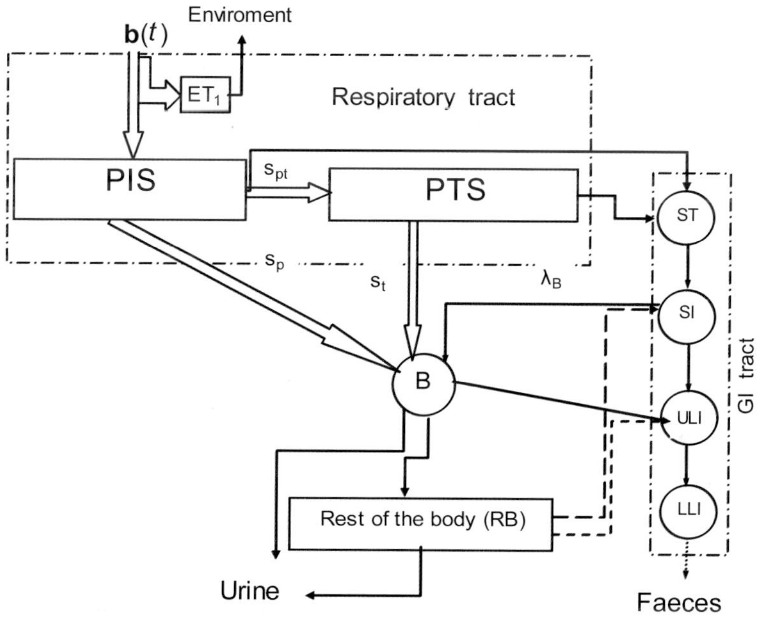

Figure 10.7 The generic ICRP 66 and 78 models applied to the intake of particles by inhalation.

According to it, the human body can be divided in three systems:

a) The human respiratory tract model (HRTM): It is applied for modeling the intake of radioactive aerosols by inhalation. The detailed description is given in ICRP 66. If a person intakes a quantity I by inhalation instantaneously, it is deposited directly in some compartments of the HRTM. The fraction deposited in each compartment is called the Initial Deposition Factor or IDF. It is a function of the Activity Median Aerodynamic Diameter (AMAD), which includes size, shape, density, anatomical and physiological parameters as well as various conditions of exposure. The IDF values may be calculated either following the procedure described in ICRP 66 (1994) or obtained from the Annex F of ICRP 66 (1994). Once the AMAD value has been determined, the program computes the IDF. Another option is to directly write the IDF values for AI, $bb_{fast+seq}$, bb_{slow}, $BB_{fast+seq}$, BB_{slow}, ET1, and ET2. The general model of the RT applies to any element except for the absorption rates $\{s_{pt}, s_p, s_t\}$ that are related to the chemical structure of the element. The ICRP gives default values for absorption rates according to types F, M or S. In Biokmod, those types can be chosen and the program will generate the default values accordingly. Another option is to directly input the absorption rate parameters.

b) The gastrointestinal tract (GI): This is used when modeling the intake of particles in the GI tract following the model provided in ICRP 30 (ICRP 1979). Particles can be introduced into the GI Tract directly by ingestion, or from the RT. The deposition occurs in the stomach (ST). Part or all the flow is transferred, through the SI (Small Intestine), to the blood (B). The transfer rate from SI to B, is given by $\lambda_B = f_1\, \lambda_{SI}/(1 - f_1)$, where f_1 is the fraction of the stable element reaching the blood (or body fluids). If $f_1 = 1$ all the flow from SI goes to B. The value of f_1 is associated to the element and its chemical structure. In Biokmod, f_1 must be introduced or a value by default (from ICRP 2001 and ICRP 1997[7]) will be applied according to the element and the absorption rate previously chosen. (The ICRP Publication 100: *Human Alimentary Tract Model for Radiological Protection*, replaces the ICRP 30. However, as of September 2016, ICRP 30 is still applied in many software programs, including Biokmod).

c) Systemic compartments: They are specific to an element or groups of elements (ICRP 2001). ICRP 78 (1997) establishes three generic groups: (i) hydrogen, cobalt, ruthenium, cesium, and californium, (ii) strontium, radium, and uranium and (iii) thorium, neptunium, plutonium, americium, and curium. For other elements not included in ICRP 78, the ICRP 30 model applies and they have the same generalized compartmental model as group (i). The same model is used for all the elements of each group although some parameters would change depending on the element. From a mathematical point of view, we can establish two groups: a) Elements whose biokinetic models do not involve recycling, this includes the group (i) and elements where ICRP 30 is still applicable, and b) elements whose biokinetic models involve recycling, this includes group (ii) and (iii).

The user can modify the respiratory and gastrointestinal tract parameters included in the application. The reference parameters are used by default. We can choose from three different types of intake (injection, ingestion or inhalation). Days (d) are used as the unit of time. The radioactive decay constant of the isotope, in days^{-1}, must be introduced by the user. More details can be found in the documentation (more than 300 pages). The next few pages provide an overview of some of the application's capabilities.

- We will use several packages included in the toolbox. Note that when launching Doses, other packages are also loaded:

```
Needs["Biokmod`Doses`"]

Resptract 1.2  2005-05-16

SysModel, version 1.5.1   2013-11-12

Humorap 3.7 2015-06-15

Biokdata 1.2.2 2004-09-3

 Doses 1.2   2015-08-27
```

- The next below uses the Resptract package to plot the lung retention at time t (in days) after an acute intake by inhalation. We need to enter: the intake in kBq, the AMAD (1–10 μm) of the radioactive aerosols and the type of metabolization (S, M or F) associated with the chemical structure of the particles inhaled, and the disintegration constant, lambda, in days^{-1} (here we assume that the radioactive aerosol intake is Pu239 but for other particles, if the disintegration constant is very small; we can use 0 instead).

```
Manipulate[met = ToExpression[metabolization];
 LogPlot[LungsRetention[int, AMADfit[amad] , met, t, lambda],
  {t, 0, tmax}, AxesLabel→{"Days", "kBq"}, PlotRange -> All], {{tmax,
   10, "Days"}, 1, 100, Appearance→"Open"}, {{int, 10, "kBq"}, 1, 100,
  Appearance→"Open"}, {{amad, 5, "AMAD (μm)"}, 1, 10, Appearance→"Open"},
 {{metabolization, "S", "Metabolization"}, {"S", "M", "F"}, ControlType→
   PopupMenu}, {{lambda, 0, "Lambda"}, ControlType→InputField},
 Initialization :> Get["Biokmod`Resptract`";], SynchronousUpdating→False]
```

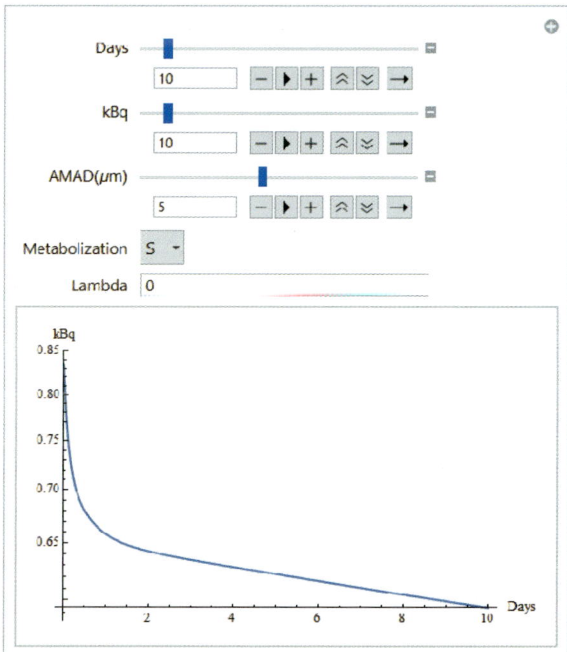

In our next example a person is going to receive an injection of 6.7×10^7 Bq of I–131. We would like to know the daily fecal and urine excretion over time, knowing that for Iodine f_1 = 1.

- We will use the function: **BiokdataReport**[*Element name* (e.g., Uranium) or *compartmental matrix* of the element, "*Intake pathway*" (e.g., "Injection"), *Type of intake* (e.g., "Acute"), *Kind of report* (e.g., "GraphicReport"), Amount *from single intake* (e.g., 6.7 10^7 or, if it is a continuous intake, {b[t], t}), value of *f1* (e.g., 1), *time* (e.g., 180 or t in days), *decay constant* (in days^{-1})].
- Before evaluating the function, we need to find out the decay constant, in days^{-1}, for ^{131}I:

```
UnitConvert[IsotopeData["I131", "Lifetime"], "Days"]
```

```
11.578 days
```

```
λI131 = 1/QuantityMagnitude[%]
```

```
0.086371
```

- The graph below displays the urine and fecal excretion for I–131 after an acute intake of $I = 6.7 \times 10^7$ Bq at $t = 0$.

```
BiokdataReport[iodine, "Injection",
  "Acute", "GraphicReport", 6.7×10^7, 1, 180, λI131]
```

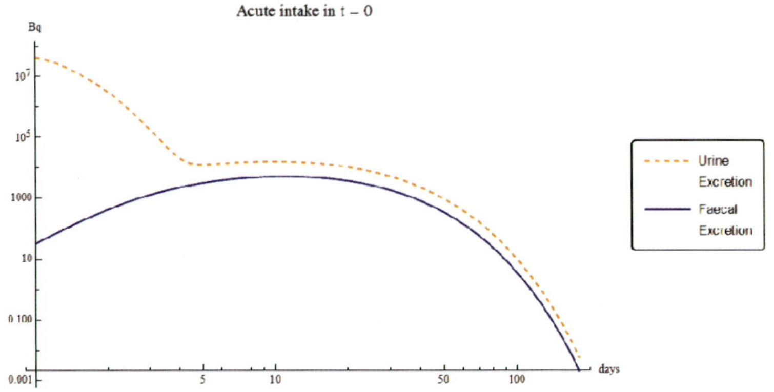

The output shows the graphical representations. However, other types of output are available such as retention functions for each compartment or the number of disintegrations accumulated in each compartment.

In this example we assume an exponential input, with t in days, of Iodine–131 by a continuous injection (through a drip):

```
b[t_] = 6 * 10^4 Exp[-10.2 t] + 2×10^3 Exp[-6.0 t];
```

- The total amount injected, in Bq, is given by:

```
Integrate[b[t], {t, 0, Infinity}]
```

```
6215.69
```

- The instruction below returns the retention function for a typical bioassay:

```
BiokdataReport[iodine, "Injection",
  "Continuous", "Automatic", {b[t] , t}, 1, t, λI131] // Chop
```

$$\{qDailyUrine[t] \to 1.10008\times 10^{9}\, e^{-12.0864\,t} - 2.46207\times 10^{8}\, e^{-10.2\,t} - 152051.\, e^{-6.\,t} + 120298.\, e^{-2.85892\,t} - 0.577698\, e^{-0.146518\,t} - 10.0907\, e^{-0.146518\,t} + 0.533278\, e^{-0.0926952\,t} + 9.35037\, e^{-0.0926952\,t},$$

$$qDailyFaecal[t] \to -0.388163\, e^{-10.2\,t} - 0.00851992\, e^{-6.\,t} + 0.552212\, e^{-2.85892\,t} - 1.33382\, e^{-1.88637\,t} + 1.35927\, e^{-1.08637\,t} - 0.221073\, e^{-0.146518\,t} - 3.86149\, e^{-0.146518\,t} + 0.191804\, e^{-0.0926952\,t} + 3.36305\, e^{-0.0926952\,t},$$

$$qWholebody[t] \to 6759.14\, e^{-12.0864\,t} - 15910.1\, e^{-10.2\,t} - 750.212\, e^{-6.\,t} + 8020.32\, e^{-2.85892\,t} - 0.26414\, e^{-1.88637\,t} + 0.791061\, e^{-1.08637\,t} - 12.8849\, e^{-0.146518\,t} - 225.061\, e^{-0.146518\,t} + 114.295\, e^{-0.0926952\,t} + 2004.02\, e^{-0.0926952\,t}\}$$

For those readers interested in learning more about the **BiokdataReport**[...] function, there's additional information in the BiokMod help files.

10.4.2 A Medical Application

For multiple single inputs: $\{b_1, \ldots, b_i, \ldots, b_n\}$ at times: $\{t_0, t_1, \ldots, t_i, \ldots, t_n\}$, where $t - t_i$ is the time since the input b_i occurred and $u(t)$ the unit impulse response, the retention function, $q_M(t)$, with $t_0 = 0$, is given by:

$$q_M(t) = b_1\, u(t) + b_2\, u(t - t_1) + \ldots + I_n\, u(t - t_{n-1}) = \sum_{i=1}^{n} b_i\, u(t - t_{i-1})$$

If we consider time to be a discrete variable measured in days, and I_j represents the intake that happened on day j, the previous equation can be written as:

$$q_M(t) = b_1\, u(t) + b_2\, u(t-1) + \ldots + b_n\, u(1) = \sum_{j=1}^{t} b_j\, u(t - j + 1)$$

We want to consider the case where multiple constant inputs $\{b_0, \ldots, b_i, \ldots, I_n\}$ occur at times: $\{t_0, t_1, \ldots, t_i, \ldots, t_n\}$ each one lasting $\{T_0, \ldots, T_i, \ldots, T_n\}$. Let's call $r(t)$ the unit function for a constant input:

$$r(t,\ T_i) = \left\{0,\ t < 0,\ \int_0^t u(t)\, dt \text{ for } 0 < t \le T_i \text{ and } \int_{t-T_i}^{t} u(t)\, dt \text{ for } t > T_i\right\}$$

The retention function for multiple constant inputs is given by:

$$q_{MC}(t) = \frac{b_0}{T_0} r(t - t_0) + \frac{b_1}{T_1} r(t - t_1) + \ldots + \frac{b_n}{T_n} r(t - t_n) = \sum_{i=1}^{n} \frac{b_i}{T_i} r(t - t_i)$$

The previous equation is the one used by the Biokmod function:

qMultiple[$\{\{b_0,\ t_0,\ T_0\},\ \{b_1,\ t_1,\ T_1\} \ldots,\ \{b_i,\ t_i,\ T_i\},\ \ldots \{b_n,\ t_n,\ T_n\}\},\ \{u[t],\ t\},\ t$].

If T_i is not included, the program assumes a single input. We can combine single and multiple inputs (e.g., $\{\{b_0,\ 0,\ T_0\},\ \{b_1,\ t_1\},\ \ldots,\ \{b_i,\ t_i,\ T_i\},\ \ldots. \{b_n,\ t_n\}\}$).

Example: A person is subjected to a medical treatment that consists of a perfusion into the blood of 10 kBq of I–131 every 5 days for 2 hours (or 1/12 day) for 20 days. We would like to know the amount retained in the thyroid function over this period of time.

- The inputs are represented as follows ($\{b_i,\ t_i,\ T_i\}$):

```
inputs = {{10, 0, 1/12}, {10, 5, 1/12},
   {10, 10, 1/12}, {10, 15, 1/12} , {10, 20, 1/12} };
```

- The unit impulse response can be obtained with the command below (we use the function `CompartNumbers` to know what compartment number is associated with Thyroid):

```
CompartNumbers[iodine]

1    Blood
2    Thyroid
3    Rest
4    Bladder
5    Urine
6    ULI
7    LLI
8    FEC
```

```
u[t_] = q2[t] /. BiokdataReport[iodine, "Injection",
    "Acute", "CompartmentContent", 1, 1, t, λI131] // Chop
```

$-0.300955\, e^{-2.85892\, t} + 0.0135875\, e^{-0.146518\, t} + 0.287368\, e^{-0.0926952\, t}$

```
Plot[u[t], {t, 0, 10}, AxesLabel → {"Days", "kBq I-131(Thyroid)"}]
```

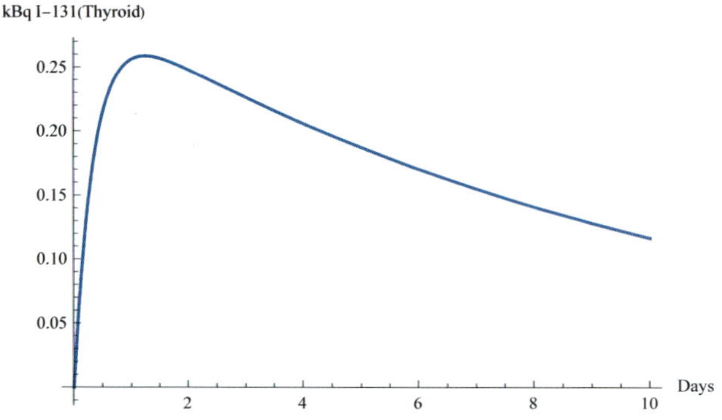

- Next, we apply the function to the multiple inputs and visualize the results:

```
Plot[qMultiple[inputs, {u[t], t}, t1], {t1, 0, 50}, PlotRange → {0, 10},
 ExclusionsStyle → {Blue, Blue}, AxesLabel → {"Days", "kBq I-131(Thyroid)"}]
```

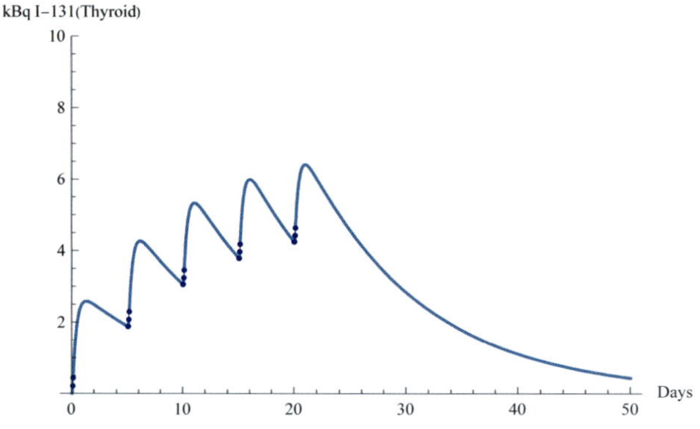

10.4.3 Bioassays Evaluation

A frequent problem in internal dosimetry is how to estimate the quantity of a certain element absorbed from bioassays. This problem is the reverse of what we've discussed so far. Let's see how to solve it with an example.

Example: As a result of the Chernobyl accident (April 26, 1986), a 39 year old civilian male weighing 80 kg was exposed to a continuous and unknown ingestion of Cs–137 (data from Ansoborlo et al., 2003).

- The measurements related to the evolution of the element retention over time in his body are given below: {time after the accident (d), activity (Bq)}:

```
completeData = {{39, 300}, {58, 671}, {75, 737},
   {130, 1661}, {156, 1846}, {170, 1882}, {198, 2247}, {234, 2493},
   {263, 2926}, {297, 3224}, {325, 3608}, {374, 3883}, {408, 3773},
   {432, 3723}, {494, 3195}, {520, 2740}, {556, 2469}, {592, 2375},
   {625, 1954}, {682, 1614}, {744, 1221}, {800, 1174}, {880, 739}};
```

```
ListPlot[completeData, Joined → True, AxesLabel → {"Days", "kBq Cs-137"}]
```

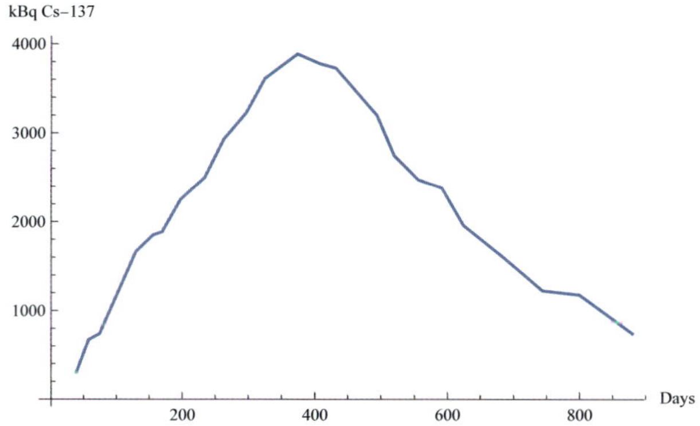

- The function below computes the overall retention for an acute intake, "1" , of Cs–137 at t = 0:

```
qWbCs137[t1_] =
 qWholebody[compartMatrix[caesium], 1, 1, t1, Log[2] / (30 * 365.24)]
```

$-0.000177381\, e^{-24.0001\, t1} + 0.00165462\, e^{-12.0001\, t1} - 0.0230333\, e^{-2.77265\, t1} + 0.0207699\, e^{-1.80006\, t1} - 0.0430482\, e^{-1.00006\, t1} + 0.139387\, e^{-0.346063\, t1} + 0.904447\, e^{-0.00636326\, t1}$

- The next command assumes a daily chronic ingestion "I" during a certain period of time t1. The ingestion of cesium stops at $t = T$ so for $t > T$, I = 0.

```
qConstant[1, {qWbCs137[t], t}, t, 2000]
```

$$\begin{cases} \text{t1 must be non negative} & t < 0 \\ 142.499 + 7.39086\times10^{-6}\, e^{-24.0001\, t} - 0.000137884\, e^{-12.0001\, t} + 0.00830732\, e^{-2.77265\, t} - 0.0115384\, e^{-1.80006\, t} + 0.0430455\, e^{-1.00006\, t} - 0.402779\, e^{-0.346063\, t} - 142.136\, e^{-0.00636326\, t} & 0 \le t \le 2000 \\ 0. - 7.39086\times10^{-6}\, e^{-24.0001\,(-2000+t)} + 0.000137884\, e^{-12.0001\,(-2000+t)} - 0.00830732\, e^{-2.77265\,(-2000+t)} + 0.0115384\, e^{-1.80006\,(-2000+t)} - 0.0430455\, e^{-1.00006\,(-2000+t)} + 0.402779\, e^{-0.346063\,(-2000+t)} + 142.136\, e^{-0.00636326\,(-2000+t)} + 7.39086\times10^{-6}\, e^{-24.0001\, t} - 0.000137884\, e^{-12.0001\, t} + 0.00830732\, e^{-2.77265\, t} - 0.0115384\, e^{-1.80006\, t} + 0.0430455\, e^{-1.00006\, t} - 0.402779\, e^{-0.346063\, t} - 142.136\, e^{-0.00636326\, t} & t > 2000 \\ 0 & \text{True} \end{cases}$$

- We can see that the retention was increasing until T. Assuming that it happened until that time and then stopped, we can fit the experimental data, taking into consideration both periods, before and after T.

```
model2[t1_?NumericQ, p_?NumericQ, tt_?NumericQ] := p qConstant[1,{qWbCs137[t],t}, t1, tt]
```

- To estimate the best fit for the intake I in Bq/day, and the period T, in days:

```
{input, timeIntake} =
 {p, tt} /. FindFit[completeData, model2[t, p, tt], {p, tt}, t]
```

{24.5952, 533.961}

- Then, the accumulated intake is:

```
input timeIntake "Bq"
```

```
13132.9 Bq
```

- The quality of the fit is shown in the plot below:

```
Plot[qConstant[input, {qWbCs137[t], t}, t1, timeIntake], {t1, 1, 1000},
 PlotRange→{Automatic, {0, 4000}}, AxesLabel→{"Days", "kBq Ca-137"},
 Epilog→ {Hue[0.], PointSize[0.02], Map[Point, completeData]}]
```

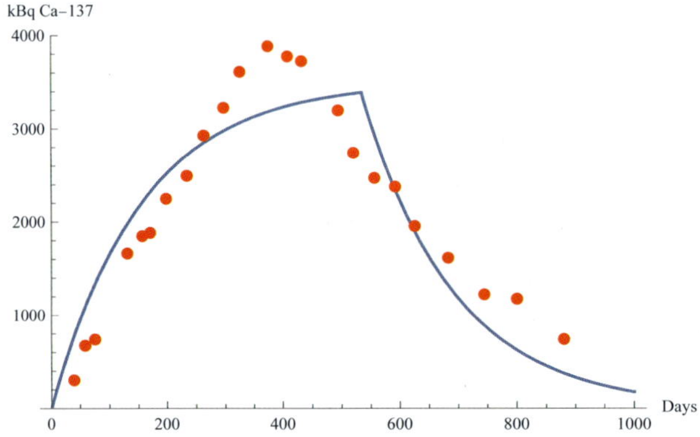

10.4.4 Optimal Experimental Design in Biokmod

- Biokmod includes a package to perform OED. In this case we solve the same example described in 10.3.1 using the Biokmod package function **Optdes**:

```
Needs["Biokmod`Optdesign`"]
```

```
Optdesign, 1.0    2007-04-09
```

```
Optdes[ inp (e^(-2.0 t-0.09 p) + e^(-0.001 t-0.2 p)),
 t, {{inp, 100}, {p, 5}} , 0.5, 1, 2, 3]
```

```
{0.16647, {t0 → 0.5, t1 → 1.98703, t2 → 8.16929, t3 → 15.083}}
```

10.4.5 Doses

Exposure to ionizing radiations may result in radiological risk. To estimate such risk, the concept of dose, measured in Sieverts (Sv, mSv), was introduced.

Almost all of the radiation that we receive on a regular basis comes from either nature or medical tests. The average annual exposure is 2.4 mSv/year per person. However, this number varies widely, ranging from 2 to 10 mSv/year, depending on the place and costumes of the specific individuals (types of food consumed, smoking habits and dwelling). There are even regions, such as Kerala in India, that amply exceed this range. As a precaution, we assume that any dose carries the risk of producing undesirable biological effects; that the relationship between the risk and the dose is linear (that is, double dose double risk); and that any dose, no matter how small, carries a risk. These criteria are used as a precaution since it's not possible to know if an effect (e.g., a certain type of cancer) is attributable to the radiations or has a different cause. For example: the linear hypothesis implies that a dose of 100 mSv may result in a 0.1% increase in the risk of cancer, a change not big enough to clearly identify the dose as the cause of the disease.

Radiation exposure can originate from either irradiation (i.e. X-rays) or assimilation (inhalation, ingestion or injection) of radioactive particles by the human body. In the first case, the exposure stops when we move away from the radiation source (for example: a radiography). However, if we assimilate the radioactive source (for example: by ingestion or injection of a radioactive isotope) the exposure continues as long as the radioactive substance remains in our body. Next, we describe how to calculate the dose in this last case.

The committed equivalent dose $H_T(\tau)$ is defined as the time integral of the equivalent dose rate in a particular tissue or organ that will be assimilated by an individual following the intake of radioactive material into the body, with τ being the integration time in years following the intake. Usually we assume $\tau = 50$ y for adults. We can calculate the doses using the Biokmod function **CommitedDose**, included in the package Doses.

Example: An adult person receives an injection of 0.01 μg of I–131. What would be the thyroid effective dose?

- First we need to convert the 0.01 μg of I–131 into Bq as follows: $A = \lambda\ n_{at} = \lambda\ n_{Av}/m_a$ (Avogadro's number n_{Av}, can be obtained directly in *Mathematica*)

```
nAv = QuantityMagnitude[UnitConvert[Quantity["AvogadroConstant"]]]
```

```
6.022141×10^23
```

```
λI131 = 1/QuantityMagnitude[IsotopeData["I131", "Lifetime"]]
```

```
9.9967×10^-7
```

```
a131 = λI131* nAv*0.01*10^-6/131
```

```
4.59553×10^7
```

The previous result indicates the extraordinary radioactive activity of such a small amount of Iodine–131, an isotope widely used in oncological treatments and medical research. Because of that, patients being treated with the isotope need to take certain precautions if they live with other people.

```
CommittedDose["I 131", "Injection", a131,
  1, 50*365.25, DisintegrationReport→ "False"]
```

```
{{Sv/Bq, 18262.5 day}, {Testes, 0.00185447}, {Ovarius, 0.00220993},
 {Red Marrow, 0.0046316}, {Colon, 0.0026153}, {Lungs, 0.00473549},
 {St Wall, 0.00196979}, {Bladder Wall, 0.0351674}, {Mama, 0.00269717},
 {Liver, 0.00218663}, {Oesophagus, 0.00706479}, {Thyroid, 20.0344},
 {Skin, 0.0031639}, {Bone Surface, 0.00607722}, {Muscle, 0.00582219},
 {Brain, 0.00667664}, {Small intestine, 0.00199493},
 {Kidneys, 0.00199287}, {Pancreas, 0.00226375}, {Spleen, 0.00212897},
 {Thymus, 0.00706479}, {Uterus, 0.00263947}, {Adrenals, 0.00220647},
 {Extrathoracic airways, 0.00683504}, {Effective, e(50), 1.00657}}
```

The effective dose is 1 Sv and the equivalent dose in the Thyroid is 20.0 Sv, a very high dose.

- In the previous section, we found that a male was exposed to an ingestion of 13,132.9 Bq of Cs–137. In term of doses, this is equivalent to:

```
Last[CommittedDose["Cs 137", "Ingestion", 13132.9,
   1, 50*365.25, DisintegrationReport→ "False"]] "Sv"
```

```
{Effective, e(50) Sv, 0.000287239 Sv}
```

This value is negligible compared to the dose received from a typical medical treatment.

10.5 Radiation Attenuation

Mathematica includes a new function with applications in radiological protection: `StoppingPowerData`. Let's take a look at a few examples based on the documentation pages.

- One way to reduce radiation exposure is through the use of shielding. For example, the thickness of a water layer needed to reduce photon radiation by a tenth is:

```
StoppingPowerData["Water", {"Particle" → "Photon",
  "Energy" → Quantity[0.1, "Megaelectronvolts"]}, "HalfValueLayer"]
```

```
4.06062 cm
```

- In the design of radiation protection products it is important to take into account the mass attenuation coefficients of the materials used. This attenuation depends on the incident particles (protons, alpha, protons, electrons, neutrons) and their energy. Here we compare the mass attenuation coefficient for various substances and energies:

```
materials =
  {"Water", lead (element), iron (element), aluminum (element)};
ListLogLogPlot[
 Table[{Quantity[x, "Megaelectronvolts"], StoppingPowerData[#,
     {"Particle" → "Photon", "Energy" → Quantity[x, "Megaelectronvolts"]},
     "MassAttenuationCoefficient"]}, {x, 0.01, 100, .5}] & /@ materials,
 Joined → True, PlotLegends → materials, AxesLabel → Automatic]
```

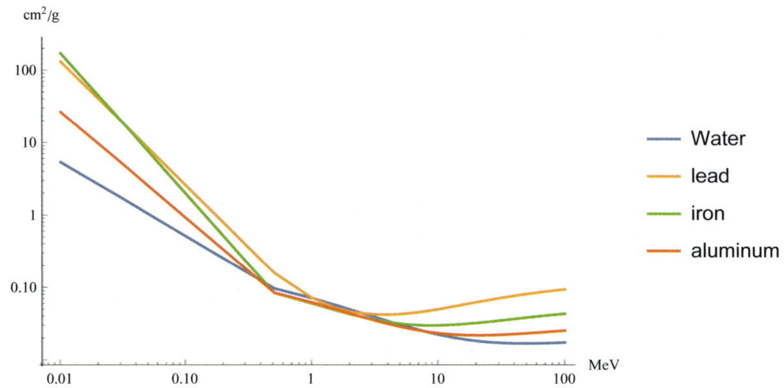

- To accurately estimate absorbed doses, we measure the amount of the energy deposited in human tissues. The example below shows the stopping power changes, dE/dx, for an alpha particle emitted from U234 as it travels through biological tissue:

```
initialKE = QuantityMagnitude@
  IsotopeData[Entity["Isotope", "Uranium234"], "BindingEnergy"]
```

```
7.600708
```

```
dEByDxVal[k_Real?Positive] :=
  QuantityMagnitude@StoppingPowerData["A150TissueEquivalentPlastic",
     {"Particle" → Entity["Particle", "AlphaParticle"], "Energy" →
       Quantity[k, "Megaelectronvolts"]}, "LinearStoppingPower"] / 10000;
dEByDxVal[k_Real] :=
 0
```

```
sol = NDSolveValue[{KiM'[x] == -dEByDxVal[KiM[x]], KiM[0] == initialKE,
 WhenEvent[KiM[x] <= 0, "StopIntegration"]}, KiM, {x, 0, 60},
 PrecisionGoal -> 3];

Plot[dEByDxVal[sol[x]], {x, 0, 60},
 Frame → True, GridLines → Automatic, FrameLabel →
  {Row[{"Distance Travelled", Quantity[None, "Micrometers"]}, "  "], Row[
    {"dE/dx", Quantity[None, "Megaelectronvolts" / "Micrometers"]}, "  "]}]
```

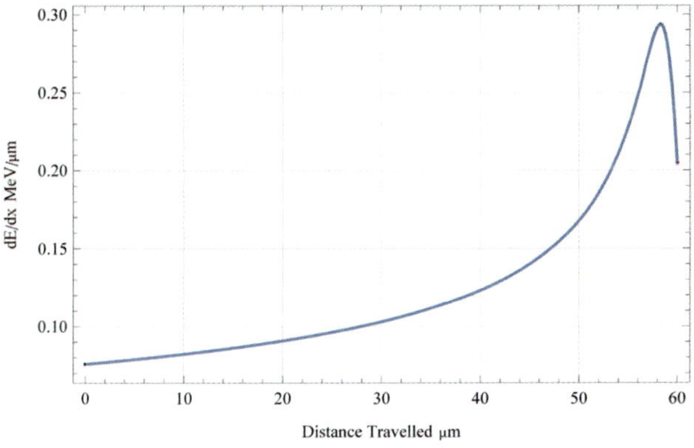

10.6 Additional Resources

About Compartmental and Physiological Modeling

J.A. Jacquez, *Compartmental Analysis in Biology and Medicine*, The University of Michigan Press, 1985.

K. Godfrey, *Compartmental Models and Their Application,* Academic Press, London, 1983.

ICRP, *Individual Monitoring for Internal Exposure of Workers*. ICRP Publication 78, 1997.

ICRP, *Human respiratory tract model for radiological protection*. ICRP Publication 66, Ann. ICRP 24(1–3), 1994a.

ICRP, *Occupational Intakes of Radionuclides: Part 1,* ICRP Publication 130, Ann. ICRP 44(2), 2015.

A. Sánchez-Navarro, C. Casquero, and M. Weiss, Distribution of Ciprofloxacin and Ofloxacin in the Isolated Hindlimb of the Rat, *Pharmaceutical Research*, 16: 587-591, 1999.

Q. Zheng, Exploring Physiologically-Based Pharmacokinetic Models, *Mathematica in Education and Research*, 6 (2): 22-28, 1997.

M. S. Roberts, M. Rowland, A dispersion model of hepatic elimination, *Journal of Pharmacokim. Biopharm 14*: pp 227-260, 1986.

About Optimal Experimental Design

J. Lopez-Fidalgo, J.M. Rodriguez-Diaz, J. M. Sanchez, G. Santos-Martin, MT, Optimal designs for compartmental models with correlated observations, *Journal of Applied Statistics*: 32(10) pp. 1075-1088, 2005.

J. M. Sanchez, J. M. Rodríguez-Díaz, Optimal design and mathematical model applied to establish bioassay programs, *Radiation Protection Dosimetry,* doi:10.1093/rpd/ncl499, 2007.

J. M. Rodríguez-Díaz, G. Sánchez-León, Design optimality for models defined by a system of ordinary differential equations, *Biometrical Journal* 56 (5), pp. 886-900, 2014.

About Biokmod

The toolbox can be downloaded from: http://diarium.usal.es/guillermo/biokmod. Extensive documentation is included.
Some of its features can be run directly over the Internet with the help of *webMathematica*: http://oed.usal.es/webMathematica/Biokmod/index.html.
G. Sanchez, Biokmod: A *Mathematica* toolbox for modeling Biokinetic Systems. *Mathematica in Education and Research* 10(2): 50-70, 2005.
G. Sanchez, Fitting bioassay data and performing uncertainty analysis with Biokmod, *Health Physics,* 92(1): 64-72, 2007.

11

Economic and Financial Applications

The latest Mathematica versions include multiple capabilities related to economics and finance. The new functionality makes it easy to perform financial visualizations and to access both historical and real time data. The commands related to stock markets and financial derivatives would be especially useful to readers interested in investment portfolio management. Additionally, the functions for optimization, with applications in economics and many other fields, have been improved significantly. A new function to solve the Traveling Salesman Problem in a very efficient way has been included. All of these topics will be discussed in the sections below using real world examples. This chapter should be read along with Chapter 6, the one covering probability and statistics.

11.1 Financial Information

11.1.1 Introduction

Mathematica, as part of its computable data functionality, includes the command `FinancialData` to access finance-related information. Although in theory the program can display thousands of indicators, in reality the accessible ones are mostly those related to the US markets. Those readers who may need reliable access to other markets or real time financial data would probably be interested in taking a look at the Wolfram Finance Platform, a new product from Wolfram Research (the developers of *Mathematica*) that focuses on finance: http://www.wolfram.com/finance-platform/.

In any case, although the information may not be directly accessible from `FinancialData`, nowadays it would be easy to access it. For example, many bank platforms enable their clients to download data in an Excel format that could then be imported into *Mathematica* and analyzed using any of the multiple functions related to finance:

http://reference.wolfram.com/mathematica/guide/Finance.html

11.1.2 FinancialData

`FinancialData["`*name*`", "`*property*`", {`*start, end, interval*`}]` gives the value of the specified property for the financial entity "name" (it could be an index, a stock, a commodity, etc.) for a given period of time (by default it returns the most recent value). We recommend to read the documentation pages to know more about this function's capabilities.

- The symbols ("name") used are the ones available in http://finance.yahoo.com/ since Yahoo Finance supplies the information to the function. In the website we can find that the symbol for the Standard & Poor's 500 index, that includes the 500 largest public companies in terms of market capitalization that trade in either NYSE or NASDAQ is: ^GSPC (we can also use SP500). We can check it as follows:

```
FinancialData["SP500", "Name" ]
```

```
S&P 500
```

- If we want to see the price (with a few minutes delay) we can type:

```
FinancialData["^GSPC"]
```

```
2269.
```

- We can also access historical data. In this example we show the trajectory of the S&P 500 index from January 1, 2000 to December 31, 2016.

```
DateListPlot[FinancialData["^GSPC",
  {"January 1 2000", "December 31 2016"}], Joined → True]
```

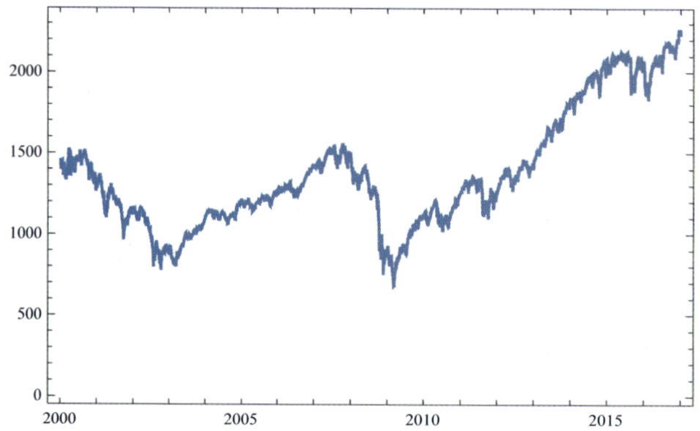

If the financial entity is trading at the time of the query, the price that we get is the one from a few minutes earlier (usually the delay is 15 minutes). The access to information in "real time" requires additional payments to 3rd party providers such as Bloomberg or Reuters. There are even people willing to pay very large amounts of money to gain an advantage of a few milliseconds when trading:

http://adtmag.com/articles/2011/07/29/why-hft-programmers-earn-top-salaries.aspx.

- The following function will dynamically update the S&P500 index (as long as the command is executed during trading hours):

```
sp500 := FinancialData["SP500", "LatestTrade"]
```

- The output changes every two seconds:

```
Dynamic[sp500, UpdateInterval → 2]
```

```
sp500
```

- We can also represent it in a self-updating graph:

```
data = {};
Dynamic[Last[AppendTo[data, sp500]],
 UpdateInterval → 2, TrackedSymbols → {}]
```

```
Dynamic[DateListPlot[data, Joined → True], SynchronousUpdating → False]
```

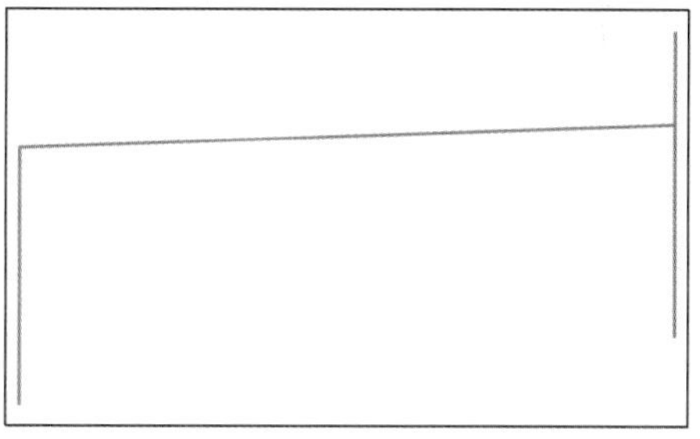

- As mentioned earlier, if the entity is not available in `FinancialData` we may still be able to import the information from specialized websites. A very interesting one is Quandl, http://www.quandl.com, that provides access to numerous economic and financial databases. Most of them are available for free (https://www.quandl.com/search?type=free) but some of them require subscription (https://www.quandl.com/search?type=premium). We can first select the data and download it in our chosen format: Excel, CSV or any other. Then we can use `Import` to analyze them using *Mathematica*. Alternatively, we can access the data directly using `Import` or `URLFetch`. If you use Quandl frequently, you may be interested in downloading QuandlLink, a free *Mathematica* package containing the function **QuandlFinancialData**, similar to `FinancialData`. The package is available in: https://github.com/bajracha71/Quandl-Mathematica-QuandlLink (Accessed on December 4, 2016).

Another interesting feature is the ability to generate email messages linked to certain events. For further information please visit: guide/AutomaticMailProcessing.

For example, we may be interested in sending a message if a stock price goes above or below a certain threshold.

- First we need to provide *Mathematica* with our email account information. This is done in **Edit ▸ Preferences... ▸ Internet Connectivity ▸ Mail Settings**. In the case of Hotmail, the configuration is the one shown in Figure 11.1. For Gmail is similar with "Server" → "smtp.gmail.com" and "PortNumber" → 587. For other popular email services such as Yahoo, you can find the relevant information on the Internet. For enterprise mail servers you may have to ask your system administrator.

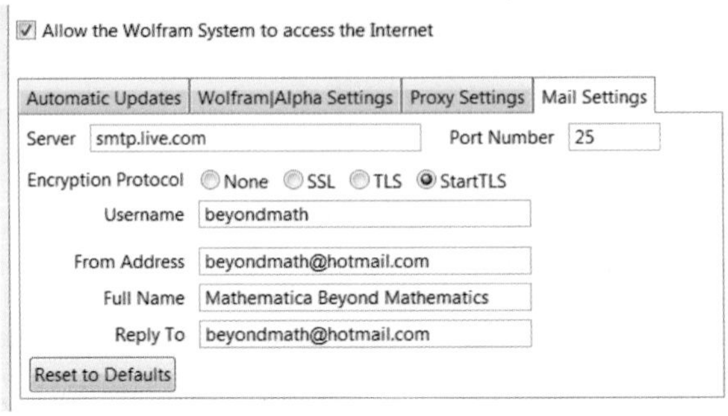

Figure 11.1 Mail Settings in Mathematica.

- Now we can send a test message:

```
SendMail["To" -> "myemail@gmail.com",
 "Subject" -> "Test message",
 "Body" -> "Test message",
  "Password" -> "your password"]
```

- If it worked correctly, we can now define the following function:

```
ss[mens_] := SendMail["To" -> "myemail@hotmail.com",
 "Subject" -> "Message via sendmail",
 "Body" -> ToString[mens],
  "Password" -> "your password"]
```

- You can use this function at your convenience. In this example, we create a stopwatch and send a message when it reaches 20. You can create similar commands to send messages when a stock goes above or below a certain threshold.

```
Dynamic[Refresh[If[Round[Clock[1000]] == 20,
   ss[Clock[1000]], Round[Clock[1000]]], UpdateInterval → 1]]
```

- In the next example we send a message at a scheduled time. The command below activates the task, the latest Euro to US Dollar exchange rate, 10 seconds from the current time, and sends an email 1 second after that.

```
RunScheduledTask[ss[FinancialData["EUR/USD"]],
  {1}, AbsoluteTime[DateString[]] + 10]
```

ScheduledTaskObject[]

11.1.3 Visualization

The number of specialized graphs for economic and financial visualization has increased significantly in the latest versions of *Mathematica*.

- All the functions below end in **Chart**. We have already seen some of them in Chapter 6. In this section, we're going to cover the ones normally used for analyzing stock prices.

```
? *Chart
```

▼ System`

BarChart	DistributionChart	PairedBarChart	RenkoChart
BoxWhiskerChart	InteractiveTradin-gChart	PieChart	SectorChart
BubbleChart	KagiChart	PointFigureChart	TradingChart
CandlestickChart	LineBreakChart	RectangleChart	

Choose any of them and click on it to access its documentation.

These graphs complement the capabilities of **FinancialData** but they can also be used with other functions. Let's see some of them:

- Let's visualize the historical closing prices for the S&P 500 using different graphs.

```
data = FinancialData["SP500", "Close", {{2016, 1, 1}, {2016, 07, 31}}];
```

- If we click on any point inside the plot, we'll see the information associated to it.

```
GraphicsRow[
 {KagiChart[data], PointFigureChart[data], LineBreakChart[data]}]
```

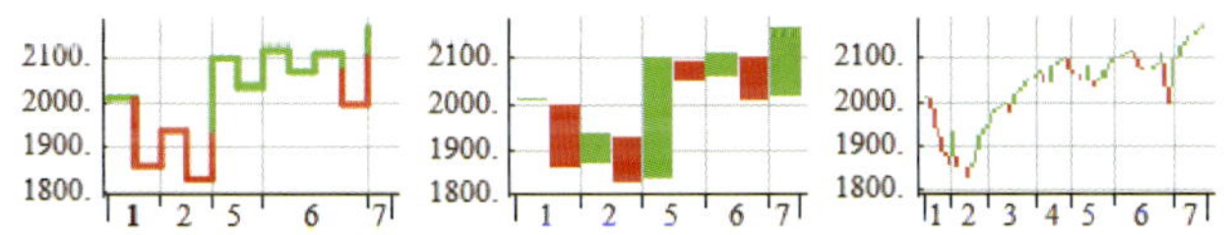

- The next graph uses the function RenkoChart with EventLabels and Placed to add symbols ("→" y "↓") indicating in what position they must be placed.

```
RenkoChart[{"NYSE:GE", {{2016, 01, 1}, {2016, 07, 31}}}, EventLabels →
  {{2015, 10, 1} → Placed["→", Before], {2016, 5, 2} → Placed["↓", Above]}]
```

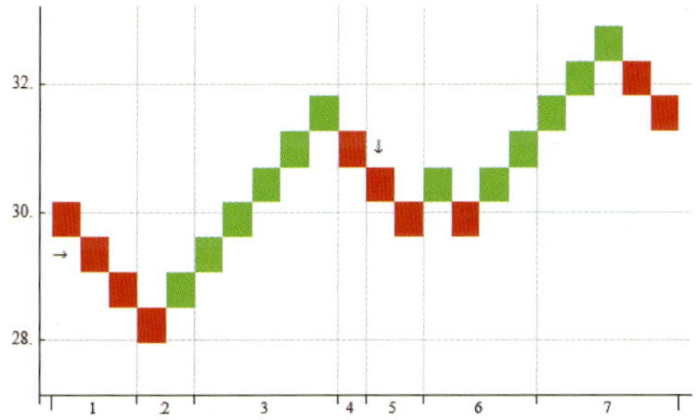

- The graph below displays the trajectory of GE in the NYSE from January 1 to July 31, 2016. For each trading day, we can see the open, high, low and close prices.

```
CandlestickChart[{"NYSE:GE", {{2016, 01, 1}, {2016, 7, 31}}}]
```

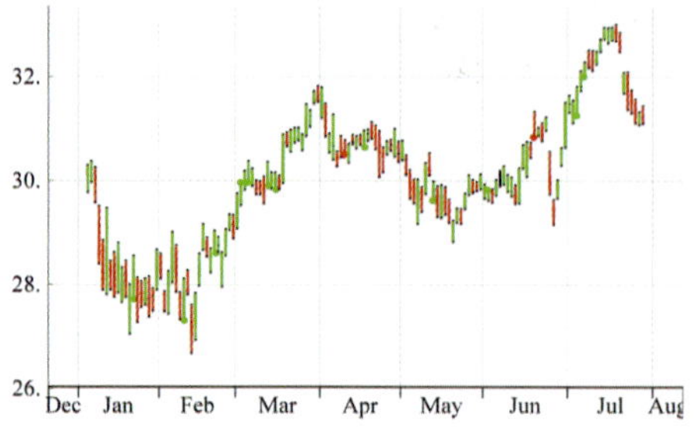

- `ChartElementFunction` is an option that when added to **`BarChart`** or **`PieChart3D`**, improves the presentation of the data.

```
GraphicsRow[{BarChart[{{1, 5, 2}, {4, 5, 7}},
   ChartElementFunction → "GradientScaleRectangle"],
  PieChart3D[{3, 2, 2, 4, 1}, ChartElementFunction → "TorusSector3D",
   SectorOrigin → {Automatic, 1}, ChartStyle → "Rainbow"]}, Frame → All]
```

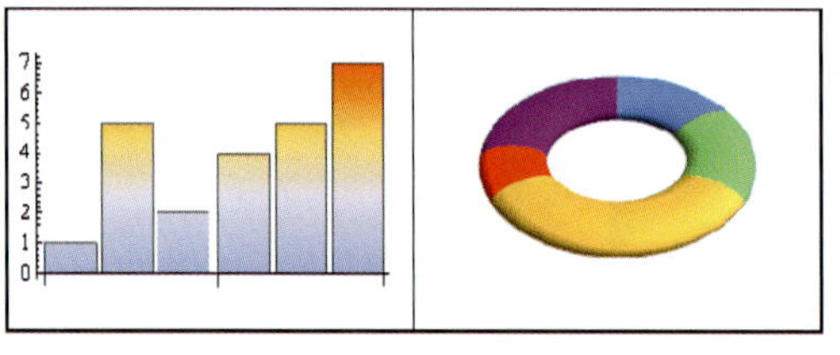

In the world of professional stock market investors, there's a group that uses chart analysis as its main tool for making investment decisions. Its members are known as chartists and their techniques are known as technical analysis. They closely study the past performance of stocks or indices to find trends and support and resistance levels. These people associate certain graph shapes to future markets ups and downs. Although there's no sound mathematical theory justifying their predictions, all economic newspapers have a section dedicated to this type of analysis.

- The powerful function `InteractiveTradingChart`, introduced in *Mathematica* 8, includes practically all the functionality that a chartist may require. In this example, we display GE's share price history for a given period. Notice that in the upper area of the graph, we can see a trimester at a time and using the bottom part we can move through the entire period. We can also choose the time interval (days, weeks or months) and even what chart type and indicators to show.

```
InteractiveTradingChart[{"NYSE:GE", {{2016, 01, 1}, {2016, 07, 31}}}]
```

11.1.4 Automatic Adjustments

With *Mathematica* we can automatically fit our financial data to a given distribution (by default the program will use the maximum likelihood method) with the function `EstimatedDistribution`.

- Let's fit Google's share price to a lognormal distribution after downloading the data.

```
googleStock =
  FinancialData["GOOG", {{2010, 3, 10}, {2016, 07, 31}, "Day"}, "Value"];
dist = EstimatedDistribution[googleStock, LogNormalDistribution[μ, σ]]
```

```
LogNormalDistribution[6.03238, 0.360796]
```

- The fit is poor on both tails. As a matter of fact, all the efforts that have been made trying to predict financial markets have produced unsatisfactory results. These markets exhibit what Mandelbrot (the father of fractal geometry) called "wild randomness" (single observations can impact the total in a very disproportionate way).

```
QuantilePlot[googleStock, dist, Filling → Automatic, FillingStyle → Red]
```

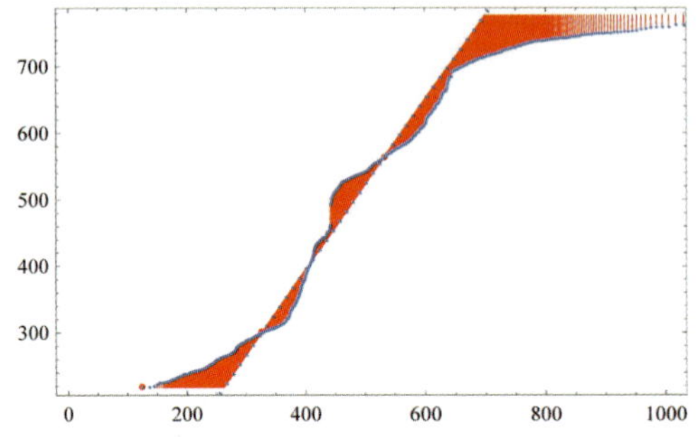

11.2 Financial Functions

The functions related to financial calculations can be found in guide/Finance. We will give an overview of some of them in this section.

11.2.1 Time Value Computations

```
Clear["Global`*"]
```

The following functions are included under the *Time value Computations* category: `TimeValue`, `EffectiveInterest`, `Annuity`, `AnnuityDue` and `Cashflow`. Let's see some examples.

`TimeValue["s", "i", "t"]` calculates the time value of a security *s* at time *t* for an interest specified by *i*.

- If we invest an amount *c* during *t* years at an annual interest rate of *i*, to find the future value *Mathematica* will apply the following formula:

```
TimeValue[c, i, t]
```

$c\,(1+i)^t$

- How much do we need to invest at 3.5% annual interest rate to get €100,000 in 10 years?

```
Solve[TimeValue[quantity, .035, 10] == 100000, quantity]
```

```
{{quantity → 70891.9}}
```

- We have two options to invest €60,000 for 10 years. Option A offers a fixed annual interest rate of 4% during the 10 years. Option B offers an interest rate of 3% during years 1 and 2, 4 % in years 3 to 5, and 4.5% until the end of the investment period. Which option is the best one?

```
{TimeValue[60000, 0.04, 10] (*product A*),
 TimeValue[60000, {{0, 0.03}, {2, 0.04}, {5, 0.045}}, 10] (*product B*)}
```

```
{88814.7, 89229.2}
```

The function `EffectiveInterest["r", "q"]` returns the effective interest rate corresponding to an interest specification *r*, compounded at time intervals *q*. It is very useful to avoid making mistakes when comparing nominal and effective interest rates. `Annuity["p", "t"]` represents an annuity of fixed payments *p* made over *t* periods. Below we show examples using both functions.

- What is the effective rate of an annual nominal interest rate of 4% compounded quarterly?

```
EffectiveInterest[.04, 1/4]
```

```
0.040604
```

- We'd like to deposit €60,000 in an account with an annual interest of 4% compounded quarterly. How much money will we be able to withdraw every quarter so the money lasts for 10 years?

```
Solve[TimeValue[Annuity[{wd, {-60000}}, 10, 1/4],
   EffectiveInterest[.04, 1/4], 10] == 0, wd]
```

```
{{wd → 1827.34}}
```

- The next set of functions creates a loan amortization table for a given period and interest rate. In this case we borrow €150,062 and make monthly payments of €1,666 during 10 years at 6% nominal annual rate. The same approach can be used with different numbers:

```
years = 10; monthlypayments = 1666; monthlyinterest = 0.06/12;
```

```
loanValue[month_] = TimeValue[
   Annuity[monthlypayments, years*12 - month], monthlyinterest, 0];
```

```
amortizationList = Table[
    {i + 1, loanValue[i], monthlypayments, loanValue[i] *monthlyinterest,
     monthlypayments - loanValue[i] *monthlyinterest,
     loanValue[i + 1]}, {i, 0, years*12 - 1}] // Round;
```

```
amortizationTable[month1_, month2_] :=
 TableForm[amortizationList[[month1 ;; month2]], TableHeadings →
   {None, {"Month", "Beginning\nPrincipal", "Monthly\nPayment",
     "Interest\nPayment", "Principal\nPayment", "Ending\nPrincipal"}}]
```

- Let's see the amortization table during the first 12 months. Try to display the entire period of the loan.

```
amortizationTable[1, 12]
```

Month	Beginning Principal	Monthly Payment	Interest Payment	Principal Payment	Ending Principal
1	150062	1666	750	916	149147
2	149147	1666	746	920	148226
3	148226	1666	741	925	147302
4	147302	1666	737	929	146372
5	146372	1666	732	934	145438
6	145438	1666	727	939	144499
7	144499	1666	722	944	143556
8	143556	1666	718	948	142607
9	142607	1666	713	953	141654
10	141654	1666	708	958	140697
11	140697	1666	703	963	139734
12	139734	1666	699	967	138767

- We can check graphically how the remaining principal goes down over time until it becomes 0 at the end of the loan:

```
ListLinePlot[Table[{i + 1, loanValue[i]}, {i, 0, years*12 - 1}],
 AxesLabel → {"Month", "Remaining principal"}, ImageSize → 300]
```

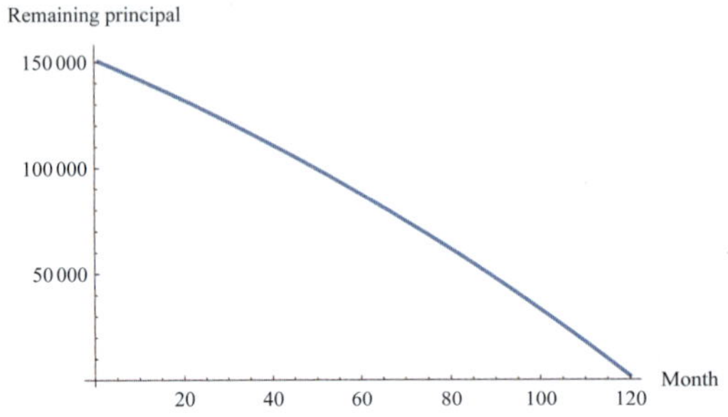

11.2.2 Bonds

```
Clear["Global`*"]
```

Bonds are financial assets that we can also analyze with *Mathematica*. When an organization needs capital, instead of borrowing from banks, they can issue bonds. Bonds can be considered a type of loan in which the lenders are the ones purchasing them. Those lenders can usually be anyone. There are different types of bonds. Normally they pay a fixed amount of interest at predetermined time intervals: annually, half-annually, quarterly or even monthly. Some bonds issued for short periods of time (less than one year), such as *Treasury Bills*, don't pay interest explicitly, they are sold at a discount from their par value.

For analyzing bonds, *Mathematica* has the function `FinancialBond`. It takes the following arguments (some of them are optional):

"FaceValue"	face value, par value
"Coupon"	coupon rate, payment function
"Maturity"	maturity or call/put date
"CouponInterval"	coupon payment interval
"RedemptionValue"	redemption value
"InterestRate"	yield to maturity or yield rate
"Settlement"	settlement date
"DayCountBasis"	day count convention

Let's see some examples.

- The yield to maturity of a €1,000 30-year bond maturing on June 31, 2018, with a coupon of 6% paid quarterly that was purchased on January 6, 2012 for €900 would be:

```
FindRoot[FinancialBond[{"FaceValue" → 1000, "Coupon" → 0.05,
   "Maturity" → {2018, 6, 31}, "CouponInterval" → 1/4},
  {"InterestRate" → y, "Settlement" → {2012, 1, 6}}] == 900, {y, .1}]
```

```
{y → 0.069269}
```

- A zero-coupon bond with a redemption value of €1,000 after 10 years is currently sold for €400. Find the implied yield rate compounded semiannually:

```
FindRoot[FinancialBond[
   {"FaceValue" → 1000, "Coupon" → 0, "Maturity" → 10, "CouponInterval" → 1/2},
   {"InterestRate" → y, "Settlement" → 0}] == 400, {y, .1}]
{y → 0.0937605}
```

11.2.3 Derivatives

```
Clear["Global`*"]
```

In Chapter 41 of the book of Genesis, the Egyptian Pharaoh asks Joseph to interpret a dream he's had about 7 sleek and fat cows followed by 7 ugly and gaunt ones. Joseph interprets the dream as periods of plenty followed by periods of famine and gives the Pharaoh the following piece of advice: "... Let Pharaoh appoint commissioners over the land to take a fifth of the harvest of Egypt during the seven years of abundance. They should collect all the food of these good years that are coming and store up the grain under the authority of Pharaoh, to be kept in the cities for food. This food should be held in reserve for the country, to be used during the seven years of famine that will come upon Egypt, so that the country may not be ruined by the famine". Some people consider this story as the earliest mention of the derivative concept based on the idea of fixing in advance the future price of a certain item. Closer to our current era, on April 3, 1848, the *Chicago Board of Trade* was established to create a marketplace for the buying and selling of several commodities (grains, cattle, hogs, etc.) at fixed prices and following certain standards. Normally, the buyer would pay in advance a small portion of the total amount to the seller. The purpose was clear: the buyer could set in advance how much things were going to cost in the future and the seller could secure a certain selling price; it could be considered as an insurance against uncertainty. Since then, this idea has been applied to many other areas: raw materials, shares, stock indexes, etc.

In finance, a derivative product is one whose price depends on the price of another product (the underlying asset). We say that the price of the former derives from the price of the latter. For example, we can create a derivative based on the price of gold without the need to physically own the gold. In principle, these instruments should play a beneficial economic role. Very often, however, people associate them with speculative transactions more similar to gambling than to uncertainty reduction. There's a kernel of truth is this opinion since investments in these products can generate either large losses or large gains. This is the reason why there have always been investors trying to (and sometimes succeeding) influence the price of the underlying for speculative purposes. Warren Buffett, the most successful investor of the 20th century and one of world's richest men, even described them as "weapons of mass destruction" although it's known that some of his investments take advantage of them, and that if used properly they help to reduce uncertainty.

When considering derivatives as an alternative to purchasing stocks, the two most important things to keep in mind are: i) For the same number of shares, derivatives require a much smaller investment. As a consequence, gains and losses are significantly higher, ii) When the share price goes up, everybody wins (more precisely, the latent value is the same for everybody). With derivatives, when someone wins somebody else loses and the other way around (it's a zero-sum game).

An example of a financial derivative is a stock option, or a well-known variant of it more commonly available to regular investors, a warrant. An option or warrant gives the owner the right to buy (*Call*) or sell (*Put*) the underlying asset at a given price usually determined at the beginning of the contract. Both have prices (premiums) that are normally much smaller than the stock associated to them. For example, we could buy a warrant to purchase Telefónica shares in a year for €15 each and pay €0.5 for it. At expiration, if the shares were trading above €15, let's say €16.5, we would make a profit of €1 per share, after subtracting the €0.5 initially paid for it (it's common to just settle the differences, without actual delivery of the underlying asset). However, if at expiration Telefónica was trading below €15, we would just lose the premium paid.

An informative overview of financial derivatives and their statistical measurement can be found in:

https://www.imf.org/external/pubs/ft/wp/wp9824.pdf.

The list below contains some of the most commonly used terms when describing derivatives transactions:

Underlying asset: the financial instrument on which a derivative's price is based.
Call option. An agreement that gives the buyer the right (but not the obligation) to buy the underlying asset at a certain time and for a certain price from the seller.
Put option. An agreement that gives the buyer the right (but not the obligation) to sell the underlying asset at a certain time and for a certain price to the seller.
Premium. The amount paid for the right to buy or sell the underlying asset.
Market price. The current price at which the underlying asset can be bought or sold.
Strike. The price at which the agreement can be exercised.

The Greek letters delta, gamma, theta and vega are used to describe the risk of changes in the underlying asset.

Delta: It's the option sensitivity to changes in the underlying asset. It indicates how the price of the option changes when the underlying price goes up or down. For an option at the money (strike price same as market price), delta is usually around 50%. This means that if the price of the underlying asset changes by €2, the price of the option will change by €1.
Gamma: It measures the sensitivity of delta to price changes in the underlying asset. Since delta is not a linear function, gamma is not linear either. Gamma decreases as the uncertainty increases.
Vega: This metric indicates how sensitive the option price is to volatility changes in the underlying asset. It represents the change in the option price if the volatility of the underlying asset changes by 1%. When volatility goes up, the price of the option also goes up. Vega may vary depending on whether the option is at the money, in the money or out of the money.
Theta: A measure of the decrease in the value of an option as a result of the passage of time (easy to remember since both start with t). The longer the time period until exercise, the higher Theta is: There's more time for the underlying asset's price to move in a direction favorable for the option holder.

Mathematica gives us a very wide array of options when using the function:

`FinancialDerivative`
["instrument", "parameters", "ambient parameters", "property (optional)"] returns the value of the specified financial instrument, computed for the stated *property*.

To make the valuation examples in this section as realistic as possible we will need certain market data.

Often, to make the purchase of an option cheaper, we can sell another one in such a way that we limit our potential gains but we also reduce the premium paid. This is the idea behind *zero cost* strategies.

An example of this strategy would be as follows: We purchase an option with a strike price equal to the exercise price (an *at the money* option) and sell another one with a higher strike price, let's suppose 18% higher. Then, this figure will be our maximum return. Obviously, the higher the potential return, the higher the potential risk and the other way around. This strategy is known as a *Call Spread*.

- In this case we use Telefónica shares trading in the NYSE (the data and functions used in the example come from Sebastián Miranda from Solventis). The stock prices for the Spanish stock exchange (IBEX) are currently not available.

```
FinancialData["NYSE:TEF", "Name"]
```

```
Telefonica SA
```

- We're going to need the 12-month EURIBOR rate. This type of interest rate is published by Euribor EBF (http://euribor-rates.eu). Let's import the 10 most recent rates (remember that in Chapter 2 we explained how to extract specific parts of a web page) as follows:

```
Import["http://www.euribor-rates.eu/euribor-rate-12-months.asp", "Data"][[
 2, 4, 1, 2, 3]]
```

```
{Rate on first day of the year,
 {{01-04-2016, 0.058%}, {01-02-2015, 0.323%},
  {01-02-2014, 0.555%}, {01-02-2013, 0.543%}, {01-02-2012, 1.937%},
  {01-03-2011, 1.504%}, {01-04-2010, 1.251%}, {01-02-2009, 3.025%},
  {01-02-2008, 4.733%}, {01-02-2007, 4.030%}}}
```

- However, unless we're creating a program to automate the process, probably the easiest thing to do is to copy the values directly. Besides, it's unusual for websites to keep their structure for a long time so, by the time you read these lines, the instruction above may not work. We'll store in memory the first rate for 2016 that we will use for our calculations later on.

```
euribor12m = 0.058 / 100
```

```
0.00058
```

- For our computations, we need to know several indicators associated to the stock. The values returned are time sensitive and depend on the time of execution of the query. We show the commands below so readers can replicate the example. Notice that after execution, the predictive interface gives us the possibility of making other queries (Figure 11.2).

```
FinancialData["NYSE:TEF", "Price"] (* stock price,
in USD, in NYSE at the time of the query*)
```

```
9.87
```

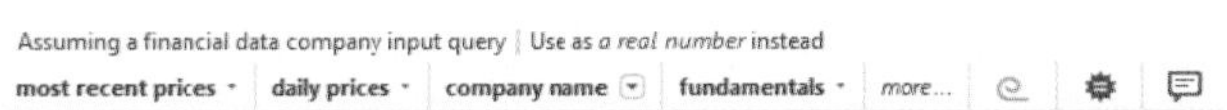

Figure 11.2 Predictive interface for financial data.

```
spotPrice = 9.87;
```

- Stock volatility during the past 50 days. (Volatility is a measure of the variability of the stock price over a specific time horizon. It's usually the standard deviation of the closing stock price during the last 50 trading sessions, or an alternative chosen time horizon):

```
FinancialData["NYSE:TEF", "Volatility50Day"]
```

```
0.528374
```

```
vol = 0.528374;
```

- Last year's dividend yield, computed as the ratio between the dividends paid and the current stock price.

```
FinancialData["NYSE:TEF", "DividendYield"]
```

```
0.0904
```

```
div = 0.0904;
```

- Now we can evaluate the price of a call option given its expiration date (in this case, in one year), and using the current stock price as the strike price:

```
FinancialDerivative[{"American", "Call"},
 {"StrikePrice" → spotPrice, "Expiration" → 1},
 {"InterestRate" → euribor12m , "Volatility" → vol,
  "CurrentPrice" → spotPrice, "Dividend" → div}]
```

```
1.67404
```

- In the next graph, we can see the evolution of the value of our call spread as a function of the underlying asset price and the remaining time. To make it easier to visualize, the solid green line indicates the current price, our strike price; and the dashed red line, the price of the underlying asset at which we maximize our profit. We need to keep in mind that our starting point in the time scale is 1. Remember that our goal is to have a maximum gain of 18%.

```
data =
  {#, 1 - #2, FinancialDerivative[{"American", "Call"}, {"StrikePrice" →
        spotPrice, "Expiration" → 1}, {"InterestRate" → euribor12m ,
       "Volatility" → 0.5, "CurrentPrice" → #, "Dividend" → div,
       "ReferenceTime" → #2}] - FinancialDerivative[{"American", "Call"},
      {"StrikePrice" → 1.18 * spotPrice, "Expiration" → 1},
      {"InterestRate" → euribor12m , "Volatility" → 0.5,
       "CurrentPrice" → #, "Dividend" → div, "ReferenceTime" → #2}]} & @@@
   Flatten[Table[{price, referencetime}, {referencetime, 0, 1, 0.05},
     {price, 0.8 spotPrice, 1.4 * spotPrice, 1}], 1];
```

```
ListPlot3D[data,
 (* Contours *)
 Mesh → {{{spotPrice, {Green, Thick}},
    {1.18 * spotPrice, {Red, Thick, Dashed}}}},
 (* Formats *)
 AxesLabel →
  (Style[#, 12] & /@ {"Underlying Price", "Remaining Time", "Profit"}),
 ColorFunction → ColorData["LightTemperatureMap"],
 ColorFunctionScaling → True]
```

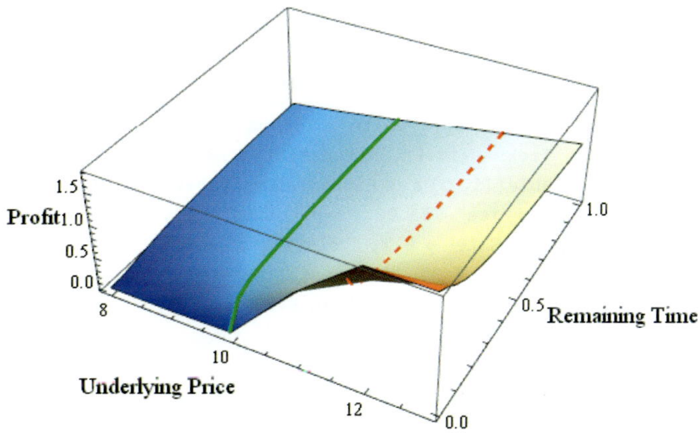

- At expiration, the profile of our profit as a function of the underlying asset price is:

```
ListPlot[Cases[data, {x_, 0., y_} → {x, y}],
 Joined → True, PlotStyle → Directive[Thickness[0.02], Red],
 AxesLabel → {"Underlying Price", "Profit"}]
```

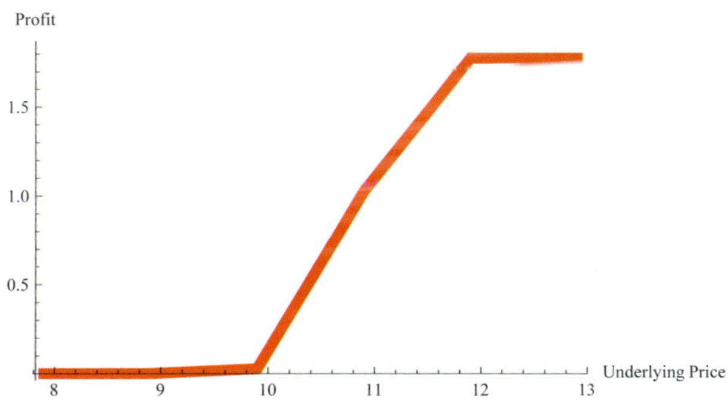

In this example, the warrants that we've used are of the American type, meaning that they can be exercised at any time until the expiration date, although it is not usually optimum to do so.

However, *Mathematica,* in addition to financial derivatives price calculations, can also compute their sensitivities, known as *Greeks*. These sensitivities are the partial derivatives of the option price with respect to different valuation parameters. For example, the *delta* of an option is the partial derivative of its price with respect to the price of the underlying asset. This measure not only gives us an idea of how sensitive the option value is to small changes in the price of the underlying asset but also, for standard options (also known as *vanilla*), it gives us the probability that the option will be exercised. G*amma,* the second derivative with respect to the underlying asset price, tells us how sensitive the *delta* is to changes to it. This metric is very important for hedging purposes. *Rho* is the partial derivative with respect to the interest rate and in the case of stock options is the least sensitive Greek. *Theta* is the derivative with respect to time to expiration. Initially, since options are not usually bought *in the money i.e.* if Telefónica trades at €12, purchasing an option for €10 is not very common, they don't have intrinsic value. They only get value from their time component. The final one, *Vega*, is the partial derivative with respect to the volatility of the underlying asset. This is such an important parameter in option valuation that for certain options, their prices are quoted in terms of the volatility of their respective underlying assets.

- The graph below shows how the Greeks change with respect to the remaining time to expiration:

```
ticks[min_, max_] := Table[
  {j, ToString[Round[12 - j 12]] <> " Months", {.03, 0}}, {j, min, max, 0.2}]
```

```
TabView[Rule @@@ Transpose[{{"Delta", "Gamma", "Rho", "Theta", "Vega"},
    ListPlot[#, Ticks → {ticks, Automatic}] & /@
     Transpose[Table[{i, #} & /@ FinancialDerivative[{"American", "Call"},
         {"StrikePrice" → 55.00, "Expiration" → 1},
         {"InterestRate" → 0.1, "Volatility" → 0.5, "CurrentPrice" → 50,
          "Dividend" → 0.05, "ReferenceTime" → i},
         "Greeks"][[All, 2]], {i, 0, 0.95, 0.05}]]}]]
```

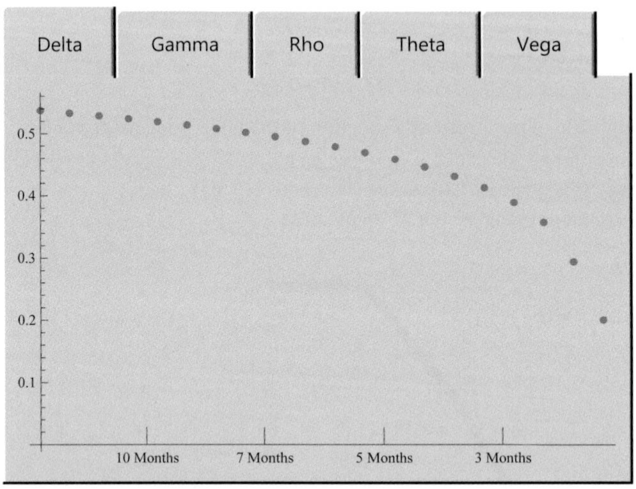

The same Greek letters are used to indicate the risks of portfolio of options due to changes in the price of the underlying assets or market conditions.

Delta: the delta of a portfolio of options with the same underlying assets is the weighted average of the delta of the individual options.

The delta factor = *the change in the price of the option as a result of changes in the price of the underlying assets*. For an at-the-money option, the delta factor is usually close to 50%, so if the underlying price changes by €1, the price of the option will change by €0.5.

11.2.4 The Black–Scholes Equation

In this example we first compute the price of a derivative using *Mathematica*'s existing function and then compare it with the result obtained by solving the Black–Scholes equation directly

(http://www.wolfram.com/language/11/partial-differential-equations/find-the-value-of-a-european-call-option.html).

- Let's start by using FinancialDerivative to compute the price of a European vanilla call option with the data from the previous section:

```
spotPrice = 9.87; euribor12m = 0.00058; vol = 0.528374;
```

```
FinancialDerivative[{"European", "Call"}, {"StrikePrice" -> spotPrice,
  "Expiration" -> 1}, {"InterestRate" -> euribor12m, "Volatility" -> vol,
  "CurrentPrice" -> spotPrice}]
```

```
2.05882
```

- Now, let's do the same computation using the Black–Scholes equation (To create more visually appealing documents you can convert the input cell to the traditional format: **Cell ▶ Convert to ▶ TraditionalForm**):

BlackScholesModel =

$$\left\{-r\, c(t,\, s) + r\, s\, \frac{\partial c(t,\, s)}{\partial s} + \frac{1}{2} s^2 \sigma^2 \frac{\partial^2 c(t,\, s)}{\partial s^2} + \frac{\partial c(t,\, s)}{\partial t} = 0,\; c(T,\, s) = \max(s - k,\, 0)\right\};$$

- After solving the boundary value problem we get:

```
(dsol = c[t, s] /.
  DSolve[BlackScholesModel, c[t, s], {t, s}][[
   1]]) // TraditionalForm
```

$$\frac{1}{2} e^{-r\,T} \left(s\, e^{r\,T} \operatorname{erfc}\!\left(\left(2 \log(k) + \left(2\, r + \sigma^2\right)(t - T) - 2 \log(s)\right) / \left(2 \sqrt{2}\, \sigma \sqrt{T - t}\right)\right) - k\, e^{r\,t} \operatorname{erfc}\!\left(\left(2 \log(k) + \left(2\, r - \sigma^2\right)(t - T) - 2 \log(s)\right) / \left(2 \sqrt{2}\, \sigma \sqrt{T - t}\right)\right)\right)$$

- Finally, we calculate the price of the derivative and compare it with the one previously obtained. We can see it's the same:

```
dsol /. {t -> 0, s → spotPrice, k -> spotPrice, σ -> vol, T -> 1,
  r -> euribor12m}
```

```
2.05882
```

11.2.5. Basel II Capital Adequacy: Internal Ratings-Based (IRB) Approach

The key cornerstone of prudential regulation of banks (and financial institutions in general) is to ensure that each bank holds sufficient equity capital to absorb unexpected losses, that is, the materialization of financial risks, mainly market (price), credit, and operational risks.

The example below, *Basel II Capital Adequacy: Internal Ratings-Based (IRB) Approach,* by Poomjai Nacaskul shows how to calculate, using the Internal Ratings-Based approach, the minimum amount of capital that a bank should keep following the standards defined by the Basel II regulatory framework on bank capital adequacy.

http://demonstrations.wolfram.com/BaselIICapitalAdequacyInternalRatingsBasedIRBApproach

- To facilitate the understanding of the example, we also include the code:

```
(*Setting defaults*)
DefaultEAD = 1000000;
MinimumPD = 0.0001; MaximumPD = 0.25;
MinimumLGD = 0.01; DefaultLGD = 0.7; MaximumLGD = 1.0;
MinimumMaturity = 1 / 250;
DefaultMaturity = 1;
MaximumMaturity = 30;

(* Create shorthands for Normal CDF & CDF inverse *)
Phi[x_] := CDF[NormalDistribution[0, 1], x];
PhiInverse[p_] := InverseCDF[NormalDistribution[0, 1], p];
```

```
(* Correlation Function, Basel Committee's way of parameterising a 2-
 parameter formula (PD, rho) with 1 parameter (PD). *)
R[PD_] := 0.12 * (1 - Exp[-50 PD]) / (1 - Exp[-50]) +
   0.24 * (1 - (1 - Exp[-50 PD]) / (1 - Exp[-50]));
ConditionalPercentageExpectedLoss[PD_, LGD_] := Phi[
   (PhiInverse[PD] + Sqrt[R[PD]] PhiInverse[0.999]) / Sqrt[1 - R[PD]]] * LGD
UnconditionalPercentageExpectedLoss[PD_, LGD_] := PD * LGD;
(* Maturity Adjustment, with a 1-year maturity as a base case *)
b[PD_] := (0.11852 - 0.05478 * Log[PD]) ^ 2;
MaturityAdjustment[Maturity_, PD_] :=
  (1 + (Maturity - 2.5) * b[PD]) / (1 - 1.5 * b[PD]);
(* Capital Requirement Calculation: note as PD --> 1,
this quantity --> 0, as credit risk is taken care of via provisioning,
i.e. E[Loss] = exposure x PD x LGD. *)
CapitalRequirement[PD_, LGD_, Maturity_] :=
  (ConditionalPercentageExpectedLoss[PD, LGD] -
     UnconditionalPercentageExpectedLoss[PD, LGD]) *
   MaturityAdjustment[Maturity, PD];
Manipulate[
 Pane[Column[{Text@Style["Basel II Capital Adequacy", Bold, 16],
    Text@"Regulatory (minimal) capital for financial institutions
       per $ exposure,\nas per Internal Ratings-Based
       Approach (IRB) Risk Weight Function (RWF)\n",
    Grid[{{Plot[CapitalRequirement[PD, LGD, Maturity], {PD, 0.0001, 0.25},
        PlotLabel -> "as a function of Probability of Default (PD)\nat
           a given Maturity (M)", ImageSize → 320],
       Plot3D[CapitalRequirement[PD, LGD, Maturity], {PD, 0.0001, 0.25},
        {Maturity, 1 / 250, 30}, PlotLabel → "as a function of PD and M\n",
        ImageSize → 280, Mesh → None, ColorFunction →
         Function[{x, y, z}, ColorData["GrayTones"][z]]]}}]},
   Alignment → Center], ImageSize → {610, 350}],
 {{LGD, 0.7, "Loss Given Default (LGD)"}, 0.01, 1, Appearance → "Labeled"},
 {{Maturity, 1, "Maturity M in years (left graph only)"},
  {1 / 250, 1 / 50, 1 / 12, 1 / 4, 1, 2, 5, 10, 15, 30},
  ControlType → RadioButton}, SaveDefinitions → True]
```

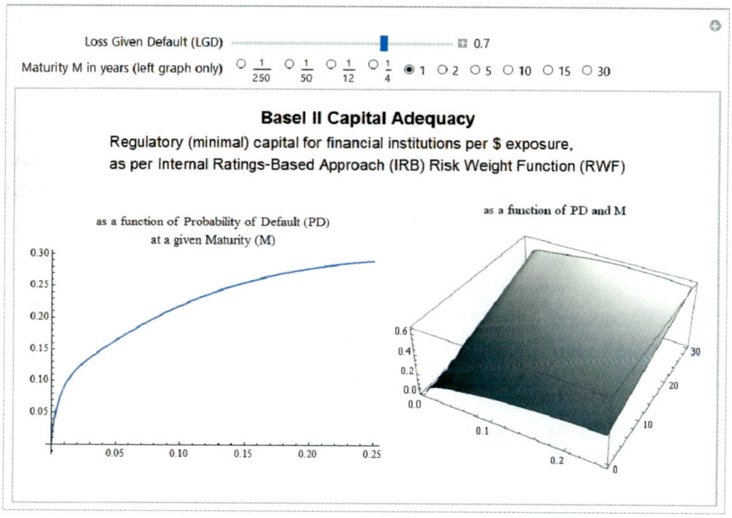

To learn more about financial modeling and risk management using *Mathematica*, Igor Hlivka has created several excellent example notebooks. You can find them in:

http://community.wolfram.com/web/ihcomm.

11.3 Optimization

11.3.1 What is Constrained Optimization?

The optimization problems that we discuss in this section are related to finding an optimum (maximum or minimum, depending on the case) for an objective function of n variables, given r constraints.

This type of problems is common in several fields such as economics, finance and engineering as we will show using different examples. You can find an excellent tutorial in the documentation pages: "Introduction to Constrained Optimization in the Wolfram Language" (tutorial/ConstrainedOptimizationIntroduction).

To solve them, *Mathematica* has the following functions:

Function	Application	Description
FindMinimum / FindMaximum	Local Numerical optimization	Linear programming methods, non - linear, interior points, and the use of second derivatives
NMinimize / NMaximize	Global Numerical optimization	Linear programming methods, Nelder - Mead, *differential evolution*, *simulated annealing*, random search
Minimize / Maximize	Exact Global optimization	Linear programming methods, algebraic cylindrical decomposition, Lagrange multipliers and other analytical methods, integer linear programming.
LinearProgramming	Linear optimization	Simplex, modified simplex, interior point

The following tree can help us decide the most appropriate function to find the minimum (To find the maximum we would use the functions `Maximize`, `NMaximize` or `FindMaximum` instead).

- The command below draws the tree. When creating a document, you may be interested in seeing only the result. To hide the input just click directly on the output marker cell.

```
TreePlot[{{"Is your problem linear?" → "LinearProgramming", "Yes"},
   {"Is your problem linear?" → "Do you want a global optimum?", "No"},
   {"Do you want a global optimum?" → "Do you want an exact solution?",
    "Yes"}, {"Do you want a global optimum?" → "Is your problem small?",
    "No"}, {"Is your problem small?" → "NMinimize", "Yes"},
   {"Is your problem small?" → "FindMinimum", "No"},
   {"Do you want an exact solution?" → "Minimize", "Yes"},
   {"Do you want an exact solution?" → "NMinimize", "No"}},
 Automatic, "Is your problem linear?", DirectedEdges -> True,
   VertexLabeling → True, VertexCoordinateRules ->
    {{0.6358, 2.049}, {-0.26, 1.3}, {1.06, 1.3}, {0.448, 0.54},
     {1.78, 0.54}, {0.76, -0.208}, {1.66, -0.208}, {0, -0.208}}] /.
  ((# → Button[TextCell[#], Inherited, BaseStyle → Dynamic[If[
           CurrentValue["MouseOver"], {"Link", FontColor →
             RGBColor[0.854902, 0.396078, 0.145098]}, {"Link"}]],
        ButtonData -> "paclet:ref/" <> #, Appearance → "None"]) & /@
    {"LinearProgramming", "NMinimize", "FindMinimum", "Minimize"})
```

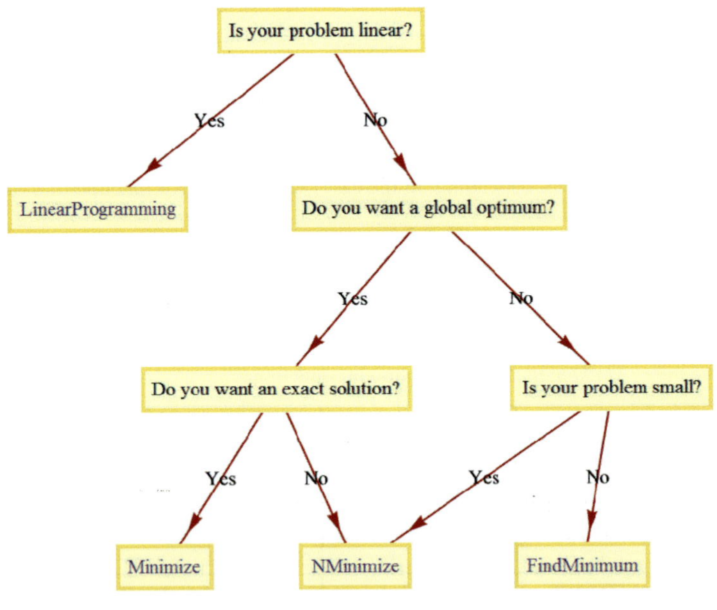

In many cases the problem can be solved using several functions. With the exception of `LinearProgramming`, the syntax for the rest of the functions is very similar: **f**[{objective function, *constraints*}, {*variables*}].

```
Clear["`Global`*"]
```

Let's see some examples.

- Given the objective function (ob) with constraints c1 and c2:

```
var = {x, y}, ob = x + y, c1 = 0 ≤ x ≤ 1, c2 = 0 ≤ y ≤ 2
```

- The maximum can be found using any of the previously mentioned commands:

```
{NMaximize[{ob,c1,c2}, var], Maximize[{ob,c1,c2}, var],
FindMaximum[{ob,c1,c2}, var]}
{{3., {x→1., y→2.}}, {3, {x→1, y→2}}, {3., {x→1., y→2.}}}
```

- We can see that the maximum occurs when {x → 1, y → 2} and is 3. We can verify it (remember that "/." is used for substitutions).

```
x + y /. {x→1, y→2}
3
```

- The minimum can be computed as follows:

```
Minimize[{ob,c1,c2}, var]
{0, {x→0, y→0}}
```

- We can interpret the problem graphically by drawing the plane $x + y$ only in the region: $\{0 \le x \le 1, 0 \le y \le 2\}$.

```
Plot3D[x + y, {x, -1, 2}, {y, -1, 2},
 RegionFunction -> Function[{x, y}, 0 ≤ x < 1 && 0 ≤ y < 2],
 AxesLabel → Automatic, Mesh → None]
```

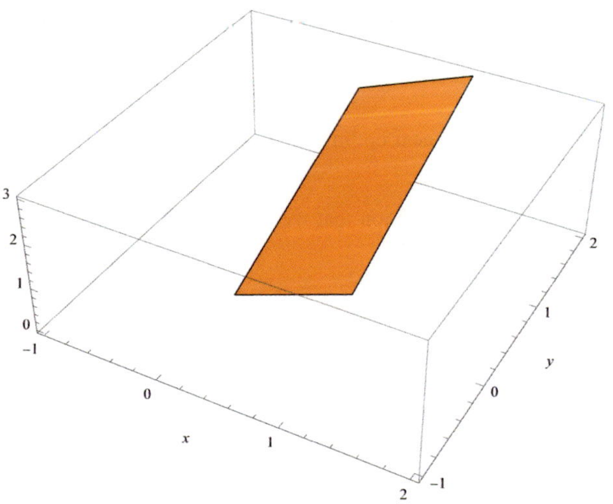

The graph clearly shows that the maximum is in (1,2) and the minimum in (0,0). Use the mouse to move the graph around to see it from different perspectives.

In certain situations we'd like the variables to take only integer values. To add this extra requirement in *Mathematica,* we use the built-in symbol **Integers**.

- Let's consider the following model in which we want to minimize the objective function (ob) (notice that equality when typing constraints is "=="):

```
Clear["`Global`*"]
```

```
var = {x, y};
ob = x + 2 y;
c1 = - 5 x + y == 7;
c2 = x + y ≥ 26;
c3 = x ≥ 3;
c4 = y ≥ 3;
NMinimize[{ob, c1, c2, c3, c4, var ∈ Integers}, var]
```

```
{58., {x → 4, y → 27}}
```

Since this is a linear problem, we can also use `LinearProgramming`. As a matter of fact, this command is the most appropriate one for linear problems, especially if the number of variables is large. The syntax is: `LinearProgramming`[*c*, *m*, *b*] . This function finds the vector **x** that minimizes the quantity **c.x** subject to the constraints **m.x** ≥ **b** and **x** ≥ 0. We can limit the values that the variables (some or all of them) can take to just integers with `LinearProgramming`[..., Integers].

- For comparison purposes, let's solve the same problem using `LinearProgramming`:

$x + 2y$	→	c: {1, 2}	
$-5x + y == 7$	→	m1 : {−5, 1}	b1 : {7, 0}
$x + y \geq 26$	→	m2 : {1, 1}	b2 : {26, 1}

Notice that the syntax to indicate the type of constraint is as follows: $\{b_i, 0\}$ if $m_i.x == b_i$; $\{b_i, 1\}$ if $m_i.x \geq b_i$ and $\{b_i, -1\}$ if $m_i.x \leq b_i$.

```
LinearProgramming[{1, 2}, {{-5, 1}, {1, 1}}, {{7, 0}, {26, 1}},
  {{3, Infinity}, {4, Infinity}}, Integers] // Quiet
```

```
{4, 27}
```

Next we'll show some examples for nonlinear optimization.

```
Clear["`Global`*"]
```

- In this problem, both, the objective function that we want to minimize and the constraints are nonlinear.

```
var = {x, y};
ob = x^2 + (y - 1) ^2;
c1 = x^2 + y^2 ≤ 4;
```

- Let's calculate the global minimum with `Minimize`:

```
Minimize[{ob, c1}, var]
```

```
{0, {x → 0, y → 1}}
```

- `NMinimize` can solve this type of problem faster (although in this example we wouldn't be able notice any difference). We add `Chop` to remove, when the solution is 0, the spurious values that may appear in the solution when using numerical approximations. Remove `Chop` from the command below and see what happens.

```
NMinimize[{ob, c1}, var] // Chop
```

```
{0, {x → 0, y → 1.}}
```

- If we are looking for a local minimum we can use `FindMinimum`. Although it's not necessary, this function is much faster if we enter an initial point from which to start the search. It's also a good idea to add `Chop` .

```
FindMinimum[{ob, c1}, {{x, 0}, {y, 0}}] // Chop
{0, {x → 0, y → 1.}}
```

- We present the previous problem graphically with `Plot3D` and `ContourPlot`, very convenient functions when trying to find optimum points visually. (The option `Contours` specifies how many contours we want to display; if not specified, *Mathematica* will choose the optimum number for the given function).

```
GraphicsRow[{Plot3D[ x^2 + (y - 1) ^2, {x, -3, 3},
   {y, -3, 3}, RegionFunction -> Function[{x, y}, x^2 + y^2 ≤ 4],
   AxesLabel → Automatic, Mesh → All], ContourPlot[x^2 + (y - 1) ^2,
   {x, -3, 3}, {y, -3, 3}, RegionFunction → Function[{x, y}, x^2 + y^2 ≤ 4],
   Contours → 100, FrameLabel → Automatic]}]
```

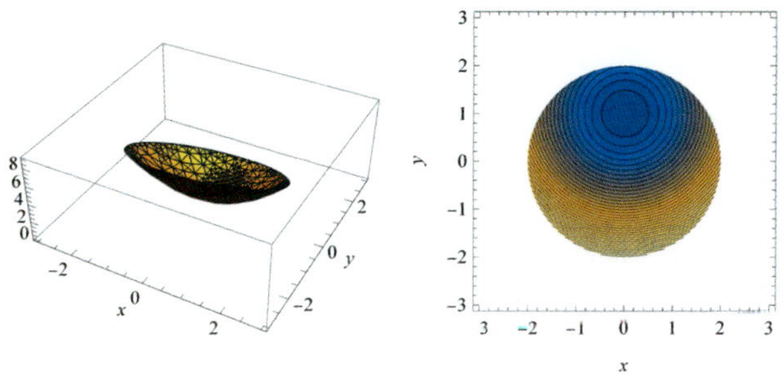

Click inside the output from `Plot3D` and move the graph around to see the minimum from different angles. In the contour plot, we can see that the minimum corresponds to the coordinates (0, 1). By default, dark colors are associated to small values and light hues are linked to bigger ones. If you place your cursor over the contour limits, you'll see the value that the function takes at that location. (All the points connected by the same line have the same function value).

- Here is another example of nonlinear optimization, in this case for finding the global minimum (`Minimize`). The objective function is Exp($-x\,y$) with the constraint that $x,\ y \in$ a Circle centered in $\{0, 0\}$ and with a radius $r = 1$. We show the result, both numerically and graphically.

```
m = Minimize[{Exp[-x y], {x, y} ∈ Disk[]}, {x, y}]
```

$$\left\{\frac{1}{\sqrt{e}},\ \left\{x \to -\frac{1}{\sqrt{2}},\ y \to -\frac{1}{\sqrt{2}}\right\}\right\}$$

```
g = Plot3D[Exp[-x y], {x, y} ∈ Disk[], Axes → True,
   Boxed → False, PlotStyle → Opacity@0.6, AxesLabel → Automatic];
```

```
Show[g,
 Graphics3D[{PointSize@Large, Red, Point[{x, y, m[[1]]} /. m[[2]]]}]]
```

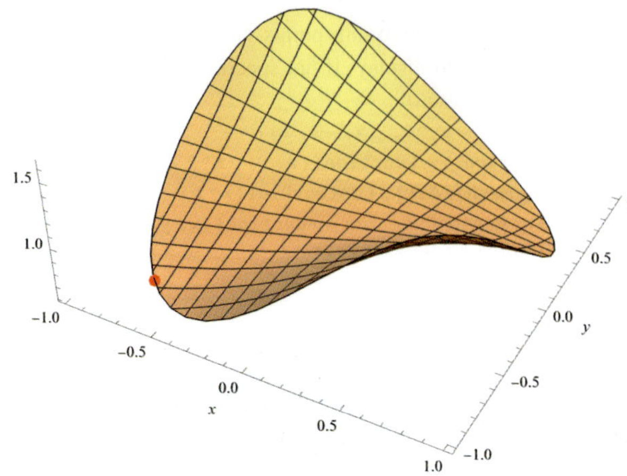

- The following example shows how the optimization algorithm approaches the optimum (minimum) point for f(x,y) in an iterative way.

```
Clear["Global`*"]

f[x_, y_] := (x - 1)^2 + 2 (y - x^2)^2

Module[{pts}, pts = Last[Last[Reap[Sow[{-1.2, 1, f[-1.2, 1]}];
     FindMinimum[f[x, y],
      {{x, -1.2}, {y, 1}}, StepMonitor :> Sow[{x, y, f[x, y]}]]]]];
 Show[Plot3D[f[x, y], {x, -1.4, 1.4}, {y, -0.6, 2}, Mesh → 5,
   PlotStyle → Automatic, PlotLabel → Style[Row[{"Finding the minimum of ",
       f[x, y], " using iteration"}], 12, Bold], AxesLabel → Automatic],
  Graphics3D[{Orange, PointSize[0.025], Point[Rest[Most[pts]]],
    Thick, Line[pts], Darker@Green, PointSize[0.05],
    Point[Last[pts]], Red, Point[First[pts]]}]]]
```

Finding the minimum of $2(y-x^2)^2+(x-1)^2$ using iteration

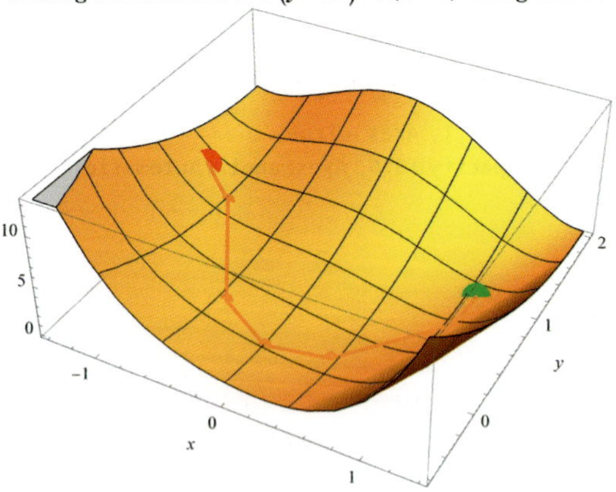

11.3.2 Application: Supplying Construction Sites

We'd like to optimize the supply of bricks from warehouses to construction sites. There are two brick warehouses $j = \{1,2\}$, where we have in stock 20 and 25 tonnes respectively. From them we supply 6 sites $i = \{1, 2, 3, 4, 5, 6\}$ with a daily demand in tonnes of: {3, 5, 4, 7, 6, 11}. We assume that the location (latitude and longitude) of the warehouses and sites is:

```
warehouses = {{7, 1}, {2, 9}};
```

```
sites = {{1.25, 1.25}, {8.75, 0.75},
   {0.5, 4.75}, {5.75, 5}, {3, 6.5}, {7.25, 7.75}};
```

We'd like to minimize the total amount of bricks transported per unit of distance, $\sum_{i,j} x_{i,j}\, d_{i,j}$, meeting the demand without exceeding the stock available in the warehouses.

For that purpose, we have to calculate the minimum of $\sum_{i,j} x_{i,j}\, d_{i,j}$, where $x_{i,j}$ corresponds to the quantities transported from the warehouses i:{1, 2} to the sites j:{1, ..., 6} with $d_{i,j}$ representing the distance between warehouse i and site j.

- To calculate the distances between the warehouses and the sites we use the function `EuclideanDistance` (in the bi-dimensional case, it's the equivalent to the Pythagorean theorem):

```
EuclideanDistance[{x1, y1}, {x2, y2}]
```

$\sqrt{\left(\text{Abs}[x1 - x2]^2 + \text{Abs}[y1 - y2]^2\right)}$

- The command below returns triplets, {i, j, $d_{i,j}$}, indicating the distance in km between the warehouse i and site j. We use `Round[value, 0.1]` to round the distance to the nearest first decimal.

```
Table[{i, j, Round[EuclideanDistance[warehouses[[i]], sites[[j]]], 0.1]},
 {i, 1, 2}, {j, 1, 6}]
```

```
{{{1, 1, 5.8}, {1, 2, 1.8}, {1, 3, 7.5}, {1, 4, 4.2},
  {1, 5, 6.8}, {1, 6, 6.8}}, {{2, 1, 7.8}, {2, 2, 10.7},
  {2, 3, 4.5}, {2, 4, 5.5}, {2, 5, 2.7}, {2, 6, 5.4}}}
```

- Therefore, the variables and the objective function, $\sum_{i,j} x_{i,j}\, d_{i,j}$, are:

```
var = {x11, x12, x13, x14, x15, x16, x21, x22, x23, x24, x25, x26};
```

```
of = 5.8 x11 + 1.8 x12 + 7.5 x13 + 4.2 x14 + 6.8 x15 +
   6.8 x16 + 7.8 x21 + 10.7 x22 + 4.5 x23 + 5.5 x24 + 2.7 x25 + 5.4 x26;
```

- The constraints are: $\sum_i x_{i,j} = \text{demand}_j$, $\sum_j x_{i,j} = \text{stock}_i$ and $x_{i,j} \geq 0$

```
constraints = {x11 + x21 == 3 && x12 + x22 == 5 && x13 + x23 == 4 && x14 + x24 == 7 &&
    x15 + x25 == 6 && x16 + x26 == 11 && x11 + x12 + x13 + x14 + x15 + x16 ≤ 20 &&
    x21 + x22 + x23 + x24 + x25 + x26 ≤ 25 && x11 ≥ 0 && x12 ≥ 0 && x13 ≥ 0 && x14 ≥ 0,
   x15 ≥ 0 && x16 ≥ 0 && x21 ≥ 0 && x22 ≥ 0 && x23 ≥ 0 && x24 ≥ 0 && x25 ≥ 0 && x26 ≥ 0};
```

- The solution is:

```
NMinimize[{of, constraints}, var]
```

```
{149.4, {x11 → 3., x12 → 5., x13 → 0., x14 → 7., x15 → 0.,
  x16 → 0., x21 → 0., x22 → 0., x23 → 4., x24 → 0., x25 → 6., x26 → 11.}}
```

This result indicates that the optimum solution, 149.4 tonnes×km, is obtained by transporting 3 tonnes from warehouse 1 to site 1, 5 tonnes from warehouse 1 to site 2 and so on.

In "Optimal Transport Scheduling", a demonstration by Yifan Hu, you can see a dynamic visualization of the resolution of this problem:
http://demonstrations.wolfram.com/OptimalTransportScheduling
If in the demonstration you use 20 and 25 tonnes, you'll see that the solution is the same (there's actually a small difference due to rounding).

11.3.3 Application: Portfolio Optimization

A bank commissions a consulting firm to analyze the most profitable way to invest €10,000,000 for two years given the options in the table below. The table includes the expected rate of return (bi-annual) and associated risk.

Product (i)	Return (%), b_i	Risk (%) r_i
Mortgages	9	3
Mutual funds	12	6
Personal loans	15	8
Commercial loans	8	2
Certificates / Bonds	6	1

The capital not invested in any of the products will be placed in government bonds (assumed to be riskless) with a bi-annual rate of return of 3%. The objective of the consulting firm is to allocate the capital to each of the products to meet the following goals:

(a) Maximize the return per € invested.
(b) Keep the possibility of loss to a maximum of 5% of the total amount invested.
(c) Invest at least 20% in commercial loans.
(d) Allocate to mutual funds and personal loans an amount no larger than the one invested in mortgages.

- Variables: x_i is the percentage of capital invested in product i. The amount placed in government bonds can be considered as a new product with an expected return, in percentage, $b_6 = 3$ and risk $r_6 = 0$. Therefore:

```
var = {x1 (*Mortgages*), x2 (*Mutual funds*),
   x3 (*Personal loans*), x4 (*Commercial loans*), x5
   (*Certificates/Bonds*), x6 (*Government debt*)};
```

- Objective function to maximize: $\sum_{i=1}^{6} b_i x_i$.

```
of = 9 x1 + 12 x2 + 15 x3 + 8 x4 + 6 x5 + 3 x6;
```

- Constraint 1: All the money is invested: $\sum_{i=1}^{6} x_i = 1$.

```
c1 = x1 + x2 + x3 + x4 + x5 + x6 == 1;
```

- Constraint 2: The average risk is $R = \sum_{i=1}^{6} r_i x_i / \sum_{i=1}^{6} x_i \leq 5$ as $r_6 = 0$ then:

```
c2 = (3 x1 + 6 x2 + 8 x3 + 2 x4 + x5) / (x1 + x2 + x3 + x4 + x5) ≤ 5

(3 x1 + 6 x2 + 8 x3 + 2 x4 + x5) / (x1 + x2 + x3 + x4 + x5) ≤ 5
```

- Constraint 3: At least 20% of the capital has to be invested in commercial loans ($x_4 \geq 0.2$).

```
c3 = x4 ≥ 0.2;
```

- Constraint 4: The percentage invested in mutual funds (x_2) and personal loans (x_3) cannot be bigger than the one invested in mortgages (x_1).

```
c4 = x2 + x3 ≤ x1;
```

- Constraint 5: No percentage invested can be negative ($x_i \geq 0$).

```
c5 = Map[# ≥ 0 &, var];
```

- Using NMaximize, *Mathematica* tells us that we'd get a return of 11.2% investing 40% in mortgages, 40% in personal loans and 20% in commercial loans.

```
sol = NMaximize[{of, c1, c2, c3, c4, c5}, var]
{11.2, {x1 → 0.4, x2 → 0., x3 → 0.4, x4 → 0.2, x5 → 0., x6 → 0.}}
"€" <> ToString@AccountingForm[10000000 sol[[1]] / 100, DigitBlock → 3]
€1,120,000.
```

11.3.4 Application: Calculating Cod Consumption

```
Clear["Global`*"]
```

Many problems in economics involve Cobb Douglas functions: $F(x, y) = A\ x^a y^b$ with $x,\ y \geq 0$ and A, a, b constants. Let's see one example:

The cod consumption in certain region follows the function $c(p,\ i) = 1.3\ p^{-1.5}\ i^{0.01}$ where p is the price per kg and i is the daily income in €/day. The previous expression is valid for daily incomes between €20 and €200 per day and the price per kg varies between €4 and €12. What would be the maximum and minimum consumption?

- We can proceed as usual defining the variables, objective function and constraints:

```
c[p_, i_] = 1.3 p^-1.05 i^0.01;
NMinimize[{c[p, i], 4 ≤ p ≤ 12 && 20 ≤ i ≤ 200}, {p, i}]
{0.0985856, {p → 12., i → 20.}}
NMaximize[{c[p, i], 0.4 < p ≤ 1.2 && 20 ≤ i ≤ 200}, {p, i}]
{3.58749, {p → 0.4, i → 200.}}
```

- We can verify the solution visually using **Plot2D** and **ContourPlot**. If we place the cursor on the second graph and move it around, we can see the different values of the function $c\ (r,\ t)$. This way we can check that the upper left corner corresponds to the largest value (maximum) and the lower right corner to the smallest one (minimum). The boundaries of the curve correspond to the constraints.

```
GraphicsRow[{Plot3D[c[p, i], {p, 4, 12}, {i, 0, 200},
   RegionFunction -> Function[{p, i}, 4 <= p <= 12 && 20 <= i <= 200],
   AxesLabel -> {"Price", "Income", "c"}, Mesh -> None],
  ContourPlot[c[p, i], {p, 4, 12}, {i, 0, 200}, RegionFunction ->
    Function[{p, i}, 4 <= p <= 12 && 20 <= i <= 200], Mesh -> None,
   Contours -> 100, FrameLabel -> Automatic, PlotLegends → Automatic]},
 Frame -> All, AspectRatio → 1/3, ImageSize → {600, 300}]
```

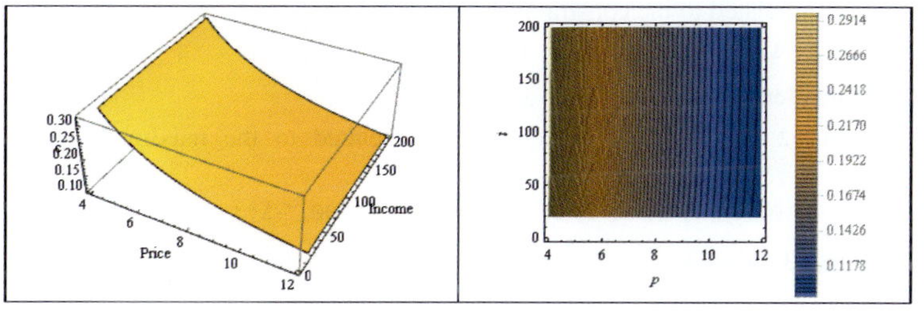

11.3.5 Application: Transportation Optimization using Linear Programming

The department store chain Carrefa would like to outsource the supply of milk cartons for its centers in Barcelona, Madrid, Seville and Valladolid. For that purpose it receives offers from three suppliers: Pascualo, Simone, and Pulova. The prices per carton are as follows: Pascualo €3.00, Simone €2.80 and Pulova €2.70. The maximum number of cartons that each supplier can send daily is shown in the table below, under the column "Maximum supply". The daily demand for each center is in the "Demand" row. The rest of the information in the table relates to the transportation costs per carton, in €.

	Maximum Supply	Barcelona	Madrid	Seville	Valladolid
Pascualo	2500	€ 0.5	€ 0.6	€ 0.7	€ 0.5
Simone	1700	€ 0.5	€ 0.6	€ 0.8	€ 0.5
Pulova	1500	€ 0.4	€ 0.5	€ 0.9	€ 0.7
	Demand	800	1100	500	300

Carrefa would like to know how much to order from each supplier to minimize the total cost.

```
Clear["Global`*"]
```

This is the type of problem that can be solved with `LinearProgramming`. For didactic purposes we're going to compare the notation in `NMinimize` with the one in `LinearProgramming`. In practice this is not necessary. As a matter of fact, the easiest way to solve the problem is by using the `LinearProgramming` syntax.

With `NMinimize`, the variables can be defined as follows: x_{ij} = number of cartons from supplier i, i:{1, 2, 3}, with {1 = Pascualo, 2 = Simone, 3 = Pulova}, to center j: {1, 2, 3, 4}, with {1 = Barcelona, 2 = Madrid, 3 = Seville, 4 = Valladolid}. With LinearProgramming (LP), it's not necessary to define the variables in an explicit way. $\{x_{11}, x_{12}, x_{13}, x_{14}, x_{21}, x_{22}, x_{23}, x_{24}, x_{31}, x_{32}, x_{33}, x_{34}\}$ is the same as {1, 1, 1, 1, 1, 1, 1, 1, 1, 1, 1, 1}.

- To define the objective function, we have to group terms by supplier. For example: the total cost of sending x units from Pascualo to Barcelona (x_{11}) would be the transportation cost plus the purchasing price: (0.5 + pa). We use the same approach for calculating the rest of the variables:

```
{pa, sim, pul} = {3, 2.8, 2.7}; LinearProgramming

of = { (0.5 + pa), (0.6 + pa), (0.7 + pa), (0.5 + pa),
    (0.5 + sim), (0.6 + sim), (0.8 + sim), (0.5 + sim),
    (0.4 + pul), (0.5 + pul), (0.9 + pul), (0.7 + pul)};
```

- The constraint **c1**, $x_{11} + x_{12} + x_{13} + x_{14} \leq 2500$, representing the maximum number of number of cartons of milk that we can receive from Pascualo is written as (the values of those variables not related to the constraint should be 0):

```
m1 = {1, 1, 1, 1, 0, 0, 0, 0, 0, 0, 0, 0}; b1 = {2500, -1};
```

  where −1 is equivalent to "≤"; 0 to "=="; and 1 to "≥".

- The constraint **c2**, $x_{21} + x_{22} + x_{23} + x_{24} \leq 1700$, related to the maximum capacity of Simone is:

```
m2 = {0, 0, 0, 0, 1, 1, 1, 1, 0, 0, 0, 0}; b2 = {1700, -1};
```

- The constraint **c3**, $x_{31} + x_{32} + x_{33} + x_{34} \leq 1500$, dealing with the supply coming from Pulova:

```
m3 = {0, 0, 0, 0, 0, 0, 0, 0, 1, 1, 1, 1}; b3 = {1500, -1};
```

- The constraint **c4**, $x_{11} + x_{21} + x_{31} == 800$, indicating the demand from Barcelona:

```
m4 = {1, 0, 0, 0, 1, 0, 0, 0, 1, 0, 0, 0}; b4 = {800, 0};
```

- The constraint **c5**, $x_{12} + x_{22} + x_{32} == 1100$, referring to the demand from Madrid:

```
m5 = {0, 1, 0, 0, 0, 1, 0, 0, 0, 1, 0, 0}; b5 = {1100, 0};
```

- The constraint **c6**, $x_{13} + x_{23} + x_{33} == 500$, about Seville:

```
m6 = {0, 0, 1, 0, 0, 0, 1, 0, 0, 0, 1, 0}; b6 = {500, 0};
```

- And finally the constraint **c7**, $x_{14} + x_{24} + x_{34} == 300$, to make sure that the needs of Valladolid are being satisfied:

```
m7 = {0, 0, 0, 1, 0, 0, 0, 1, 0, 0, 0, 1}; b7 = {300, 0};
```

- Now we can solve the problem (`LinearProgramming` assumes that all the variables are nonnegative so we don't have to include those constraints).

```
sol = LinearProgramming[of,
  {m1, m2, m3, m4, m5, m6, m7}, {b1, b2, b3, b4, b5, b6, b7}]
```

```
{0., 0., 0., 0., 400., 0., 500., 300., 400., 1100., 0., 0.}
```

- **LinearProgramming** returns the optimal value for each variable. If we'd like to calculate the minimum total transportation cost, we just need to multiply the objective function by the solution (we use "." for matrix multiplication).

```
of.sol
```

```
8870.
```

11.3.6 Application: Analyzing Enriched Uranium

A key aspect in the operation of nuclear power plants is the cost of the fuel. Virtually all nuclear reactors use uranium as their main source of energy input. The minerals containing it are widely distributed in nature, but profitable deposits under current economic conditions are concentrated in a limited number of countries. According to estimates, these deposits will last for another 70 to 80 years which reduces the amount of money invested in finding new sources. Fast reactors that reuse uranium would extend the availability of the nuclear fuel to thousands of years. However, the cost and the immaturity of the technology don't justify the development of this type of reactor for the time being. Nevertheless, the current ones, although inefficiently, can reuse part of the spent fuel. Some power plants already do that.

- Uranium found in nature mostly consists of three isotopes. The command below displays their respective abundance (fraction of atoms of an isotope in the element). We can see that almost everything is either U238 or U235:

```
DeleteCases[Transpose[{IsotopeData["U", "Symbol"],
   IsotopeData["U", "IsotopeAbundance"] // Chop}], {_, _Integer}]
```

$\{\{^{234}U, 0.000054000\}, \{^{235}U, 0.0072040\}, \{^{238}U, 0.992742\}\}$

Most nuclear reactors use uranium enriched in the U–235 isotope. The proportion of this isotope in heavy metals can reach up to 5%. The process to enrich uranium consists of two steps: turning natural uranium into gas (UF6) and then putting the gas into centrifuge tubes (there are other enrichment techniques but this one is the most commonly used). The centrifuge tubes are cylinders that spin at very high speeds pulling the heavier U–238 into the walls of the cylinder and leaving the lighter U–235 closer to the center where it can be sucked out. The result is the extraction of uranium with a higher proportion of U–235 than the one found in nature.

Summarizing: In the enrichment process we start with natural uranium, that has a mass fraction $X_f = 0.00711$ (equivalent to 0.711% of Uranium–235, or 0.711% in isotopic abundance) and we get enriched uranium with a fraction X_p of U–235 higher than X_f. The process creates depleted uranium as a by-product with a proportion X_w of U–235, lower than the natural one.

The equation below returns the initial number of kilograms needed (kgU of natural uranium, known as *feed*) to obtain 1 kgU with enrichment X_p, generating as a by-product depleted uranium with a fraction X_w of U–235:

$$\text{factorfeed} = \frac{X_p - X_w}{X_f - X_w}$$

From this equation we can see that to get 1 kgU with an enrichment X_p, the initial feed depends on the desired content, X_w, of the depleted uranium. The value of X_w expressed as a percentage of U–235 is known as the "**tail**". The lower the required X_w, the lower the feed needed. In principle, it may seem that the best outcome would be to obtain a value of X_w as low as technologically possible. However, there's a problem: the energy consumption needed to generate a target X_w increases nonlinearly as X_w gets smaller. The process is described next.

The method for calculating the energy consumption uses an internationally recognized measurement unit named **SWU** (Separative Work Unit). This unit is defined as the minimum energy required to obtain 1 kgU enriched by a fraction X_p from uranium with an isotopic fraction X_f (usually natural uranium), generating as a by-product uranium with an isotopic fraction X_w. The actual calculation is done with the following equation:

$$\text{factorSWU} = \left((2 X_p - 1) \text{Log}\left[\frac{X_p}{1 - X_p}\right] - (2 X_w - 1) \text{Log}\left[\frac{X_w}{1 - X_w}\right]\right) - \left(\left((2 X_f - 1) \text{Log}\left[\frac{X_f}{1 - X_f}\right] - (2 X_w - 1) \text{Log}\left[\frac{X_w}{1 - X_w}\right]\right)(X_p - X_w)\right) \Big/ (X_f - X_w)$$

(Log is the natural logarithm, that is, in base e)

The previous equation doesn't refer to the real energy consumption that depends on the technology used. It's based on the thermodynamical notion of consumption and there's an agreement to use it regardless of the real consumption.

- The two previous equations can be put together in the following function that calculates the feed and SWU necessary to get 1 kgU enriched by 100 X_p and a fixed percentage 100 X_w of depleted U–235. In this function we use W and P to refer to X_w and X_p respectively. F represents X_f and its value is 0.00711 or 0.711% (natural uranium).

```
swufeed[P1_, W1_] =
  Module[{F, swu, feed, P, W}, P = P1/100; W = W1/100; F = 0.00711;
swu = ((2*P - 1)*Log[P/(1 - P)] - (2*W - 1)*Log[W/(1 - W)]) -
(((2*F - 1)*Log[F/(1 - F)] - (2*W - 1)*Log[W/(1 - W)])*(P - W))/
    (F - W);
feed = (P - W)/(F - W); Thread[{"factorSWU", "factorFeed"} ->
    {swu, feed}]];
```

- In the next example we calculate the feed and SWU needed to obtain 5% enriched uranium with a tail of 0.3%.

```
swufeed[5, 0.3]
```

```
{factorSWU → 7.19832, factorFeed → 11.4355}
```

- For a given tail, in the command below we use 0.2%, the function calculates the SWU required for 1 kgU with enrichment *enr*:

```
func[enr_] = "factorSWU" /. swufeed[enr, 0.2]
```

$$-6.18776 + 258.096\left(-0.002 + \frac{enr}{100}\right) + \left(-1 + \frac{enr}{50}\right) \text{Log}\left[\frac{enr}{100\left(1 - \frac{enr}{100}\right)}\right]$$

$$-6.18776 + 258.096\left(-0.002 + \frac{enr}{100}\right) + \left(-1 + \frac{enr}{50}\right) \text{Log}\left[\frac{enr}{100\left(1 - \frac{enr}{100}\right)}\right]$$

- We fit the previous function to a line in the range *enr* 1.0% - 3.5%.

```
linear1[x_] = Fit[Table[{i, func[i]}, {i, 1.0, 3.5, 0.5}], {1, x}, x]
```

```
-1.76531 + 2.02534 x
```

- Now we compare the exact calculation and its numerical approximation. We can see that both lines, although close to each other, are different.

```
Plot[{linear1[x], func[x]}, {x, 1.0, 3.5},
 AxesLabel → {"Enrichment level (%)", "SWU"}]
```

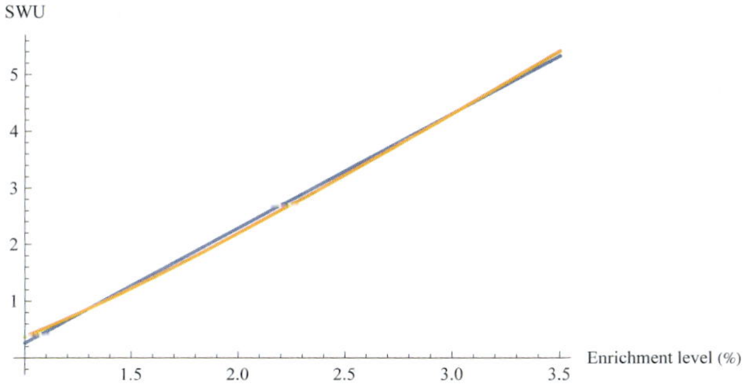

For the uranium buyer, the setting of the tail is very important and that choice will be determined by the cost of the feed and the SWU.

The enriched uranium cost can be computed using the following expression:

$priceEu = factorfeed \times pFeed + factorSWU \times pSWU$, with *pFeed* being the price of the feed and *pSWU* the price of the SWU.

- The next function calculates the price of 1 kgU for a given tail, enrichment level and feed and SWU prices:

```
priceEu[enr_, tail_, pFeed_, pSWU_] :=
 pFeed "factorFeed" + pSWU "factorSWU" /. swufeed[ enr, tail]
```

Using the import capabilities discussed in Chapter 2, we can get the spot feed (natural UF6 in kgU) and SWU prices from http://www.uxc.com/review/UxCPrices.aspx. If we open the link, the information will be inside the table: *Ux Month-End Spot Prices as of MM, YYYY.* The command below will work as long as the website has not been modified since the last time it was accessed (updated on December 31, 2016):

```
priceuranium = Import[
    "http://www.uxc.com/review/uxCPrices.aspx", "Data"][[2, 2, 3;;8]];
```

Note: To avoid potential problems we have also copied the output directly in *Mathematica*:

```
priceuranium =
  {{"U 3 O 8 Price (lb)", "$20.25", "( +2.00 )", "€19.35", "( +1.91 )"},
   {"NA Conv. (kgU)", "$5.85", "( Unch. )", "€5.59", "( Unch. )"},
   {"EU Conv. (kgU)", "$6.40", "( Unch. )", "€6.12", "( Unch. )"},
   {"NA UF 6 Price (kgU)", "$58.00", "( +4.60 )", "€55.43", "( +4.40 )"},
   {"NA UF 6 Value § (kgU)", "$58.76", "( +5.23 )",
    "€56.15", "( +5.00 )"}, {"EU UF 6 Value § (kgU)",
    "$59.31", "( +5.23 )", "€56.68", "( +5.00 )"}};
```

- We can even choose a specific value for later use. The next command extracts the price of the feed (EU UF6 value). Nevertheless, we copy it so that the reader can reproduce the example using the same data.

```
{priceuranium[[5, 2]], priceuranium[[6, 2]]}
```

```
{$58.76, $59.31}
```

```
{pFeed, pSWU} = {58.76, 59.31};
```

- Given the spot prices for the feed and SWU, the cost of 1 kgU enriched by 5% and with a 0.3% tail will be:

```
priceEu[5, 0.3, pFeed, pSWU]
```

```
1098.88
```

- The graph below displays the price as a function of the tail. We can clearly see that there's a minimum economic value associated to the tail.

```
ListPlot[Table[{i, priceEu[5, i, pFeed, pSWU]}, {i, 0.10, 0.40, 0.025}],
 AxesOrigin → {0, 2500}, Joined → True, AxesLabel → {"Tail(%)", "Price"}]
```

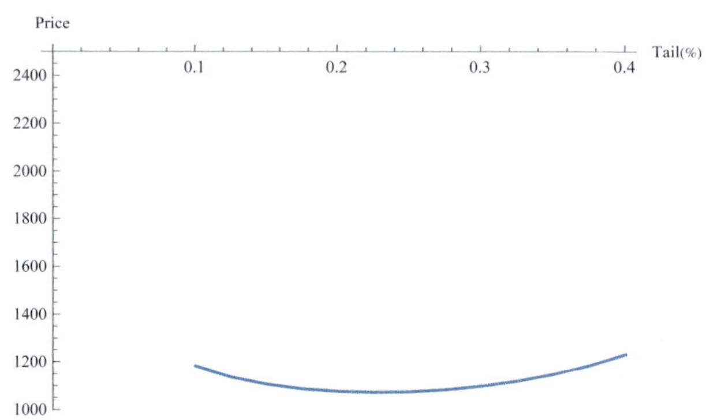

- The following function calculates the optimum purchasing prices for different enrichment levels. Notice that for the given feed and SWU prices, the optimum tail doesn't depend on the enrichment target.

```
Table[{enr, NMinimize[{priceEu[enr, tail, pFeed, pSWU] , 0.15 ≤ tail ≤ 0.5},
    tail]}, {enr, 1, 5}]
```

```
{{1, {114.273, {tail → 0.22767}}},
 {2, {335.907, {tail → 0.22767}}}, {3, {575.243, {tail → 0.22767}}},
 {4, {821.986, {tail → 0.22767}}}, {5, {1072.88, {tail → 0.22767}}}}
```

In practice, the tail value is rounded to two decimals.

The next example shows how to calculate the cost of the enriched uranium needed to generate 1 kWh in European Union.

- The normal annual consumption of a 1,000 MW power plant, the most common one, is of 20,000 kgU 4.5% enriched. Using the optimum tail from the previous data, the cost of this uranium expressed in euros would be:

```
eurodollar = FinancialData["EUR/USD"]
```

```
1.0601
```

```
annualcomsumption = eurodollar 20000 priceEu[4.5, 0.23, pFeed, pSWU] "euros"
```

2.00794×10^7 euros

- A 1,000 MW power plant working non-stop for one year will generate the following amount of energy in kWh (kilowatt-hour):

```
annualkwh = 1000000 × 24 × 365 "kWh"
```

```
8760000000 kWh
```

- Therefore, the cost per kWh would be:

```
annualcomsumption / annualkwh
```

$\frac{0.00229217 \text{ euros}}{\text{kWh}}$

- The previous cost refers exclusively to the enriched uranium. To that number we must add the cost of its production, transportation and insurance among others. In total, around 20% of the uranium price. Therefore, the cost of fuel per kWh expressed in euro cents would be:

```
nuclearfuelkwh = 100 × 1.2 annualcomsumption / annualkwh / "euros" "cents"
```

$\frac{0.27506 \text{ cents}}{\text{kWh}}$

This figure doesn't reflect the true cost of nuclear kWh, that is influenced by additional factors, especially the initial construction investment.

The cost of the electricity in Europe is higher than 10 c/kWh. If we compare it with the nuclear fuel cost, we can see that the latter represents only a small portion of the former. Therefore, big swings in fuel costs will have a small impact in the cost per nuclear kWh. In the case of other fuels such as natural gas, the most commonly used one for generating electricity, their costs can represent up to 80% of the total production expense.

11.4 The Shortest Path Problem

11.4.1 The Traveling Salesman Problem

There are problems in mathematics that look very specialized and easy to solve at first sight but in reality they are very complicated and their resolution can have applications in a wide variety of fields. One of the most famous examples is the *Traveling Salesman Problem* (TSP). This problem can be defined as follows: A traveler would like to visit *N* cities starting and finishing in the same arbitrarily chosen one, minimizing the total distance traveled and passing through each city only once. If you think about it, there are many real-world problems that are actually equivalent to the TSP: route optimization (transportation, logistics), optimum network layout (trains, roads, electricity), even the connections inside microprocessors to minimize calculation time.

A graphical demonstration of the problem created by Jon McLoone can be found in: (http://demonstrations.wolfram.com/TravelingSalesmanProblem/)

Mathematically, the problem consists of finding a permutation P = $\{c_0, c_1, ..., c_{n-1}\}$ such that $d_P = \sum_{i=1}^{n-1} d[c_i, c_{i+1}]$ would be a minimum, with d_{ij} representing the distance from city i to city j.

The seemingly simplest solution would be to enumerate all the possible routes, calculate the distance traveled for each route and select the shortest one. The problem with this approach is that computing time increases dramatically when the number of cities gets bigger. For example, a computer would take 3 seconds to solve the problem if there were 10 cities, a bit more than half a minute for 11 cities and almost 80,000 years! to solve the problem for 20 cities.

Conclusion: This solving method is not feasible. Therefore, several alternative algorithms have been developed that use a variety of strategies to simplify the problem (Something similar to computer chess programs that instead of calculating all possible moves, use different criteria to choose the most appropriate ones).

For this type of problem, *Mathematica* has the function:

`FindShortestTour[{`*e1*, *e2*, ...`}]`. This function takes a list of coordinates: {*e1*, *e2*, ...} and attempts to find the optimum ordering so that the total distance traveled is minimized and all the points have been visited just once. The output returns both, the total distance and the order of traveling.

Example: Given a set of points p = {1, 2, 3, 4, 5, 6} each with coordinates{x,y}:{{4, 3},{1, 1},{2, 3},{3, -5},{-1, 2},{3, 4}}, find the shortest path that will visit all the points only once.

```
d = {{4, 3}, {1, 1}, {2, 3}, {3, -5}, {-1, 2}, {3, 4}};
```

```
{dist, order} = FindShortestTour[d]
```

$\{2\sqrt{2} + \sqrt{5} + 3\sqrt{10} + \sqrt{65}, \{1, 4, 2, 5, 3, 6, 1\}\}$

- We sort the points following the order from the output above and represent them graphically:

```
d[[order]]
```

```
{{4, 3}, {3, -5}, {1, 1}, {-1, 2}, {2, 3}, {3, 4}, {4, 3}}
```

```
Graphics[Line[%]]
```

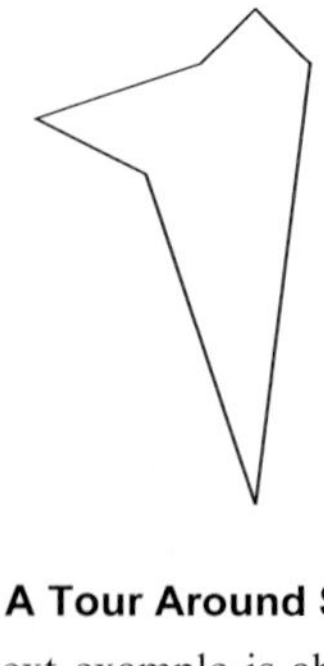

11.4.2 A Tour Around South American Cities

The next example is about organizing a trip to several cities in South America. In what order should we visit them to minimize the total distance traveled?

- The cities we're visiting are the following (we indicate both, the cities and their respective regions and countries to avoid ambiguity):

```
cities={Entity["City", {"Asuncion", "Asuncion", "Paraguay"}],
 Entity["City", {"Bogota", "DistritoCapital", "Colombia"}],
 Entity["City", {"RioDeJaneiro", "RioDeJaneiro", "Brazil"}],
 Entity["City", {"BuenosAires", "BuenosAires", "Argentina"}],
 Entity["City", {"Caracas", "DistritoCapital", "Venezuela"}],
 Entity["City", {"LaPaz", "LaPaz", "Bolivia"}],
 Entity["City", {"Lima", "Lima", "Peru"}],
 Entity["City", {"Montevideo", "Montevideo", "Uruguay"}],
 Entity["City", {"Quito", "Pichincha", "Ecuador"}],
 Entity["City", {"Santiago", "Metropolitana", "Chile"}]};
```

- Now we can calculate the best visiting order:

```
order = Last[FindShortestTour[GeoPosition[cities]]]
```

```
{1, 8, 4, 10, 6, 7, 9, 2, 5, 3, 1}
```

- Finally, we can display the route on a map:

```
GeoListPlot[cities[[order]], Joined → True]
```

11.4.3 Moon Tourism

It's the year 2035 and we are planning to organize sightseeing tours around the Moon. We want to find the shortest route for visiting all the Apollo landing sites.

- "Apollo landings" is one of the available entity classes that have been included in the function `MannedSpaceMissionData`.

```
MannedSpaceMissionData["Classes"]
```

```
{Apollo landings, Apollo manned mission, Chinese manned mission,
 Earth orbit manned mission, lunar manned mission,
 manned space mission, Mercury Atlas manned mission,
 Mercury Redstone manned mission, pre-flight test manned mission,
 project Gemini manned mission, project Mercury manned mission,
 Russian manned mission, Shenzhou manned mission,
 Soyuz manned mission, STS manned mission,
 sub-orbital flight manned mission, United States manned mission,
 Voskhod manned mission, Vostok manned mission}
```

```
MannedSpaceMissionData[Apollo manned mission (manned space missions)]
```

```
{Apollo 1, Apollo 7, Apollo 8, Apollo 9, Apollo 10, Apollo 11,
 Apollo 12, Apollo 13, Apollo 14, Apollo 15, Apollo 16, Apollo 17}
```

```
apollolandings =
  MannedSpaceMissionData[Apollo landings (manned space missions),
   {"Position"}, "EntityAssociation"]
```

```
<|Apollo 11 → {GeoPosition[{0.6741, 23.47}, Moon]},
  Apollo 12 → {GeoPosition[{-3.012, -23.42}, Moon]},
  Apollo 14 → {GeoPosition[{-3.645, -17.47}, Moon]},
  Apollo 15 → {GeoPosition[{26.13, 3.634}, Moon]},
  Apollo 16 → {GeoPosition[{-8.973, 15.50}, Moon]},
  Apollo 17 → {GeoPosition[{20.19, 30.77}, Moon]}|>
```

■ The names of the manned landing missions are:

```
missionnames = Keys[apollolandings]
```

```
{Apollo 11, Apollo 12, Apollo 14, Apollo 15, Apollo 16, Apollo 17}
```

■ Notice that Apollo 13 is missing. Let's find out why:

```
MannedSpaceMissionData[Apollo 13 (manned space mission), "Description"]
```

```
{intended to be the third manned mission to land on the moon,
 mid-mission onboard explosion forced landing to be aborted,
 returned safely to Earth}
```

■ Create a map showing the locations of the landings:

```
GeoListPlot[missionnames, GeoRange → All,
 GeoProjection → "Orthographic", GeoLabels → True, LabelStyle → White]
```

- To extract the coordinates of those landings:

```
landingcoords = Values[apollolandings] // Flatten
```

```
{GeoPosition[{0.6741, 23.47}, Moon],
 GeoPosition[{-3.012, -23.42}, Moon],
 GeoPosition[{-3.645, -17.47}, Moon],
 GeoPosition[{26.13, 3.634}, Moon],
 GeoPosition[{-8.973, 15.50}, Moon],
 GeoPosition[{20.19, 30.77}, Moon]}
```

- Next, we find out that the shortest route starts where Apollo 11 landed:

```
order = Last[FindShortestTour[landingcoords]]
```

```
{1, 6, 4, 2, 3, 5, 1}
```

- Finally, we create a plot showing the path of the tour:

```
GeoListPlot[landingcoords[[order]], Joined → True, GeoLabels → Automatic]
```

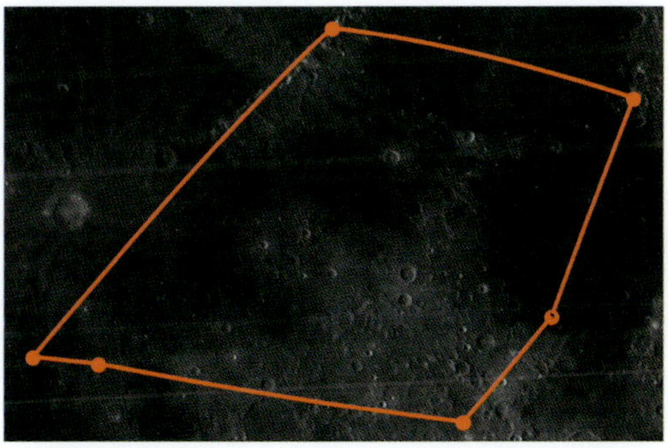

11.5 Optimum Flows

In this example we want to maximize the total number of freight cars that can be sent from Vancouver to Winnipeg given the following network characteristics:

The code generating graph g is not shown.

g = 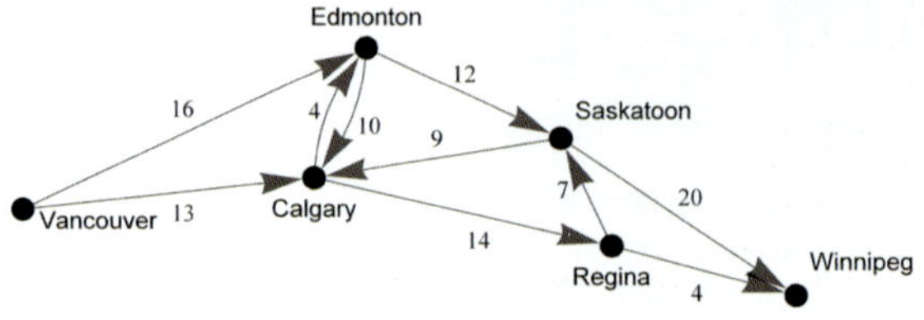

;

- The previous illustration has been created using the `Graph` function (for further details see Section 4.4). For simplifying purposes, let's just focus on 3 cities: { (1) Vancouver, (2) Calgary, (3) Edmonton}. The flow directions (i→j) and capacities (`EdgeCapacity`) are: {From (1) Vancouver to (2) Calgary, 13, from (1) Vancouver to (3) Edmonton, 4, from (3) Edmonton to (2) Calgary, 10 and from (2) Edmonton to (3) Calgary, 4}. We display that information using `DirectedEdge` and with the help of `VertexLabels` and `Placed` position the labels in the desired locations. The style is defined using `Style`[#, "Style"]&. You can use the documentation to learn more about the syntax for each function.

```
g1 = Graph[{
  "Vancouver", "Calgary", "Edmonton"}, {{{
  1, 2}, {1, 3}, {2, 3}, {3, 2}}, Null}, {
  EdgeCapacity -> {13, 16, 4, 10},
  EdgeLabels -> {
    DirectedEdge["Vancouver", "Calgary"] -> Placed[13, {0.5, {2, 1}}],
    DirectedEdge["Vancouver", "Edmonton"] -> Placed[16, {0.5, {-1.5, 2}}],
    DirectedEdge["Edmonton", "Calgary"] -> Placed[10, {0.5, {-0.2, 1.5}}],
    DirectedEdge["Calgary", "Edmonton"] -> Placed[4, {0.5, {-0.7, 1.5}}] },
  GraphStyle -> "BasicBlack", VertexLabels -> {
   "Edmonton" -> Placed["(3) Ed.", Above, Style[#, Red] & ],
   "Vancouver" -> Placed["(1) Vanc.", After, Style[#, Blue] & ],

    "Calgary" -> Placed["(2) Cal.", Above, Style[#, FontFamily -> "Helvetica"] & ]},
  VertexSize -> {Small}}, ImagePadding → All, ImageSize → Small]
```

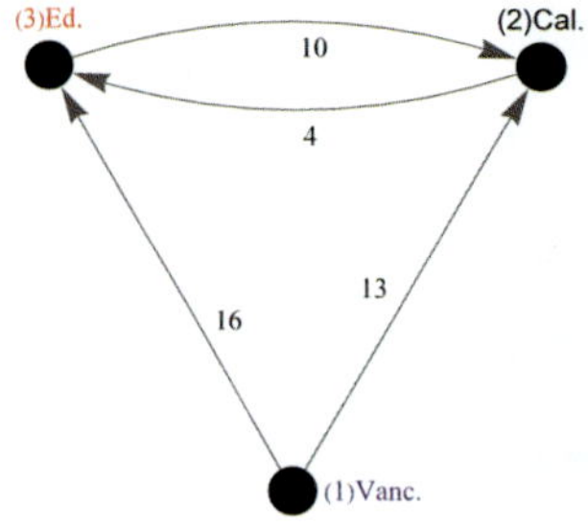

■ In the next example we add `VertexCoordinates` to specify the coordinates for the center of the vertices. We also modify several options to make it easier to overlay the graph on a map of Canada later.

```
Graph[{
  "Vancouver", "Calgary", "Edmonton"}, {{{
  1, 2}, {1, 3}, {2, 3}, {3, 2}}, Null}, {
  EdgeCapacity -> {13, 16, 4, 10},
  EdgeLabels -> {
   DirectedEdge["Edmonton", "Calgary"] -> Placed[10, {0.5, {-0.2, 1}}],
   DirectedEdge["Vancouver", "Edmonton"] -> Placed[16, {0.5, {1, 0}}],
  DirectedEdge["Vancouver", "Calgary"] -> Placed[13, {0.6, {1, 1.5}}]},
  GraphStyle -> "BasicBlack", ImagePadding -> {{0, 40}, {0, 10}},
  VertexCoordinates -> {{-0.253, 0.112}, {-0.1419, 0.1239}, {-0.122,
   0.1735}}, VertexLabels -> {
   "Edmonton" ->
   Placed["Edmonton", Above, Style[#, FontFamily -> "Helvetica"] & ],
   "Vancouver" ->
   Placed["Vancouver", {{1.5, -1}, {0, 0}},
    Style[#, FontFamily -> "Helvetica"] & ],
   "Calgary" ->
   Placed["Calgary", Below, Style[#, FontFamily -> "Helvetica"] & ]},
  VertexSize -> {Small}}]
```

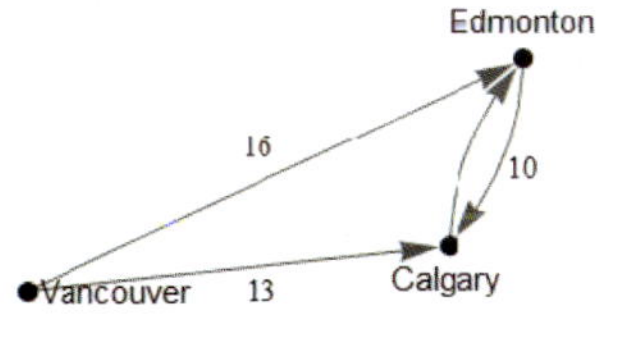

■ To find the optimum flow we use the function `FindMaximumFlow`.

```
F = FindMaximumFlow[g, "Vancouver", "Winnipeg", "OptimumFlowData"]
```

OptimumFlowData[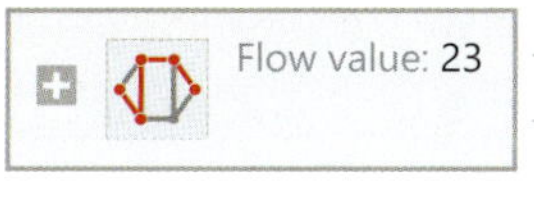]


```
F["FlowValue"]
```

23

■ We can see the optimum flow per edge as follows:

```
Text@Grid[Prepend[{#, F[#]} & /@ F["EdgeList"],
   {Style["Railroad", Bold], Style[ "Edge Flow", Bold]}]]
```

Railroad	**Edge Flow**
Vancouver ⟷ Calgary	7
Vancouver ⟷ Edmonton	16
Calgary ⟷ Regina	11
Edmonton ⟷ Calgary	4
Edmonton ⟷ Saskatoon	12
Regina ⟷ Saskatoon	7
Regina ⟷ Winnipeg	4
Saskatoon ⟷ Winnipeg	19

- Here we display the result on top of the map of Canada, downloaded using **CountryData** (for further details see Chapter 5).

```
Panel[Show[{CountryData["Canada", "Shape"], F["FlowGraph"]},
  PlotRange → {{-0.31, 0.083}, {0.04, 0.25}}]]
```

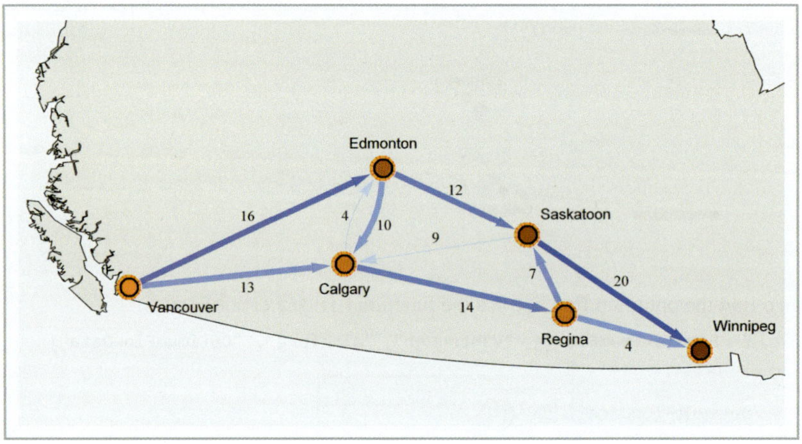

11.6 Additional Resources

The Wolfram Finance Platform: http://www.wolfram.com/finance-platform

Mathematica's finance related functions:
http://reference.wolfram.com/mathematica/guide/Finance.html

Financial engineering: http://www.wolfram.com/solutions/industry/financial-engineering-and-mathematics

Economics: http://www.wolfram.com/solutions/industry/economics/

Financial risk management: http://www.wolfram.com/solutions/industry/financial-risk-management/

Constrained optimization tutorial: tutorial/ConstrainedOptimizationOverview

12

Faster, Further

This chapter discusses how to make computationally intensive tasks more efficient by distributing the calculations among several cores in a single computer or even among several computers in a grid. All of this can be done using Wolfram Research products such as: gridMathematica and Wolfram Lightweight Grid Manager (WLGM). If our computer has a NVidia graphics card, we also have the option of using the card processors through the Compute Unified Device Architecture (CUDA) programming interface, integrated in Mathematica, for extra processing power. The chapter also describes complementary tools such as Workbench, an excellent Integrated Development Environment (IDE) for large Mathematica-based projects, and webMathematica, perfectly suited to develop programs that run in a web server and can be executed from any browser. Finally we will refer to some new functionality added to newest Mathematica version (wavelet analysis, control systems, etc.)

This chapter has been written using several multicore computers.

12.1 Parallel Computing

Computer programs are usually run sequentially: A problem is broken down into a series of instructions performed one by one in order. Only one instruction is carried out at any time, with its output being passed to the next one until the entire program has been executed.

Parallel computing consists of breaking down programs into discrete components which can be executed simultaneously. In *Mathematica,* this is done by distributing the individual program parts among multiple kernels. This is possible when the computer we're using has several processors, if we distribute the computation among several machines or a combination of both. (B. Blais, *Introduction to Parallel Computing,* Lawrence Livermore National Laboratory: https://computing.llnl.gov/tutorials/parallel_comp).

To do parallel computing with several processors, one of them must act as the master distributing and synchronizing the operations with the rest of the processors (located in the same computer and/or in other computers connected to the grid). This task implies an extra amount of time (communication, synchronization, etc.) that must be taken into account. The objective is to reduce the total computation time compared to how long it would take to perform the calculation sequentially. There are many scientific and technical fields where parallel computing is very useful or even the only realistic way to perform certain type of calculations. For example: in Meteorology, there used to be predictive models that took longer to evaluate than the actual prediction window.

In the past, parallel programming was a niche area only available for expensive hardware. However, nowadays even PCs have more than one processor. This computation power gets even bigger if we connect computers to a grid. One of the best examples of this approach is the well-known SETI program, a series of scientific activities undertaken in search of extraterrestrial intelligence. The program uses small radiotelescopes to capture signals from outer space. Since analyzing those signals requires a very large amount of computation time, the SETI@home (http://setiathome.berkeley.edu) project was created. It offers us the possibility of installing in our PCs a small application that SETI can use to analyze the signals when the computer is idle. There are other projects that also use PCs from volunteers all over the world for different purposes such as the analysis of proteins (Folding@home: http://folding.stanford.edu).

In summary, *Mathematica* has the capability to perform calculations in parallel right out of the box. In addition, it's possible to set up a grid to distribute computations to several computers using WLGM and/or grid *Mathematica* (that includes WLGM). If you have an NVidia graphics card, you can also use CUDA (included in *Mathematica*) for parallel computing.

12.2 Parallel Programming

12.2.1 How to Use More than One Kernel

Usually, computer programs are executed sequentially in a single CPU. Although not exactly the same, we can say that in *Mathematica* a sequential program will be executed in a single kernel, while a program that includes parallel calculations will be distributed among all the available kernels and its instructions executed simultaneously.

- Let's start with a list of numbers that we'd like to factorize (transform them into a product of prime numbers).

```
numbers = {342895, 423822, 1546,
   168972, 123555, 23440, 1378, 957, 348300, 2987};
```

Factorization plays a crucial role in cryptography. As a matter of fact, the transmission of encrypted information is often done using numbers obtained from multiplying two prime numbers. This technique is the most commonly used one when doing secure commercial transactions over the Internet.

- For factorization purposes, *Mathematica* has the `FactorInteger` function that when used along with `Map` (or `/@`) can be applied to more than one number at once:

```
list = Map[FactorInteger, numbers]
```

```
{{{5, 1}, {7, 1}, {97, 1}, {101, 1}}, {{2, 1}, {3, 1}, {7, 1}, {10091, 1}},
 {{2, 1}, {773, 1}}, {{2, 2}, {3, 1}, {14081, 1}},
 {{3, 1}, {5, 1}, {8237, 1}}, {{2, 4}, {5, 1}, {293, 1}},
 {{2, 1}, {13, 1}, {53, 1}}, {{3, 1}, {11, 1}, {29, 1}},
 {{2, 2}, {3, 4}, {5, 2}, {43, 1}}, {{29, 1}, {103, 1}}}
```

- The previous result is formatted as a list. Next, we display a table where we can see each number and its corresponding prime factors.

```
Grid[Transpose[{numbers, CenterDot @@ (Superscript @@@ #) & /@ list}],
 Frame → All]
```

342 895	$5^1 \cdot 7^1 \cdot 97^1 \cdot 101^1$
423 822	$2^1 \cdot 3^1 \cdot 7^1 \cdot 10091^1$
1546	$2^1 \cdot 773^1$
168 972	$2^2 \cdot 3^1 \cdot 14081^1$
123 555	$3^1 \cdot 5^1 \cdot 8237^1$
23 440	$2^4 \cdot 5^1 \cdot 293^1$
1378	$2^1 \cdot 13^1 \cdot 53^1$
957	$3^1 \cdot 11^1 \cdot 29^1$
348 300	$2^2 \cdot 3^4 \cdot 5^2 \cdot 43^1$
2987	$29^1 \cdot 103^1$

Operationally, the kernel or logical core of *Mathematica* (not to be mistaken with the physical core of the computer) factorized each number sequentially. That is, it first factorized 342,895, then 423,822 and so on. In this case, the factorization time was very small and so wouldn't justify the use of parallel computations. However, when the factorization involves big numbers, the calculation time increases dramatically so we'd be very interested in using methods to reduce it, such as parallel computing.

For parallel computing purposes, *Mathematica* has certain functions that are added to the usual ones to parallelize calculations but before we can take advantage of them, we need to set up *Mathematica* properly.

- The standard installation of *Mathematica* includes the possibility of using more than one kernel in the host machine with the actual number depending on the license. We can access the computer configuration by going to: **Edit ▶ Preferences ▶ Parallel**. We will see a screen similar to the one in Figure 12.1 (for further details you can watch the video http://www.wolfram.com/broadcast/screencasts/parallelcomputing).

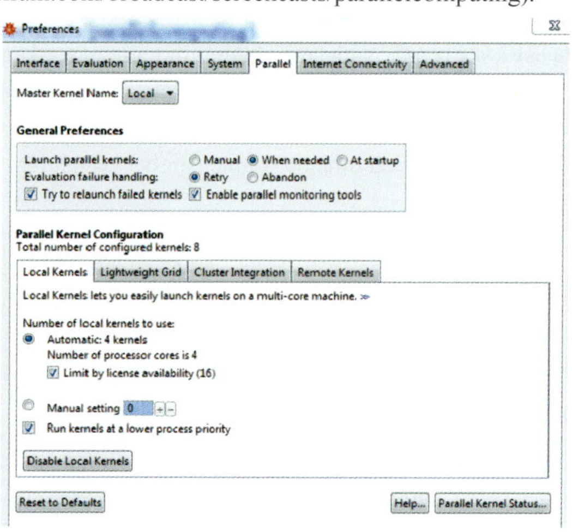

Figure 12.1 Parallel Settings in *Mathematica*.

In this case, the number of kernels that will be launched automatically is 4, the same as the number of computer cores. Notice that the kernels are set by default to low priority, giving preference to other computer processes.

- To launch the available kernels we click on the **Parallel Kernel Status...** button. We can also do the same by going the menu bar and choosing: **Evaluation ▶ Parallel Kernel Status**. In both cases, a window similar to the one in Figure 12.2 will appear:

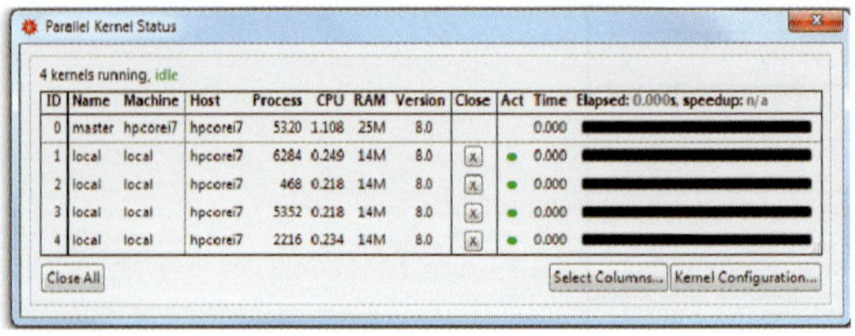

Figure 12.2 Parallel Kernel Status.

- An alternative way to launch the kernels directly is to write down the following command (for this to work, the kernels should not have been launched previously using any of the methods described above):

```
LaunchKernels[]
```

```
{KernelObject[1, local], KernelObject[2, local],
 KernelObject[3, local], KernelObject[4, local]}
```

If you launch more cores than the ones available in the computer, you'll get a $Failed message for each core not found (this can happen, for example, if you have configured your machine for WLGM, to which we will refer later, but the application has not been activated yet).

- We can always find out our available kernels using the following command:

```
Kernels[]
```

```
{KernelObject[1, local], KernelObject[2, local],
 KernelObject[3, local], KernelObject[4, local]}
```

To execute a program in parallel, in addition to having more than one kernel available, the instructions have to explicitly state that their execution should be done in parallel. Usually this is done using `Parallelize`[*exp*], with *exp* being the regular *Mathematica* command. `Parallelize` will launch the kernels first, if they haven't been launched already using any of the methods described in the previous paragraphs.

- You can also use the commands that directly enable parallel evaluation:

```
?Parallel*
```

▼ System`

ParallelArray	ParallelEvaluate	ParallelNeeds	ParallelSum
ParallelCombine	Parallelization	Parallelogram	ParallelTable
ParallelDo	Parallelize	ParallelProduct	ParallelTry
Parallelepiped	ParallelMap	ParallelSubmit	

- To factorize the numbers previously defined, we put the command we used earlier inside `Parallelize` (Instead of **`Parallelize`[`Map`[*f,expr*]]** we can also use the equivalent expression `ParallelMap`[*f,expr*]).

```
Parallelize[Map[FactorInteger, numbers]]
```

```
{{{5, 1}, {7, 1}, {97, 1}, {101, 1}}, {{2, 1}, {3, 1}, {7, 1}, {10091, 1}},
 {{2, 1}, {773, 1}}, {{2, 2}, {3, 1}, {14081, 1}},
 {{3, 1}, {5, 1}, {8237, 1}}, {{2, 4}, {5, 1}, {293, 1}},
 {{2, 1}, {13, 1}, {53, 1}}, {{3, 1}, {11, 1}, {29, 1}},
 {{2, 2}, {3, 4}, {5, 2}, {43, 1}}, {{29, 1}, {103, 1}}}
```

In this case we have used all the available kernels to simultaneously factorize the numbers.

We can try to use this command with other functions, although as mentioned before, the execution may not be faster. That will depend on many factors such as the type of calculation, the number of kernels available, etc. If the computation time is short (a few seconds or less) we will not notice any difference or the parallel calculation may even take longer.

- In the example below we show the computation time with and without parallelization. When measuring computation times for comparison purposes, it's recommended to start a new session (you can use `Quit []` to exit the current one) since, as we have seen in previous chapters, *Mathematica* will automatically reuse previous calculations reducing the execution time significantly. We use `AbsoluteTiming` instead of `Timing` because `AbsoluteTiming` takes into account the entire calculation time while `Timing` only considers the CPU time. The ";" tells *Mathematica* not to print the output of the command since we're only interested in its computation time.

$$\sum_{i=1}^{10000} i!\text{; // AbsoluteTiming}$$

```
{1.2817, Null}
```

$$\text{Parallelize}\left[\sum_{i=1}^{10000} i!\right]\text{; // AbsoluteTiming}$$

```
{0.782273, Null}
```

- Sometimes is not possible to parallelize a computation:

$$\text{Parallelize}\left[\int \frac{1 + x\hat{\ }2}{x}\, dx\right]$$

··· Parallelize: $\int \frac{1+x^2}{x} dx$ cannot be parallelized; proceeding with sequential evaluation.

$$\frac{x^2}{2} + \text{Log}[x]$$

12.2.2 Defining Functions

When doing parallel programming, it's very important to remember that if we define a function and we'd like it to be used by all the kernels, we have to explicitly state it, as shown in the next example.

- We define a function that generates a random number between 1 and 10.

```
rand := RandomInteger[10]
```

- Next, we use `DistributeDefinitions` to parallelize the previous function so it can be used by all the available kernels.

```
DistributeDefinitions[rand];
```

- Finally, we execute the function in all the different kernels.

```
ParallelEvaluate[rand]
```

```
{6, 10, 4, 9}
```

- Once a function has been parallelized, it will remain so even after closing the kernels. In this example, we close the kernels and open them again.

```
CloseKernels[]; LaunchKernels[];
```

- The definition still works with the new kernels:

```
ParallelEvaluate[rand]
```

```
{3, 2, 4, 1}
```

- Commands in *Mathematica* that start with `Parallel`, such as `ParallelTable` (equivalent to **`Parallelize`**[**`Table`**[...]]), will automatically parallelize the definitions, without the need to use `DistributeDefinitions`, as the following example shows:

```
fl[n_] := 2^n
```

```
ParallelTable[fl[k], {k, 1, 6}]
```

```
{2, 4, 8, 16, 32, 64}
```

- An alternative way to execute the previous command is by using `With` to give values to the local variables:

```
WaitAll[Table[With[{i = i}, ParallelSubmit[2^i]], {i, 6}]]
```

```
{2, 4, 8, 16, 32, 64}
```

- Sometimes, we may be interested in performing the calculation in a specific kernel. In this example we choose kernel 2 to execute the computations:

```
CloseKernels[]; LaunchKernels[];
```

```
kernel2 = Kernels[][[2]]
```

```
KernelObject[10, local]
```

Remember that if we don't parallelize variables, we will find unexpected behavior.

- In this example the equality **`a === 2`** (remember that **`SameQ`** (===) checks whether two expressions are identical) returns false even though **`a = 2`**, because **`a`** has not been defined for all the kernels:

```
a = 2;
ParallelEvaluate[a === 2, kernel2]
```

```
True
```

- We can fix the problem using `With`:

```
With[{a = 2}, ParallelEvaluate[a === 2, kernel2] ]
```

```
True
```

```
CloseKernels[];
```

12.2.3 ParallelTry

```
LaunchKernels[];
```

With `ParallelTry` we can evaluate a command in different kernels and get the result from the kernel that executes it first. This can be useful, for example, when comparing different computation methods.

- In the example below we compare different methods for evaluating the minimum of a function. The command returns the fastest one:

```
AbsoluteTiming[ParallelTry[
  {#, NMinimize[{ⅇ^Sin[50 x] - Sin[10 (x + y)] + 1/4 (x^2 + y^2), y ≥ 0 && x ≥ 0},
     {x, y}, Method → #]} &, {"DifferentialEvolution",
   "NelderMead", "RandomSearch", "SimulatedAnnealing"}]]
```

```
{1.06826, {SimulatedAnnealing, {-0.318523, {x → 0.345965, y → 1.06244}}}}
```

12.2.4 ParallelCombine

`ParallelCombine` distributes part of a computation among all the available kernels combining the partial results. It's normally used with list elements. Its syntax is: **`ParallelCombine`**[*f*, {e_1, ..., e_n}] that applies *f* to the elements of a list and groups the partial results.

- The command below calculates the first 9 primes. It's the same as **`Prime`**[{1, ..., 9}] except that the calculation in this case is done in parallel:

```
ParallelCombine[Prime, {1, 2, 3, 4, 5, 6, 7, 8, 9}]
```

```
{2, 3, 5, 7, 11, 13, 17, 19, 23}
```

12.2.5 ParallelEvaluate

The function `ParallelEvaluate` evaluates an expression in different kernels and returns the result from each one of them. By default it uses all the available kernels but we can specify which ones to use. This function is commonly employed in calculations involving Monte Carlo simulation. The next example shows how to calculate π through simulation (an illustrative but quite inefficient method).

- First, we generate *m* pairs of real numbers {*x*, *y*} between 0 and 1. Then we find out which ones fall inside the circle with radius $r = 1$ with **`Norm`**[x y] < 1 equivalent to $x^2 + y^2 < 1$. Finally, we multiply by 4 since we're only generating positive {*x*, *y*}, simulating just 1/4 of the circle:

```
With[{n = 10^5}, ParallelEvaluate[
  4. Count[RandomReal[{0, 1}, {n, 2}], xy_ /; Norm[xy] < 1] / n]]
```

```
{3.13996, 3.13508, 3.15372, 3.15392}
```

- The result is an estimate from each kernel. We calculate the mean to get a final value:

```
Mean[%]
```

```
3.14567
```

12.2.6 ParallelMap

`ParallelMap` is another commonly used function that, as previously mentioned, is equivalent to `Map` for sequential computations.

- We are going to solve again the previous example of approximating π. We generate pairs of points $\{x, y\}$ with values between -1/2 and 1/2. Next we select those points inside a circle with radius 1/2 centered at (0,0) using again **Norm**$[r] < r$. We then proceed to calculate π using the output given by the command below:

```
Solve[(pi r^2)/1 == interiorPoints/total, pi]
```

$$\left\{\left\{\text{pi} \to \frac{\text{interiorPoints}}{r^2\ \text{total}}\right\}\right\}$$

```
r = 1/2;

DistributeDefinitions[r]

{r}

pts = ParallelTable[RandomReal[{-r, r}, {25000, 2}], {Length[Kernels[]]}];

interiorpts = ParallelMap[Select[#, (Norm[#] < r &)] &, pts];
```

- By combining the results, we get the following approximation:

```
DistributeDefinitions[pts, interiorpts];
N[Mean[ParallelTable[Length[interiorpts[[i]]] / (Length[pts[[i]]] * r^2),
    {i, Length[Kernels[]]}]]]

3.14076
```

12.2.7 Finding the Roots of an Equation

We'd like to find the solutions to the equation $\cos(t^2) = e^{-t} + 0.3$ with `FindRoot`, a function that uses numerical algorithms requiring an initial guess.

- The graph below displays the multiple solutions (actually an infinite number of them) that correspond to the intersection points of both curves:

```
Plot[{Cos[t^2], Exp[-t] + .3}, {t, 0, 10}]
```

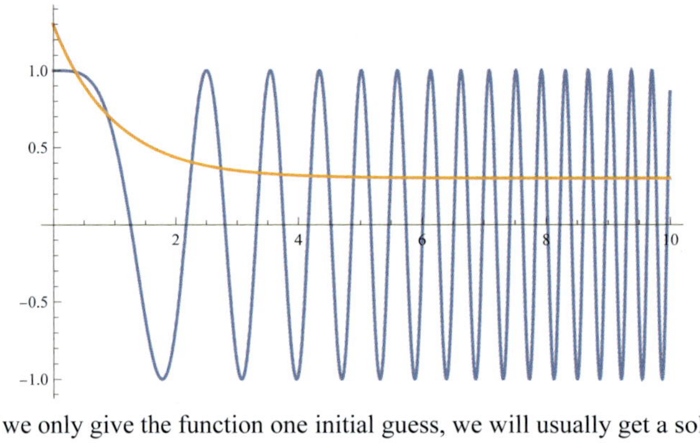

- If we only give the function one initial guess, we will usually get a solution close to it:

```
FindRoot[{Cos[t^2] == Exp[-t] + .3}, {t, 3}]

{t → 2.26454}
```

We are going to find the solution with and without parallelization. We just want to compare the computation times so we'll not show the results.

- Using multiple initial guesses, we can find all the equation roots for a given domain. We have used Map since what we're really doing is solving the equation for different starting points:

```
initval = Table[x, {x, 0, 4 Pi, .1}];
```

```
vals = Map[FindRoot[{Cos[t^2] == Exp[-t] + .3}, {t, #}] &, initval]; //
 AbsoluteTiming
```

```
{0.0323881, Null}
```

- The command below solves the problem replacing Map with ParallelMap.

```
pvals = ParallelMap[FindRoot[{Cos[t^2] == Exp[-t] + .3}, {t, #}] &,
    initval]; // AbsoluteTiming
```

```
{0.0904666, Null}
```

In this case the difference is negligible since, as mentioned before, there are many factors that may influence the computation time.

- We visually represent the solutions:

```
Show[Plot[{Cos[t^2], Exp[-t] + .3}, {t, 0, 12}], Graphics[
  {Red, Point[MapThread[{t /. #1, #2} &, {pvals, Cos[t^2 /. pvals]}]]}]]
```

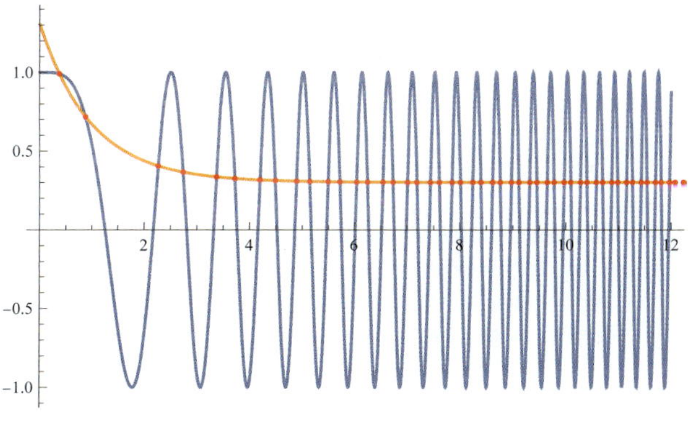

12.3 Application: The Mandelbrot Set

12.3.1 Defining the Mandelbrot Set

The Mandelbrot set (http://mathworld.wolfram.com/MandelbrotSet.html) is arguably the most famous fractal set. It can be created by recursion as follows :

$$z_0 = C$$

$$z_{n+1} = z_n^2 + C$$

where C is any complex number.

If we take C such that z_{n+1} doesn't tend to infinity, the points generated by this recursion belong to the Mandelbrot set.

- The previous series can be easily computed in *Mathematica* with FixedPointList.

```
FixedPointList[#^2 + c &, c, 5]
```

$\{c, c+c^2, c+(c+c^2)^2, c+(c+(c+c^2)^2)^2,$
$c+(c+(c+(c+c^2)^2)^2)^2, c+(c+(c+(c+(c+c^2)^2)^2)^2)^2\}$

- With $C = 1$, the series diverges (in the example we display only the first 6 terms). It tends to infinity in absolute terms, so it doesn't belong to it.

```
FixedPointList[#^2 + 1 &, 0, 5]
```

```
{0, 1, 2, 5, 26, 677}
```

- However if $C = -1$, the series is bounded and therefore belongs to the set.

```
FixedPointList[#^2 - 1 &, 0, 5]
```

```
{0, -1, 0, -1, 0, -1}
```

12.3.2 Visualizing the Mandelbrot Set

Often, the Mandelbrot set is represented using the escape-time coloring algorithm based on applying different colors depending on the number of iterations required to prove if a series belongs to the set. Darker colors (in our example bright red) indicate that only a few iterations are needed while lighter ones mean that many iterations were required. There's always a limit for the maximum number of iterations allowed. Since *Mathematica* 10 the build-up function `MandelbrotSetPlot` is available, however, we will build a function step by step to represent the Mandelbrot set.

- `FixedPointList` with the option `SameTest` enables us to set a logical test to stop the iteration as soon as it returns true starting with the first two values of the series. The command below returns the number of iterations needed to get an absolute value greater than 2 (the limit traditionally used) and if after 50 iterations, the absolute value still doesn't meet the condition, it returns the number of iterations +1. The higher the number of iterations, the higher the certainty that the series belongs to the Mandelbrot set:

```
Length[FixedPointList[#^2 - 0.2 + 0.1 I &, 0, 50, SameTest → (Abs[#] > 2 &)]]
```

```
51
```

- We define a function that implements the previous approach for a given complex number. In this case we use a maximum of 20 iterations:

```
mf1[z_] :=
  Length[FixedPointList[#^2 + z &, z, 20, SameTest → (Abs[#] > 2 &)]]
```

- We apply it to find out what points in the complex plane $x + i\,y$ belong to it. We add `AbsoluteTiming` to compare the computation times using different methods.

```
pts1 = Table[mf1[x + I y], {x, -2, 1, 0.002}, {y, -1, 1, 0.002}]; //
  AbsoluteTiming
```

```
{26.8221, Null}
```

- We use `ArrayPlot` to visualize the previous points.

```
ArrayPlot[pts1, ColorFunction -> Hue]
```

12.3.3 Faster

In this section we show several techniques, apart from parallelization, to reduce computation time.

These techniques require the presence of a C compiler in the computer. (CCompilerDriver/tutorial/Overview). For 32-bit systems, an option is to install Visual C++ Express, a Microsoft application that can be downloaded for free from: http://www.microsoft.com/visualstudio. In case of 64-bit systems, one option is to install the 64-bit version of MinGW (http://mingw-w64.sourceforge.net). In our case we have downloaded mingw-w64-bin_i686-mingw_20111220 and copied it to the folder C:\mingw.

- To run the C compiler we need to load the following package first.

```
Needs["CCompilerDriver`"]
```

- The command below will display all the C compilers available in your computer:

```
CCompilers[]
```

```
{{Name → Visual Studio,
  Compiler → CCompilerDriver`VisualStudioCompiler`VisualStudioCompiler,
  CompilerInstallation →
   C:\Program Files (x86)\Microsoft Visual Studio 14.0,
  CompilerName → Automatic}}
```

- If the command above returns an empty list it may be that either you don't have any compiler installed or, as in our case, you have installed MinGW in folder: C:/MinGW. To let *Mathematica* know that you have the open-source software in your computer, execute the following commands:

```
Needs["CCompilerDriver`GenericCCompiler`"]

$CCompiler =
  {"Compiler" -> GenericCCompiler, "CompilerInstallation" → "C:/MinGW",
   "CompilerName" → "x86_64-w64-mingw32-gcc.exe"};
```

Now we can launch the available kernels as explained before. To use kernels over a grid we need to make sure that we have access to them through WLGM.

- Here we repeat the calculation from the previous section but in parallel:

```
mf2[z_] :=
 Length[FixedPointList[#^2 + z &, z, 20, SameTest → (Abs[#] > 2 &)]]

DistributeDefinitions[mf2];
```

```
pts2 = ParallelTable[mf2[x + I y],
     {x, -2, 1, 0.002}, {y, -1, 1, 0.002}]; // AbsoluteTiming
```

```
{14.3025, Null}
```

- We can still significantly reduce the computation time with the command `Compile` (not limited to parallel calculations). We use it to indicate that we're going to evaluate a function only for numeric inputs. This way the program saves time by not performing the checks associated to symbolic calculations and can apply its own routines optimized for numerical calculations.

```
mf3 = Compile[{{c, _Complex}},
   Length[FixedPointList[#^2 + c &, c, 50, SameTest → (Abs[#2] > 2.0 &)]]];
DistributeDefinitions[mf3];
pts3 = ParallelTable[mf3[x + y ⅈ],
     {x, -2, 1, 0.002}, {y, -1, 1, 0.002}]; // AbsoluteTiming
```

```
{2.35448, Null}
```

- We can perform the computation even faster by taking advantage of some of the options of `Compile`. Using this function we can convert the Mathematica defined function into C code that is compiled and dynamically linked:

```
cfun = Compile[ {{x}}, x Sin[x^2],
   CompilationTarget → "C", RuntimeAttributes → {Listable}];
cfun[
 {3,
  4}]
```

```
{1.23636, -1.15161}
```

- Now, we can define the fastest function so far:

```
mf4 = Compile[{{c, _Complex}}, Module[{num = 1}, FixedPoint[(num++;
        #1^2 + c) &, 0, 99, SameTest → (Re[#1]^2 + Im[#1]^2 ≥ 4 &)];
    num], CompilationTarget → "C",
   RuntimeAttributes → {Listable}, Parallelization → True];

pts4 = mf4[Table[x + ⅈ y, {x, -2, 1, 0.002}, {y, -1, 1, 0.002}]]; //
  AbsoluteTiming
```

```
{2.31239, Null}
```

- We display the results from the three functions using different color templates and compare them with the built-in function `MandelbrotSetPlot`:

```
GraphicsRow[{ArrayPlot[pts2, ColorFunction → "GreenPinkTones"],
  ArrayPlot[pts3, ColorFunction → "BlueGreenYellow"],
  ArrayPlot[pts4 , ColorFunction → "Rainbow"]}]
```

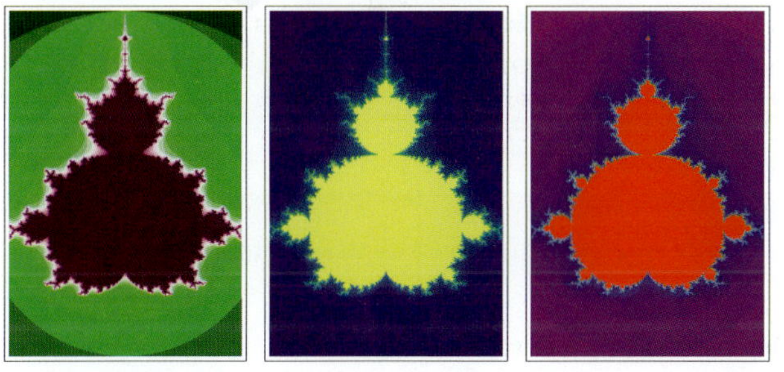

```
GraphicsRow[{MandelbrotSetPlot[ColorFunction → "GreenPinkTones"],
  MandelbrotSetPlot[ColorFunction → "BlueGreenYellow"],
  MandelbrotSetPlot[ColorFunction → "Rainbow"]}]
```

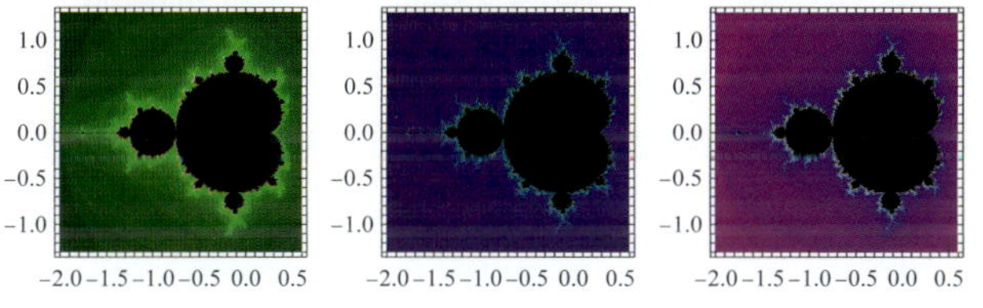

If you are a programmer you can even generate C code that can be converted into an executable file (CCodeGenerator/tutorial/Overview) or you can call external programs (guide/CallingExternalPrograms).

- To generate C code we need to load the CCodeGenerator package. In the example below, we load it after clearing all the variables from memory:

```
Clear["Global`*"]
```

```
Needs["CCodeGenerator`"]
```

- The next command generates optimized C code for a given expression:

```
s = ExportString[Compile[x, Sin[x^2]], "C"];
```

- We use `Panel` in the example below to show the output of the previous instruction inside a frame displaying the generated code along with all the required headings and libraries necessary to compile an executable file.

```
Panel[z[StringReplace[s, "\r\n" → "\n"],
   {Full, 300}, Scrollbars → {Automatic, True}],
  BaseStyle → {FontFamily → "Courier", FontSize → 8}]
```

```
z[#include "math.h"

#include "WolframRTL.h"

static WolframCompileLibrary_Functions funStructCompile;

static mbool initialize = 1;

#include "m-c048be80-539a-43dc-a4ee-59b1b37a8292.h"

DLLEXPORT int Initialize_m-c048be80-539a-43dc-a4ee-59b1b37a8292(WolframLibraryData libData)
{
if( initialize)
{
funStructCompile = libData->compileLibraryFunctions;
initialize = 0;
}
return 0;
}

DLLEXPORT void Uninitialize_m-c048be80-539a-43dc-a4ee-59b1b37a8292(WolframLibraryData libData)
{
if( !initialize)
{
initialize = 1;
}
}

DLLEXPORT int
    m-c048be80-539a-43dc-a4ee-59b1b37a8292(WolframLibraryData libData, mreal A1, mreal *Res)
{
mreal R0_0;
mreal R0_1;
mreal R0_2;
R0_0 = A1;
R0_1 = R0_0 * R0_0;
R0_2 = sin(R0_1);
*Res = R0_2;
funStructCompile->WolframLibraryData_cleanUp(libData, 1);
return 0;
}

, {Full, 300}, Scrollbars → {Automatic, True}]
```

12.4 Application: Comparing Organisms Genetically

Let’s see how parallel computing is used in the real world. In this case, we are trying to determine the genetic proximity of organisms using a string metric known as the Levenshtein distance.

The Levenshtein distance is used in information theory to measure the similarity between two sequences. We can define it as the minimum number of single character edits necessary to convert one sequence into the other one.

For example: To go from kitten to sitting, we proceed as follows:

kitten → sitten (we substitute 'k' with 's')
sitten → sittin (we substitute 'e' with 'i')
sittin → sitting (we insert 'g' at the end).

This means that the Levenshtein distance is 3 since we have made 3 edits to turn kitten into sitting.

- This metric can be calculated in *Mathematica* with `EditDistance`:

```
EditDistance["kitten", "sitting"]
```

```
3
```

Let's apply this metric to compare the genomes of three organisms using the files located in the Data directory that, along with other examples in this book, can be downloaded from http://diarium.usal.es/guillermo.

- We first set the working directory to the one where the files are located (Data). This way we can use Import directly without the need to state the entire path:

```
SetDirectory[FileNameJoin[{NotebookDirectory[], "Data"}]]
```

```
C:\Users\guill\Documents\MathematicaBeyond\Data
```

```
picrogenes = Import["picrophilus_torridus_gene.fna"];
acidogenes = Import["thermoplasma_acidophilum_gene.fna"];
volcaniumgenes = Import["thermoplasma_volcanium_gene.fna"];
```

In the previous example, the computation time was very short but, when the number of characters to compare increases, the time required for the calculations increases drastically.

- Let's see the partial contents of one of the files containing the genome of the picrogenes. The command below shows some of their DNA bases (A,C, G, T):

```
picrogenes
```

```
{ATGAATTCGGATAACTTTCAATCCTATTTAGGTTATTATTTAACCTGGATTTCCTTGGCCAT
  ... CAATCCTATTTAGGTTATTATTTAACCCGTTTCTAATAAACAATATAATAAATGATTTTTAA, <<1596>>, ...}
large output   show less   show more   show all   set size limit...
```

- These three organisms are very simple and their genome size (total number of genes) is respectively:

```
Length /@ {picrogenes, acidogenes, volcaniumgenes}
```

```
{1598, 1575, 1610}
```

- Let's load the imported data to analyze them using parallelization:

```
With[{ picrogenes = picrogenes, acidogenes = acidogenes,
   volcaniumgenes = volcaniumgenes}, ParallelEvaluate[picro = picrogenes;
    acido = acidogenes; volcanium = volcaniumgenes;];];
```

- The command below shows the data for one gene that will be used as an input for the first kernel:

```
ParallelEvaluate[picro[[1]], Kernels[][[1]]]
```

```
ATGAATTCGGATAACTTTCAATCCTATTTAGGTTATTATTTAACCTGGATTTCCTTGGCCATATTTGCCGAGA\
  TACTATACTTTCAATCCTATTTAGGTTATTATTTAACATATATCGGATTTCAAACCCTTATAGATAGACT\
  AGACTTTCAATCCTATTTAGGTTATTATTTAACGCCAGAGGCATTTCTCCGTGCGTGGTCCTTTAATGAC\
  TTTCAATCCTATTTAGGTTATTATTTAACACAACAGGCAAAAACAGAATTAAGACTGGCCATTGAGCTTT\
  CAATCCTATTTAGGTTATTATTTAACCCGTTTCTAATAAACAATATAATAAATGATTTTTAA
```

For the next example, we want to compare the first eight genes of *thermoplasma acidophilum* with those of *thermoplasma volcanium.* However, before doing it, let's discuss a simpler case first. In the example below we compare the initial 4 bases of the first 2 genes of volcaniumgenes, with their respective counterparts in *thermoplasma acidophilum.*

- We start by extracting the relevant bases and genes from both organisms:

```
{vo1, vo2, aci1, aci2} =
  {StringTake[volcaniumgenes[[1 ]], 4], StringTake[volcaniumgenes[[2]], 4],
   StringTake[acidogenes[[1]], 4], StringTake[acidogenes[[2]], 4] }
```

```
{ATGA, ATGA, ATGT, ATGG}
```

Notice how the first two bases of *thermoplasma volcanium* are actually the same.

- To make the comparison we'll use a combination of the `EditDistance` and `Min` functions. However, in the command below we use **f1** and **f2** first to gain a better understanding of how the actual instruction works. We show the logic in four steps:

```
f2[Map[f1[x, #] &, {aci1, aci2}]]
```

```
f2[{f1[x, ATGT], f1[x, ATGG]}]
```

```
Map[Function[{x}, f2[Map[f1[x, #] &, {aci1, aci2}]]], {vo1, vo2}]
```

```
{f2[{f1[ATGA, ATGT], f1[ATGA, ATGG]}],
 f2[{f1[ATGA, ATGT], f1[ATGA, ATGG]}]}
```

- Here **f1** and **f2** are replaced with `EditDistance` and `Min` respectively:

```
Map[Function[{x}, f2[Map[EditDistance[x, #] &, {aci1, aci2}]]],
 {vo1, vo2}]
```

```
{f2[{1, 1}], f2[{1, 1}]}
```

```
Map[Function[{x}, Min[Map[EditDistance[x, #] &, {aci1, aci2}]]],
 {vo1, vo2}]
```

```
{1, 1}
```

- Finally we apply the previous pattern using `ParallelMap` to compare the first eight genes of *thermoplasma acidophilum* with those of *thermoplasma volcanium*:

```
acidovolcanium =
   ParallelMap[Function[{x}, Min[Map[EditDistance[x, #] &, acido]]],
    volcaniumgenes[[1 ;; 8]]]; // AbsoluteTiming
```

```
{45.7372, Null}
```

Depending on the computer being used, the calculation time may take up to several minutes.

```
acidovolcanium
```

```
{304, 700, 265, 227, 497, 243, 122, 104}
```

- We proceed the same way to compare the genes between *picrophilus torridus* and *thermoplasma volcanium*:

```
picrovolcanium =
   ParallelMap[Function[{x}, Min[Map[EditDistance[x, #] &, picro]]],
    volcaniumgenes[[1 ;; 8]]]; // AbsoluteTiming
```

```
{46.4057, Null}
```

```
picrovolcanium
```

```
{399, 968, 442, 353, 550, 240, 180, 124}
```

- The graph below shows the genetic proximity of the first eight genes in both cases: *thermoplasma acidophilum* vs. *thermoplasma volcanium* and *picrophilus torridus* vs. *thermoplasma volcanium*:

```
ListLinePlot[{acidovolcanium, picrovolcanium},
 PlotLegends → {"acido vs. volcanium", "picro vs. volcanium"}]
```

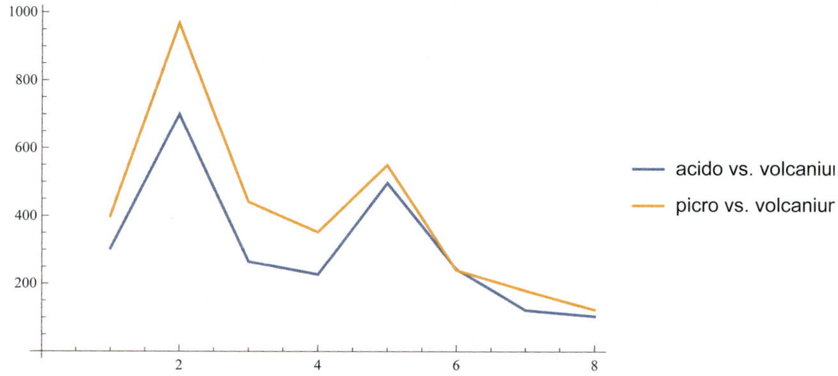

- Comparing *thermoplasma acidophilum* vs. *thermoplasma volcanium* and *picrophilus torridus* vs. *thermoplasma volcanium* in terms of their mean distances we get:

```
{Mean[acidovolcanium], Mean[picrovolcanium]} // N
```

```
{307.75, 407.}
```

- A more efficient alternative calculation method using `WaitAll` is:

```
(acidovolcanium = Map[Function[{gene}, ParallelSubmit[
      Min[EditDistance[gene, #] & /@ acido]]], volcaniumgenes[[1 ;; 8]]];
picrovolcanium = Map[Function[{gene}, ParallelSubmit[
      Min[EditDistance[gene, #] & /@ picro]]], volcaniumgenes[[1 ;; 8]]];
acidovolcanium;
picrovolcanium;
acidovolcaniumresult = WaitAll[acidovolcanium];
picrovolcaniumresult = WaitAll[picrovolcanium];) // AbsoluteTiming
```

```
{74.0176, Null}
```

The calculation time is shorter than the sum of the times of the previous two computations.

- The results are identical to the previously obtained ones:

```
Mean[acidovolcaniumresult] // N
```

```
307.75
```

```
Mean[picrovolcaniumresult] // N
```

```
407.
```

12.5 Grid Computing with Wolfram Lightweight Grid Manager (WLGM)

We start a new session. If you haven't exited the previous one you can do it now.

```
Quit[]
```

Wolfram Lightweight Grid Manager (WLGM) enables the connection between computers in a grid to perform parallel computations. WLGM is included in some *Mathematica* licenses while grid*Mathematica* is a stand alone program.

The computational power of parallelization comes to life when using a large number of processors. This is often possible when there are many available PCs connected in a grid eliminating the need to purchase expensive specialized machines. In universities or firms we can leverage the existing resources for performing parallel computations when they're not in use or are being underutilized as is often the case.

The connection of computers in a network to execute parallel calculations with *Mathematica* can be done in several ways.

12.5.1 Network with Cluster Integration

If your network has a cluster management tool such as: Computer Cluster Server, Windows HPC Server 2008 or Platform™ LSF® among others, you can use it with the appropriate settings. Choose in the main menu toolbar: **Edit ▸ Preferences ▸ Parallel ▸ Cluster Integration**, and after pressing **Enable Cluster Integration**, you will see a menu similar to Figure 12.3. The right configuration will depend on the type of tool present in your network (for further details see your system documentation).

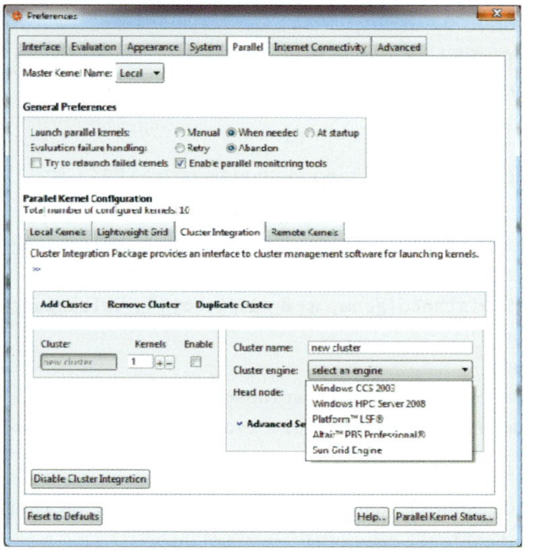

Figure 12.3 Cluster Integration Settings.

Another option is to access the kernels of a stand alone machine from your local PC. In this case, you will need to configure your connection with **Edit ▸ Preferences ▸ Parallel ▸ Remote Kernels**.

12.5.2 WLGM Installation

Nevertheless, the most recommended and simplest way to connect computers in a grid is with the Wolfram Lightweight Grid Manager (WLGM: http://www.wolfram.com/lightweight-grid-manager).

Since version 7, *Mathematica* by default installs the Wolfram Lightweight Grid Client (WLGC). However, to be able to use this resource we have to install the Wolfram Lightweight Grid Manager (WLGM) in each computer as well. This program is included at no additional cost as part of Wolfram Premier Service. Usually, we will receive a link to download the setup file.

We need to install the program in all the computers running *Mathematica* that we want to connect in a grid network. **It's recommended that you keep in mind the following advice during the installation of WLGM**:

1. Run the setup file either as administrator or with administrator privileges.

2. During the installation process (Figures 12.4 and 12.5), we will be asked to configure two accounts: a user account (by default wolframgrid but you can choose your own account) and an administrator account "admin" (the name assigned by default) for the administrator account. We can modify the configuration using the administrator account. Normally, this will not be necessary so we will use WLGM default configuration.

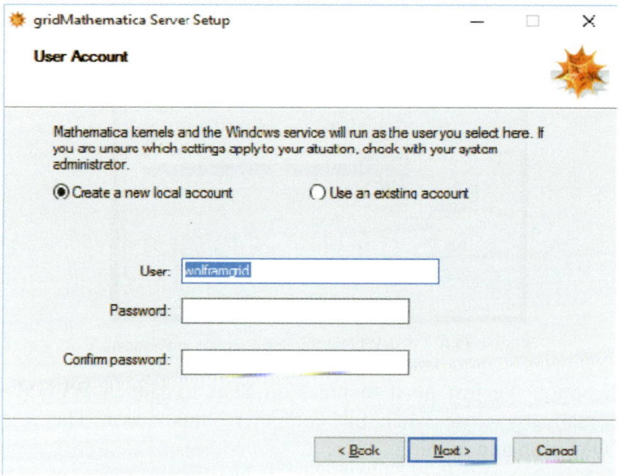

Figure 12.4 Creating a new account for WLGM.

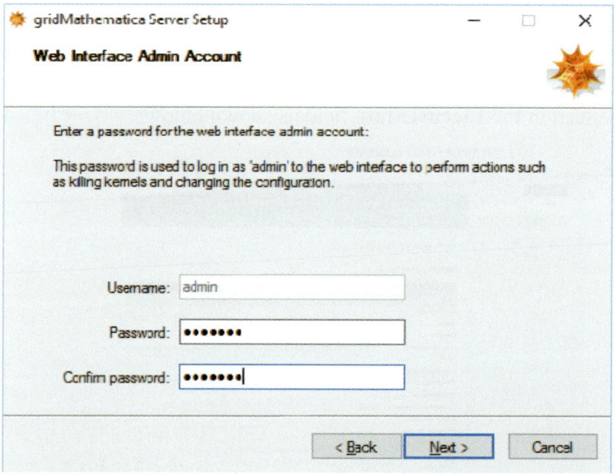

Figure 12.5 Setting up the password for the admin account in WLGM.

3. If your network is protected by a firewall, you may have to open (in each computer) ports 3737 and 5353 (mDNS) to enable traffic in both directions.

4. If you use the firewall provided by Windows, go to the **Windows Firewall** item located in the **Control Panel** and after clicking on it, in the *Advanced settings* window configure ports 3737 and 5353 to allow inbound and outbound communication.

The WLGM installation creates two icons: WLGM Web Manager and WLGM Service Console. Through WLGM Service Console we can enable the automatic execution of WLGM (default configuration) every time the computer boots up or set it to manual. Most of the time, the automatic configuration is the most convenient one. To set it to manual (for example: when we only use WLGM sporadically) we will open the WLGM Service Console with administrator privileges and choose manual for the *Startup type* dropdown box (Figure 12.6):

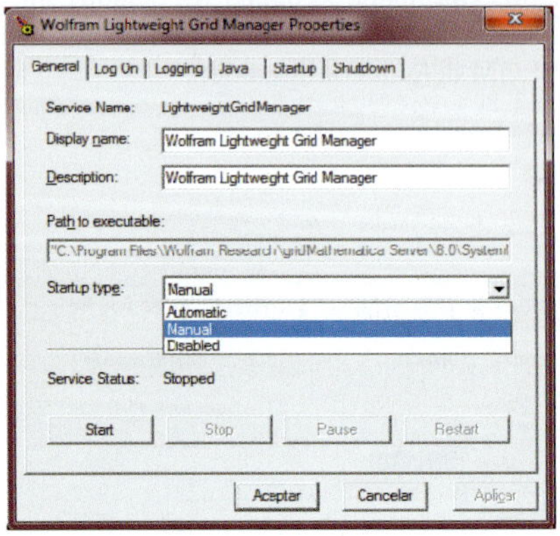

Figure 12.6 The WLGM Service Console in Windows.

With the manual option we just need to press on Start to launch WLGM or Stop to quit. WLGM will automatically close when the computer shuts down. The rest of this section assumes that WLGM is up and running.

If we click on the WLGM Web Manager icon when WLGM is running, a web-based interface will appear (Figure 12.7). From there we can manage the computers available to run *Mathematica* kernels, deciding who has access to what machine and to how many kernels. There are also tools to monitor use, anticipate problems or fix them.

If we experience technical difficulties, it is important to check that the status of the license is valid. That can be seen in the **LicenseState** field as shown below:

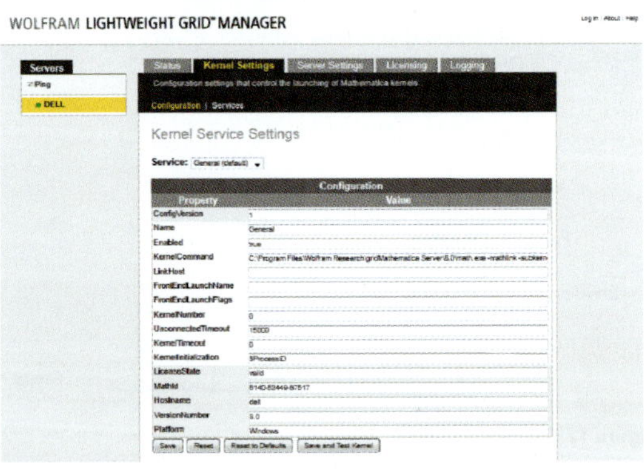

Figure 12.7 WLGM Web Manager.

We can also access the WLGM documentation by clicking on Help (upper-right corner in the previous screenshot).

If the LicenseState doesn't show valid but *Mathematica* is working properly you may have to copy the *mathpass* file in the BaseDirectory removing it from the UserBaseDirectory. This can be done as follows (http://support.wolfram.com/kb/topic/gridmathematica):

- Execute the command below:

```
SystemOpen[$UserBaseDirectory]
```

- Look inside the subfolder Licensing for a file named *mathpass*. If it's there, move it to the subdirectory Licensing located in:

```
SystemOpen[$BaseDirectory]
```

  Then, after making sure that *Mathematica* and WLGM are not running, launch WLGM again from the WLGM Service Console.

12.5.3 WLGM Configuration

Start WLGM and open a new session in *Mathematica* or even better, close the program and open it again.

Now go to **Edit ▸ Preferences ▸ Parallel ▸ Lightweight Grid**

You should be able to see how many connected computers are available (Figure 12.8). Select the ones you want to use (enable: ✓). In the example below we have installed WLGM in two computers at home connected to a Wi-Fi network.

- Choose the number of kernels that you want to use in each of the computers. In our case there are two computers identified as HPCorei7 and Pepon, each of them with 4 kernels. HPCorei7 is the master one.

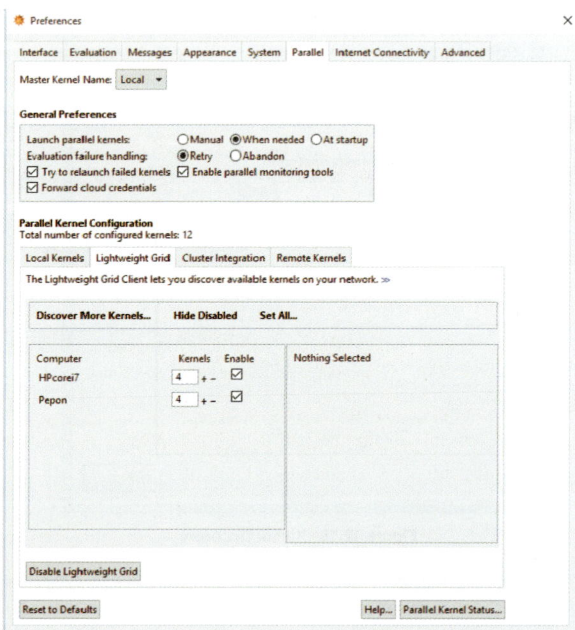

Figure 12.8 Lightweight Grid Preferences Panel.

- You need to make sure that WLGM is running in each computer by checking that the status of the service (Figure 12.9).

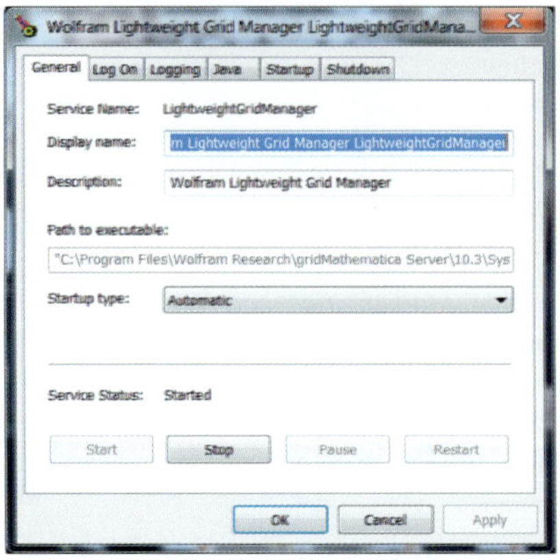

Figure 12.9 Checking the service status of WLGM (Service Status: Started).

- Besides the computers in our grid, we can also connect to machines in other networks (most likely we will have to configure the firewall to allow the connection) that have *Mathematica* and WLGM installed on them. For that purpose we can click on Discover More Kernels and type the address or addresses of the networks where those computers are located (Figure 12.10):

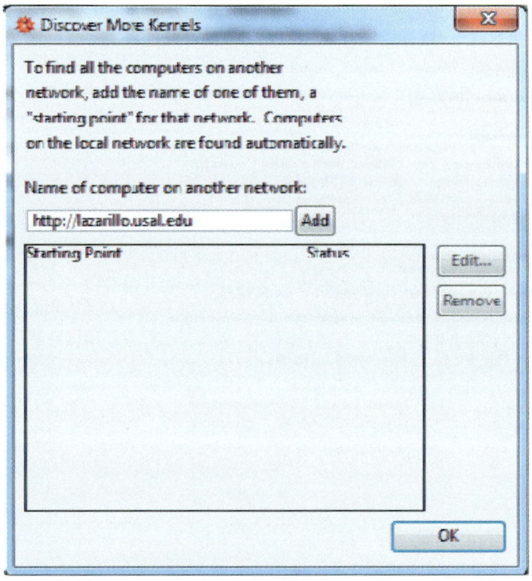

Figure 12.10 Kernel Discovery.

The total number of available kernels will be limited by the license type. Normally it will be 16, but in enterprise licenses that number may be higher.

One of the advantages of WLGM is that it makes it possible to connect to many types of computers, even those with different operating system.

- We can get very detailed information of our system as follows:

```
SystemInformation[]
```

Try clicking on the different tabs. Notice that they all include dropdown menus and interactive buttons.

Once WLGM has been successfully configured, its regular use is very simple and most of the time it will be limited to just setting up the number of kernels for each session. It's recommended that the number of cores launched in each computer is the same as the number of CPU cores.

- We can also use the command below to know our network status at any time. `ParallelEvaluate` will launch the kernels if they haven't been launched yet:

```
TableForm[ParallelEvaluate[{$KernelID, $MachineName, $SystemID,
   $VersionNumber, $ProcessID, TimeUsed[], MaxMemoryUsed[],
   "ProcessPriority" /. SystemOptions["ProcessPriority"]}],
 TableHeadings -> {None, {"ID", "host", "OS", "Version", "Process",
    "CPU Time", "Memory", "Priority"}}, TableDepth → 2]
```

ID	host	OS	Version	Process	CPU Time	Memory
1	pepon	Windows–x86–64	10.3	6724	0.969	40 848 200
2	pepon	Windows–x86–64	10.3	11 240	0.985	40 886 224
3	pepon	Windows–x86–64	10.3	10 512	0.984	40 885 528
4	pepon	Windows–x86–64	10.3	6044	0.954	40 847 616
5	hpcorei7	Windows–x86–64	10.3	2776	2.704	40 747 320
6	hpcorei7	Windows–x86–64	10.3	13 024	2.5	40 709 376
7	hpcorei7	Windows–x86–64	10.3	12 396	2.562	40 710 184
8	hpcorei7	Windows–x86–64	10.3	5080	2.625	40 747 072
9	pepon	Windows–x86–64	10.3	8404	0.937	40 764 888
10	pepon	Windows–x86–64	10.3	6140	0.937	40 803 328
11	pepon	Windows–x86–64	10.3	7060	1.015	40 764 888
12	pepon	Windows–x86–64	10.3	7876	1.047	40 803 328

- If we now perform a parallel computation all the available kernels will be used. However, we also need to take into consideration that although the computing power has increased there will be some overhead related to the management of the communication among all the computers in the grid. It's always a good idea to run some tests to make sure that the tasks we want to perform would benefit from using the computers in the grid. Naturally, grid computing would always be recommended for time intensive calculations taking more than just a few minutes.

```
Parallelize[∑_{i=1}^{10000} i!]; // AbsoluteTiming
```

{0.386827, Null}

12.6 Compute Unified Device Architecture (CUDA)

- We start a new session

```
Quit[]
```

NVidia graphics cards, present in many computers, include multiple processors that can be used for parallel computations using a programming language named Compute Unified Device Architecture (CUDA) (http://www.nvidia.com/object/what_is_cuda.html). CUDA takes advantage of the power of the GPU (Graphical Processing Unit) to allocate computer instructions among the kernels in the card, boosting the overall performance of the system.

Mathematica since Version 7 includes the possibility of using CUDA.

- To program in CUDA we must load the following package first:

```
<< CUDALink`
```

- CUDALink requires the presence of a CUDA-enabled NVIDIA graphics card in the system (since 2009, almost all the graphics cards have that capability). We can check it by running the following command:

```
CUDAInformation[]

{1 →
  {Name → GeForce GTS 450, Clock Rate → 1750000, Compute Capabilities → 2.1,
   GPU Overlap → 1, Maximum Block Dimensions → {1024, 1024, 64},
   Maximum Grid Dimensions → {65535, 65535, 65535},
   Maximum Threads Per Block → 1024,
   Maximum Shared Memory Per Block → 49152, Total Constant Memory → 65536,
   Warp Size → 32, Maximum Pitch → 2147483647,
   Maximum Registers Per Block → 32768, Texture Alignment → 512,
   Multiprocessor Count → 4, Core Count → 128, Execution Timeout → 1,
   Integrated → False, Can Map Host Memory → True, Compute Mode → Default,
   Texture1D Width → 65536, Texture2D Width → 65536,
   Texture2D Height → 65535, Texture3D Width → 2048, Texture3D Height → 2048,
   Texture3D Depth → 2048, Texture2D Array Width → 16384,
   Texture2D Array Height → 16384, Texture2D Array Slices → 2048,
   Surface Alignment → 512, Concurrent Kernels → True,
   ECC Enabled → False, TCC Enabled → False, Total Memory → 1073741824}}
```

Now we can use the CUDA functions. For further information, please see the tutorial: CUDALink/tutorial/Overview.

- One area where CUDA can be very useful is in image processing. The command below creates an interface where the user can process input images from a web camera in real time.

```
Manipulate[fun[CurrentImage[], r], {fun,
  {CUDAErosion, CUDADilation, CUDAOpening, CUDAClosing}}, {r, 1, 10, 1}]
```

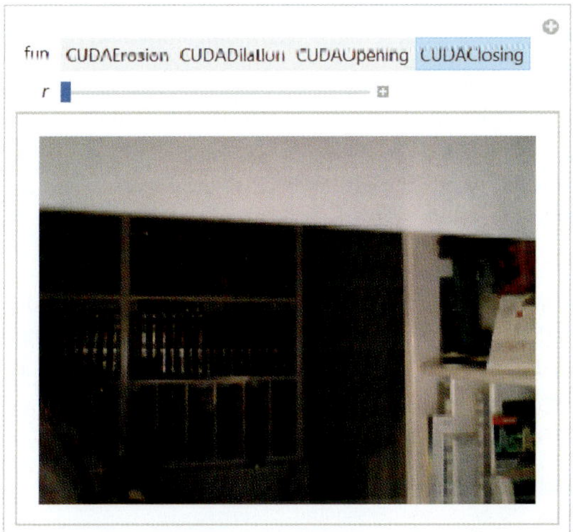

12.7 *Mathematica* for the Web: web*Mathematica*

12.7.1 What Is web*Mathematica*?

Nowadays, people expect that computer applications can be run over the Internet. In the first chapter we saw that *Mathematica* offers that capability using http://www.wolfram.com/cloud/ but Wolfram Research also has another tool, *webMathematica*, especially useful if you want to develop sophisticated Internet-based applications that can be accessed from anywhere with a browser. With web*Mathematica* and very little knowledge of Java, you can build web applications in *Mathematica*. The user will interact with the program using a browser, the calculations will be done in the server where *Mathematica* runs and the results will be sent back to the user. For non-commercial purposes, *Mathematica* premier service users can obtain a free license for the program.

Internally, web*Mathematica* uses Java Servlet and JavaServer Pages (JSPs). But before we can run it, we need to first install the Java Runtime Environment (JRE), a program that can be downloaded for free from http://java.com in case we don't have it yet in our computers. We will also need to be able to run servlets, small Java programs that usually run in servlet containers. There are containers for different architectures and operating systems and they all can be integrated into web servers such as Apache. Probably, the most well-known servlet container is Apache-Tomcat, the one used in this book. It can be downloaded for free from http://tomcat.apache.org.

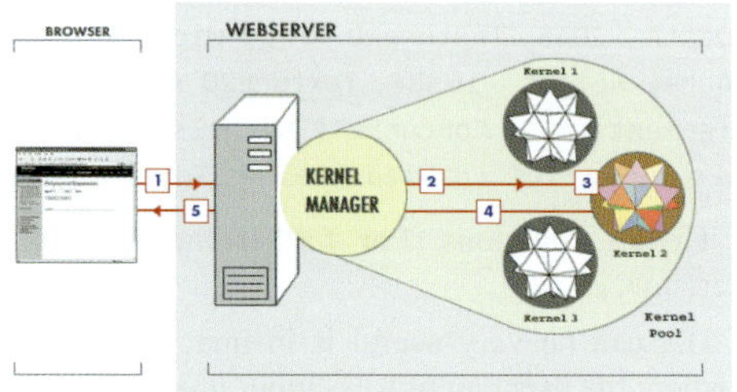

Figure 12.11 Web*Mathematica* in action.

The way web*Mathematica* works is as follows (Figure 12.11):

1. The browser (e.g., Firefox, Chrome) sends a request to the web*Mathematica* server.

2. The web*Mathematica* server calls the *Mathematica* kernel. Given that several users can interact simultaneously, each new open kernel will be assigned to a specific request.

3. The *Mathematica* kernel is then initialized with the input parameters given by the user. *Mathematica* performs the calculations and sends the results back to the server where they can be seen by the user using the browser.

4. The web*Mathematica* server closes the kernel, making it available to other users.

The input requests are usually sent directly through the browser but when the size of the input is big, it can be uploaded as a file (different formats are allowed). The outputs are normally shown as images in the browser but there are also other alternative formats such as HTML, *Mathematica* notebooks, MathML, SVG, XML, PostScript, and PDF.

You can try some examples in:

http://www.wolfram.com/products/webmathematica/examples/examples.html

12.7.2 Installation and Testing

In this section we are going to use web*Mathematica* 3.4. The complete installation process (JRE + Tomcat + web*Mathematica*) can be quite time consuming. Before installing web*Mathematica* we need to make sure that JRE and Tomcat are running properly. To do that, type the following in your browser address bar: http://localhost:8080/examples/servlets. You should see a screen similar to Figure 12.2 when pressing **Execute**:

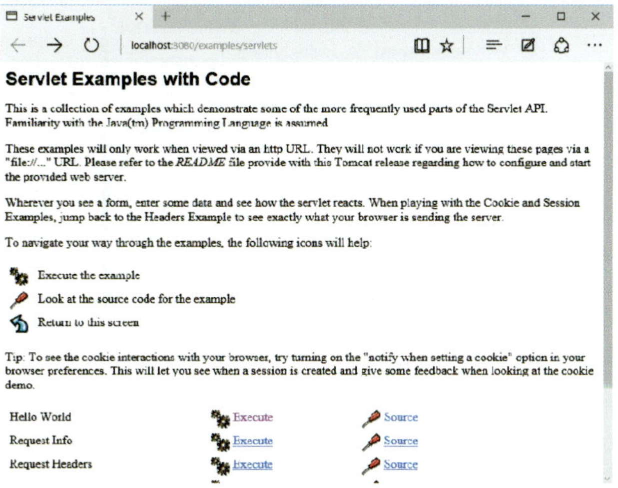

Figure 12.12 Servelet Examples Page.

If instead of Figure 12.2 you see the message: “web not available”, Tomcat may not be working or it may not have been configured correctly. In particular, you need to make sure that the path to the Java Virtual Machine is indicated properly by pointing at the jvm.dll file located in Program Files and not the one in Program Files (x86) (Figure 12.13),

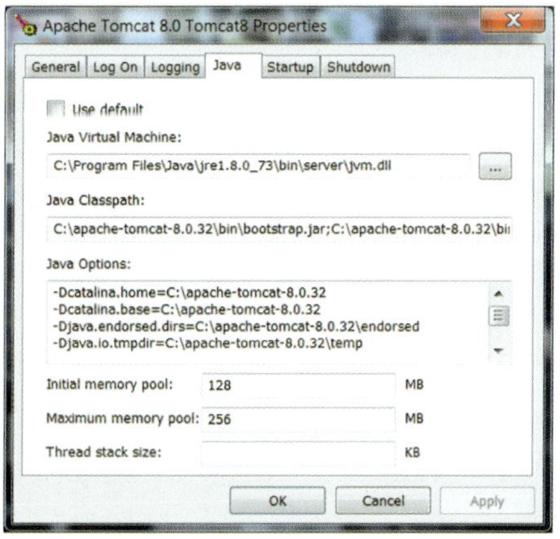

Figure 12.13 Troubleshooting Apache Tomcat.

- Make sure that Tomcat is running. You can do it using the Monitor Tomcat icon. The Service Status should be “Started” (Figure 12.14).

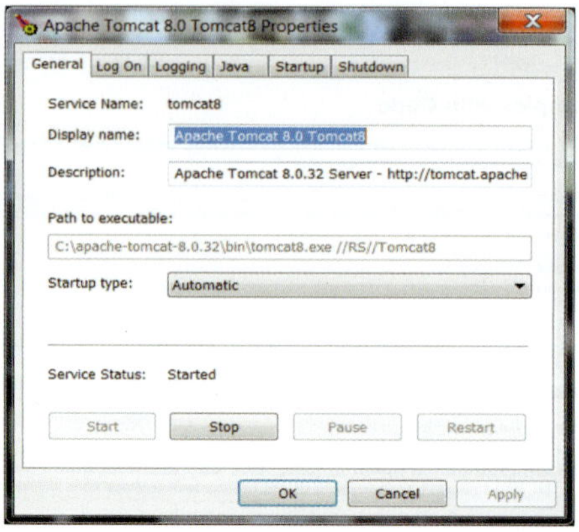

Figure 12.14 Checking the status of the Tomcat service.

Once we know Tomcat is working properly, we can proceed with the web*Mathematica* installation. Basically we just need to copy the folder web*Mathematica* in the Tomcat subdirectory: \webapps (usually in C:\Program Files\Apache Software Foundation\Tomcat N\webapps), with N=7.x, 8.x or 9.x depending on the installed Tomcat version.

- Locate the file: "...\webapps\webMathematica\WEB-INF*MSPConfiguration.xml*", open it and insert the path to the *MathKernel.exe* in your machine. Usually, it would be like this (change the path to *MathKernel.exe* if necessary):

```
<MSPConfiguration version =" 3.0">
 <DataStore>
    <DataStoreClass> com.wolfram.msp.datastore.FileStore </DataStoreClass>
    </DataStore>
    <KernelPool>
    <KernelExecutable> C:\Program Files\Wolfram
Research\Mathematica\10.3\MathKernel.exe </KernelExecutable>
        <KernelPoolName> General </KernelPoolName>
        <URLPattern> /* </URLPattern>
    </KernelPool >
   </MSPConfiguration>
```

- Check that it is working properly by typing in your browser:
  http://localhost:8080/webMathematica
- You should see Figure 12.15:

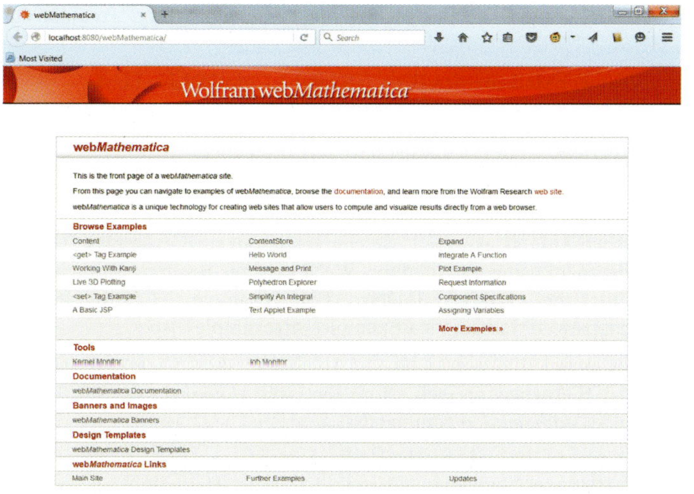

Figure 12.15 Web*Mathematica* home page.

- Next, under "Browse Examples", if we click on **Expand** (http://localhost:8080/webMathematica/BrowseExamples/Expand.html) and press **EVALUATE**, web*Mathematica* will return Figure 12.16:

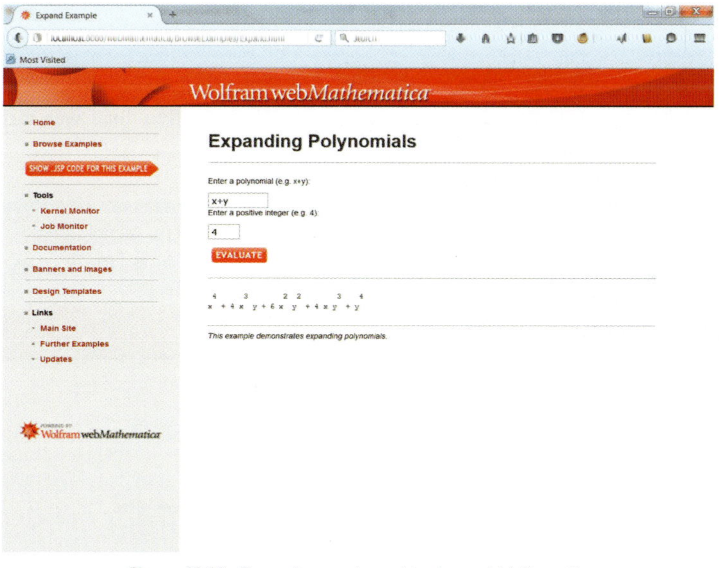

Figure 12.16 Expanding a polynomial using web*Mathematica*.

In Windows 10/8 if when running the example you get an error message of the type "FrontEndError:..." you need to increase the memory that is assigned by default (Select: "Problems Launching the Front End (Windows only)"):

http://reference.wolfram.com/webMathematica/tutorial/TroubleShootingSpecificProblems.html

Another possible issue could be related to Java not working properly. To make sure that's not the case, try:

http://localhost:8080/webMathematica/BrowseExamples/Plot3DLive.html.

If it works, your browser is compatible with web*Mathematica* and it has been configured properly. In our experience, Firefox doesn't have any problems.

12.7.3 Applications Development

Once everything is working as expected, we can create our own examples. The easiest way to get started is to type them using a text editor. We'll save the resulting file with a jsp extension and copy it to the servlet folder (For example: C:\Program Files\apache-tomcat-8.0.32\webapps\webMathematica\Examples\myexample1.jsp).

The basic examples included in web*Mathematica* have a "Show Code" button that will display the contents of the file when pressed. Note that web*Mathematica* uses HTML tags with web*Mathematica* and *Mathematica* commands.

- Below we show part of the code for the Hello example: http://localhost:8080/webMathematica/BrowseExamples/Hello.html.

```
<html><head><title>Hello World</title>...</head>
<body>
...
<h4>Date[]</h4>
<msp:evaluate>Date::usage </msp:evaluate>
<p>Its current value is:</p>
<msp:evaluate>Date[]</msp:evaluate>
</div>
...
</body></html>
```

Notice that the HTML file has two tags specifically for web*Mathematica*: `<msp:evaluate>` **Date[]** </msp : evaluate >

- The output is shown in Figure 12.17:

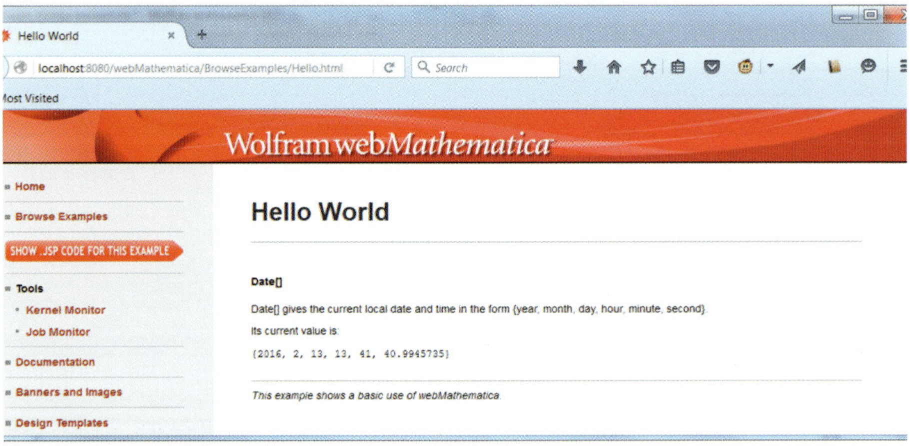

Figure 12.17 Our first program in web*Mathematica*.

12.7.4 Building Your Own Example

The example that follows is based on:
http://localhost:8080/webMathematica/BrowseExamples/Expand.html. We'll use it as a template and modify it accordingly (remember: use the Show Code button to see its contents).

We're going to create a file that uses the function **SizeSample**[N_Integer, alpha_, f_] included in the "StatisticalPackage" file, developed in section 6.10.2., to calculate the sample size.

```
Needs["StatisticalPackage`"]
```

```
?"StatisticalPackage`*"
```

▼ StatisticalPackage`

ProbAlpha	SizeSample

For instance, to calculate the sample size required to be 95% confident that in a population of 1,000 units the fraction of defective elements is at most 1%, we would type the following:

```
SizeSample[200, 0.05, 0.01]
```

```
155
```

Let's see how we can build an application using this function.

- Create a new file with a text editor, such as Notepad in Windows, and name it *Statisticalexample.jsp* (you can use another name but the extension must be jsp).
- Next, copy all the code extracts in the next few paragraphs to the new file.

```
<!DOCTYPE HTML PUBLIC "-//W3C//DTD HTML 4.01 Transitional//EN" "http://www.w3.org/TR/html4/loose.dtd">
<<!DOCTYPE HTML PUBLIC "-//W3C//DTD HTML 4.01 Transitional//EN" "http://www.w3.org/TR/html4/loose.dtd">
<%@ taglib uri="http://www.wolfram.com/msp" prefix="msp" %>
<html>
<head>
<title>Statistical example</title>
<meta http-equiv="Content-Type" content="text/html; charset=iso-8859-1"/>
<link rel="shortcut icon" href="../Resources/Images/favicon.ico" type="image/x-icon"/>
<link rel="stylesheet" type="text/css" href="../Resources/CSS/webMathematica.css"/>
<script type="text/javascript" src="../Resources/JavaScript/webMathematica.js"></script>
</head>
<body>
<div class="container_noborder">
<div class="section">
<h1>Statistical example</h1>
</div>
<div class="section">
<form action="statisticalexample.jsp" method="post">
```

- The next three lines tell web*Mathematica* to load the package. Obviously, it should have been placed previously in the Applications folder inside the $BaseDirectory (FileNameJoin[{$BaseDirectory,"Applications"}]):

```
<msp:evaluate>
Needs["StatisticalPackage`"];
```

```
</msp:evaluate>
```

- This section of the code contains the instructions related to the inputs. Notice the addition of default values:

```
<div>
Enter the population size (A positive integer >100): <br/>
<input type="text" name="num"  size="4" value="<msp:evaluate>MSPValue[ $$num, "200"]</msp:evaluate>"/>
</div>
<div>
Enter the fraction of defective elements : <br/>
<input type="text" name="ff"  size="4" value="<msp:evaluate>MSPValue[ $$ff, "0.01"]</msp:evaluate>" />
</div>
<div>
Enter the alpha value (usually a number between 0.01- 0.1) : <br/>
<input type="text" name="alpha"  size="4" value="<msp:evaluate>MSPValue[ $$alpha, "0.05"]</msp:evaluate>" />
</div>
```

- At this stage we include an **Evaluate** button that will call up a graphics file located in a subdirectory in the folder where our file is located:

```
<div>
<input type="image" src="../Resources/Images/Buttons/evaluate.gif" alt="Evaluate"/>
</div>
</form>
</div>
<div class="section">
```

- The next 3 lines correspond to the function that the package is going to use. Look carefully at the syntax being used:
  MSPBlock[{$$num,$$alpha,$$ff },SizeSample[$$num,$$alpha,$$ff]].
  Notice the use of $$ to refer to arguments previously defined. After them, we give a brief description of the application and close the remaining open HTML commands.

```
<msp:evaluate>
MSPBlock[{$$num,$$alpha,$$ff },SizeSample[$$num,$$alpha,$$ff]]
</msp:evaluate>
</div>
<p class="description">
The example we just created returns the sample size n required to have a confidence level 1-alpha that the population N contains a fraction of defective elements lower or equal to f.
</p>
</div>
</body>
</html>
```

- Next we save the file S*tatisticalexample.jsp* to: ...\Apache Software Foundation\Tomcat 8.0\webapps\webMathematica\Examples.
- Finally, with Tomcat running, go to:
  http://localhost:8080/webMathematica/Examples/Statisticalexample.jsp
  and you should see Figure 12.18 after pressing the "EVALUATE" button:

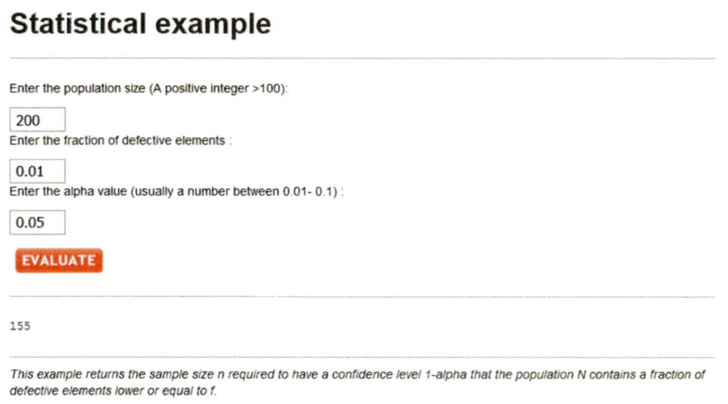

Figure 12.18 Running the StatisticalPackage in web*Mathematica*.

The development of web*Mathematica* applications like the previous one is easier with Workbench, a tool that we're going to discuss in the next section.

- You can download *StatisticalPackage.m* and *Statisticalexample.jsp* file from: http://diarium.usal.es/guillermo.

12.8 Software Development with Wolfram Workbench

If you are going to build large software systems in *Mathematica* (tutorial/BuildingLargeSoftwareSystemsInTheWolframLanguage) you may be interested in using Wolfram Workbench (http://www.wolfram.com/products/workbench), an Integrated Development Environment (IDE) based on software from the Eclipse Foundation (http://www.eclipse.org). Workbench can help you edit, test and debug code. It can also generate help files than can be integrated within the *Mathematica* documentation. It's highly recommended if you'd like to develop a high-quality enterprise application.

A large project will require two things: Code (consisting normally of several packages) and extensive documentation. Workbench will help you get both.

Wolfram Workbench requires the following software:

1. *Mathematica*

2. Java Development Kit (JDK) 8 or later

3. Eclipse 4.6 (Neon) or later

Note that the Wolfram Workbench plugin is compatible with the various standard Eclipse IDEs.

Wolfram recommend having the latest version of these programs installed before setting up the Wolfram Workbench plugin.

- To set up Wolfram Workbench, install the plugin inside Eclipse and configure it to point to your *Mathematica* installation as it is shown in the following figures (more details in http://support.wolfram.com/kb/27221):

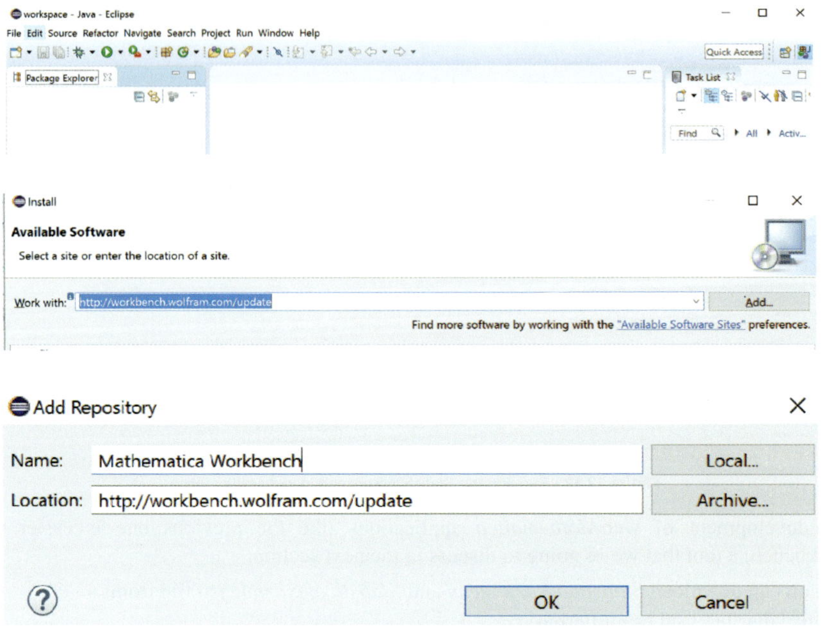

Figure 12.19 Step1 to configure Workbench.

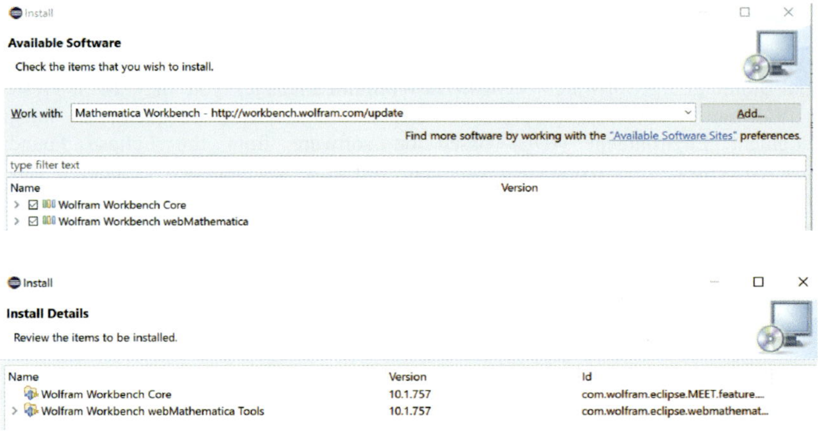

Figure 12.20 Step 2 to configure Workbench.

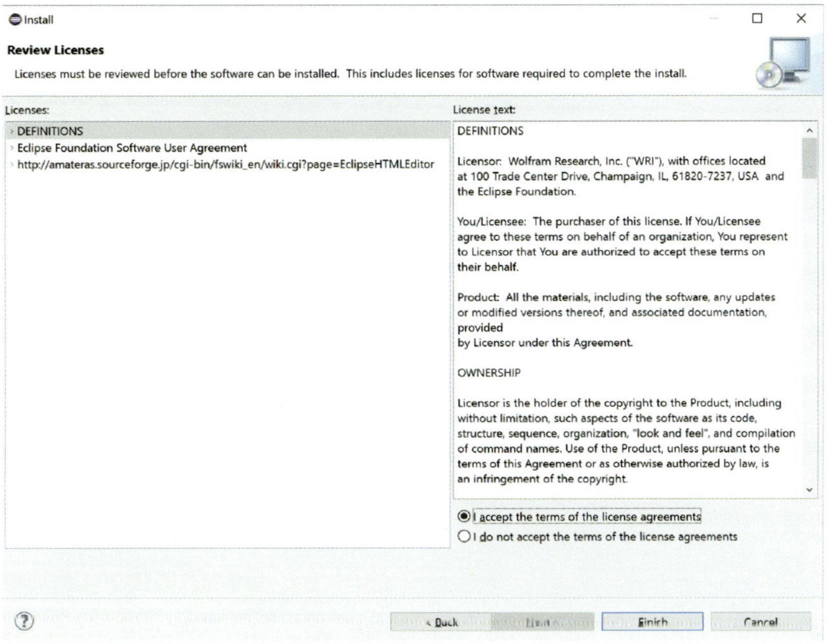

Figure 12.21 Step 3 to configure Workbench.

Restart Eclipse for the changes to take effect.

- The Wolfram Workbench plugin is now installed. To verify, choose **Help ▶ About Eclipse** and identify the Wolfram Workbench logo toward the bottom of the window.

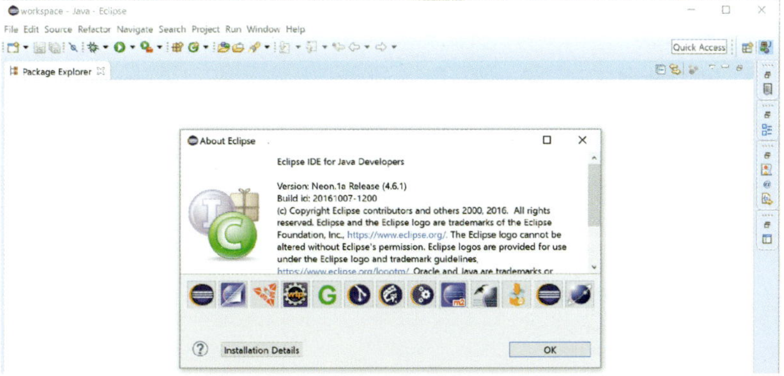

Figure 12.22 The Wolfram Workbench plugin is now installed.

- Configuring the Wolfram Workbench Plugin: Open the Eclipse Preferences (Windows/Linux: **Window ▶ Preferences**, Mac: **Eclipse ▶ Preferences**) and select the Wolfram category on left-hand side of the Preferences window.

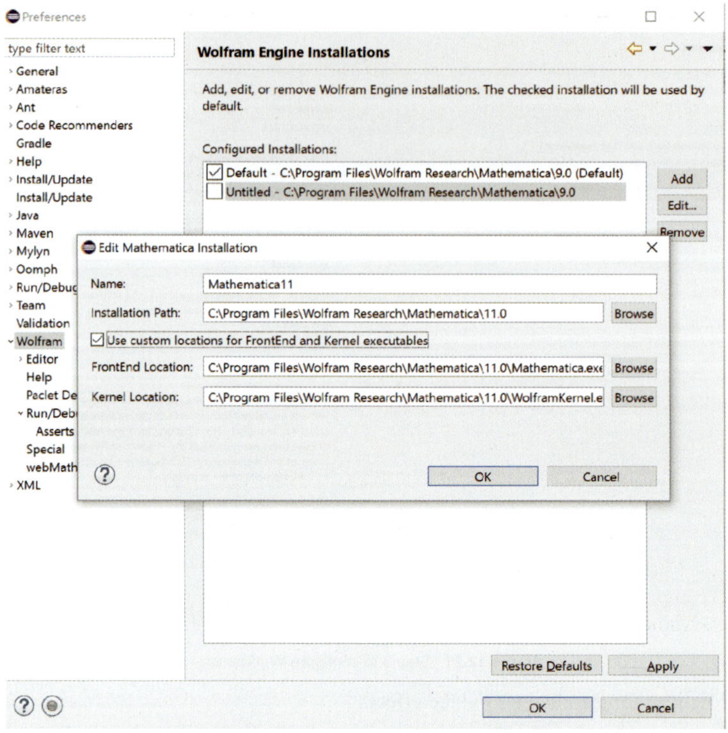

Figure 12.23 Workbench Configuration Page.

- To create a new application project, go to **File ▶ New ▶ Project ▶ Application Project** and type the name of your project.

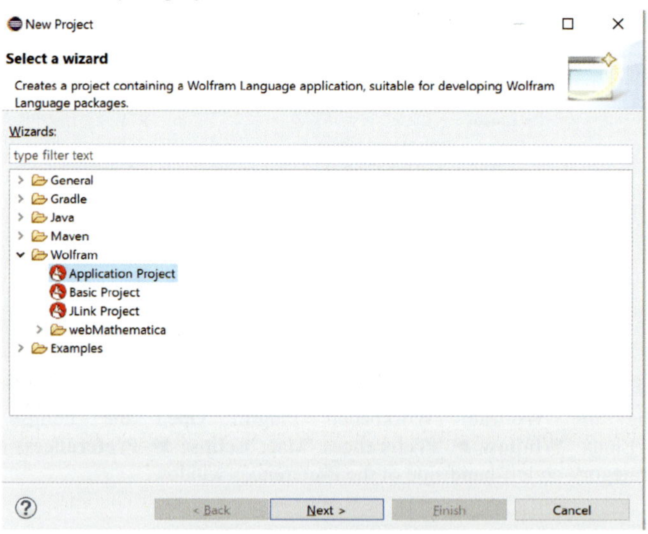

Figure 12.24 Creating a new project.

- After clicking on **Next**, we can assign a name to the application (in this example: Project1).

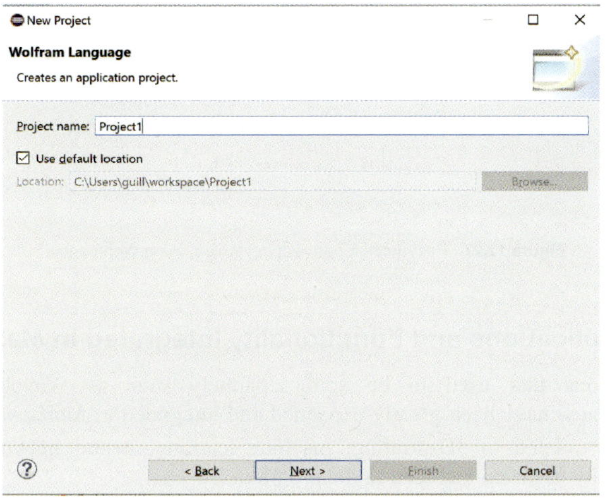

Figure 12.25 Creating an application project.

- Click on **Finish**. You'll see in the Package Explorer tab the projects that you have. Select Project1 and press on PacletInfo.m. Figure 12.26 will appear with basic details about your project: Author, project version, *Mathematica* version, and a brief description.

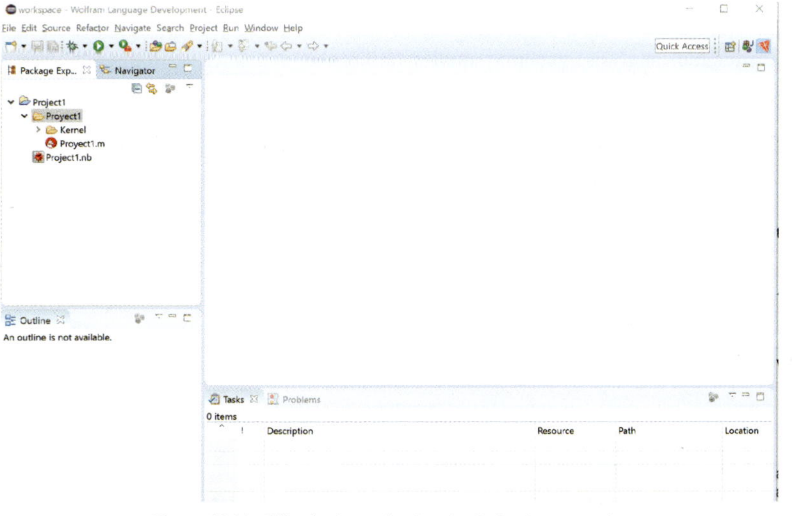

Figure 12.26 Filling in the application details for the new project.

To learn more you can always visit http://www.wolfram.com/workbench/. At the bottom of the webpage (see Figure 12.27) you will find a link to many useful videos. Additionally, you can also go through examples of completed projects by downloading some of them. It will help you to build your own projects and their associated documentation.

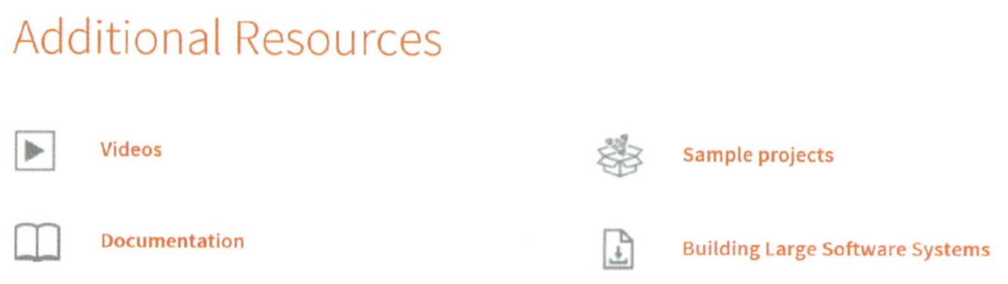

Figure 12.27 Resources for developing project using Workbench.

12.9 New Applications and Functionality Integrated in *Mathematica*

Several applications that used to be sold separately such as Wavelet Analysis and Control Systems now have been greatly expanded and integrated in *Mathematica*. In addition, new functionality related to 3D-printing, machine learning, neural networks and external devices connectivity has been included in the program.

Here is a short list:

12.9.1 Wavelet Analysis

Wavelet analysis (guide/Wavelets) has applications in many different fields that require the study of signals. In many cases they have actually replaced Fourier transforms as the preferred tool for performing such analysis. Some of those fields are: image processing, electrocardiogram analysis, genomic studies, molecular dynamics, astrophysics, seismology, optics, quantum mechanics, meteorology, voice recognition, etc.

12.9.2 Control Systems

Control systems (guide/ControlSystems), now part of *Mathematica*, includes numerous functions for the analysis and design of control systems. To learn more visit:

http://www.wolfram.com/solutions/industry/control-systems.

12.9.3 3D-Printing

Since Version 11 Mathematica provides capabilities to directly 3D print geometric models, using either an online printing service or your own printer. More information available in:

http://www.wolfram.com/language/11/3d-printing.

12.9.4 Machine Learning and Neural Networks

Machine learning and neural networks are some the areas of interest of Stephen Wolfram. He has been directly involved in developing *Mathematica*'s functionality in both areas. Additional information in:

http://www.wolfram.com/language/11/neural-networks
http://www.wolfram.com/language/11/improved-machine-learning

12.9.5 External Devices

Mathematica's connectivity capabilities have been greatly enhanced with the availability of a Wolfram Language framework for connecting external devices. The list of all the devices that can be connected to the program can be found in:

guide/UsingConnectedDevices and http://devices.wolfram.com.

If you happen to be an amateur astronomer, you may enjoy the following blog entry:

http://blog.wolfram.com/2014/12/29/serial-interface-control-of-astronomical-telescopes/.

It covers examples of how to use the Wolfram Language to control telescopes using the LX200 or Celestron NexStar protocols, very popular among astronomy fans.

12.9.6 WolframAlpha API and Widgets

Another area that you may be interested in, is the development of applications and widgets for mobile devices such as smartphones and tablets that make using WolframAlpha easier. You can find tools to develop them in:

http://products.wolframalpha.com/developers/.

12.10. Additional Resources

To access the following resources, if they refer to the help files, write their locations in a notebook (e.g., CUDALink/guide/CUDALink), select them and press <F1>. In the case of external links, copy the web addresses in a browser:

Parallelization tutorial:
reference.wolfram.com/mathematica/ParallelTools/tutorial/Overview.html.
CUDALink: CUDALink/guidc/CUDALink
OpenCLLink: OpenCLLink/guide/OpenCLLink
WebMathematica overview: http://www.wolfram.com/products/webmathematica
web*Mathematica* examples: http://www.wolfram.com/products/webmathematica/examples
Web development in general: http://www.wolfram.com/solutions/industry/web-development/
Information about Workbench: http://www.wolfram.com/products/workbench
Examples created by the author: http://diarium.usal.es/guillermo/mathematica/

A summary of the new functionality included in the latest Mathematica versions can be found in:

http://www.wolfram.com/mathematica/new-in-10/
http://www.wolfram.com/mathematica/new-in-11/
http://www.wolfram.com/mathematica/new-in-11/for-existing-users/

Index